MW00983671

Lubrication Fundamentals

Second Edition, Revised and Expanded

EXXON ESSO Mobil

MECHANICAL ENGINEERING

A Series of Textbooks and Reference Books

1. *Spring Designer's Handbook*, Harold Carlson
2. *Computer-Aided Graphics and Design*, Daniel L. Ryan
3. *Lubrication Fundamentals*, J. George Wills
4. *Solar Engineering for Domestic Buildings*, William A. Himmelman
5. *Applied Engineering Mechanics: Statics and Dynamics*, G. Boothroyd and C. Poli
6. *Centrifugal Pump Clinic*, Igor J. Karassik
7. *Computer-Aided Kinetics for Machine Design*, Daniel L. Ryan
8. *Plastics Products Design Handbook, Part A: Materials and Components; Part B: Processes and Design for Processes*, edited by Edward Miller
9. *Turbomachinery: Basic Theory and Applications,* Earl Logan, Jr.
10. *Vibrations of Shells and Plates*, Werner Soedel
11. *Flat and Corrugated Diaphragm Design Handbook,* Mario Di Giovanni
12. *Practical Stress Analysis in Engineering Design,* Alexander Blake
13. *An Introduction to the Design and Behavior of Bolted Joints,* John H. Bickford
14. *Optimal Engineering Design: Principles and Applications,* James N. Siddall
15. *Spring Manufacturing Handbook,* Harold Carlson
16. *Industrial Noise Control: Fundamentals and Applications,* edited by Lewis H. Bell
17. *Gears and Their Vibration: A Basic Approach to Understanding Gear Noise,* J. Derek Smith
18. *Chains for Power Transmission and Material Handling: Design and Applications Handbook,* American Chain Association
19. *Corrosion and Corrosion Protection Handbook,* edited by Philip A. Schweitzer
20. *Gear Drive Systems: Design and Application,* Peter Lynwander
21. *Controlling In-Plant Airborne Contaminants: Systems Design and Calculations,* John D. Constance
22. *CAD/CAM Systems Planning and Implementation,* Charles S. Knox
23. *Probabilistic Engineering Design: Principles and Applications,* James N. Siddall
24. *Traction Drives: Selection and Application,* Frederick W. Heilich III and Eugene E. Shube
25. *Finite Element Methods: An Introduction,* Ronald L. Huston and Chris E. Passerello
26. *Mechanical Fastening of Plastics: An Engineering Handbook,* Brayton Lincoln, Kenneth J. Gomes, and James F. Braden
27. *Lubrication in Practice: Second Edition,* edited by W. S. Robertson
28. *Principles of Automated Drafting,* Daniel L. Ryan
29. *Practical Seal Design,* edited by Leonard J. Martini
30. *Engineering Documentation for CAD/CAM Applications,* Charles S. Knox
31. *Design Dimensioning with Computer Graphics Applications,* Jerome C. Lange
32. *Mechanism Analysis: Simplified Graphical and Analytical Techniques,* Lyndon O. Barton
33. *CAD/CAM Systems: Justification, Implementation, Productivity Measurement,* Edward J. Preston, George W. Crawford, and Mark E. Coticchia
34. *Steam Plant Calculations Manual,* V. Ganapathy
35. *Design Assurance for Engineers and Managers,* John A. Burgess

36. *Heat Transfer Fluids and Systems for Process and Energy Applications,* Jasbir Singh
37. *Potential Flows: Computer Graphic Solutions,* Robert H. Kirchhoff
38. *Computer-Aided Graphics and Design: Second Edition,* Daniel L. Ryan
39. *Electronically Controlled Proportional Valves: Selection and Application,* Michael J. Tonyan, edited by Tobi Goldoftas
40. *Pressure Gauge Handbook,* AMETEK, U.S. Gauge Division, edited by Philip W. Harland
41. *Fabric Filtration for Combustion Sources: Fundamentals and Basic Technology,* R. P. Donovan
42. *Design of Mechanical Joints,* Alexander Blake
43. *CAD/CAM Dictionary,* Edward J. Preston, George W. Crawford, and Mark E. Coticchia
44. *Machinery Adhesives for Locking, Retaining, and Sealing,* Girard S. Haviland
45. *Couplings and Joints: Design, Selection, and Application,* Jon R. Mancuso
46. *Shaft Alignment Handbook,* John Piotrowski
47. *BASIC Programs for Steam Plant Engineers: Boilers, Combustion, Fluid Flow, and Heat Transfer,* V. Ganapathy
48. *Solving Mechanical Design Problems with Computer Graphics,* Jerome C. Lange
49. *Plastics Gearing: Selection and Application,* Clifford E. Adams
50. *Clutches and Brakes: Design and Selection,* William C. Orthwein
51. *Transducers in Mechanical and Electronic Design,* Harry L. Trietley
52. *Metallurgical Applications of Shock-Wave and High-Strain-Rate Phenomena,* edited by Lawrence E. Murr, Karl P. Staudhammer, and Marc A. Meyers
53. *Magnesium Products Design,* Robert S. Busk
54. *How to Integrate CAD/CAM Systems: Management and Technology,* William D. Engelke
55. *Cam Design and Manufacture: Second Edition; with cam design software for the IBM PC and compatibles, disk included,* Preben W. Jensen
56. *Solid-State AC Motor Controls: Selection and Application,* Sylvester Campbell
57. *Fundamentals of Robotics,* David D. Ardayfio
58. *Belt Selection and Application for Engineers,* edited by Wallace D. Erickson
59. *Developing Three-Dimensional CAD Software with the IBM PC,* C. Stan Wei
60. *Organizing Data for CIM Applications,* Charles S. Knox, with contributions by Thomas C. Boos, Ross S. Culverhouse, and Paul F. Muchnicki
61. *Computer-Aided Simulation in Railway Dynamics,* by Rao V. Dukkipati and Joseph R. Amyot
62. *Fiber-Reinforced Composites: Materials, Manufacturing, and Design,* P. K. Mallick
63. *Photoelectric Sensors and Controls: Selection and Application,* Scott M. Juds
64. *Finite Element Analysis with Personal Computers,* Edward R. Champion, Jr., and J. Michael Ensminger
65. *Ultrasonics: Fundamentals, Technology, Applications: Second Edition, Revised and Expanded,* Dale Ensminger
66. *Applied Finite Element Modeling: Practical Problem Solving for Engineers,* Jeffrey M. Steele
67. *Measurement and Instrumentation in Engineering: Principles and Basic Laboratory Experiments,* Francis S. Tse and Ivan E. Morse
68. *Centrifugal Pump Clinic: Second Edition, Revised and Expanded,* Igor J. Karassik
69. *Practical Stress Analysis in Engineering Design: Second Edition, Revised and Expanded,* Alexander Blake
70. *An Introduction to the Design and Behavior of Bolted Joints: Second Edition, Revised and Expanded,* John H. Bickford
71. *High Vacuum Technology: A Practical Guide,* Marsbed H. Hablanian
72. *Pressure Sensors: Selection and Application,* Duane Tandeske
73. *Zinc Handbook: Properties, Processing, and Use in Design,* Frank Porter
74. *Thermal Fatigue of Metals,* Andrzej Weronski and Tadeusz Hejwowski
75. *Classical and Modern Mechanisms for Engineers and Inventors,* Preben W. Jensen
76. *Handbook of Electronic Package Design,* edited by Michael Pecht
77. *Shock-Wave and High-Strain-Rate Phenomena in Materials*, edited by Marc A. Meyers, Lawrence E. Murr, and Karl P. Staudhammer
78. *Industrial Refrigeration: Principles, Design and Applications*, P. C. Koelet

79. *Applied Combustion*, Eugene L. Keating
80. *Engine Oils and Automotive Lubrication*, edited by Wilfried J. Bartz
81. *Mechanism Analysis: Simplified and Graphical Techniques, Second Edition, Revised and Expanded*, Lyndon O. Barton
82. *Fundamental Fluid Mechanics for the Practicing Engineer*, James W. Murdock
83. *Fiber-Reinforced Composites: Materials, Manufacturing, and Design, Second Edition, Revised and Expanded*, P. K. Mallick
84. *Numerical Methods for Engineering Applications*, Edward R. Champion, Jr.
85. *Turbomachinery: Basic Theory and Applications, Second Edition, Revised and Expanded*, Earl Logan, Jr.
86. *Vibrations of Shells and Plates: Second Edition, Revised and Expanded*, Werner Soedel
87. *Steam Plant Calculations Manual: Second Edition, Revised and Expanded*, V. Ganapathy
88. *Industrial Noise Control: Fundamentals and Applications, Second Edition, Revised and Expanded*, Lewis H. Bell and Douglas H. Bell
89. *Finite Elements: Their Design and Performance*, Richard H. MacNeal
90. *Mechanical Properties of Polymers and Composites: Second Edition, Revised and Expanded*, Lawrence E. Nielsen and Robert F. Landel
91. *Mechanical Wear Prediction and Prevention*, Raymond G. Bayer
92. *Mechanical Power Transmission Components*, edited by David W. South and Jon R. Mancuso
93. *Handbook of Turbomachinery*, edited by Earl Logan, Jr.
94. *Engineering Documentation Control Practices and Procedures*, Ray E. Monahan
95. *Refractory Linings: Thermomechanical Design and Applications*, Charles A. Schacht
96. *Geometric Dimensioning and Tolerancing: Applications and Techniques for Use in Design, Manufacturing, and Inspection*, James D. Meadows
97. *An Introduction to the Design and Behavior of Bolted Joints: Third Edition, Revised and Expanded*, John H. Bickford
98. *Shaft Alignment Handbook: Second Edition, Revised and Expanded*, John Piotrowski
99. *Computer-Aided Design of Polymer-Matrix Composite Structures*, edited by S. V. Hoa
100. *Friction Science and Technology*, Peter J. Blau
101. *Introduction to Plastics and Composites: Mechanical Properties and Engineering Applications*, Edward Miller
102. *Practical Fracture Mechanics in Design*, Alexander Blake
103. *Pump Characteristics and Applications*, Michael W. Volk
104. *Optical Principles and Technology for Engineers*, James E. Stewart
105. *Optimizing the Shape of Mechanical Elements and Structures*, A. A. Seireg and Jorge Rodriguez
106. *Kinematics and Dynamics of Machinery*, Vladimír Stejskal and Michael Valášek
107. *Shaft Seals for Dynamic Applications*, Les Horve
108. *Reliability-Based Mechanical Design*, edited by Thomas A. Cruse
109. *Mechanical Fastening, Joining, and Assembly*, James A. Speck
110. *Turbomachinery Fluid Dynamics and Heat Transfer*, edited by Chunill Hah
111. *High-Vacuum Technology: A Practical Guide, Second Edition, Revised and Expanded*, Marsbed H. Hablanian
112. *Geometric Dimensioning and Tolerancing: Workbook and Answerbook*, James D. Meadows
113. *Handbook of Materials Selection for Engineering Applications*, edited by G. T. Murray
114. *Handbook of Thermoplastic Piping System Design*, Thomas Sixsmith and Reinhard Hanselka
115. *Practical Guide to Finite Elements: A Solid Mechanics Approach*, Steven M. Lepi
116. *Applied Computational Fluid Dynamics*, edited by Vijay K. Garg
117. *Fluid Sealing Technology*, Heinz K. Muller and Bernard S. Nau
118. *Friction and Lubrication in Mechanical Design*, A. A. Seireg
119. *Influence Functions and Matrices*, Yuri A. Melnikov
120. *Mechanical Analysis of Electronic Packaging Systems*, Stephen A. McKeown

121. *Couplings and Joints: Design, Selection, and Application, Second Edition, Revised and Expanded,* Jon R. Mancuso
122. *Thermodynamics: Processes and Applications*, Earl Logan, Jr.
123. *Gear Noise and Vibration*, J. Derek Smith
124. *Practical Fluid Mechanics for Engineering Applications,* John J. Bloomer
125. *Handbook of Hydraulic Fluid Technology*, edited by George E. Totten
126. *Heat Exchanger Design Handbook*, T. Kuppan
127. *Designing for Product Sound Quality*, Richard H. Lyon
128. *Probability Applications in Mechanical Design*, Franklin E. Fisher and Joy R. Fisher
129. *Nickel Alloys,* edited by Ulrich Heubner
130. *Rotating Machinery Vibration: Problem Analysis and Troubleshooting*, Maurice L. Adams, Jr.
131. *Formulas for Dynamic Analysis*, Ronald Huston and C. Q. Liu
132. *Handbook of Machinery Dynamics*, Lynn L. Faulkner and Earl Logan, Jr.
133. *Rapid Prototyping Technology: Selection and Application*, Ken Cooper
134. *Reciprocating Machinery Dynamics: Design and Analysis*, Abdulla S. Rangwala
135. *Maintenance Excellence: Optimizing Equipment Life-Cycle Decisions,* edited by John D. Campbell and Andrew K. S. Jardine
136. *Practical Guide to Industrial Boiler Systems*, Ralph L. Vandagriff
137. *Lubrication Fundamentals: Second Edition, Revised and Expanded,* D. M. Pirro and A. A. Wessol

Additional Volumes in Preparation

Reliability Verification, Testing, and Analysis in Engineering Design, Gary S. Wasserman

Mechanical Life Cycle Handbook: Good Environmental Design and Manufacturing, Mahendra Hundal

Micromachining of Engineering Materials, edited by Joseph McGeough

Mechanical Engineering Software

Spring Design with an IBM PC, Al Dietrich

Mechanical Design Failure Analysis: With Failure Analysis System Software for the IBM PC, David G. Ullman

Lubrication Fundamentals

Second Edition, Revised and Expanded

D. M. Pirro
Exxon Mobil Corporation
Fairfax, Virginia

A. A. Wessol
Lubrication Consultant
Manassas, Virginia

MARCEL DEKKER, INC. NEW YORK • BASEL

The first edition of this book was written by J. George Wills (Marcel Dekker, 1980).

ISBN: 0-8247-0574-2
This book is printed on acid-free paper.

Headquarters
Marcel Dekker, Inc.
270 Madison Avenue, New York, NY 10016
tel: 212-696-9000; fax: 212-685-4540

Eastern Hemisphere Distribution
Marcel Dekker AG
Hutgasse 4, Postfach 812, CH-4001 Basel, Switzerland
tel: 41-61-261-8482; fax: 41-61-261-8896

World Wide Web
http://www.dekker.com

The publisher offers discounts on this book when ordered in bulk quantities. For more information, write to Special Sales/Professional Marketing at the headquarters address above.

Current printing (last digit):
10 9 8 7 6 5 4 3 2

PRINTED IN THE UNITED STATES OF AMERICA

Preface

Lubrication and the knowledge of lubricants not only are subjects of interest to all of us but they are also critical to the cost effective operation and reliability of machinery that is part of our daily lives. Our world, and exploration of regions beyond our world, depends on mechanical devices that require lubricating films. Whether in our homes or at work, whether knowingly or unknowingly, we all need lubricants and some knowledge of lubrication. Fishing reels, vacuum cleaners, and lawn mowers are among the devices that require lubrication. The millions of automobiles, buses, airplanes, and trains depend on lubrication for operation, and it must be effective lubrication for dependability, safety, and minimization of environmental impact.

Many changes in the field of lubrication have occurred since the first edition of *Lubrication Fundamentals* was published more than 20 years ago. Today intricate and complex machines are used to make paper products; huge rolling mills turn out metal ingots and sheets; metalworking machines produce close-tolerance parts; and special machinery is used to manufacture cement, rubber, and plastic products. New metallurgy, new processes, and never before used materials are often part of these machines that require lubrication. The newer machinery designs have taken advantage of these as well as other technologies, which often involve computers to assist in producing ultra-high precision parts at production rates that were once only dreamed of. These advances have led to faster machine speeds, greater load-handling capability, higher machine temperatures, smaller capacity lubricant reservoirs, and less frequent lubrication application up to and including fill-for-life lubrication. As a result, there has been an explosion in both higher performance and specialty application oils and greases. The impact of these lubricants on our natural environment has also been a driver for new lubricant technology.

This second edition of *Lubrication Fundamentals* builds upon the machinery basics discussed in the first edition, much of which is still applicable today. The second edition also addresses many of the new lubricant technologies that were introduced or improved upon in the last 20 years to meet the needs of modern machinery. As we progress through this century, lubricant suppliers will be faced with many challenges. Critical activities along the lubricant value chain that are impacted by technology include new lubrication requirements, petroleum crude selection, base stock manufacture, product formulation and evaluation, lubricant application, and environmental stewardship. These will be exciting times for industry, especially for those participating in the quest to develop the new lubricant molecule for the future.

D. M. Pirro
A. A. Wessol

ACKNOWLEDGMENTS

Lubrication Fundamentals: Second Edition, Revised and Expanded, like all technical publications of this magnitude, is not the work of one or two people. It is the combined effort of hundreds, even thousands, of engineers, designers, chemists, physicists, writers, and artists—the compendium of a broad spectrum of talent working over a long period of time. The field of lubrication fundamentals starts with the scientists who study the basic interaction of oil films with bearings, gears, and cams under various stresses and loads. It then takes the unique cooperation that exists between the machine designer and equipment builders, on one side, and the lubricant formulators and suppliers, on the other, along with the cooperation that takes place in the many associations such as STLE, SAE, ACEA, ASTM, ISO, DIN, NLGI, AGMA, and API, to name but a few. It culminates in the mating of superior lubricants properly applied with the requirements of the most efficient machines operating today.

The lubricants industry is most grateful to lubrication pioneers such as J. George Wills, the author of the first edition. More than 20 years ago, Wills, an acknowledged expert in the field of lubrication in the nuclear power industry, identified the need for a practical resource on lubrication. He developed a vision, secured the support and resources to undertake such a monumental effort, and then dedicated the effort to turn his vision into reality. We are privileged to be able to build upon this effort and share the many technological advances in industry.

It would be impossible to list the host of people who have helped to put this second edition together. The book compiles the many technical publications of Exxon Mobil Corporation and the cooperative offerings of the foremost international equipment builders. Impossible though it may be to acknowledge the contributions of everyone, the following must be singled out for thanks:

Our lubricant business leaders at ExxonMobil—John Lyon, Jeff Webster, Don Salamack, J. Ian Davidson, and George Siragusa—first for their acceptance of the idea and then for their encouragement to complete the project

The following engineers, researchers, and technologists at ExxonMobil, who made significant contributions to this edition—W. Russ Murphy, S. Levi Pearson, Marcia Rogers, Charles Baker, Mary McGuiness, Tim McCrory, John Doner, Betsey Varney, Carl Gerster, and Elena Portoles

The many original equipment manufacturers we have worked with for many years, for sharing their knowledge and technology

The many other marketers, engineers, formulators, and researchers (past and present) from Mobil and ExxonMobil for their contributions and comments

Contents

1

Introduction

Petroleum is one of the naturally occurring hydrocarbons that frequently include natural gas, natural bitumen, and natural wax. The name ''petroleum'' is derived from the Latin *petra* (rock) and *oleum* (oil). According to the most generally accepted theory today, petroleum was formed by the decomposition of organic refuse, aided by high temperatures and pressures, over a vast period of geological time.

I. PREMODERN HISTORY OF PETROLEUM

Although petroleum occurs, as its name indicates, among rocks in the earth, it sometimes seeps to the surface through fissures or is exposed by erosion. The existence of petroleum was known to primitive man, since surface seepage, often sticky and thick, was obvious to anyone passing by. Prehistoric animals were sometimes mired in it, but few human bones have been recovered from these tar pits. Early man evidently knew enough about the danger of surface seepage to avoid it.

The first actual use of petroleum seems to have been in Egypt, which imported bitumen, probably from Greece, for use in embalming. The Egyptians believed that the spirit remained immortal if the body was preserved.

About the year 450 B.C., Herodotus, the father of history, described the pits of Kir ab ur Susiana as follows:

> At Ardericca is a well which produces three different substances, for asphalt, salt and oil are drawn up from it in the following manner. It is pumped up by means of a swipe; and, instead of a bucket, half a wine skin is attached to it. Having dipped down with this, a man draws it up and then pours the contents into a reservoir, and being poured from this into another, it assumes these different forms: the asphalt and salt immediately become solid, and the liquid oil is collected. The Persians call it Phadinance; it is black and emits a strong odor.

Pliny, the historian, and Dioscorides Pedanius, the Greek botanist, both mention ''Sicilian oil,'' from the island of Sicily, which was burned for illumination as early as the beginning of the Common Era.

The Scriptures contain many references to petroleum, in addition to the well-known story of Moses, who as an infant was set afloat on the river in a little boat of reeds waterproofed with pitch, and was found by Pharaoh's daughter. Some of these biblical references include the following:

> Make thee an ark of gopher wood; rooms shalt thou make in the ark, and shalt pitch it within and without with pitch. (Genesis VI.14)
> And they had brick for stone, and slime (bitumen) had they for mortar. (Building the Tower of Babel, Genesis XI.3)
> And the Vale of Siddim was full of slime (bitumen) pits; and the kings of Sodom and Gomorrah fled, and fell there. . . . (Genesis XIV.10)

Other references are found in Strabo, Josephus, Diodorus Siculus, and Plutarch, and in more recent times much evidence has accumulated that petroleum was known in almost every part of the world.

Marco Polo, the Venetian traveler and merchant, visited the lands of the Caspian Sea in the thirteenth century. In an account of this visit, he stated:

> To the north lies Zorzania, near the confines of which there is a fountain of oil which discharges so great a quantity as to furnish loading for many camels. The use made of it is not for the purpose of food, but as an unguent for the cure of cutaneous distempers in men and cattle, as well as other complaints; and it is also good for burning. In the neighboring country, no other is used in their lamps, and people come from distant parts to procure it.

Sir Walter Raleigh, while visiting the island of Trinidad off the coast of Venezuela, inspected the great deposit of bitumen there. The following is taken from *The Discoveries of Guiana* (1596):

> At this point called Tierra de Brea, or Piche, there is that abundance of stone pitch that all the ships of the world may be therewith loden from thence, and wee made triall of it in trimming our ships to be most excellent good, and melteth not with the sunne as the pitch of Norway, and therefore for ships trading the south partes very profitable.

II. PETROLEUM IN NORTH AMERICA

On the North American continent, petroleum seepages were undoubtedly known to the aborigines, but the first known record of the substance was made by the Franciscan Joseph de la Roche D'Allion, who in 1629 crossed the Niagara River from Canada and visited an area later known as Cuba, New York. At this place, petroleum was collected by the Indians, who used it medicinally and to bind pigments used in body adornments.

In 1721, Charlevois, the French historian and missionary who descended the Mississippi River to its mouth, quotes a Captain de Joncaire as follows: "There is a fountain at the head of a branch of the Ohio River (probably the Allegheny) the waters of which like oil, has a taste of iron and serves to appease all manner of pain."

The *Massachusetts Magazine,* Volume 1, July 1789, contains this account under the heading "American Natural Curiosities":

> In the northern parts of Pennsylvania, there is a creek called Oil Creek, which empties into the Allegheny River. It issues from a spring, on the top of which floates an oil similar to that called Barbadoes tar; and from which one man may gather several gallons in a day. The troops sent to guard the western posts halted at this spring, collected some of the oil, and

> bathed their joints with it. This gave them great relief from the rheumatic complaints with which they were affected. The waters, of which the troops drank freely, operated as a gentle purge.

Although the practice of deriving useful oils by the distillation of bituminous shales and various organic substances was generally known, it was not until the nineteenth century that distillation processes were widely used for a number of useful substances, including tars for waterproofing, gas for illumination, and various chemicals, pharmaceuticals, and oils.

In 1833 Dr. Benjamin Silliman contributed an article to the *American Journal of Science* that contained the following report:

> The petroleum, sold in the Eastern states under the name of Seneca Oil, is a dark brown color, between that of tar and molasses, and its degree of consistency is not dissimilar, according to temperature; its odor is strong and too well known to need description. I have frequently distilled it in a glass retort, and the naphtha which collects in the receiver is of a light straw color, and much lighter, more odorous and inflammable than the petroleum; in the first distillation, a little water usually rests in the receiver, at the bottom of the naphtha; from this it is easily decanted, and a second distillation prepares it perfectly for preserving potassium and sodium, the object which led me to distil it, and these metals I have kept under it (as others have done) for years; eventually they acquire some oxygen, from or through the naphtha, and the exterior portion of the metal returns, slowly, to the condition of alkali—more rapidly if the stopper is not tight.
>
> The petroleum remaining from the distillation is thick like pitch; if the distillation has been pushed far, the residuum will flow only languidly in the retort, and in cold weather it becomes a soft solid, resembling much the maltha or mineral pitch.

Along the banks of the Kanawha River in West Virginia, petroleum was proving a constant source of annoyance in the brine wells; and one of these wells, in 1814, discharged petroleum at periods of from 1 to 4 days, in quantities ranging from 30 to 60 gallons at each eruption. A Pittsburgh druggist named Samuel M. Kier began bottling the petroleum from these brine wells around 1846 and selling the oil for medicinal purposes. He claimed it was remarkably effective for most ills and advertised this widely. In those days, many people believed that the worse a nostrum tasted, the more powerful it was. People died young then, and often did not know what killed them. In the light of today's knowledge, we would certainly not recommend drinking such products. Sales boomed for awhile; but in 1852 there was a falling off in trade. Therefore, the enterprising Mr. Kier began to distill the substance for its illuminating oil content. His experiment was successful and was a forerunner, in part, of future commercial refining methods.

In 1853 a bottle of petroleum at the office of Professor Crosby of Dartmouth College was noticed by Mr. George Bissel, a good businessman. Bissel soon visited Titusville, Pennsylvania, where the oil had originated, purchased 100 acres of land in an area known as Watsons Flats, and leased a similar tract for the total sum of $5000. Bissel and an associate, J. D. Eveleth, then organized the first oil company in the United States, the Pennsylvania Rock Oil Company. The incorporation papers were filed in Albany, New York, on December 30, 1854. Bissel had pits dug in his land in the hope of obtaining commercial quantities of petroleum, but was unsuccessful with this method. A new company was formed, which was called the Pennsylvania Rock Oil Company of Connecticut, with New Haven as headquarters. The property of the New York corporation was transferred to the new company, and Bissel began again.

In 1856 Bissel read one of Samuel Kier's advertisements on which was shown a drilling rig for brine wells. Suddenly it occurred to him to have wells drilled, as was being done in some places for brine. A new company, the Seneca Oil Company, succeeded the Connecticut firm, and an acquaintance of some of its partners, E. L. Drake, was selected to conduct field experiments in Titusville. Drake found that to reach hard rock in which to try the drilling method, some unusual form of shoring was needed to prevent a cave-in. It occurred to him to drive a pipe through the loose sand and shale; a plan afterward adopted in oil well and artesian well drilling.

Drilling then began under the direction of W. A. Smith, a blacksmith and brine well driller, and went down 69½ ft. On Saturday, August 27, 1859, the drill dropped into a crevice about 6 in. deep, and the tools were pulled out and set aside for the work to be resumed on Monday. However, Smith decided to visit the well that Sunday to check on it, and upon peering into the pipe saw petroleum within a few feet of the top. On the following day, the well produced the incredible quantity of 20 barrels a day.

III. DEVELOPMENT OF LUBRICANTS

During the period from 1850 to 1875, many men experimented with the products of petroleum distillation then available, attempting to find uses for them, in addition to providing illumination. Some of the viscous materials were investigated as substitutes for the vegetable and animal oils previously used for lubrication, mainly those derived from olives, rapeseed, whale, tallow, lard, and other fixed oils.

As early as 1400 B.C., greases, made of a combination of calcium and fats, were used to lubricate chariot wheels. Traces of this grease were found on chariots excavated from the tombs of Yuaa and Thuiu. During the third quarter of the nineteenth century, greases were made with petroleum oils combined with potassium, calcium, and sodium soaps and placed on the market in limited quantities.

Gradually, as distillation and refining processes were improved, a wider range of petroleum oils as produced to take the place of the fatty oils. These mineral oils could be controlled more accurately in manufacture and were not subject to the rapid deterioration of the fatty oils.

Some of the fatty oils continued to be used in special services as late as the early part of the twentieth century. Tallow was fairly effective in steam cylinders as a lubricant. However, it was not always pleasant to handle, since maggots often appeared in the tallow particularly in hot weather. Lard oil was used for cutting of metals, and castor oil was used to lubricate the aircraft engines of World War I. Even, today, some fatty oils are still used as compounding in small percentages with mineral oils, but chemical additives have taken their place for the majority of users.

As machinery has increased in complexity and applications have expanded to more severe climatic conditions such as operation of gas and crude oil producing equipment in Alaska, mining in Siberia, high altitude jet aircraft, and equipment in space programs, so has the technology in research and development of lubricants. One example is the fast developing field of synthetic lubricants to provide a full range of lubricants to meet the requirements of extremes of temperatures and operating conditions. Another would be a class of lubricants designed to be less damaging to the environment where there is potential for inadvertent spills or leakage.

IV. FUTURE PROSPECTS

The twenty-first century will continue to see advancements in equipment technology. As equipment is designed to achieve higher production levels, this will result in higher operating speeds, increased temperatures and higher system pressures that will place greater demands on the lubricants. These demands, coupled with the trends of reduced or maintenance-free operation, increased environmental awareness and regulations, and greater attention to safety issues, will continue to challenge lubricant technology and associated research and development activities.

2

Refining Processes and Lubricant Base Stocks

Petroleum, or crude oil, is refined to make many essential products used throughout the world in homes and industry. These products include gas burned as fuel, gasoline, kerosene, solvents, fuel oil, diesel fuel, lubricating products, and industrial specialty products, (waxes, chemicals, asphalt, and coke). Usually, crude oil is refined in two stages: refining of light products and refining of lubricating oils and waxes. The refining of light products, which is concerned with all these substances except lubricants, specialty products, waxes, asphalts, and coke, is accomplished at or slightly above atmospheric pressure. Although all the products discussed are not actually light in weight or color (e.g., the heavy fuels, oils, and asphalts), their production is grouped with that of the light products because they are all made in the same or similar equipment.

At approximately 700°F (371°C), the residuum from light products refining has a tendency to decompose. Thus, the refining of lubricating oils and waxes takes place under vacuum conditions and at temperatures under the decomposition point.

There are two basic refining processes; separation and conversion. The separation process selects certain desirable components by distillation, solvent extraction, and solvent dewing. The conversion process involves changing the chemical structure of certain undesirable crude oil components into desirable components. Conversion processes also include a degree of removal of nondesirable species. The types of refining process are discussed in this chapter following brief general discussions of crude oil handling and its initial fractionation into light products, vacuum gas oil, and residuum.

I. CRUDE OIL

A. Origin and Sources

The petroleum that flows from our wells today was formed many millions of years ago. It is believed to have been formed from the remains of tiny aquatic animals and plants that settled with mud and silt to the bottoms of ancient seas. As successive layers built up, those remains were subjected to high pressures and temperatures and underwent chemical transformations, leading to the formation of the hydrocarbons and other constituents of crude oil described herein. In many areas, this crude oil migrated and accumulated in porous rock formations overlaid by impervious rock formations that prevented further travel. Usually a layer of concentrated salt water underlies the oil pool.

The states of Alaska, Texas, California, and Louisiana, with their offshore areas, are the largest producers of crude oil in the United States, although petroleum was first produced in Pennsylvania. Today, a major portion of this nation's needs is supplied from Canada, Mexico, South America, and the oil fields in the Middle East.

B. Production

Crude oil was first found as seepages, and one of the earliest references to it may have been the "fiery furnace" of Nebuchadnezzar. This is now thought to have been an oil seepage that caught fire. Currently, holes are drilled as deep as 5 miles to tap oil-bearing strata located by geologists. The crude oil frequently comes to the surface under great pressure and in combination with large volumes of gas. Present practice is to separate the gas from the oil and process the gas to remove from it additional liquids of high volatility to form what is called "natural gasoline," for addition to motor gasoline. The "dry" gas is sold as fuel or recycled back to the underground formations to maintain pressure in the oil pool and, thus, to increase recovery of crude oil. Years ago much of this gas was wasted by burning it in huge flares.

C. Types and Composition

Crude oils are found in a variety of types ranging from light-colored oils, consisting mainly of gasoline, to black, nearly solid asphalts. Crude oils are very complex mixtures containing very many individual hydrocarbons or compounds of hydrogen and carbon. These range from methane, the main constituent of natural gas with one carbon atom, to compounds containing 50 or more carbon atoms (Figure 2.1). The boiling ranges of the compounds increase roughly with the number of carbon atoms.

Typical boiling point ranges for various crude oil fractions are as follows:

Far below 0°F (−18°C) for the light natural gas hydrocarbons with one to three carbon atoms
About 80–400°F (27–204°C) for gasoline components
400–650°F (204–343°C) for diesel and home heating oils
Higher ranges for lubricating oils and heavier fuels

The asphalt materials cannot be vaporized because they decompose when heated and their molecules either *"crack"* to form gas, gasoline, and lighter fuels, or unite to form even heavier molecules. The latter form carbonaceous residues called "coke," which as discussed later, can be either a product or a nuisance in refining.

Crude oils also contain varying amounts of compounds of sulfur, nitrogen, oxygen, various metals such as nickel and vanadium, and some entrained water-containing dissolved salts. All these materials can cause trouble in refining or in subsequent product applications, and their reduction or removal increases refining costs appreciably. In addition, some of the materials must be removed or substantially reduced to meet ecological

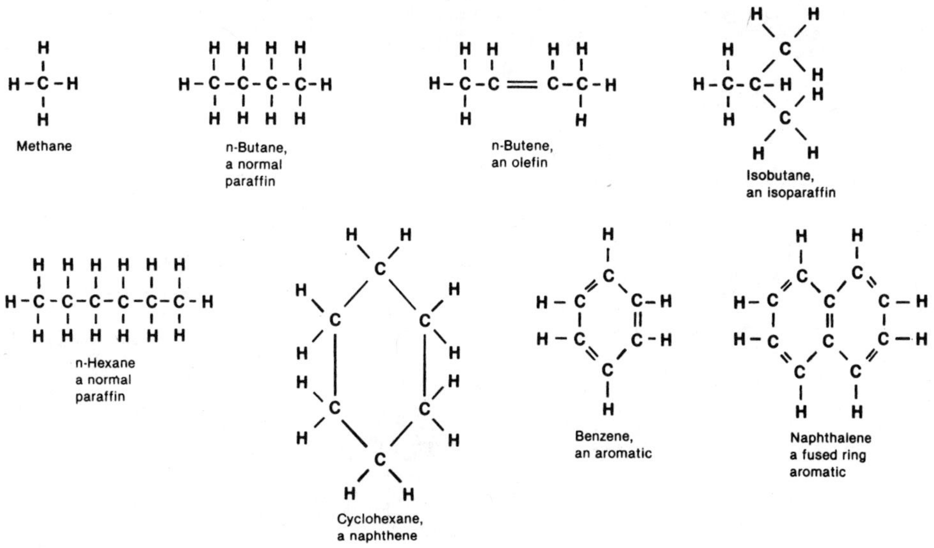

Methane

n-Butane, a normal paraffin

n-Butene, an olefin

Isobutane, an isoparaffin

n-Hexane a normal paraffin

Cyclohexane, a naphthene

Benzene, an aromatic

Naphthalene a fused ring aromatic

Figure 2.1 Typical hydrocarbon configurations.

or environmental regulations. For example, federal, state, and local regulatory agencies have passed laws limiting sulfur content in fuels.

The carbon atom is much like a four-holed Tinker Toy piece, and even more versatile. Reference to Figure 2.1 shows that for molecules containing two or more carbon atoms, a number of configurations can exist for each number of carbon atoms; in fact, the number of such possible shapes increases with the number of carbon atoms. Each configuration has distinct properties. Compounds with the carbon atoms in a straight line (normal paraffins) have low octane ratings when in the gasoline boiling range but make excellent diesel fuels. They consist of waxes when they are in the lubricating oil boiling range. Branched chain and ring compounds with low hydrogen content like benzene may cause knocking in diesel engines, but can act as an antiknock additive in gasoline.

II. REFINING

A. Crude Distillation

Crude oil is sometimes used in its unprocessed form as fuel in power plants and in some internal combustion engines; but in most cases, it is separated or converted into different fractions, which in turn require further processing to supply the large number of petroleum products needed. In many cases, the first step is to remove from the crude certain inorganic salts suspended as minute crystals or dissolved in entrained water. These salts break down during processing to form acids that severely corrode refinery equipment, plug heat exchangers and other equipment, and poison catalysts used in subsequent processes. Therefore, the crude is mixed with additional water to dissolve the salts and the resultant brine is removed by settling.

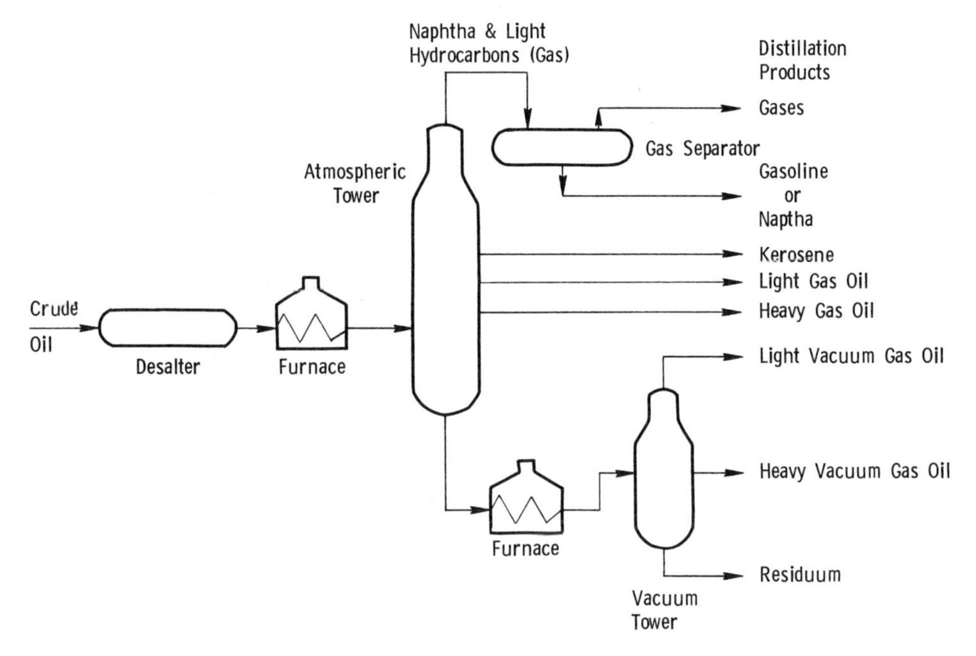

Figure 2.2 Crude distillation unit.

After desalting, the crude is pumped through a tubular furnace (Figure 2.2) where it is heated and partially vaporized. The refinery furnace usually consists of connected lengths of pipe heated externally by gas or oil burners. The mixture of hot liquid and vapor from the furnace enters a fractionating column. This is a device that operates at slightly above atmospheric pressure and separates groups of hydrocarbons according to their boiling ranges. The fractionating column works because there is a gradation in temperature from bottom to top so that, as the vapors rise toward the cooler upper portion, the higher boiling components condense first. As the vapor stream moves up the column, lower boiling vapors are progressively condensed. Trays are inserted at various levels in the column to collect the liquids that condense at those levels. Naphtha, an industry term for raw gasoline that requires further processing, and light hydrocarbons are carried over the top of the column as vapor and are condensed to liquid by cooling. Kerosene, diesel fuel, home heating fuels, and heavy oils (called gas oils) are withdrawn as side cuts from the successively lower and hotter levels of the tower.

A heavy black atmospheric residuum is drawn from the bottom of the column. The combination of furnace and atmospheric tower is sometimes called a ''pipe still.''

Because of the tendency of residuum to decompose at temperatures about 700°F (371°C), heavier (higher boiling) oils such as lubricating oils must be distilled off in a separate vacuum fractionating tower. The greatly reduced pressure in the tower markedly lowers the boiling points of the desired hydrocarbon compounds. Bottom materials from the vacuum tower are used for asphalt, or are further processed to make other products.

The fractions separated by crude distillation are sometimes referred to as ''straight run'' products. The character of their hydrocarbon constituents is not changed by distillation. If all the separated fractions were reassembled, we would recover the original crude.

B. Lubricating Oils

The term ''lubricating oils'' is generally used to include all the classes of lubricating materials that are applied as fluids. Lubricating oils are composed of base oils plus additives to enhance specific characteristics. In the remainder of this section, the term ''base stock'' replaces ''base oil''; the two are synonymous. On a volume basis, the vast majority of the world's lubricating base stock is obtained by refining distillate or residual fractions obtained directly from crude oil. This section is concerned primarily with ''mineral'' oils. Because of their growing importance, synthetic lubricants, both oils and greases, are discussed in Chapter 5.

Lubricating base stocks are made from the more viscous portion of the crude oil, which remains after removal by distillation of the gas oil and lighter fractions. They have been prepared from crude oils obtained from most parts of the world. Although crude oils from various parts of the world differ widely in properties and appearance, there is relatively little difference in their elemental analysis. Thus, crude oil samples will generally show carbon content ranging from 83 to 87%, and hydrogen content from 11 to 14%. The remainder is composed of elements such as oxygen, nitrogen, and sulfur, and various metallic compounds. An elemental analysis, therefore, gives little indication of the extreme range of physical and chemical properties that actually exists, or of the nature of the lubricating base stocks that can be produced from a particular crude oil through conventional refining techniques.

An idea of the complexity of the lubricating base stock refining problem can be obtained from a consideration of the variations that can exist in a single hydrocarbon molecule with a specific number of carbon atoms. For example, the paraffinic molecule containing 25 carbon atoms (a compound falling well within the normal lubricating oil range) has 52 hydrogen atoms. This compound can have about 37 million different molecular arrangements. When it is considered that there are also naphthenic and aromatic hydrocarbon molecules (Figure 2.1) containing 25 carbon atoms, it will be seen that the possible variations in molecular arrangement for a 25-carbon molecule are immense. The possible variations are increased still further when heteroatoms (e.g., sulfur, nitrogen, oxygen) are considered. This accounts for much of the variation in physical characteristics and performance qualities of base stocks prepared from different crude sources.

Increasing quality demands on base stocks and the finished lubricants of which they are an integral part require that lubricant refiners have access to advanced tools to help predict crude compositions and match those with the optimum processes that yield the best overall products. Traditionally, the process of approving a specific crude for base stock manufacturing consisted of a lengthy trial-and-error process that involved a costly refinery test runs, extensive product testing, and evaluation periods of up to a year. When approvals were finally issued, they would be limited to the specific operating conditions of the test. ExxonMobil has replaced this approach with a system based on hydrocarbon characterization and compositional modeling of the crude to give the refiner the ability to select crudes and match those with process parameters to provide the best products at the lowest costs. By using the compositional modeling approach, it is possible to evaluate the feasibility and economics of any crude, for any specific lubricant refinery and predict refinery yields and finished product performance. This approach integrates all aspects of production using detailed composition analysis of crudes, resides, distillates, raffinates, and dewaxed stocks. It links all aspects to a common denominator—*composition.*

As a result, to minimize variations and produce products that will provide consistent performance in specific applications, the refiner follows four main stages in the manufacture of base stocks from the various available crudes:

1. Hydrocarbon characterization and compositional modeling of the available crudes
2. Selection and segregation of crudes according to the principal types of hydrocarbon present in them
3. Distillation of the crude to separate it into fractions containing hydrocarbons in the same general boiling range
4. Processing to remove undesirable constituents from the various fractions, or conversion of these materials to more desirable materials.

C. Crude Oil Selection

One way to understand the extreme differences that can exist among crude oils is to examine some of the products that are made from different types of crude. Crudes range from "paraffin" types, which are high in paraffin hydrocarbons, through the "intermediate" or "mixed base" types to the "naphthenic" types, which are high in hydrocarbons containing ring structures. Asphalt content varies in crudes of different types.

Table 2.1 shows two base stocks that are similar in viscosity, the most important physical property of a lubricant. The base stock on the left is made from a naphthenic crude. This type of crude is unusual because it contains essentially no wax. In fact, the very low pour point, −50°F (−46°C), of this stock results from the unique composition of the compounds in the crude—no processing has been employed to reduce the pour point. In contrast, the stock on the right required dewaxing to reduce its pour point from about 80°F (27°C) to 0°F (−18°C). One other important difference between these base stocks is shown by the differences in viscosity index. While both oils have similar viscosities at 100°F (38°C), the viscosity of the naphthenic oil will change with temperature much more than the viscosity of the paraffin stock. This is reflected in the lower viscosity index (VI) of the naphthenic oil. For products that operate over a wide temperature range, such as automotive engine oils, the naphthenic stock would be less desirable. Generally, naphthenic base stocks are used in products that have a limited range of operating temperature and call for the unique composition of naphthenic crudes—with the resultant low pour point. Long-term supply of naphthenic crudes is uncertain, and alternatives are being sought to replace these base stocks as the supply diminishes. Recognizing the large differ-

Table 2.1 Lube Base Stocks

Crude type	Naphthenic	Paraffinic
Viscosity SUS at 100°F (cSt at 38°C)[a]	100 (20.53)	100 (20.53)
Pour point, °F (°C)	−50 (−45.5)	0 (−18)
Viscosity index	15	100
Flash point, °F (°C)	340 (171)	390 (199)
Gravity, API	24.4	32.7
Color (ASTM)	1.5	0.5

[a] SUS, Saybolt universal seconds.

ences that can exist between different crudes, the major factors that must now be considered in lube crude selection are supply, refining, finished product quality, and marketing.

1. Supply Factors

Supply factors in crude selection are the quantities available, the constancy of composition from shipment to shipment, and the cost and ease of segregating the particular crude from other shipments.

Since, on average, about 10 barrels of crude oil are needed to make a barrel of lube base stock by means of the conventional refining processes, relatively large volume crudes are desirable for processing. Crude oil with variable composition will cause problems in the refinery because of the rather limited ability to adjust processing to compensate for crude changes. Since only some crudes are suitable for lube base stock manufacture by conventional refining processes, segregation of crude oils is essential. If the cost of segregation is too high, the crude will not be used for lube oil manufacture. For example, if one of the Alaskan North Slope crudes were suggested for lube processing, the inability to segregate the crude at reasonable cost would clearly eliminate it from consideration, since many of the crudes with which it would be mixed in the Trans-Alaskan Pipeline are extremely poor for conventional lube manufacture. Alternate refining processes, discussed later in this chapter, allow more flexibility in crude selection owing to the ability to convert undesirable components of the crude to desirable components.

2. Refining Factors

Refining factors important in selecting a crude oil for lube base stocks are the ratio of distillate to residuum, the processing required to prepare suitable lube base stock, and the final yield of finished lube base stocks. For a crude to be useful for lube manufacture, it must contain a reasonable amount of material in the proper boiling range. For instance, very light crudes (such as condensates) would not be considered as lube crudes because they contain only a few percent of material in the higher boiling range needed for lube base stocks.

Once it has been established that the crude contains a reasonable amount of material in the lube boiling range, the response of the crude to available processes must be examined. If the crude requires very severe refining conditions or exhibits low yields on refining, it will be eliminated. A example of this is Gippsland crude, an offshore Australian crude that met the supply criteria, had reasonable distillate yields, and even responded well to furfural extraction. However, in the dewaxing process, Gippsland, because of its very high wax content, showed an extremely low yield of dewaxed oil. The dewaxing yield was so far out of line that it was not possible to process this crude economically by conventional extractive processing.

3. Product Quality Factors

Product factors concern the quality aspects of all the products refined from the crude—not only the lube base stock. These product qualities include the base stock quality and its response to presently available additives and, also, the quality of light products and by-products extracted from the crude.

Since almost 90% of the crude will end up in nonlube products, this portion cannot be ignored. In some situations, the quality of a certain by-product (e.g., asphalt) can be of overriding importance in the evaluation of a crude.

Table 2.2 Examples of Satisfactory Crudes for Lube Base Stock Manufacture

Arabian
Extra Light
Light
Medium
Heavy
Basrah
Citronelle
Iranian Light
Kirkuk
Kuwait
Lago Medio
Louisiana Light
Luling/Lyton/Karnes
Mid-Continent Sweet
Raudhatain
West Texas Bright

The lube base stock produced from a crude must not only be satisfactory in chemical and physical characteristics but must respond to the additives that are readily available on the market. Lube base stocks produced from different crudes and/or different refining processes may respond differently to specific additives and resultant finished lubricant performance characteristics could be effected. If new or different additives are needed because of a new crude or refining process, the economics of using this crude for lubes must support the cost of finding these additives and implementing them within the system.

4. Marketing Factors

Marketing factors to be considered in evaluating a crude oil for lube base stock manufacture include the viscosity range and overall product quality required by the lube oil market, and the operating and investment costs of manufacturing a lube oil based on the market product sales price.

The location (market) in which the crude oil will be used can have a major impact on the economics of its use for lubes. A crude that contains a great deal of low viscosity material would be ideal (in this respect) for use in the United States. Although the U.S. product demand requires a lot of low viscosity product, this requirement might be totally unsuitable for use in a different market in which larger amounts of high viscosity oil were necessary. Likewise, the product quality required depends very strongly on the market being served, as do operating and investment costs. Therefore, in addition to the physical and chemical composition of a crude oil, selection finally becomes an economic business decision.

Having discussed the factors that must be considered in selecting a crude oil for lube base stock manufacture, we list examples of satisfactory crudes in Table 2.2.

III. LUBRICANT BASE STOCKS

Lube base stocks make up a significant portion of the finished lubricants, ranging from 70% of automotive engine oils to 99% of some industrial oils. The base stocks contribute

Table 2.3 API Base Stock Categories

Category	Amount (%) of		VI
	Saturates	Sulfur	
API group I (solvent-refined)	< 90	> 0.03	80–120
API group II (hydroprocessed)	≥ 90	≤ 0.03	80–120
API group III	≥ 90	≤ 0.03	≥ 120
API group IV is polyalphaolefins			
API group V is for ester and other base stocks not included in groups I–IV			

significant performance characteristics to finished lubricants in areas such as thermal stability, viscosity, volatility, the ability to dissolve additives and contaminants (oil degradation materials, combustion by-products, etc.), low temperature properties, demulsibility, air release/foam resistance, and oxidation stability. This list, indicates the importance of it base stock processing and selection, along with the use of proper additives and blending procedures, in achieving balanced performance in the finished lubricant.

As mentioned earlier, the two basic refining processes for obtaining lubricant base stocks are those for separation and conversion. Sometimes the base stocks produced by these methods are referred to as *conventional base oils* and *unconventional base oils,* respectively. Conventional refining technology involves the separation of the select desirable components of the crude by distillation, solvent extraction, and solvent dewaxing. Some additional steps or modifications such as hydrofinishing can be added to this process but would still be classified as conventional. This process is used in about two-thirds of the world's production of paraffinic base stocks.

The American Petroleum Institute (API) has defined five categories of lubricant base stocks to try to separate conventional, unconventional, synthetic, and other classifications of base stocks. Of these five categories, groups I, II, and III are mineral oils and are classified by the amounts of saturates and sulfur and by the viscosity index of each. Group IV is reserved for polyalphaolefins (see Chapter 5, Synthetics) and group V is ester and other base stocks not included under groups I–IV. The API classification system is based on the base stock characteristics as just mentioned, not on the refining process used. Group III base stocks are very high VI products that are typically achieved through a hydrocracking process. The categories in the API system are given in Table 2.3.

If a given base stock falls under a group I classification, it does not necessarily mean that it is better or worse than a base stock that falls under a group II classification. Although the group II base stock would have lower levels of sulfur and aromatics, increased potential for improved oxidation stability, and a higher viscosity index, it may provide poorer solubility of additives and contaminants than a conventionally refined base stock that falls under group I. The real measurement of the base stock suitability for formulating finished lubricants is in the performance of the finished lubricants.

IV. LUBE REFINING PROCESSES

The most common processes used to produce lube base stocks in refineries worldwide involve separation processes that is, processes that operate by dividing feedstock, which

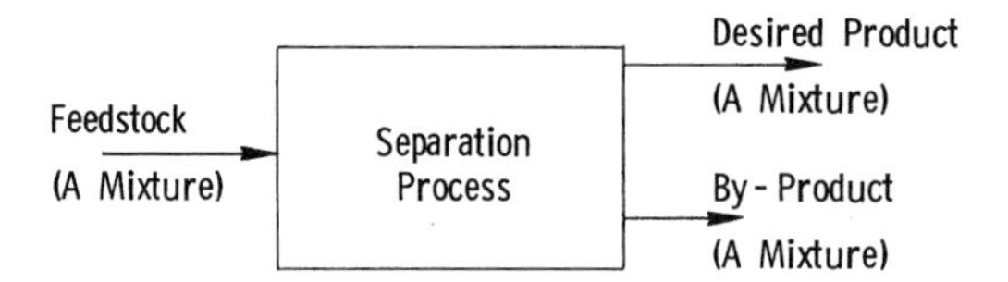

Figure 2.3 Lube separating process.

is a complex mixture of chemical compounds, into products. Usually, this results in two sets of products: the desired lube product and by-products. Thus, although the products themselves are complex mixtures, the compounds in each of the products are similar in either physical or chemical properties. On the other hand, the fastest-growing method for lube manufacture is by the alternate conversion process, which involves converting undesirable structures to desirable lube molecules under the influence of hydrogen pressure and selected catalysts.

The concept of a separation process is basic to understanding lube base stock manufacture. Figure 2.3 is a simple diagram of a separation process. By comparison, Figure 2.4 shows a simple schematic of a hydroprocessing conversion process. While the desired lube products from the two processes have many similarities, the respective by-products of the two processes are quite different, because of the different processes. However, while the basic properties (e.g., viscosity) of the desired products from the two processes are similar, there are differences in hydrocarbon structure and heteroatom (S, N, O) content that can be important in final quality. This is discussed further in this chapter.

The two processes are compared in Figure 2.5, which shows the (alternate) paths, with approaches starting with distillation processes (extraction). Following vacuum distillation, the extraction approach includes solvent extraction (propane deasphalting and removal of aromatics with furfural or other solvent), removal of waxy components by solvent extraction with methyl ethyl ketone (MEK) or other solvent, and finally a clay "finishing" process, which removes some heteroatoms. For the conversion approach, the primary upgrade is through catalytic hydrotreatment, which results in conversion of hydrocarbons to more desirable structures (as well as some removal of heteroatoms as gases). Conversion uses a separate catalytic hydrogen process for conversion of waxy paraffins and employs a final hydrotreatment step as finishing step. Also, ExxonMobil has pioneered the use of hydrodewaxing with solvent-upgraded stocks. It is also possible to employ solvent extraction and hydrotreatment in combination for primary upgrading. Such approaches are known as "hybrid" processing.

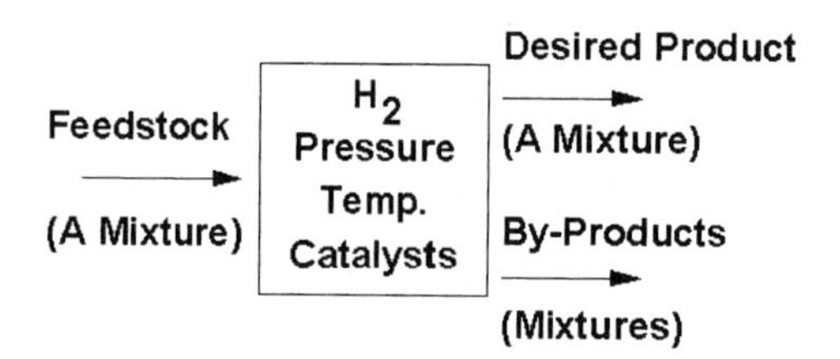

Figure 2.4 Lube conversion process.

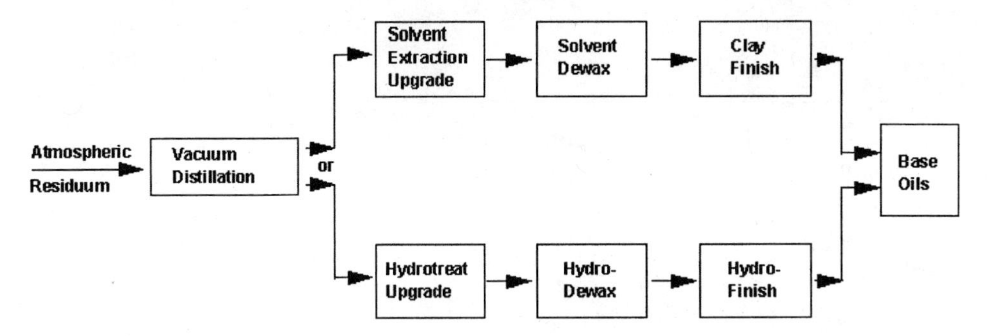

Figure 2.5 Lube processing schemes extraction (top) and conversion (bottom).

Sections IV.A and IV.B briefly discuss the two primary processes employed to produce lube base stocks; a more detailed discussion of all the processes used by the various refining techniques is given in Section V.

A. Lube Separation (Extractive) Process

A simplified block flow diagram (Figure 2.6) indicates the five processes in conventional lube oil refining:

1. Vacuum distillation
2. Propane deasphalting
3. Furfural extraction (solvent extraction)
4. Methyl ethyl ketone (MEK) dewaxing/hydrodewaxing
5. Hydrofinishing

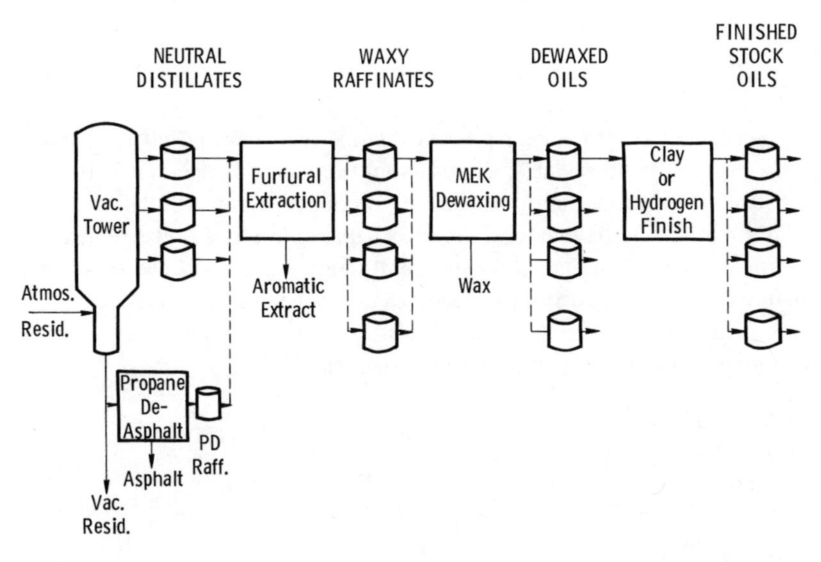

Figure 2.6 Lube separation process.

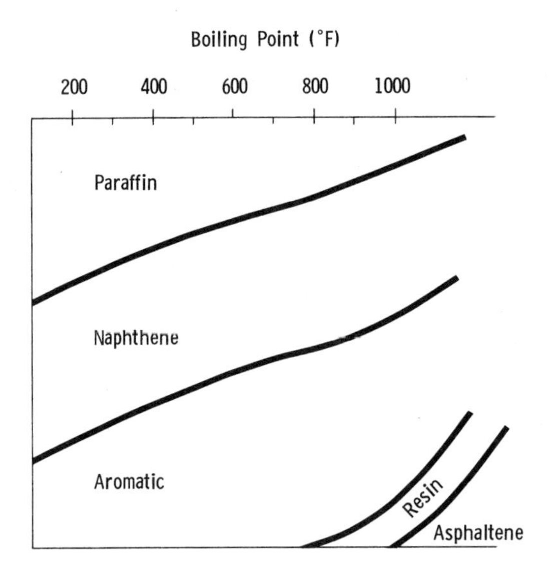

Figure 2.7 Simplified diagram of crude oil composition.

The first four items are separation processes. The fifth, hydrofinishing, is a catalytic reaction with hydrogen to decolorize the base stocks and further remove or convert some of the undesirable components to desirable components (isoparaffins). The purpose of these processes is to remove or convert materials that are undesirable in the final product.

Before discussing these processes in detail, a brief discussion may be useful in clarifying their interrelationship. A simplified representation of a crude oil (Figure 2.7) shows that the crude consists not only of compounds (paraffins, naphthenes, aromatics) that are chemically different but also of compounds that are chemically similar (e.g., paraffins) but differ in boiling point.

1. Vacuum Distillation

Assuming that an acceptable crude has been processed properly in the atmospheric distillation column for recovery of light products, the residuum from the atmospheric distillation column is the feedstock for the vacuum distillation column. Vacuum distillation is the first step in refining lubricating base stocks. This is a separation process that segregates crude oil into products that are similar in boiling point range. In terms of the simplified picture of crude oil in Figure 2.7, distillation can be represented as a vertical cut, as in Figure 2.8, where distillation divides the feedstock (crude oil) into products that consist of materials with relatively narrow ranges of boiling points.

2. Propane Deasphalting

Propane deasphalting (PD) operates on the very bottom of the barrel—the residuum. This is the product shown in the simplified distillation on the right of Figure 2.8, the highest boiling portion of the crude. Note that the residuum in Figure 2.8 contains some types of compound not present in the other products from distillation—resins and asphaltenes. PD,

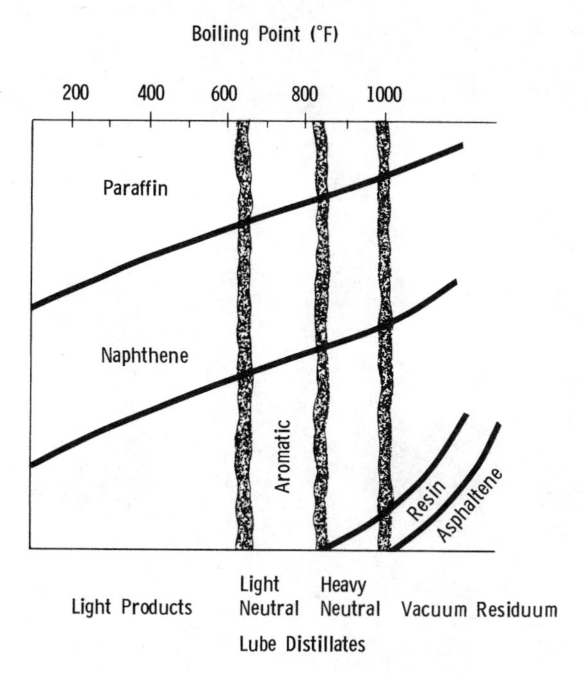

Figure 2.8 Lube distillation.

which removes these materials, can also be represented as a horizontal cut, as in Figure 2.9. The residuum has now been divided into two products, one containing almost all the resins and asphaltenes, called PD tar, and the other containing compounds that are similar chemically to those in the lube distillates but of higher boiling point. This material is called deasphalted oil and is refined in the same way as lube distillates.

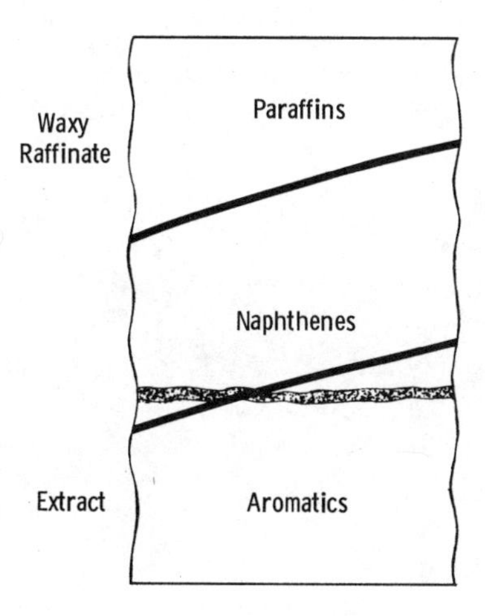

Figure 2.9 Propane deasphalting.

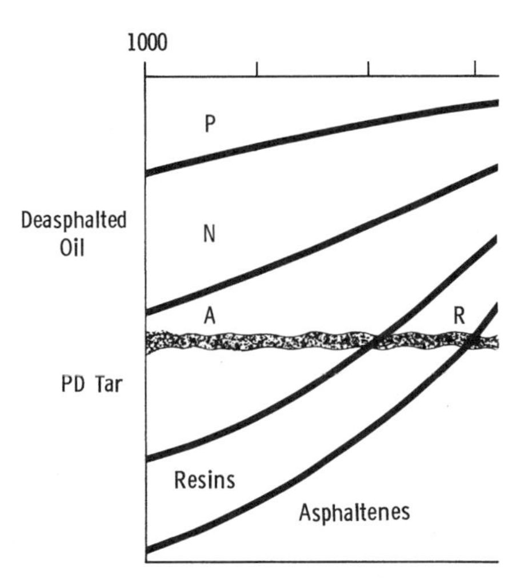

Figure 2.10 Furfural extraction.

3. Furfural Extraction

Furfural extraction, the next process in the sequence, can be represented as a horizontal cut (Figure 2.10), dividing the distillate or deasphalted oil into a raffinate, which is the desired material, and a by-product called an extract, which is mostly aromatic compounds. Solvents other than furfural can be used for the extraction process.

4. MEK Dewaxing

Methyl ethyl ketone (MEK) dewaxing of the raffinate from the furfural extraction is another horizontal cut (Figure 2.11), producing a by-product wax that is almost completely paraffins and a dewaxed oil that contains paraffins, naphthenes, and some aromatics. This dewaxed oil is the base stock for many fluid lubricants. For certain premium applications, however, a finishing step is needed.

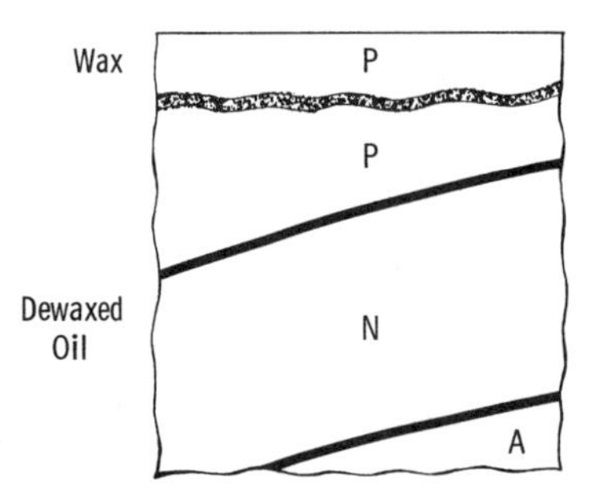

Figure 2.11 Dewaxing.

5. Hydrofinishing

Hydrofinishing by chemical reaction of the oil with hydrogen changes the polar compounds slightly but retains them in the oil. In this process, the most obvious result is oil of a lighter color. From the simplified description of crude oil, the product is indistinguishable from the dewaxed oil feedstock. Hydrofinishing has superseded the older clay processing of conventional stocks because of simpler, lower cost operation.

The purpose of this simplified look at lube processing is to emphasize that it is a series of processes to remove undesirable materials or conversion of the undesirable materials to isoparaffins. Thus, for a crude oil to be used for lube base stock manufacture, it must contain some good base oil. The residuum from the atmospheric distillation column is the feedstock for the lube vacuum distillation column.

B. Lube Conversion Process

Hydroprocessing offers methods to further aid in the removal of aromatics, sulfur, and nitrogen from the lube base stocks. Two of the hydroprocessing methods use the feedstock from the solvent refining process (hydrofinishing and hydrotreating) and a third, hydrocracking, uses the vacuum gas oil (VGO) from the crude distillation unit. The lube base stocks produced by hydrofinishing and hydrotreating would be classified as API group I and II base stocks; the base stocks produced using the hydrocracking process would generally be classified in groups II and III. As mentioned earlier in this chapter, classification into an API group is based on the base stock specifications, not the process for refining.

A simplified block flow diagram (Figure 2.12) for hydrocracking indicates four primary processes:

1. Vacuum distillation
2. Hydrocracking
3. Hydrodewaxing
4. Hydrotreating

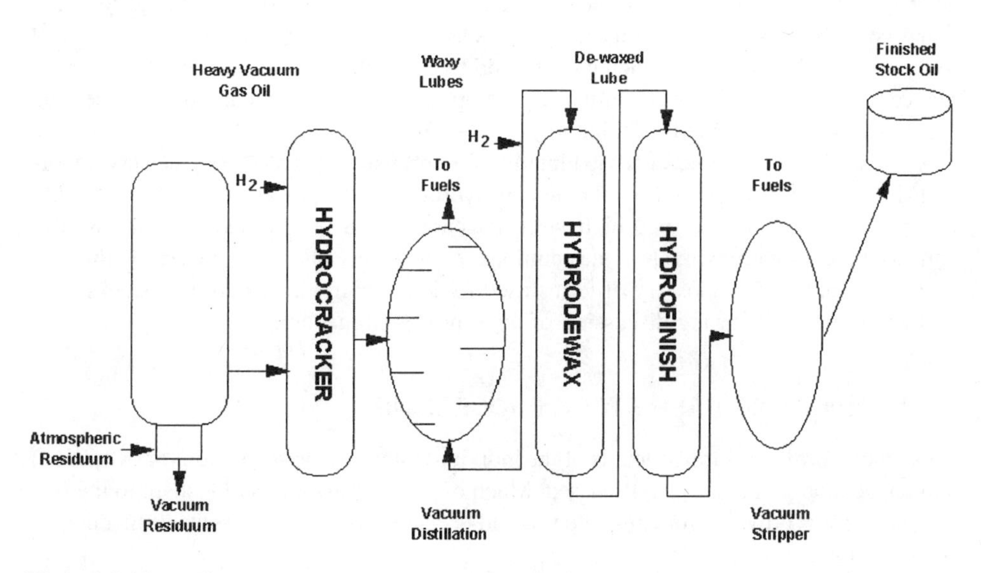

Figure 2.12 Lube conversion process.

In addition to the four primary steps, additional vacuum distillation and vacuum stripping are necessary to produce the needed viscosity grades and volatility characteristics.

1. Vacuum Distillation

While the vacuum distillation process is the same one used in extractive processing, in this case the propane deasphalter is omitted because the vacuum tower residuum is normally not hydroprocessed; rather, it is processed conventionally by solvent treatment.

2. Hydrocracking

The hydrocracking unit is a catalytic processing unit that converts less desirable hydrocarbon species to more desirable species in the presence of hydrogen at pressures up to 3000 psi. Typically, aromatic and naphthene rings are opened to produce a higher portion of saturated paraffinic molecules. The hydrogen also removes heteroatoms as gases—hydrogen sulfide, ammonia, and water.

3. Hydrodewaxing

The hydrodewaxing unit, like the hydrocracker, is also a catalytic hydrogenation unit, but in this case the catalyst employed is specific to converting waxy normal paraffins to more desirable isoparaffin structures.

4. Hydrotreating

Because the hydrocracking and hydrodewaxing processes involve breaking of carbon–carbon bonds, it is necessary to have a final hydrotreating stage to induce saturation of any remaining unsaturated molecules, which could cause thermal or oxidative instability in the base oil and finished products blended from it.

5. Properties of Hydroprocessed Stocks

ExxonMobil views the use of hydroprocessed base stocks as another component of its arsenal aimed at achieving finished lubricants that exhibit superior performance while being cost-effective to the end user. Proper selection of base stocks—conventional and hydroprocessed—and the balancing of the additives are the key to achieving total performance. This set of procedures continues to comprise the best overall approach in meeting the needs of customers and machinery.

Hydroprocessed base stocks exhibit different properties (Table 2.4). One very important difference is the general availability of hydroprocessed stocks above ISO VG 100, except for some hydrofinished high viscosity base oils. Selective additives are used to enhance these properties and provide additional performance characteristics in the finished lubricant. Table 2.5, a glossary of terms used to define hydroprocessed base stocks, will be helpful in the following discussion of hydroprocessing techniques.

V. LUBRICATING BASE STOCK PROCESSING

This section provides further details of the individual steps of the separation (or extractive) and conversion processes just discussed. Much of the information is redundant to the two processes but is provided for completeness and because this section presents further details on the following processes:

Table 2.4 Comparing Hydroprocessed Base Stocks

Base stock type	Hydrofinished	Hydrotreated VHQ	Hydrocracked HDC, VHVI, or XHQ	Solvent-refined (reference)
Processing severity	Low	Moderate	High	—
Process purpose	Saturate olefins, aromatics; remove S and N	Saturate olefins, aromatics; remove S and N	Saturate olefins; remove S and N Open Ring structures	—
Feedstock	Solvent-refined	Solvent-refined	Distillate HDC and VHVI distillate wax XHQ	—
Inspection properties	Little change from feedstock	Lower aromatics, S and N	Very low aromatics, S, and N	
Viscosity index	90–105	105 +	95–105 (HDC), 115–130 (VHVI), 130 + (XHQ)	90–105
ISO viscosity grade range	22–460	22–100	22–100 (HDC/VHVI) 22–46 (XHQ)	22–460
Performance attributes	Improved color, demulsibility, foam stability	Improved oxidation stability	Much improved oxidation stability	—
Solubility	Excellent	Very good	Moderate	Excellent

1. Vacuum distillation
2. Propane deasphalting
3. Furfural extraction
4. Solvent extraction
5. Hybrid processing
6. MEK dewaxing
7. Catalytic dewaxing
8. Wax isomerization
9. Hydrofinishing
10. Hydrotreating
11. Hydrocracking

Because increasing quality demands are being placed on lubricating oils for some critical applications, base stocks must be able to provide much higher performance levels. Moreover, because additives cannot always overcome deficiencies in base stock characteristics, lube refiners must have access to or develop sophisticated models such as the hydrocarbon characterization and compositional models discussed earlier in this chapter. Making high quality lubes at reasonable costs requires the most up-to-date process design and the best use of catalysis.

Table 2.5 Glossary of Base Oil Terms

Aromatics. Highly unsaturated cyclic hydrocarbon with a high level of reactivity.

Atmospheric resid. High boiling bottoms from atmospheric pressure distillation.

Distillate. Cut or fraction taken from vacuum distillation tower having viscosity suitable for lubricants.

HDC. Hydrocracked base stock. Produced from distillate, HDC typically exhibits a viscosity index (VI) of 95–105.

HDP. Hydroprocessed; adjective for base stock manufacturing by hydroprocessing.

Hydrocracking. Processing that requires more more severe conditions for hydroprocessing to convert aromatics to naphthenes and paraffins.

Hydrodewaxing (or catalytic dewaxing). Converts normal paraffins (wax) to fuel components or iso paraffins.

Hydrofinishing. Mild Hydroprocessing, usually following conventional solvent refining to saturate olefins. Results improve color, demulsibility, and foam characteristics. Little change in general inspection properties.

Hydroprocessing. Use of hydrogen and catalysts to remove and/or convert less desirable components of crude into more desirable lubricant components. This is a general term for the various lube hydrogen processes.

Hydrotreating. Moderate severity hydroprocessing that converts aromatics to naphthenes and reduces sulfur and nitrogen levels.

Iso-paraffins. Preferred hydrocarbon type for lubricant base stocks, giving excellent oxidation stability and low temperature fluidity.

MLDW. Mobil Lube (oil) DeWaxing, a patented hydrodewaxing process.

MSDW, Mobil Selective DeWaxing, a patented process specifically targeted for hydroprocessed stocks.

MWI. Mobil Wax Isomerization, another patented process.

Normal paraffins. Straight chain hydrocarbons, exhibiting wax characteristics.

Olefins. Hydrocarbon with areas of unsaturation.

PAO. Polyalphaolefin, a pure isoparaffin with excellent lubricant characteristics.

Polars. Sulfur-, nitrogen-, and oxygen-containing compounds that can promote oxidation instability.

Raffinate. Product of solvent extraction after aromatics removal.

Saturate. To add hydrogen to aromatics or olefins to increase resistance to chemical reactivity.

Solvent dewaxing. Use of solvents to remove normal paraffins (wax) from base stocks for improved low-temperature fluidity.

Solvent extraction. Use of solvents to remove aromatics and sulfur from vacuum distillates to increase VI and oxidation stability.

SUS. Saybolt Universal Seconds, a standard measure of viscosity determined by using testing equipment known as a Saybolt Universal Viscometer.

TAN. Total acid number, a measure of acid level in a lubricating oil or hydraulic fluid.

VGO. Vacuum gas oil the output of the atmospheric distillation phase after normal of low boiling point fractions.

VHQ. ExxonMobil's acronym for a very high quality base for severe turbine and other applications. Produced by hydrotreating of solvent-refined base stocks.

VHVI. Very high viscosity index base stocks, produced by more severe hydrocracking. VIs are in the range of 115–130.

VI. Viscosity index, a standard measure of the rate change of viscosity with temperature. The higher the VI, the lower is the rate of change with changes in temperature.

XHQ. Extrahigh quality base stocks produced by hydrocracking wax; VI 130. ExxonMobil uses the MWI process to produce XHQ stocks.

A. Vacuum Distillation

Vacuum distillation is considered to be the first process in lube base stock manufacturing. This assumes that an acceptable crude has been processed properly in an atmospheric distillation column for recovery of light products. The portion of the residuum from the atmospheric distillation column that boils between 650°F (340°C) and 700°F (371°C) is not used as light product because of its adverse specifications (pour point or cloud point of diesel or heating fuel). An additional upper limitation to which an atmospheric column can be operated is imposed by thermal cracking. If a crude is heated excessively, the molecules will be broken up into smaller, lower boiling point molecules. These thermally cracked materials can cause product quality problems as well as operating problems in the atmospheric distillation unit.

In vacuum distillation, the feedstock is separated into products of similar boiling point to control the physical properties of the base stocks that will be produced from the vacuum tower. The major properties that are controlled by vacuum distillation are viscosity, flash points, and carbon residue. The viscosity of the oil base stock, the most important physical property for lubrication, is determined by the viscosity of the distillate.

Figure 2.13 illustrates vacuum distillation in terms of viscosity separation. The heavy curve represents the distribution of viscosity in a specific crude oil. The figure shows three vacuum tower distillates and a residuum (boxed areas). Notice that each distillate contains material displaying a range of viscosity both higher and lower than the average viscosity. In the example, some overlap is shown, and, in reality, overlap is present in the products of commercial vacuum towers. However, to avoid downstream processing problems (e.g., a light neutral with too much viscous material, which will be difficult to dew), a vacuum tower must be run properly, to prevent excessive overlap. High viscosity material in a heavy neutral (i.e., too much overlap with the vacuum residuum) not only causes processing problems but results in inferior product quality.

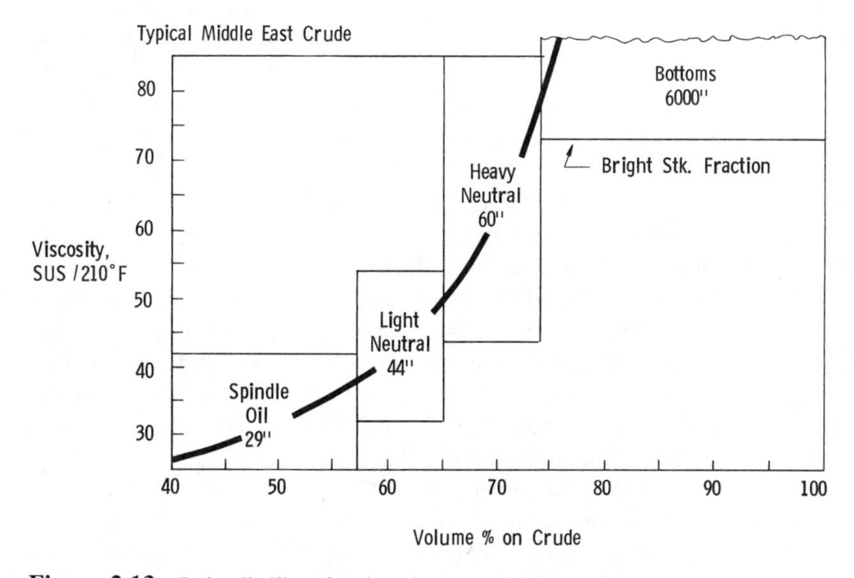

Figure 2.13 Lube distillate fractionation viscosities.

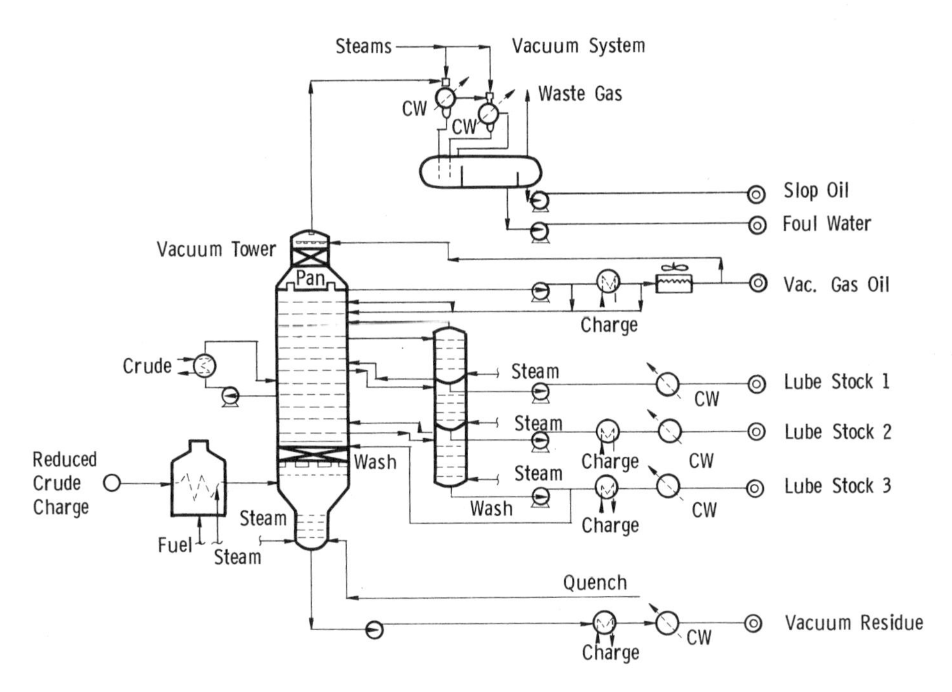

Figure 2.14 Vacuum tower fractionation for lube distillates.

Figure 2.14 is a simplified flow diagram of a lube vacuum column. The crude feed (a reduced crude charge consisting of the residuum from the atmospheric distillation) is heated in a furnace and flows into the flash zone of the column, where the vapor portion begins to rise and the liquid falls. Temperatures of about 750°F (399°C) are used in the heater, and steam is added to assist in the vaporization. A vacuum is maintained in the flash zone by a vacuum system connected to the top of the column. By reducing the pressure to less than one-tenth of atmospheric pressure, materials boiling up to 1000°F (538°C) at atmospheric pressure can be vaporized without thermal cracking. As the hot vapors rise through the column, they are cooled by removing material from the column, cooling it, and returning it to the column.

The liquid in the column wets the packing and starts to flow back down the column, where it is revaporized by contacting the rising hot vapors. Special trays, called draw trays, are installed at various points in the column; these permit the collection and removal of the liquid from the column. The liquid that is withdrawn contains not only the material that is normally liquid at the temperature and pressure of the draw tray but also a small amount of lower boiling material dissolved in the liquid. To remove this lower boiling material, the distillate after removal from the vacuum tower is charged to a stripping column, where steam is introduced to strip out these low boiling materials.

In terms of the physical properties of the distillate, the stripping column adjusts the flash point by removing low boiling components. The low boiling materials and the steam are returned to the vacuum column. Other lube distillates are stripped similarly. The vacuum residuum is also steam stripped; however, this is generally done internally in the vacuum tower in a stripping section below the flash zone.

B. Propane Deasphalting

The highest boiling portions of most crude oils contain resins and asphaltenes. To provide an oil with acceptable performance, these materials must be removed. Traditionally, deasphalting of vacuum residuum has been used to remove these materials. Figure 2.15 is a simplified illustration of deasphalting. As shown here, a solvent (generally propane) is mixed with the vacuum residuum. The paraffins, naphthenes, and aromatics are more soluble in the solvent than the resins and asphaltenes. After mixing, the system is allowed to settle, and two liquid phases form (the analogy of a system of oil and water may be helpful in visualizing the two liquid layers). The top layer is primarily propane containing the soluble components of the residuum. The bottom layer is primarily the asphaltic material with some dissolved propane. The two phases can be separated by decanting, and upon removal of the solvent from each, the separation is accomplished.

The separation is carried out in the refinery in one continuous operation (Figure 2.16). The residuum (usually diluted with a small amount of propane) is pumped to the middle of the extraction column. Propane (usually about 6–8 volumes per volume of residuum) is charged to the bottom of the column. Since the residuum is more dense than the propane, the residuum will flow down the column, with the propane rising up in a counterflow. The mixing is provided within the column, by either perforated plates or a rotor with disks attached. The rising propane dissolves the more soluble components, which are carried out the top of the column with the propane. The insoluble, asphaltic material is removed from the bottom of the column. Temperatures used in the column range from approximately 120°F (50°C) to 180°F (80°C). To maintain the propane as a liquid at the temperatures used, the column must be operated under pressure (about 500 psig, or 35 atm). Propane is vaporized from the products and is then recovered and liquefied for recycling by compressing and cooling it.

Figure 2.17 shows the types of chemical compound separated in deasphalting: the residuum composition is defined in terms of component groups measured by a solid–liquid chromatographic method. The components are saturates (paraffins and naphthenes); mononuclear, dinuclear, and polynuclear aromatics; resins; and asphaltenes. The tar is composed mostly of resins and asphaltenes.

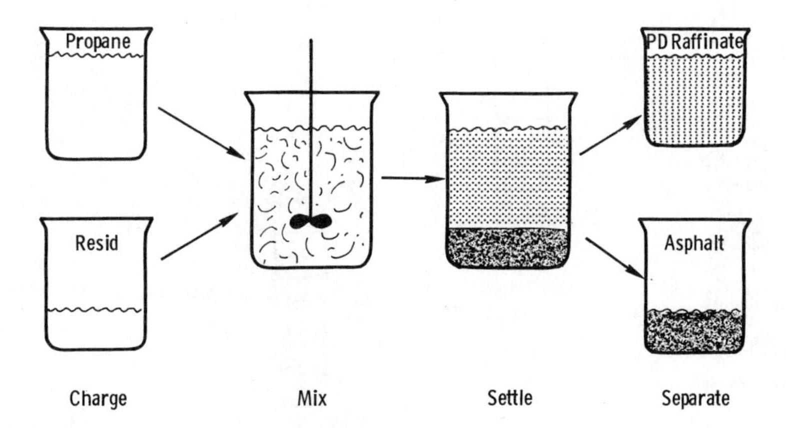

Figure 2.15 Simplified propane deasphalting.

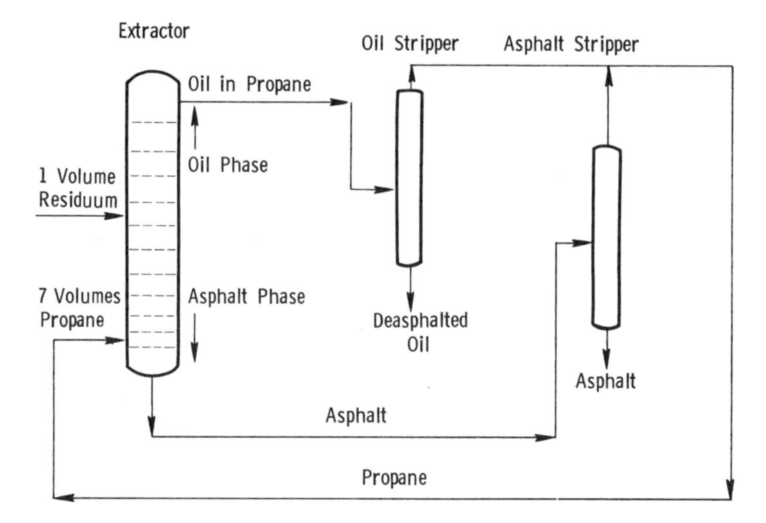

Figure 2.16 Continuous propane deasphalting process.

The propane-deasphalted (PD) oil is composed primarily of saturates and mononuclear, dinuclear, and polynuclear aromatics. However, since the separation is not ideal, a small amount of these components is lost to the tar, while the PD oil is left with small amounts of resins and asphaltenes. After deasphalting, the PD oil is processed the same manner as the lube distillates.

While no widely accepted substitute for propane deasphalting has been available, there is a combination process called Duo-Sol that combines in one process both propane

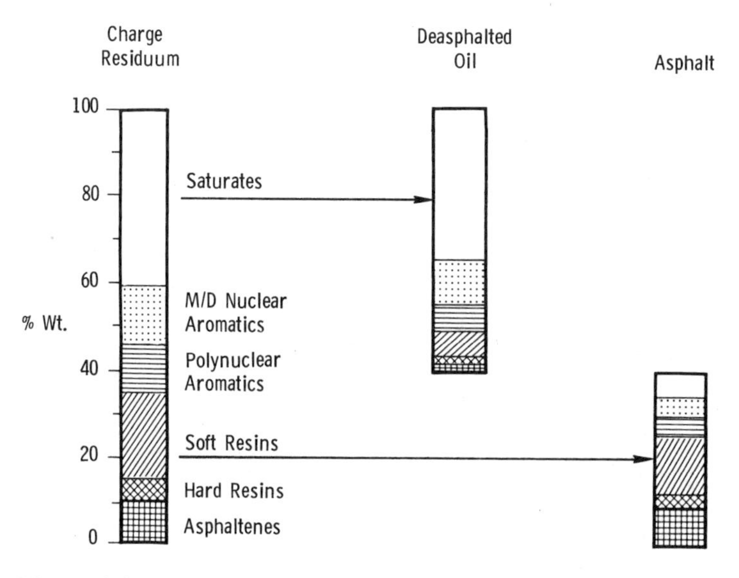

Figure 2.17 Propane deasphalting of mid-continent sweet residuum.

deasphalting and extraction. The solvent extraction portion of the Duo-Sol process employs either phenol or cresylic acid as the extraction solvent.

C. Furfural Extraction

Before discussing furfural extraction and the subsequent processing, it should be pointed out that from this stage on, a lube refinery differs from most fuel refineries in another important aspect. Lube units process very different feedstocks at various times. For example, a furfural extraction unit will process not only the deasphalted oil from the PD unit but the distillate feeds from the vacuum tower as well. These different feedstocks are processed in ''blocked out'' operation. This means that while one of the feeds is being processed, the others are collected in intermediate tankage and are processed later. Runs of various lengths, from days to weeks, are scheduled so that demands are met and intermediate tankage requirements are not excessive. During the switching from one stock to another, some transition oil is produced and must be disposed of. One other consequence of this blocked-out type of operation is that each section of the process units must be designed to handle the most demanding service. As a unit's capacity for various stocks may be limited by the requirements imposed on its different sections.

Furfural extraction separates aromatic compounds from nonaromatic compounds. In its simplest form, the process consists of mixing furfural with the feedstock, allowing the mixture to settle into two liquid phases, decanting, and removing the solvent from each phase. This can be demonstrated in a glass graduated cylinder, where the more dense furfural dissolves the dark-colored aromatic materials from a distillate, leaving a lighter-colored raffinate product. The resultant product from the furfural extraction shows an increase in thermal and oxidative stability as well as an improvement in viscosity and temperature characteristics, as measured by a higher viscosity index (VI).

Figure 2.18 is a simplified flow diagram of a commercial furfural extraction unit. The feedstock is charged to the middle of the extraction column, the furfural near the top. The density difference causes a counterflow in the column; the downward-flowing furfural

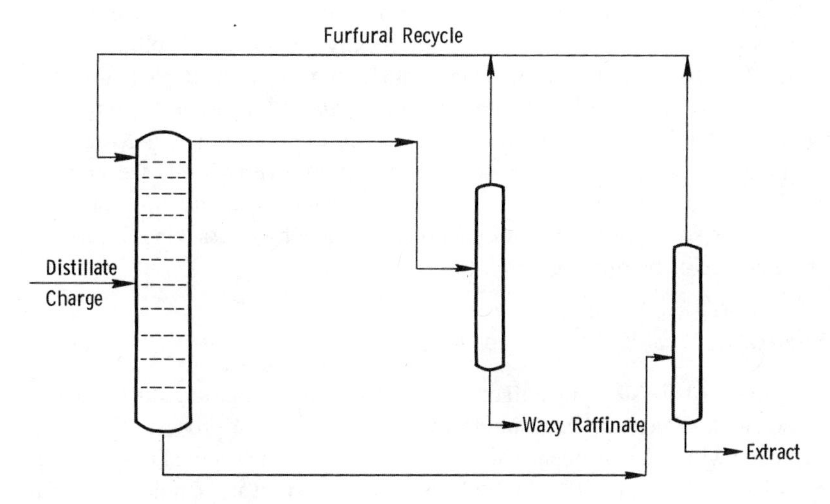

Figure 2.18 Furfural extraction.

dissolves the aromatic compounds. The furfural raffinate rises and is removed from the top of the column. The furfural extract is removed from the bottom of the column. Each of the products is passed to the solvent recovery system, with the furfural being recycled as feed to the extractor. The solvent recovery system, which is not discussed here in any detail, is much more complicated than this treatment suggests.

The major effect of furfural extraction on the physical properties of a base stock is an increase in viscosity index—an improvement in the viscosity–temperature relationship of the oil. However, equally important, but less obvious, changes in the base stock result. Although oxidation and thermal stability are improved, there is no physical property of the base stock that is easily measurable that can be related to these characteristics. Thus, while viscosity index is sometimes used to monitor the day-to-day operations of a furfural extraction unit, VI is only an indication of the continuity of the operation, not an absolute criterion of quality. The quality of base stocks for a given product (and the refining conditions needed to produce the base stocks) is arrived at by extensive testing, ranging from bench-scale to full-fleet testing programs. Careful control of the operation of the refining units is essential to assure the continuous production of base stocks meeting all the quality criteria needed in the final products.

D. Solvent Extraction

Another solvent extraction process that has been used employs phenol as the solvent. The capacity and the number of phenol extraction units in operation (if Duo-Sol units are included with phenol) exceed those available for furfural extraction. However, recent trends indicate that the use of phenol extraction will decline. ExxonMobil's recent solvent extraction units are based on furfural extraction. The action of these solvents on the oil charge is quite similar, although different operation conditions are used with each solvent. Some lube refineries use *N*-methylpyrrolidone (NMP) as the extraction solvent.

E. Hybrid Processing

Certain lubricants, used in critical applications, require base stocks with high viscosity indices (> 105 VI) and very low sulfur contents. While solvent extraction can achieve this VI level with some crudes, raffinate yields are typically very low and sulfur removal is limited to about 80%. An alternative processing route for high VI is to couple hydrotreating with extraction. This is commonly called *hybrid processing*. While hydrotreating helps to preserve yield, the chemical reactions that occur result in some additional viscosity loss. By properly targeting the correct distillate viscosity, extraction severity, and hydrotreating severity, high VI base stocks can be produced at economic yields. Another application of *hybrid processing* consists of mild furfural extraction and lube hydrocracking to produce base stocks from crudes of marginal quality.

F. MEK Dewaxing

The next process in lube base stock manufacture is the removal of wax to reduce the pour point of the base stock. Figure 2.19, a simple representation of the process, shows the waxy oil being mixed with MEK–toluene solvent. The mixture is then cooled to a temperature between 10°F (−12°C) and 20°F (−6°C) below the desired pour point. The wax crystals that form are kept in suspension by stirring during the cooling. The wax is then

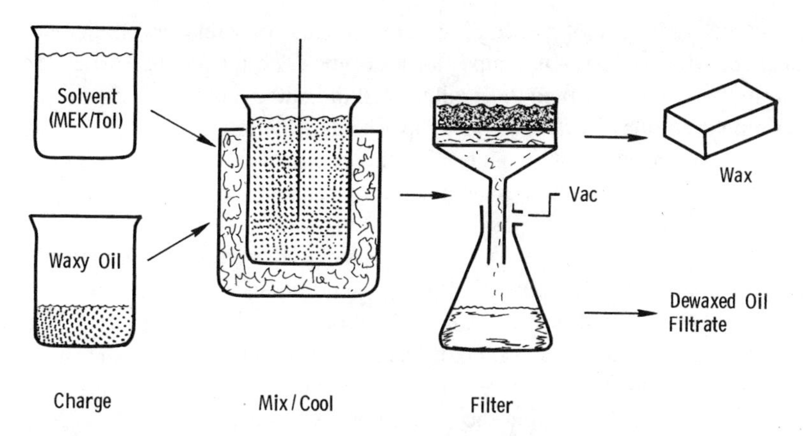

Figure 2.19 Simplified MEK–toluene dewaxing for lubes.

removed from the oil by filtration. Solvent is removed from both the oil and the wax and recycled for reuse.

A simplified flow diagram for a commercial dewaxing unit is shown in Figure 2.20. In this unit, the waxy oil is mixed with solvent and heated sufficiently to dissolve all the oil and wax. The purpose of this step is to destroy all the crystals that are in the oil so that the crystals that will be separated at the filter are formed under carefully controlled conditions. The solution is then cooled, first with cooling water, then by heat exchange with cold product, and finally by a refrigerant. In some cases, more solvent is added at various points in the heat exchange train.

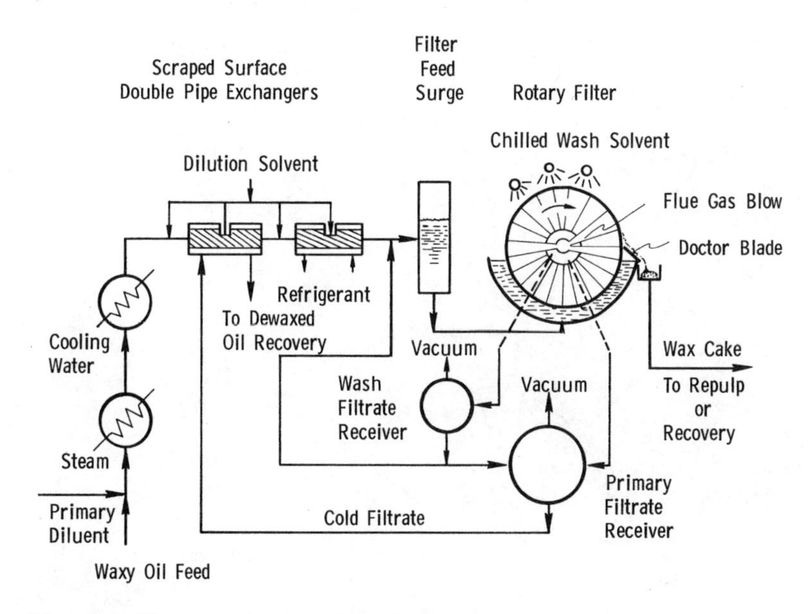

Figure 2.20 Dewaxing for lubes: MEK process.

One other distinctive feature of this cooling train is the use of scraped-wall, double-pipe heat exchangers, which consist of a pipe inside a pipe. The inner pipe carries the solvent–oil–wax mixture; the outer pipe the cooling medium, either cold product or refrigerant. The inner pipe is equipped with a set of scraper blades that rotate and scrape away any wax that plates out on the walls of the inner pipe. This action is necessary to maintain a reasonable rate of heat transfer. Although the method is efficient, scraped-wall, double-pipe heat exchangers are expensive and costly to maintain.

The cooled slurry is passed to filter feed surge drum and then to the filter itself, where the actual separation is accomplished. Rotary vacuum filters used in dewaxing plants are large drums covered with a filter cloth, which prevents the wax crystals from passing through to the inside of the drum as the drum rotates in a vat containing the slurry of wax, oil, and solvent. A vacuum applied inside the drum pulls the solvent–oil mixture (filtrate) through the cloth, thus separating the oil from the wax.

Figures 2.21–2.24 illustrate the operation of this type of filter: by looking at the end of the drum in each one, we can follow a single segment as the drum rotates through one revolution. In Figure 2.21, the segment is submerged in the slurry vat and is building up a wax cake on the filter cloth. The filtrate is being pulled into the interior of the drum. As the drum rotates through the slurry, the wax cake will increase in thickness and, as the segment leaves the vat, the vacuum state is maintained for a short period to dry the cake and remove as much oil and solvent as possible for the wax cake. Figure 2.22 shows the cold wash solvent being applied to the cake, displacing more of the oil in the cake. The wash portion of the cycle is followed by another short period of drying. The wax cake (Figure 2.23) is lifted from the filter by means of flue gas. This is accomplished by applying a positive pressure to the inside of the drum. As the drum rotates, the wax cake is guided from the drum by means of a blade (Figure 2.24), which directs it to a conveyor and then to the solvent recovery system. This segment of the drum is then ready to reenter the vat and continue with another cycle of pickup, dry, wash, dry, cake lift, and wax removal.

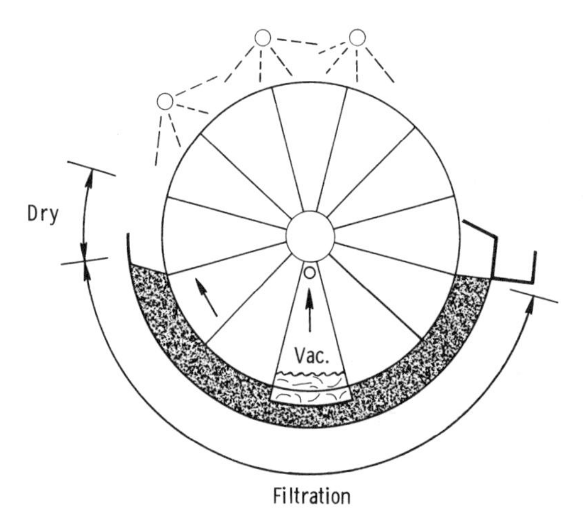

Figure 2.21 Dewaxing cycle filtration.

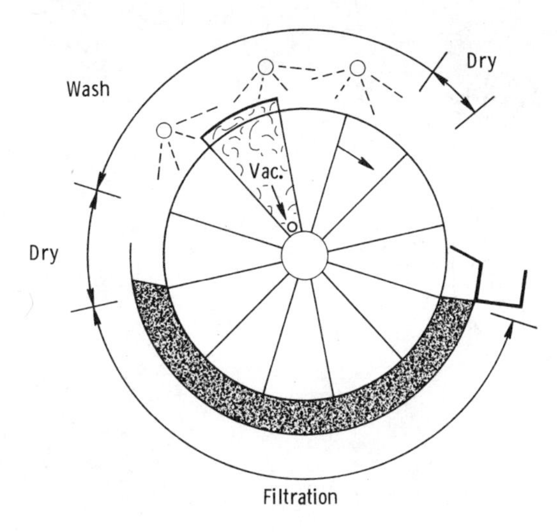

Figure 2.22 Dewaxing cycle wash.

The oil and solvent are removed continuously from the inside of the drum through a complicated valving system on one end of the drum, and then are pumped to a solvent recovery system. After removal of solvent for recycle, the base stock is ready for use in many applications. The wax from the dewaxing filter, after its solvent is removed, is the starting material for wax manufacture.

In addition to MEK–toluene, several other dewaxing solvents are used. Propane, the same solvent used for deasphalting, may be used for dewaxing. The solubility characteristics of propane are unique, since at the temperatures needed for dewaxing, wax is insolu-

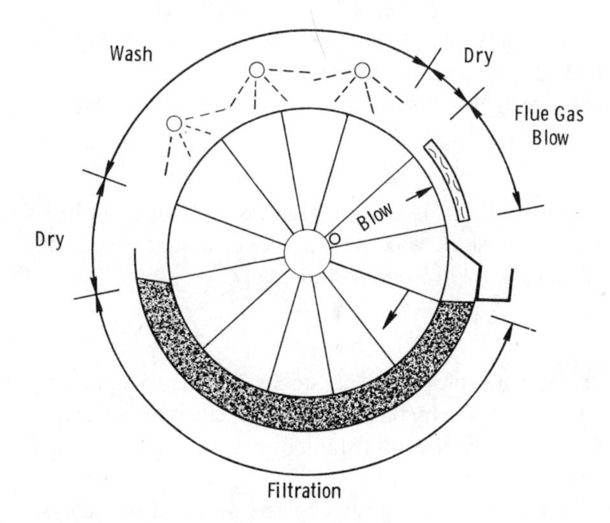

Figure 2.23 Dewaxing cycle flue gas blow.

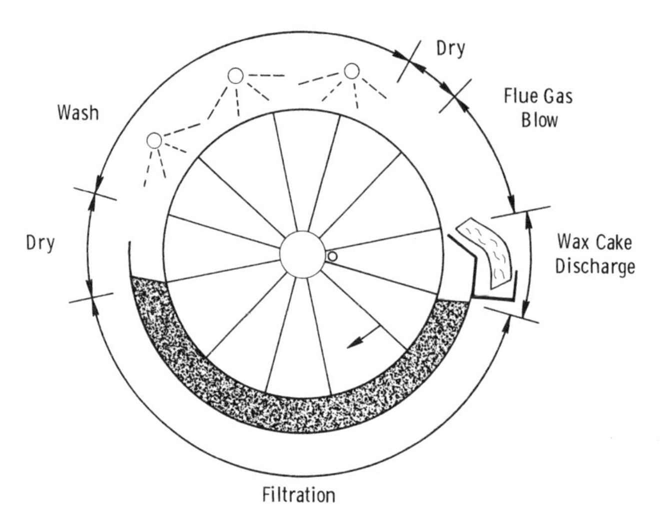

Figure 2.24 Dewaxing cycle wax cake discharge.

ble and may be filtered from the propane–oil mixture. Methyl isobutyl ketone (MIBK) also has been used as a dewaxing solvent. The operation of these solvents is generally similar to that of MEK–toluene.

G. Catalytic Dewaxing

Catalytic dewaxing is a cost-effective and flexible altnerative to solvent dewaxing for lube base stock manufacture. Unlike solvent dewaxing, which physically separates wax crystals, catalytic dewaxing transforms the wax to either nonwaxy isoparaffinic lube molecules or light fuels by-products, depending on the dewaxing catalyst. The first-generation process, Mobil Lube DeWaxing Process (MLDW), produced specification pour point lubes by shape selectively cracking waxy feed molecules to naphtha and LPG. The second-generation process and catalyst developed by ExxonMobil, Mobil Selective DeWaxing (MSDW), is specifically targeted for hydrocracked and severely hydrotreated stocks. Figure 2.25 shows the schematic of the catalytic dewaxing process developed and used at ExxonMobil.

H. Wax Isomerization

Was isomerization is a process for making base stocks of extrahigh viscosity index (140+ VI) from concentrated wax streams such as slack wax. The process sequence typically includes a hydrocracking step followed by a hydroisomerization step.

I. Hydrofinishing

For many base stocks, dewaxing is the final process. The stocks are then shipped to blending plants, where the final products are made by blending base stocks with additives. Some stocks—particularly premium stocks—require a finishing process to improve color, oxidation, or thermal stability of the base stock.

As shown in Figure 2.26, the hydrofinishing process consists of a bed of catalyst through which hot oil and hydrogen are passed. The catalyst slightly changes the molecular

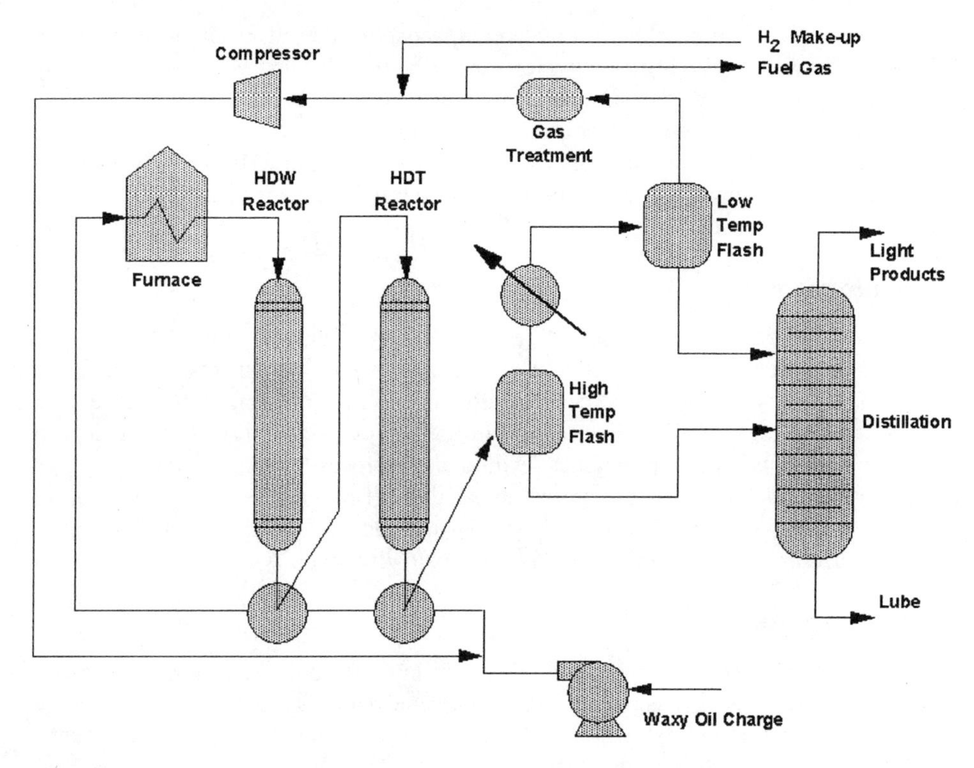

Figure 2.25 Catalytic dewaxing process flow diagram.

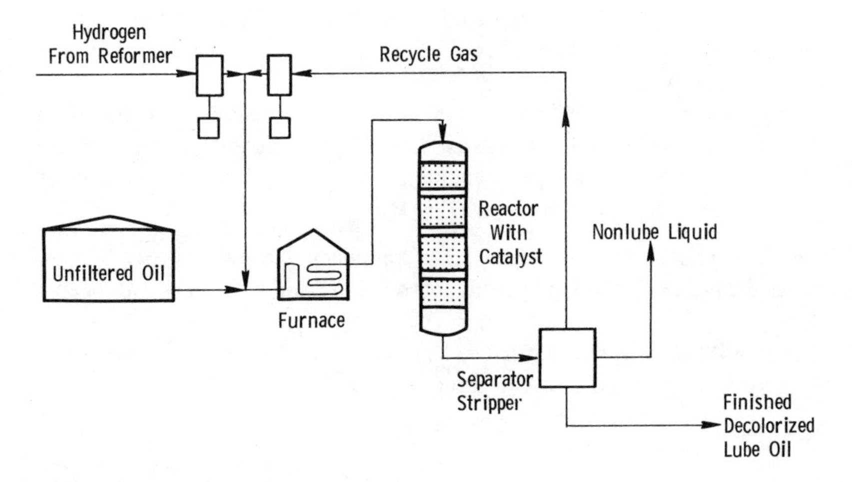

Figure 2.26 Lube hydrofinishing process.

structure of the color bodies and unstable components in the oil, resulting in a lighter colored oil that is improved in certain performance qualities. The process operates similarly to processes that are used to desulfurize kerosenes and diesel fuels.

Hydrofinishing represents relatively mild operation at relatively low temperatures and pressures. It will also saturate residual olefins to form paraffins. This process stabilizes base stock color and improves demulsibility and air release characteristics. Slight improvement in oxidation resistance may also result.

J. Hydrotreating

Hydrotreating represents a more severe set of operating conditions than hydrofinishing. At higher pressures and with selected catalysts, the aromatic rings become saturated to become naphthenes. In addition to converting aromatics, hydrotreating can remove most of the sulfur and nitrogen. This also has a positive effect on oxidation stability and deposit control. ExxonMobil uses a proprietary hydrotreating process to manufacture very high quality (VHQ) stocks for severe turbine applications. This process allows retention of selected aromatic compounds to provide the desired additive and contaminant solvency in the finished product, while maintaining good oxidative stability.

K. Hydrocracking

Lube hydrocracking is an alternative to solvent refining for producing lube base stocks. In solvent lube processing, the main objective is to remove undesirable low VI components in crude via liquid–liquid extraction. Lube hydrocracking may be employed to convert the undesirable components into valuable lube molecules.

Hydrocracking is the most severe hydroprocessing operation and is less dependent on feedstock than solvent refining. However, feedstock can have a significant impact on the product properties. Hydrocracking vacuum distillate usually targets base stocks in the 95–105 VI range. Higher operating severities can increase this up to 115+ VI, but with loss of yield. Use of a high wax (paraffin) content feed will result in an even higher VI (130+). Hydrocracking produces predominantly lower viscosity base stocks (80–500 SUS) owing to the cracking of larger, heavier molecules. Thus hydrocracked base stocks cannot be used in many heavy industrial and engine oil products. They must be blended with solvent-refined base stocks and/or other thickening agents to achieve the higher viscosity required in some products.

Hydrocracking is a well-established process in the refining industry. It has been widely used for many years in fuels manufacture and its use in lube manufacture is currently expanding. This growing interest for lube manufacture stems from several advantages hydrocracking offers versus solvent processing:

1. **Higher lube yields:** converts undesirable components to lubes.
2. **Broader feedstock flexibility:** permits production from lower quality, cheaper crudes.
3. **Higher quality base oils:** can produce base stocks meeting emerging standards for high performance such as API groups II and III.

3
Lubricating Oils

I. ADDITIVES

The preceding chapter discussed refining processes and base stock manufacturing. The lube oil base stock is the building block with respect to which appropriate additives are selected and properly blended to achieve a delicate balance in performance characteristics of the finished lubricant. It is important to mention again that the various base stock manufacturing processes can all produce base stocks with the necessary characteristics to formulate finished lubricants with the desirable performance levels. The key to achieving the highest levels of performance in finished lubricants is in the understanding of the interactions of base stocks and additives and matching those to requirements of machinery and operating conditions to which they can be subjected.

Additives are chemical compounds added to lubricating oils to impart specific properties to the finished oils. Some additives impart new and useful properties to the lubricant, some enhance properties already present, while some act to reduce the rate at which undesirable changes take place in the product during its service life.

Additives, in improving the performance characteristics of lubricating oils, have aided significantly in the development of improved prime movers and industrial machinery. Modern passenger car engines, automatic transmissions, hypoid gears, railroad and marine diesel engines, high speed gas and steam turbines, and industrial processing machinery, as well as many other types of equipment, would have been greatly retarded in their development were it not for additives and the performance benefits they provide.

Additives for lubricating oils were used first during the 1920s, and their use has since increased tremendously. Today, practically all types of lubricating oil contain at least one additive, and some oils contain additives of several different types. The amount of additive used varies from a few hundredths of a percent to 30% or more.

In addition to their beneficial effects, additives can have detrimental side effects, especially if the dosage is excessive or if interactions with other additives occur. It is the responsibility of the oil formulator to achieve a balance of additives for optimum performance, and to ensure by testing that this combination is not accompanied by undesirable side effects. When this is achieved, it is usually unnecessary and undesirable for the oil user to add oil additive supplements.

The more commonly used additives are discussed in the following sections. Although some are multifunctional, as in the case of certain VI improvers that also function as pour point depressants or dispersants or antiwear agents that also function as oxidation inhibitors, they are discussed in terms of their primary function only.

A. Pour Point Depressants

Certain high molecular weight polymers function by inhibiting the formation of a wax crystal structure that would prevent oil flow at low temperatures. Two general types of pour point depressant are used:

1. Alkylaromatic polymers adsorb on the wax crystals as they form, preventing them from growing and adhering to each other.
2. Polymethacrylates cocrystallize with wax to prevent crystal growth.

The additives do not entirely prevent wax crystal growth, but rather lower the temperature at which a rigid structure is formed. Depending on the type of oil, pour point depression of up to 50°F (28°C) can be achieved by these additives, although a lowering of the pour point by about 20–30 F° (11–17 C°) is more common.

B. Viscosity Index Improvers

VI improvers are long chain, high molecular weight polymers that function by causing the relative viscosity of an oil to increase more at high temperatures than at low temperatures. Generally this result is due to a change in the polymer's physical configuration with increasing temperature of the mixture. It is postulated that in cold oil the molecules of the polymer adopt a coiled form so that their effect on viscosity is minimized. In hot oil, the molecules tend to straighten out, and the interaction between these long molecules and the oil produces a proportionally greater thickening effect. *Note:* Although the oil–polymer mixture still decreases in viscosity as the temperature increases, the decrease is not as great as it would have been in the oil alone.

Among the principal VI improvers are methacrylate polymers and copolymers, acrylate polymers, olefin polymers and copolymers, and styrene butadiene copolymers. The degree of VI improvement from these materials is a function of the molecular weight distribution of the polymer.

The long molecules in VI improvers are subject to degradation due to mechanical shearing in service. Shear breakdown occurs by two mechanisms. Temporary shear breakdown occurs under certain conditions of moderate shear stress and results in a temporary loss of viscosity. Apparently, under these conditions the long molecules of the VI improver align themselves in the direction of the stress, thus reducing resistance to flow. When the stress is removed, the molecules return to their usual random arrangement and the temporary viscosity loss is recovered. This effect can be beneficial in that it can temporarily reduce oil friction to permit easier starting, as in the cranking of a cold engine. Permanent shear breakdown occurs when the shear stresses actually rupture the long molecules, converting them into lower molecular weight materials, which are less effective VI improvers. This results in a permanent viscosity loss, which can be significant. It is generally the limiting factor controlling the maximum amount of VI improver that can be used in a particular oil blend.

VI improvers are used in engine oils, automatic transmission fluids, multipurpose tractor fluids, and hydraulic fluids. They are also used in automotive gear lubricants. Their use permits the formulation of products that provide satisfactory lubrication over a much wider temperature range than is possible with straight mineral oils alone.

C. Defoamants

The ability of oils to resist foaming varies considerably depending on type of crude oil, type and degree of refining, and viscosity. In many applications, there may be considerable tendency to agitate the oil and cause foaming, while in other cases even small amounts of foam can be extremely troublesome. In these cases, a defoamant may be added to the oil.

Silicone polymers used at a few parts per million are the most widely used defoamants. These materials are essentially insoluble in oil, and the correct choice of polymer size and blending procedures is critical if settling during long-term storage is to be avoided. Also, these additives may increase air entrainment in the oil. Organic polymers are sometimes used to overcome these difficulties with the silicones, although much higher concentrations are generally required.

It is thought that the defoamant droplets attach themselves to the air bubbles and can either spread or form unstable bridges between bubbles, which then coalesce into larger bubbles, which in turn rise more readily to the surface of the foam layer where they collapse, thus releasing the air.

D. Oxidation Inhibitors

When oil is heated in the presence of air, oxidation occurs. As a result of this oxidation, both the oil viscosity and the concentration of organic acids in the oil increase, and varnish and lacquer deposits may form on hot metal surfaces exposed to the oil. In extreme cases, these deposits may be further oxidized to form hard, carbonaceous materials.

The rate at which oxidation proceeds is affected by several factors. As the temperature increases, the rate of oxidation increases exponentially. A rule of thumb is that for each 10°C (18°F) rise in temperature, the oxidation rate of mineral oil will double. Greater exposure to air (and the oxygen it contains), or more intimate mixing with it, will also increase the rate of oxidation. Many materials, such as metals, particularly copper and iron and organic and mineral acids, may act as catalysts or oxidation promoters.

Although the complete mechanism of oil oxidation is not too well defined, it is generally recognized as proceeding by free radical chain reaction. Reaction chain initiators are formed first from unstable oil molecules, and these react with oxygen to form peroxy radicals, which in turn attack the unoxidized oil to form new initiators and hydroperoxides. The hydroperoxides are unstable and divide, forming new initiators to expand the reaction. Any materials that will interrupt this chain reaction will inhibit oxidation. Two general types of oxidation inhibitor are used: those that react with the initiators, peroxy radicals, and hydroperoxides to form inactive compounds, and those that decompose these materials to form less reactive compounds.

At temperatures below 200°F (93°C), oxidation proceeds slowly and inhibitors of the first type are effective. Examples of this type are hindered (alkylated) phenols such as 2,6-ditertiary-butyl-4-methylphenol (also called 2,6 ditertiary-butylparacresol, DBPC), and aromatic amines such as *N*-phenyl-α-naphthylamine. These are used in products such as turbine, circulation, and hydraulic oils, which are intended for extended service at moderate temperatures.

When the operating temperature exceeds about 200°F (93°C), the catalytic effects of metals become important factors in promoting oil oxidation. Under these conditions, inhibitors that reduce the catalytic effect of the metals must be used. These materials usually react with the surfaces of the metals to form protective coatings and for that reason are sometimes called metal deactivators. Typical of additives of this type are the dithiophosphates, primarily zinc dithiophosphate. Since the dithiophosphates also act to decompose hydroperoxides at temperatures above 200°F (93°C), they inhibit oxidation by this mechanism as well.

Oxidation inhibitors may not entirely prevent oil oxidation when conditions of exposure are severe, and some types of oil are inhibited to a much greater degree than others. Oxidation inhibitors are not, therefore, cure-alls, and the formulation of a satisfactorily stable oil requires proper refining of a suitable base stock combined with careful selection of the type and concentration of oxidation inhibitor. It should also be pointed out that other additives can reduce oxidation stability in performing their design functions. Proper formulation requires the balancing of all the additive reactions to achieve the desired total performance characteristics.

E. Rust and Corrosion Inhibitors

A number of kinds of corrosion can occur in systems served by lubricating oils. Probably the two most important types are corrosion by organic acids that develop in the oil itself and corrosion by contaminants that are picked up and carried by the oil.

Corrosion by organic acids can occur, for example, in the bearing inserts used in internal combustion engines. Some of the metals used in these inserts, such as the lead in copper-lead or lead-bronze, are readily attacked by organic acids in oil, as illustrated in Figure 3.1. The corrosion inhibitors form a protective film on the bearing surfaces that

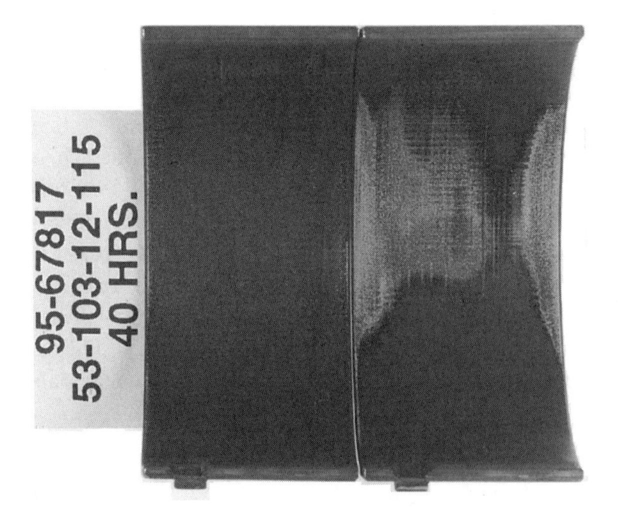

Figure 3.1 Heavily corroded copper-lead bearing.

Figure 3.2 Satisfactorily protected copper-lead bearing.

prevents the corrosive materials from reaching or attacking the metal (Figure 3.2). The film may be either adsorbed on the metal or chemically bonded to it.

During combustion in gasoline or diesel engines, certain materials in the fuel, such as sulfur and antiknock scavengers, can burn to form strong acids. These acids can then condense on the cylinder walls and be carried to other parts of the engine by the lubricant. Corrosive wear of rings and cylinder walls, and corrosion of crankshafts, rocker arms, and other engine components can then occur.

It has been found that the inclusion of highly alkaline materials in the oil will help to neutralize these strong acids as they are formed, greatly reducing this corrosion and corrosive wear. These alkaline materials are also used to provide detergency. See the detailed discussion in Section I.F, Detergents and Dispersants.

Rust inhibitors are usually compounds having a high polar attraction toward metal surfaces. By physical or chemical interaction at the metal surface, they form a tenacious, continuous film that prevents water from reaching the metal surface. Typical materials used for this purpose are amine succinates and alkaline earth sulfonates. The effectiveness of a properly selected rust inhibitor is illustrated in Figure 3.3, where specimen 9 is rust free and the other specimens display varying degree of corrosion.

Rust inhibitors can be used in most types of lubricating oil, but the selection must be made carefully to avoid problems such as corrosion of nonferrous metals or the formation of troublesome emulsions with water. Because rust inhibitors are adsorbed on metal surfaces, an oil can be depleted of rust inhibitor in time.

F. Detergents and Dispersants

In internal combustion engine service, a variety of effects tends to cause oil deterioration and the formation of harmful deposits. These deposits can interfere with oil circulation, build up behind piston rings to cause ring sticking and rapid ring wear, and affect clearances and proper functioning of critical components, such as hydraulic valve lifters. Once formed, such deposits are generally hard to remove except by mechanical cleaning. The use of

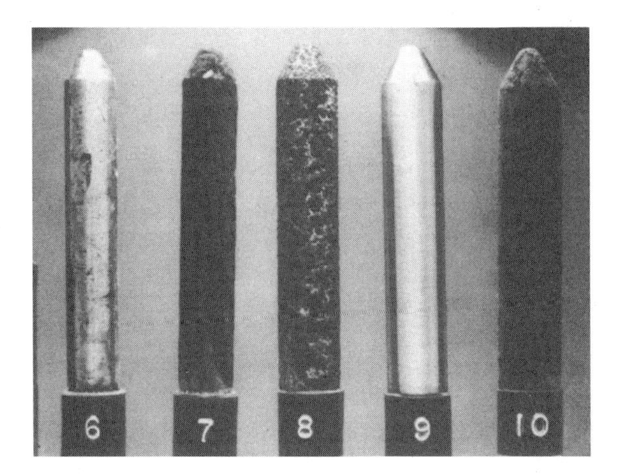

Figure 3.3 ASTM rust test specimens.

detergents and dispersants in the oil can delay the formation of deposits and reduce the rate at which they accumulate on metal surfaces. An essential factor with this approach is regular draining and replacement of the oil so that the contaminants in it are removed from the engine before the oil's capacity to hold them is exceeded.

Detergents are generally considered to be chemical compounds that chemically neutralize deposit precursors that form under high temperature conditions or as the result of burning fuels with high sulfur content or other materials that form acidic combustion by-products. Dispersants, on the other hand, are chemical compounds that disperse or suspend in the oil potential sludge- or varnish-forming materials, particularly those formed during low temperature operation when condensation and partially burned fuel find their way into the oil. These contaminants are removed from the system when the oil is drained. There is no sharp line of demarcation between detergents and dispersants. Detergents have some ability to disperse and suspend contaminants, while dispersants have some ability to prevent the formation of high temperature deposits.

The principal detergents used today are organic soaps and salts of alkaline earth metals such as barium, calcium, and magnesium. These materials are often referred to as metallo-organic compounds. Calcium and magnesium sulfonates, and calcium phenates (or phenol sulfides) are widely used. The sulfonates and phenates may be neutral or overbased; that is, they may contain more of the alkaline metal than is required to neutralize the acidic components used in diesel engine oils to neutralize the strong acids formed from combustion of the sulfur in the fuel. This neutralization reduces corrosion and corrosive wear and minimizes the tendency of these acids to cause oil degradation.

Overbased materials are generally used at lower concentration in gasoline engine oils, where the fuel sulfur is much lower. The overbased materials are included in the formulation to help reduce corrosion in low temperature operation. Both neutral and overbased materials also act to disperse and suspend potential varnish forming materials resulting from oil oxidation, preventing these materials from depositing on engine surfaces.

Metallo-organic detergents, on combustion, leave an ashy residue (see Section II.I, Sulfated Ash). In some cases, this may be detrimental in that the ash can contribute to combustion chamber deposits. In other cases, it may be beneficial in that the ash provides wear-resistant coatings for surfaces such as valve faces and seats.

Typical dispersants (also called polymeric dispersants and ashless dispersants) in use today are described as polymeric succinimides, olefin/P_2S_5 reaction products, polyesters, and benzylamides. These are based on long chain hydrocarbons that are acidified and then neutralized with a compound containing basic nitrogen. The hydrocarbon portion provides oil solubility, while the nitrogen portion provides an active site that attracts and holds potential deposit-forming materials to keep them suspended in the oil.

While the primary use of detergents and dispersants is in engine oils, they are also being used in products such as automatic transmission fluids, hydraulic oils, and circulation oils for high temperature service. In these applications, the detergents and dispersants help to prevent the deposition of lacquer and varnish resulting from oil oxidation, thus supplementing the effects of the oxidation inhibitors.

G. Antiwear Additives

Antiwear additives are used in many lubricating oils to reduce friction, wear, and scuffing and scoring under boundary lubrication conditions, that is, when full lubricating films cannot be maintained. As the oil film becomes progressively thinner as a result of increasing loads or temperatures, contact through the oil film is first made by minute surface irregularities or asperities. As these opposing asperities make contact, friction increases and welding can occur. As sliding continues, the welds break immediately, but the process can form new roughness through metal transfer, as well as wear particles, which can cause scuffing and scoring. Two general classes of materials are used to prevent metallic contact, depending on the severity of the requirements.

Mild antiwear and friction-reducing additives, sometimes called boundary lubrication additives, are polar materials such as fatty oils, acids, and esters. They are long chain materials that form an adsorbed film on the metal surfaces with the polar ends of the molecules attached to the metal and the molecules projecting more or less normal to the surface. Contact is then between the projecting ends of the layers of molecules on the opposing surfaces. Friction is reduced, and the surfaces move more freely relative to each other. Wear is reduced under mild sliding conditions, but under severe sliding conditions the layers of molecules can be rubbed off, with the result that their wear-reducing effect is lost.

H. Extreme Pressure Additives

At high temperatures or under heavy loads where more severe sliding conditions exist, compounds called extreme pressure (EP) additives are required to reduce friction, control wear, and prevent severe surface damage. These materials function by chemically reacting with the sliding metal surfaces to form relatively oil insoluble surface films. The kinetics of the reaction are a function of the surface temperatures generated by the localized high temperatures that result from rubbing between opposing surface asperities, and breaking of junctions between these asperities.

Even with extreme pressure additives in the lubricant, wear of new surfaces may be high initially. In addition to the normal break-in wear, nascent metal (freshly formed, chemically reactive surfaces), time, and temperature are required to form the protective surface films. After the films have formed, relative motion is between the layers of surface films rather than the metals. The sliding process can lead to some film removal, but since replacement by further chemical reaction is rapid, the loss of metal is extremely low. This process gradually depletes the amount of EP additive available in the oil, although the rate of depletion is usually very slow. Thus, there will be sufficient additive left to provide adequate protection for the metal surfaces except possibly under severe operating conditions, where makeup rates are low and normal drain intervals are exceeded.

The severity of the sliding conditions dictates the reactivity of the EP additives required for maximum effectiveness. The optimum reactivity occurs when the additives minimize the adhesive or metallic wear without leading to appreciable corrosive or chemical wear. Additives that are too reactive lead to the formation of excessively thick surface films, which have less resistance to attrition, so some metal is lost by the sliding action. Since a particular EP additive may have different reactivity with different metals, it is important to match additive metal reactivity to the additives not only with the severity of the sliding system but also with the specific metals involved. For example, some additives that are excellent for steel-on-steel systems may not be satisfactory for bronze-on-steel systems operating at similar sliding severity because they are too reactive with the bronze.

Another important function of EP additives is that because the chemical reaction is greatest on the asperities where contact is made and localized temperatures are highest, they lead to polishing of the surfaces. The load is then distributed more uniformly over a greater contact area, which allows for a reduction in sliding severity, more effective lubrication, and a reduction in wear.

Extreme pressure agents are usually compounds containing sulfur, chlorine, or phosphorus, either alone or in combination. The compounds used depend on the end use of the lubricant and the chemical activity required in it. Sulfur compounds, sometimes with chlorine or phosphorus compounds, are used in many metal-cutting fluids. Sulfur–phosphorus combinations are used in most industrial and automotive gear lubricants. These materials provide excellent protection against gear tooth scuffing and have the advantages of better oxidation stability, lower corrosivity, and often lower friction than other combinations that have been used in the past.

II. PHYSICAL AND CHEMICAL CHARACTERISTICS

A multitude of physical and chemical tests yield useful information on the characteristics of lubricating oils. However, the quality or the performance features of a lubricant cannot be adequately described on the basis of physical and chemical tests alone. Thus, major purchasers of lubricating oils, such as the military, equipment builders, and many commercial consumers, include performance tests as well as physical and chemical tests in their purchase specifications. Physical and chemical tests are of considerable value in maintaining uniformity of products during manufacture. Also, they may be applied to used oils to determine changes that have occurred in service, and to indicate possible causes for those changes.

Some of most commonly used tests for physical or chemical properties of lubricating oils are outlined in the following sections, with brief explanations of the significance of the tests from the standpoint of the refiner and consumer. For detailed information on methods of test, the reader is referred to the American Society for Testing and Materials handbooks (Annual Standards for Petroleum Products and Lubricants) the British Institute of Petroleum handbook (Standard Methods for Testing Petroleum and Its Products), the U.S. Federal Test Method Standard No. 791, and similar publications used in a number of other countries.

A. Carbon Residue

The carbon residue of a lubricating oil is the amount of deposit, in percentage by weight (wt %), left after evaporation and pyrolysis of the oil under prescribed conditions. In the test, oils from any given type of crude oil show lower values than those of similar viscosity containing residual stocks. Oils of naphthenic type usually show lower residues than those of similar viscosity made from paraffinic crudes. The more severe the refining treatment—whether an oil is subjected to solvents, hydroprocessing, filtration, or acid treatment—the lower the carbon residue value will be. Although many finished lubricating oils now contain additives that may contribute significantly to the amount of residue in the test, their effect on performance is distinctly beneficial.

Originally the carbon residue test was developed to determine the carbon-forming tendency of steam cylinder oils. In the years that followed, unsuccessful attempts were made to relate carbon residue values to the amount of carbon formed in the combustion chambers and on the pistons of internal combustion engines. Since such factors as fuel composition and engine operation and mechanical conditions, as well as other lubricating oil properties, are of equal or greater importance, carbon residue values alone have only limited significance. The carbon residue determination is now made mainly on base oils used for engine oil manufacture, straight mineral engine oils such as aircraft engine oils, and some products of the cylinder oil type used for reciprocating air compressors. In these cases, the determination is an indication of the degree of refining to which the base stock has been subjected.

B. Color

The color of lubricating oils as observed by light transmitted through them varies from practically clear or transparent to opaque or black. Usually, the various methods of measuring color are based on a visual comparison of the amount of light transmitted through a specified depth of oil with the amount of light transmitted through one of a series of colored glasses. The color is then given as a number corresponding to the number of the colored glass.

Color variations in lubricating oils result from differences in crude oils, viscosity, and method and degree of treatment during refining, and in the amount and nature of the additives included. During processing, color is a useful guide to the refiner to indicate whether processes are operating properly. In finished lubricants, color has little significance except in the case of medicinal and industrial white oils, which are often compounded into, or applied to, products in which staining or discoloration would be undesirable. Although color changes in used lube oils should not be taken as a condemning criterion, a color change accompanied by other physical changes such as odor or oxidation would possibly signal a need for action.

C. Density and Gravity

The density of a substance is the mass of a unit volume of it at a standard temperature. The specific gravity (relative density) is the ratio of the mass of a given volume of a material at a standard temperature to the mass of an equal volume of water at the same temperature. API gravity is a special function of specific gravity (sp gr), which is related to it by the following equation:

$$\text{API gravity} = \frac{141.5}{\text{sp gr } 60/60^\circ\text{F}} - 131.5$$

The API gravity value, therefore, increases as the specific gravity decreases. Since both density and gravity change with temperature, determinations are made at a controlled temperature and then corrected to a standard temperature by use of special tables.

Density and gravity can be determined by means of hydrometers (Figure 3.4). The hydrometer can be calibrated to read any of the three properties: density, specific gravity, or API gravity.

Gravity determinations are quickly and easily made. Because products of a given crude oil, having definite boiling ranges and viscosities, will fall into definite ranges, this property is widely used for control in refinery operations. It is also useful for identifying oils, provided the distillation range or viscosity of the oils is known. The primary use of API gravity, however, is to convert weighed quantities to volume and measured volumes to weight.

In testing used oils, particularly used engine oils, a decrease in specific gravity (increase in API gravity) may indicate fuel dilution, whereas an increase in specific gravity might indicate the presence of contaminants such as fuel soot or oxidized materials. Additional test information is necessary to fully explain changes in gravity, since some effects tend to cancel others. For hydrocarbon-based materials, the API gravity can also be used

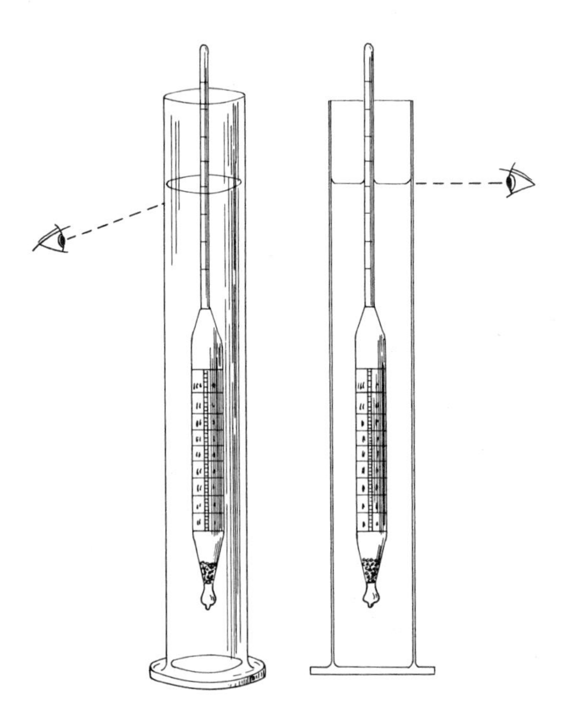

Figure 3.4 Use of hydrometer to determine density and gravity.

to determine the heating value of the material (Btu/gal). Tables are readily available to read these values directly once the API gravity has been determined.

D. Flash and Fire Points

The flash point of an oil is the temperature at which the oil releases enough vapor at its surface to ignite when an open flame is applied. For example, if a lubricating oil is heated in an open container, ignitable vapors are released in increasing quantities as the temperature rises. When the concentration of vapors at the surface becomes great enough, exposure to an open flame will result in a brief flash as the vapors ignite. When a test of this type is conducted under certain specified conditions, as in the Cleveland open cup method (Figure 3.5), the bulk oil temperature at which this happens is reported as the flash point. The release of vapors at this temperature is not sufficiently rapid to sustain combustion, so the flame immediately dies out. However, if heating is continued, a temperature will be reached at which vapors are released rapidly enough to support combustion. This temperature is called the fire point. For any specific product, flash and fire points will vary depending on the apparatus and the heating rate. Temperatures are raised in 5°F increments for this test.

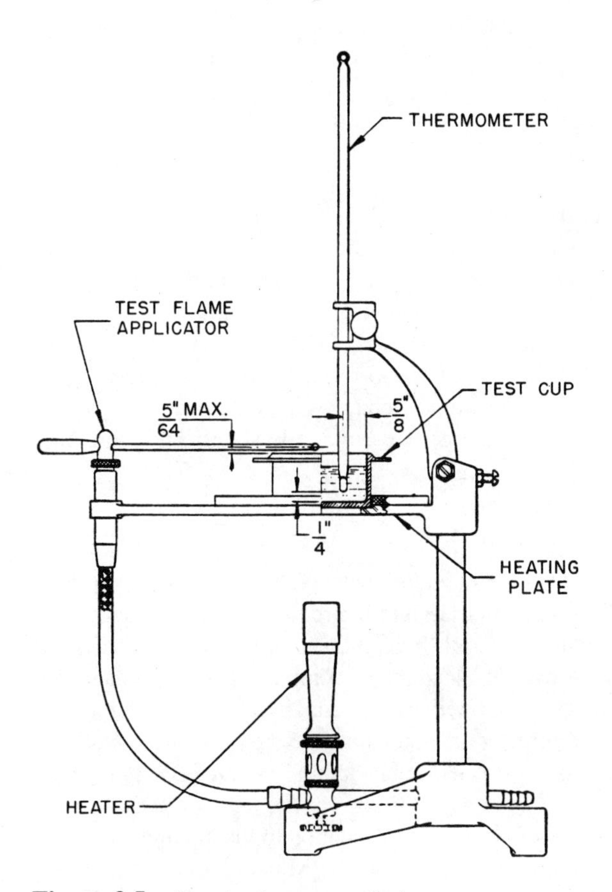

Figure 3.5 Cleveland open cup flash tester.

The flash point of new oils varies with viscosity: higher viscosity oils have higher flash points. Flash points are also affected by the type of crude and by the refining process. For example, naphthenic oils generally have lower flash points than paraffinic oils of similar viscosity.

Flash and fire tests are of value to refiners for control purposes and are significant to consumers under certain circumstances for safety considerations. Also, in certain high temperature applications, use of an oil with a low flash point, indicating higher volatility, may result in higher oil consumption rates. Flash and fire points are of little value in determining whether fire-resistant fluids are safe near possible points of ignition.

In the inspection of a used oil, a significant reduction in the flash point usually indicates contamination with lower flash material. An exception to this occurs when certain products are used for long periods at high temperatures and undergo thermal cracking with the formation of lighter hydrocarbons and a reduction in the original flash point.

E. Neutralization Number (NN)

Tests were developed to provide a quick determination of the amount of acid in an oil by neutralizing it with a base. The amount of acid in the oil was expressed in terms of the amount of a standard base required to neutralize a specified volume of oil. This quantity of base came to be called the neutralization number (NN) of the oil.

Some of the products of oxidation of petroleum hydrocarbons are organic acids. An oil oxidizes in service, caused by exposure to elevated temperatures, and there is a tendency for it to become more acidic. Thus, measuring the acidity of an oil in service was one way of following the progress of oxidation. This technique is still used with a number of products, primarily those intended for extended service, such as steam turbine oils and electrical insulating oils.

Many of the additives now used to improve performance properties may make an oil acidic or basic, depending on their composition. In other cases, the additives may contain weak acids and weak bases that do not react with each other in the oil solution, but do react with both the strong acid and the strong base used in the neutralization number tests to give both acidic and basic neutralization numbers. Other additives also undergo exchange reactions with the base used for neutralization to give false acid values in the test. These effects of the additives tend to obscure any changes in acidity occurring in the oil itself, so neutralization number determinations may have little significance for some oils containing additives.

F. Total Acid Number (TAN)

The total acid number (TAN) of an oil is synonymous with neutralization number. The TAN of an oil is the weight in milligrams of potassium hydroxide required to neutralize one gram of oil and is a measure of all the materials in an oil that will react with potassium hydroxide under specified test conditions. The usual major components of such materials are organic acids, soaps of heavy metals, intermediate and advanced oxidation products, organic nitrates, nitro compounds, and other compounds that may be present as additives. It is worth mentioning that new and used oil can exhibit both TAN and TBN (total base number) values.

Organic acids may form as a result of progressive oxidation of the oil, and the heavy metal soaps result from reaction of these acids with metals. Mineral acids (i.e., strong inorganic acids), if present in an oil sample, are neutralized by potassium hydroxide and would, therefore, affect the TAN determination. However, such acids are seldom present except in internal combustion engines using high sulfur fuels or in cases of contamination.

Since a variety of degradation products contribute to the TAN value, and since the organic acids present vary widely in corrosive properties, the test cannot be used to predict corrosiveness of an oil under service conditions.

G. Total Base Number (TBN)

The total base number of an oil is the quantity of acid, expressed in terms of the equivalent number of milligrams of potassium hydroxide, that is required to neutralize all basic constituents present in one gram of oil. This test is normally used with oils that contain alkaline, acid-neutralizing, additives. The rate of consumption of these alkaline materials (TBN depletion) is an indication of the projected serviceable life of the oil. With used oils, it indicates how much acid-neutralizing additive remains in the oil. Typical oils of this nature include diesel engine oils for internal combustion engines that use fuels containing acid-producing constituents such as sulfur or chlorine. As long as any significant amount of TBN remains in the oil, there should not be any *strong acids* present. However, the nature of high alkaline and metallic antioxidant additives sometimes allow for both TAN and TBN values to be obtained on the same sample. This occurs for both new and used oils.

H. Pour Point

The pour point of a lubricating oil is the lowest temperature at which it will pour or flow when it is chilled without disturbance under prescribed conditions. Most mineral oils contain some dissolved wax and, as an oil is chilled, this wax begins to separate as crystal that interlock to form a rigid structure that traps the oil in small pockets in the structure. When this wax crystal structure becomes sufficiently complete, the oil will no longer flow under the conditions of the test. Since, however, mechanical agitation can break up the wax structure, it is possible to have an oil flow at temperatures considerably below its pour point. Cooling rates also affect wax crystallization; it is possible to cool an oil rapidly to a temperature below its pour point and still have it flow.

While the pour point of most oils is related to the crystallization of wax, certain oils, which are essentially wax free, have viscosity-limited pour points. In these oils the viscosity becomes progressively higher as the temperature is lowered until at some temperature no flow can be observed. The pour points of such oils cannot be lowered with pour point depressants, since these agents act by interfering with the growth and interlocking of the wax crystal structure.

Untreated lubricating oils show wide variations in pour points. Distillates from waxy paraffinic or mixed base crudes typically have pour points in the range of 80–120°F (27–49°C), while raw distillates from naphthenic crudes may have pour points on the order of 0°F (−18°C) or lower. After solvent dewaxing, the paraffin distillates will have pour points on the order of 20–0°F (−7 to −18°C); where lower pour points oils are required, pour point depressants are normally used.

From the consumer's viewpoint, the importance of the pour point of an oil is almost entirely dependent on its intended use. For example, the pour point of a winter-grade engine oil must be low enough to permit the oil to be dispensed readily and to flow to the pump suction in the engine at the lowest anticipated ambient temperatures. On the other hand, there is no particular need for low pour points for oils to be used inside heated plants or in continuous service such as steam turbines, or for many other applications.

I. Sulfated Ash

The sulfated ash of a lubricating oil is the residue, in percent by weight, remaining after three processes: burning the oil, treating the initial residue with sulfuric acid, and burning the treated residue. It is a measure of the noncombustible constituents (usually metallic materials) contained in the oil.

New, straight mineral lubricating oils contain essentially no ash-forming materials. Many of the additives used in lubricating oils, such as detergents, contain metallo-organic components, which will form a residue in the sulfated ash test, and so the concentration of such materials in an oil is roughly indicated by the test. Thus, during manufacture, the test gives a simple method of checking to ensure that the additives have been incorporated in approximately the correct amounts. However, since the test combines all metallic elements into a single residue, additional testing may be necessary to determine if the various metallic elements are in the oil in the correct proportions.

Several manufacturers now include a maximum limit on sulfated ash content in their specifications for internal combustion engine oils of certain types. This is done in the belief that while the sulfated ash content results from the incorporation of materials intended to improve overall oil performance, excessive quantities of some of these materials may contribute to such problems as combustion chamber deposits, top ring wear, or port deposits in engines with two-stroke cycles.

With used oils, an increase in ash content usually indicates a buildup of contaminants such as dust and dirt, wear debris, wrong oil used as makeup, and possibly other contamination.

J. Viscosity

Probably the most important single property of a lubricating oil is its viscosity. A factor in the formation of lubricating films under both thick and thin film conditions, viscosity affects heat generation in bearings, cylinders, and gears; it governs the sealing effect of the oil and the rate of consumption or loss; and it determines the ease with which machines may be started under cold conditions. For any piece of equipment, the first essential for satisfactory results is to use an oil of proper viscosity to meet the operating conditions.

The basic concept of viscosity is shown in Figure 3.6, which shows a plate being drawn at a uniform speed over a film of oil. The oil adheres to both the moving surface and the stationary surface. Oil in contact with the moving surface travels with the same velocity U as that surface, while oil in contact with the stationary surface is at zero velocity. In between, the oil film may be visualized as being made up of many layers, each drawn by the layer above it at a fraction of velocity U that is proportional to its distance above the stationary plate (Figure 3.6, lower view). A force F must be applied to the moving plate to overcome the friction between the fluid layers. Since this friction is the result of viscosity, the force is proportional to viscosity. Viscosity can be determined by measuring the force required to overcome fluid friction in a film of known dimensions. Viscosity determined in this way is called *dynamic* or *absolute* viscosity.

Dynamic viscosities are usually reported in poise (P) or centipoise (cP; 1 cP = 0.01 P), or in SI units in pascal-seconds (Pa·s; 1 Pa·s = 10 P). Dynamic viscosity, which is a function only of the internal friction of a fluid, is the quantity used most frequently in

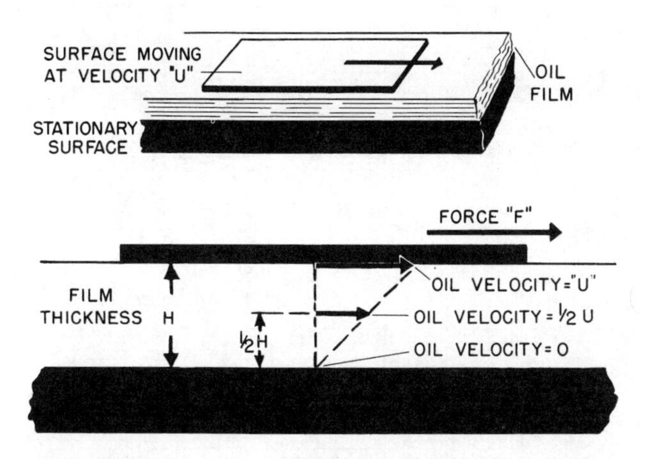

Figure 3.6 Concept of dynamic viscosity.

bearing design and oil flow calculations. Because it is more convenient to measure viscosity in a manner such that the measurement is affected by the density of the oil, *kinematic* viscosities normally are used to characterize lubricants.

The kinematic viscosity of a fluid is the quotient of its dynamic viscosity divided by its density, both measured at the same temperature and in consistent units. The most common units for reporting kinematic viscosities now are the stokes (St) or centistokes (cSt; 1 cSt = 0.01 St), or in SI units, square millimeters per second (mm^2/s; 1 mm^2/s = 1 cSt). Dynamic viscosities, in centipoise, can be converted to kinematic viscosities, in centistokes, by dividing by the density in grams per cubic centimeter (g/cm^3) at the same temperature. Kinematic viscosities, in centistokes, can be converted to dynamic viscosities, in centipoise, by multiplying by the density in grams per cubic centimeter. Kinematic viscosities, in square millimeters per second (mm^2/s), can be converted to dynamic viscosities, in pascal-seconds, by multiplying by the density, in grams per cubic centimeter, and dividing the result by 1000.

Other viscosity systems, including those of Saybolt, Redwood, and Engler, are in use and will probably continue to serve for many years because of their familiarity to some people. However, the instruments developed to measure viscosities in these systems are rarely used. Most actual viscosity determinations are made in centistokes and converted to values in the other systems by means of published SI conversion tables.

The viscosity of any fluid changes with temperature—increasing as the temperature is decreased, and decreasing as the temperature is increased. Thus, it is necessary to have some method of determining the viscosities of lubricating oils at temperatures other than those at which they are measured. This is usually accomplished by measuring the viscosity at two temperatures, then plotting these points on special viscosity–temperature charts developed by ASTM. A straight line can then be drawn through the points and viscosities at other temperatures read from it with reasonable accuracy, as illustrated in Figure 3.7. The line should not be extended below the pour point or above approximately 300°F (for most lubricating oils) since in these regions it may no longer be straight.

The two temperatures most used for reporting viscosities are 40°C (104°F) and 100°C (212°F).

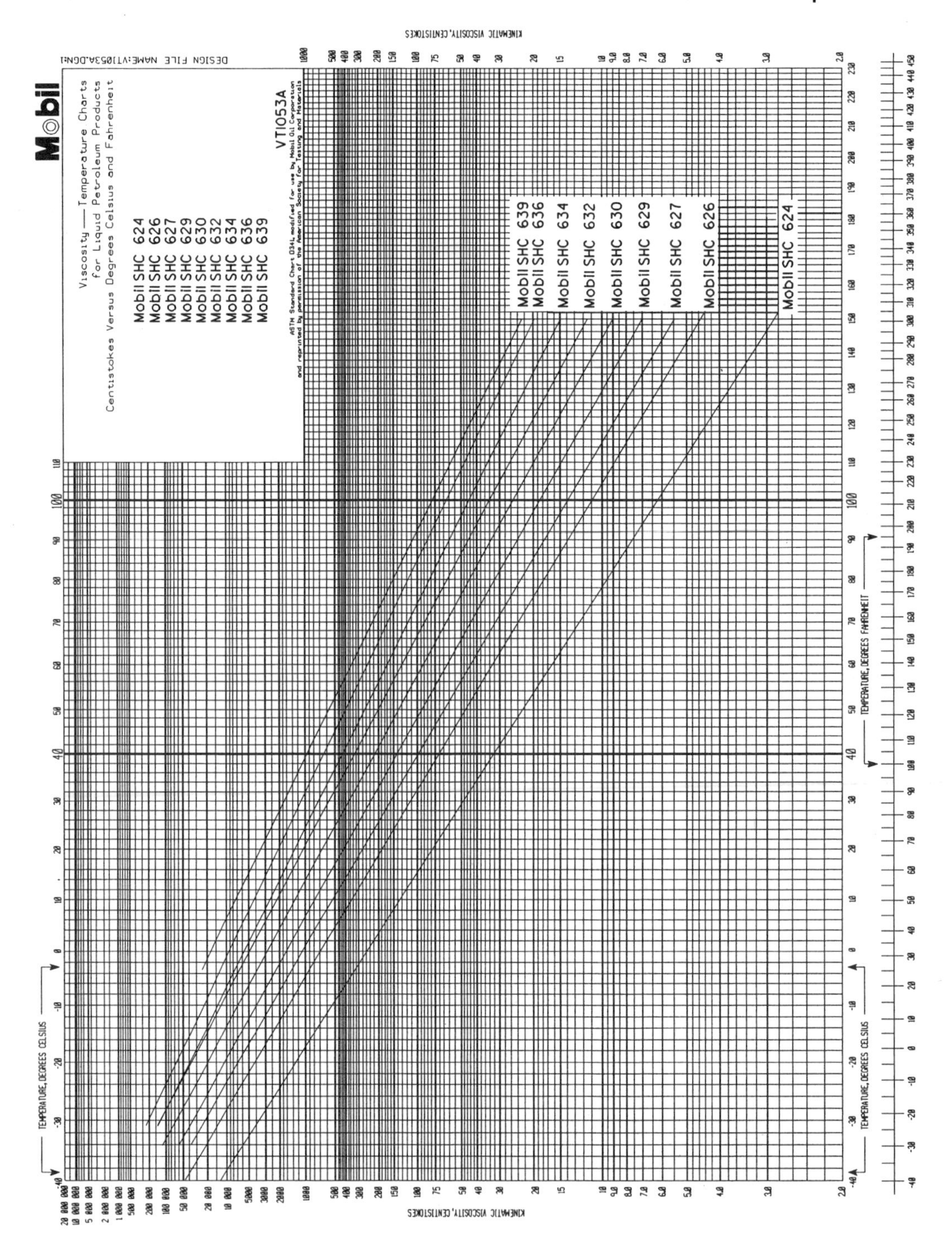

Figure 3.7 ASTM viscosity–temperature chart.

In selecting the proper oil for a given application, viscosity is a primary consideration. It must be high enough to provide proper lubricating films but not so high that friction losses in the oil will be excessive. Since viscosity varies with temperature, it is necessary to consider the actual operating temperature of the oil in the machine. Other considerations, such as whether a machine must be started at low ambient temperatures, must also be taken into account.

Three viscosity numbering systems are in use to identify oils according to viscosity ranges. Two of these are for automotive lubricants and one for industrial oils.

1. Engine Oil Viscosity Classification

SAE (Society of Automotive Engineers) Standard J300 classifies oils for use in automotive engines by viscosities determined at low shear rates and high temperature (100°C), at high shear rate and high temperature (150°C), and at both low and high shear rates at low temperature (−5°C to −40°C). The ranges for grades in this classification are shown in Table 3.1.

In these classifications, grades with the suffix letter W are intended primarily for use at low ambient temperatures, while grades without the suffix letter are intended for

Table 3.1 Engine Oil Viscosity Classification: SAE J300 (April 1997)[a]

SAE viscosity grade	Low temperature (°C) cranking viscosity (cP, max)[b]	Low temperature (°C) pumping viscosity (cP, max, with no yield stress)[c]	Kinematic viscosity (cSt) at 100°C min[d]	Kinematic viscosity (cSt) at 100°C max[d]	High shear viscosity (cP), at 150°C and 10^6 s^{-2}, min[e]
0W	3250 at −30	60 000 at −40	3.8	—	—
5W	3500 at −25	60 000 at −35	3.8	—	—
10W	3500 at −20	60 000 at −30	4.1	—	—
15W	3500 at −25	60 000 at −25	5.6	—	—
20W	4500 at −20	60 000 at −20	5.6	—	—
25W	6000 at − 5	60 000 at −15	9.3	—	—
20	—	—	5.6	< 9.3	2.6
30	—	—	9.3	< 12.5	2.9
40	—	—	12.5	< 16.3	2.9[f]
40	—	—	12.5	< 16.3	3.7[g]
50	—	—	16.3	< 21.9	3.7
60	—	—	21.9	< 26.1	3.7

[a] All values are critical specifications as defined by ASTM D 3244.
[b] ASTM D 5293.
[c] ASTM D 4684. Note that the presence of any yield stress detectable by this method constitutes a failure regardless of viscosity.
[d] ASTM 445.
[e] ASTM D 4683 or ASTM D 4741.
[f] 0W-40, 5W-40, and 10W-40 grades.
[g] 15W-40, 20W-40, 25W-40 and 40 grades.

use where low ambient temperatures will not be encountered. Multigrade oils are used for year-round service in automotive engines except for some 2 two-stroke diesel engines. An example of a multigrade oil is as follows: oils can be formulated that will meet the low temperature limits of the 5W grade and the 100°C limits for the 30 grade. Such a product can then be designated an SAE 5W-30 grade and is referred to as a multigrade or multiviscosity oil. Oils of this type generally require the use of VI improvers (see Section I: Additives) in conjunction with petroleum or synthetic lubricating oil base stocks.

High Temperature–High Shear (HT/HS) viscosities are measured at 150°C by ASTM Standard D 4683 or D 4741. The HT/HS is measured under shear stresses similar to those experienced in engine bearings. The HT/HS value indicates the temporary shear stability of the viscosity index (VI) improvers used to formulate multigrade engine oils. The SAE J300 classification shows that a SAE 40 grade oil must have a HT/HS rate viscosity exceeding 3.69 cP at 150°C or it will not perform as a 40 grade oil under engine operating conditions.

At low temperature, the high shear rate viscosity is measured by ASTM D 5293, a multitemperature cold cranking simulator method. The low temperature, low shear rate viscosity is measured by ASTM D 4684. Results of both tests have been shown to correlate with engine starting and engine oil pumpability at low temperatures.

This SAE system is widely used by engine manufacturers to determine suitable oil viscosities for their engines, and by oil marketers to indicate the viscosities of internal combustion engine oils.

2. Axle and Manual Transmission Lubricant Viscosity Classification

SAE Recommended Practice J306 classifies lubricants for use in automotive manual transmissions and drive axles by viscosity measured at 100°C (212°F), and by maximum temperature at which they reach a viscosity of 150,000 cP (150 Pa·s) when cooled and measured in accordance with ASTM Standard D 2983 (Method of Test for Apparent Viscosity at Low Temperature Using the Brookfield Viscometer).

The limits for lubricant oil viscosity are given in Table 3.2. Multigrade oils such as 80W-90 or 85W-140 can be formulated under this system. This limiting viscosity of 150,000 cP was selected on the basis of test data indicating that lubrication failures of pinion bearings of a specific axle design could be experienced when the lubricant viscosity exceeded this value. Since other axle designs, as well as transmissions, may have higher

Table 3.2 Axle and Manual Transmission Lubricant Viscosity Classification-SAE J306*

SAE viscosity grade	Maximum temperature for viscosity of 150,000 cP (°C)	(cSt) Viscosity at 100° Celsius	
		Minimum	Maximum
70W	−55	4.1	—
75W	−40	4.1	—
80W	−26	7.0	—
85W	−12	11.0	—
90	—	13.5	24.0
140	—	24.0	41.0
250	—	41.0	—

Table 3.3 Viscosity System for Industrial Fluid Lubricants

Viscosity system grade identification	Midpoint viscosity [cSt (mm^2/s)] at 40.0°C	Kinematic viscosity limits [cSt (mm^2/s)] at 40.0°C	
		Minimum	Maximum
ISO VG 2	2.2	1.98	2.42
ISO VG 3	3.2	2.88	3.52
ISO VG 5	4.6	4.14	5.06
ISO VG 7	6.8	6.12	7.48
ISO VG 10	10	9.00	11.0
ISO VG 15	15	13.5	16.5
ISO VG 22	22	19.8	24.2
ISO VG 32	32	28.8	35.2
ISO VG 46	46	41.4	50.6
ISO VG 68	68	61.2	74.8
ISO VG 100	100	90.0	110
ISO VG 150	150	135	165
ISO VG 220	220	198	242
ISO VG 320	320	288	352
ISO VG 460	460	414	506
ISO VG 680	680	612	748
ISO VG 1000	1000	900	1100
ISO VG 1500	1500	1350	1650

or lower limiting viscosities, it is the responsibility of the gear manufacturer to specify the actual grades that will provide satisfactory service under different ambient conditions.

3. Viscosity System for Industrial Fluid Lubricants

A system was developed jointly by ASTM and the Society of Tribologists and Lubrication Engineers (STLE) to establish a series of definite viscosity levels that could be used as a common basis for specifying or selecting the viscosity of industrial fluid lubricants, and to eliminate unjustified intermediate grades. The system was originally based on viscosities measured at 100°F but was converted to viscosities measured at 40°C in the interests of international standardization. In this form, the system now appears as ASTM Standard D 2422, American National Standard Z11.232, (British Standards Institution Standard BS 4231, Deutsches Institut für Normung) (DIN) No. 51519, and International Standards Organization (ISO) Standard 3448. The ISO viscosity ranges and identifying grade numbers are shown in Table 3.3. Figure 3.8 is an easy conversion chart to other viscosity classification systems.

K. Viscosity Index

Different oils have different rates of change of viscosity with temperature. For example, a distillate oil from a naphthenic base crude would show a greater rate of change of viscosity with temperature than would a distillate oil from a paraffin crude. The viscosity index is a method of applying a numerical value to this rate of change, based on comparison

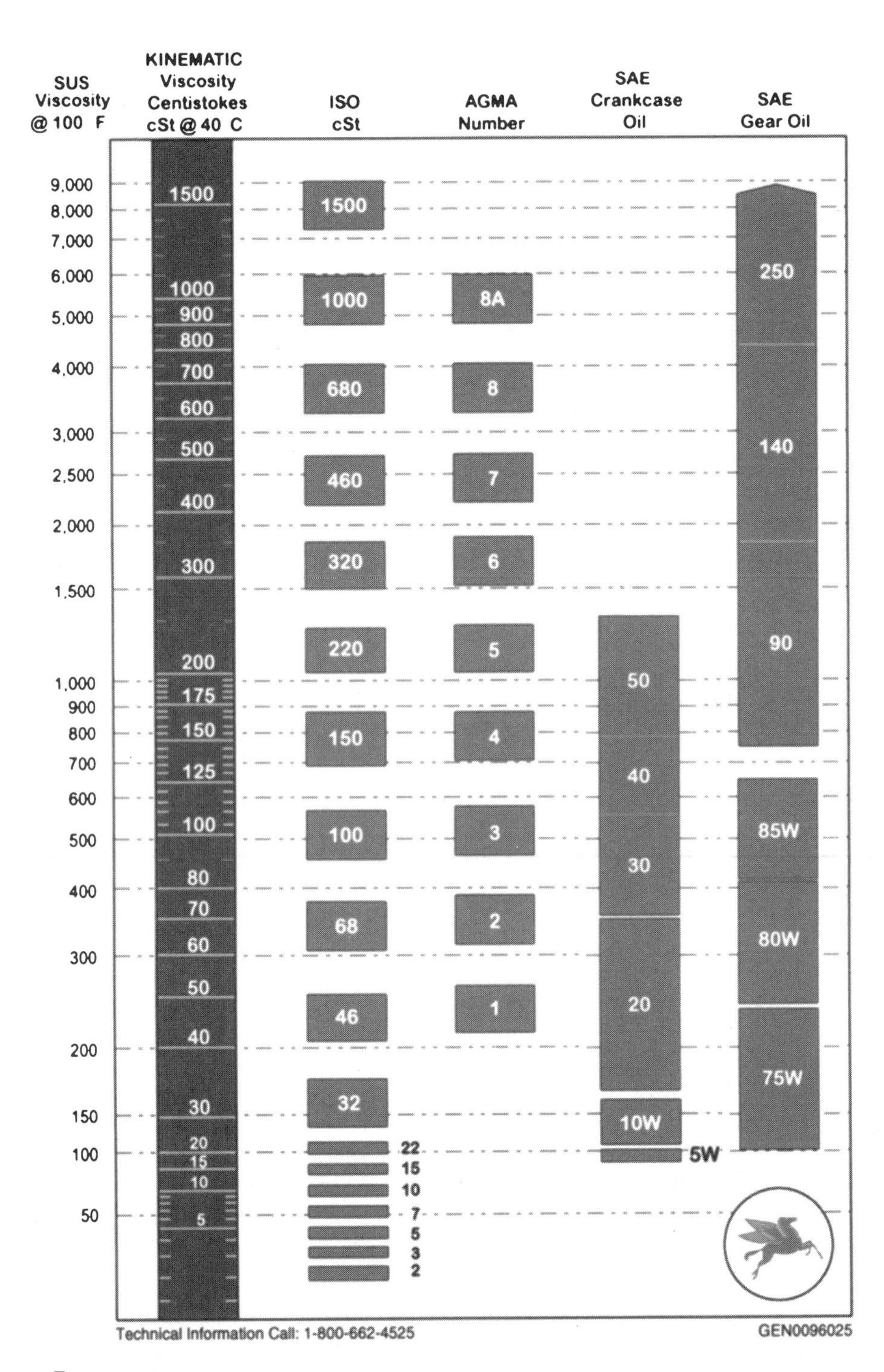

Figure 3.8 Viscosity equivalents.

with the relative rates of change of two arbitrarily selected types of oil that differ widely in this characteristic. A high VI indicates a relatively low rate of change of viscosity with temperature; a low VI indicates a relatively high rate of change of viscosity with temperature. For example, consider a high VI oil and a low VI oil having the same viscosity at, say, room temperature: as the temperature increased, the high VI oil would thin out less and, therefore, would have a higher viscosity than the low VI oil at higher temperatures.

The VI of an oil is calculated from viscosities determined at two temperatures by means of tables published by ASTM. Tables based on viscosities determined at both 100°F and 212°F, and 40°C and 100°C are available.

Finished mineral-based lubricating oils made by conventional methods range in VI from somewhat below 0 to slightly above 100. Mineral oil base stocks refined through special hydroprocessing techniques can have VIs well above 100. Some synthetic lubricating oils have VIs both below and above this range. Additives called VI improvers can be blended into oils to increase VIs; however, VI improvers are not always stable in lubricating environments exposed to shear or thermal stressing. Accordingly, these additives must be used with due care to assure adequate viscosity over the anticipated service interval for the application for which they are intended.

In many types of service, where the operating temperature remains more or less constant, VI is of little concern. However, in services where the operating temperature may vary over a wide range, such as in passenger car engines, it is probably safe to say that the VI of the oil used should be as high as practicable, consistent with other performance properties.

III. EVALUATION AND PERFORMANCE TESTS

Testing under actual service conditions is the best way to evaluate the performance of a lubricant and should be the final evaluation used before a new product or new formulation is placed on the market. Service testing is not always practical, however, during the early phases of lubricant development because of the time and cost involved. Therefore, shorter and less costly tests are used in the laboratory during development of both commercial and experimental lubricants. The correlation of results from these laboratory bench and rig tests with field performance has been a major effort of petroleum laboratories for many years. Many tests have been proposed, but a large proportion of them have been discarded for various reasons. Most of the tests that have survived have done so because they have shown the ability to predict with a fair degree of reliability some of the aspects of how a lubricant will perform in service.

Selecting lubricants based solely on meeting a set of laboratory bench tests will not always guarantee meeting the requirements of an application. Most standard laboratory test procedures were developed for evaluating *new* oil under laboratory conditions with some expectations of correlation to field application. In a lab test that is completed in minutes to months, it is extremely difficult to duplicate operating conditions and contamination for equipment that operates over a period of years. This is why extensive field testing of new products must be an integral part of new product development.

Since the process of test development and refinement is continuing, the following discussion treats the tests in a fairly general manner.

A. Oxidation Tests

Oxidation of lubricating oils depends on the temperature, amount of oxygen contacting the oil, and the catalytic effects of metals. If the oil's service conditions are known, these three variables can be adjusted to provide a test that closely represents actual service. However, oxidation in service is often an extremely slow process, so the test may be time-consuming. To shorten the test time, the test temperature is usually raised and catalysts added to accelerate the oxidation. Unfortunately, these measures tend to make the test a less reliable indication of expected field performance. As a result, very few oxidation tests have received wide acceptance, although a considerable number are used by specific laboratories that have developed satisfactory correlations for them.

One oxidation test that is widely used is ASTM D 943, Oxidation Characteristics of Inhibited Steam-Turbine Oils. This test is commonly known as TOST (turbine oil stability test). While TOST is intended mainly for use on inhibited steam turbine oils, it has been used for hydraulic and circulation oils, and for base oils for use in the manufacture of turbine, hydraulic, and circulation oils. The test is operated at a moderate temperature (95°C, 203°F). Iron and copper catalyst wires are immersed in the oil sample, to which water is added. Oxygen is bubbled through the sample at a prescribed rate. The test is run either for a prescribed number of hours, after which the neutralization number of the oil is determined, or until the neutralization number reaches a value of 2.0. The result in the latter case is than reported as the hours to a neutralization number of 2.0.

Objections to ASTM D 943 are that extremely long test times, often on the order of several thousand hours, are required for stable oils, and that the only criterion for acceptability is the neutralization number. Severe sludging and deposits on the catalyst wires can occur with some oils without excessive increase in the neutralization number. A modification of the procedure, called procedure B, overcomes some of these latter objections by requiring a determination of the sludge content.

Two additional oxidation tests that are used primarily in Europe are the IP 280 procedure for turbine oils and the PNEUROP procedure of the Comité Européen des Constructeurs de Compresseurs et d'Outillage for compressor oils. Because of ISO activities in international standardization, these tests are also receiving attention in the United States.

In the IP 280 procedure, often referred to as the CIGRE test (for **C**onférence **I**nternationale des **G**randes **R**eseaux **E**lectriques à Haute Tension), oxygen is passed through a sample of oil containing soluble iron and copper catalysts. The sample is held at 120°C (248°F) and test time is 164 h. During the test, volatile acids formed are absorbed in an absorption tube. At the end of the test, the acid numbers of the oil and the absorbent are determined and combined to give the total acidity; the sludge is determined as a weight percent. These may then be further combined to give the total oxidation products. Compared to ASTM D 943, the CIGRE test requires a short, fixed test time, and the amount of sludge formed during the test is an important criterion of the evaluation. Where the limits for satisfactory performance in the test are properly set, some believe that good correlation with performance in modern turbines is obtained. A concern of the procedure is the use of an oil soluble catalyst which does not recognize the benefits of oils that specifically resist catalyst dissolution in service.

In the PNEUROP test, a sample of oil containing iron oxide as a catalyst is aged by being held at 200°C (392°F) for 24 h while air is bubbled through it. At the end of the test, the evaporation loss and the Conradson carbon residue (CCR) of the remaining sample are determined. The evaporation loss is significant only in that it must not exceed 20%, so the main criterion is the CCR value. The test is believed to correlate to some extent with the tendency of oils to form carbonaceous deposits on compressor valves.

B. Thermal Stability

Thermal stability, as opposed to oxidation stability, is the ability of an oil or additive to resist decomposition under prolonged exposure to high temperatures with minimal oxygen present. Decomposition may result in thickening or thinning of the oil, increasing acidity, the formation of sludge, or any combination of these.

Thermal stability tests usually involve static heating, or circulation over hot metal surfaces. Exposure to air is usually minimized, but catalyst coupons of various metals may be immersed in the oil sample. While no tests have received wide acceptance, a number of proprietary tests are used to evaluate the thermal stability of products such as hydraulic fluids, gas engine oils, and oils for large diesel engines.

C. Rust Protection Tests

The rust-protective properties of lubricating oils are difficult to evaluate. Rusting of ferrous metals is a chemical reaction that is initiated almost immediately when a specimen is exposed to air and moisture. Once initiated, the reaction is difficult to stop. Thus, when specimens are prepared for rust tests, extreme care must be taken to minimize exposure to air and moisture so that rusting will not start before the rust-protective agent has been applied and the test begun. Even with proper precautions, rust tests do not generally show good repeatability or reproducibility.

Most laboratory rust tests involve polishing or sandblasting a test specimen, coating it with the oil to be tested, then subjecting it to rusting conditions. Testing may be in a humidity cabinet, by atmospheric exposure, or by some form of dynamic test. In the latter category is ASTM D 665, Rust Preventive Characteristics of Steam Turbine Oil in the Presence of Water. In this test, a steel specimen is immersed in a mixture of distilled or synthetic seawater and the oil under test. The oil and water moisture is stirred continuously during the test, which usually lasts for 24 h. The specimen is then examined for rusting.

The IP 220 ''Emcor'' method is used widely in Europe for oils and greases. This procedure employs actual ball bearings, which are visually rated for rust on the outer races at the end of the test. This is a dynamic test that can be run with a given amount of water flowing into the test bearing housings. Severity may be increased with the use of salt water or acid water in place of the standard distilled water.

Other dynamic tests may involve testing of actual mechanisms lubricated with test oil, as in the CRC L-33 test described in Automotive Gear Lubricant Tests. Additional rust tests for automotive engines are discussed later on in this chapter (Section IV, Engine Tests for Oil Performance).

D. Foam Tests

The most widely used foam test is ASTM D 892, Foaming Characteristics of Lubricating Oils. Air is blown through an oil sample held at a specified temperature for a specified period of time. Immediately after blowing is stopped, the amount of foam is measured and reported as the foaming tendency. After a specified period of time to allow for the foam to collapse, the volume of foam is again measured and reported as the foam stability.

This test gives a fairly good indication of the foaming characteristics of new, uncontaminated oils, but service results may not correlate well if contaminants such as moisture or finely divided rust, which can aggravate foaming problems, are present in a system. For a number of applications, such as automatic transmission fluids, special foam tests involving severe agitation of the oil have been developed. These are generally proprietary tests used for specification purposes.

E. EP and Antiwear Tests

One of the main functions of a lubricant is to reduce mechanical wear. Closely related to wear reduction is the ability of lubricants of the extreme pressure (EP) type to prevent scuffing, scoring, and seizure as applied loads are increased. As a result, a considerable number of machines and procedures have been developed to try to evaluate EP and antiwear properties. In a number of cases, the same machines are used for both purposes, although different operating conditions may be used.

Wear can be divided into four classifications based on the cause:

Abrasive wear
Corrosive (chemical) wear
Adhesive wear
Fatigue wear

1. Abrasive Wear

Abrasive wear is caused by abrasive particles, either contaminants carried in from outside or wear particles formed as a result of adhesive wear. In either case, oil properties do not have much direct influence on the amount of abrasive wear that occurs, except through their ability to carry particles to filtering systems that remove them from the circulating oils.

2. Corrosive or Chemical Wear

Corrosive or chemical wear results from chemical action on the metal surfaces combined with rubbing action that removes corroded metal. A typical example is the wear that may occur on cylinder walls and piston rings of diesel engines burning high sulfur fuels. The strong acids formed by combustion of the sulfur can attack the metal surfaces, forming compounds that can be fairly readily removed as the rings rub against the cylinder walls. Direct measurement of this wear requires many hours of test unit operation, which is often done as the final stage of testing new formulations. However, useful indications have been obtained in relatively short periods of time by means of sophisticated electronic techniques. Rusting is a chemical action that may result from long periods of inoperation, high moisture conditions, and inability of the oil to protect nonwetted surfaces. On equipment start-up, the rust is removed in contact areas, not only resulting in loss of surface metal but also generating particles that cause abrasive wear.

3. Adhesive Wear

Adhesive wear in lubricated systems occurs when, owing to load, speed, or temperature conditions, the lubricating film becomes so thin that opposing surface asperities can make contact. If adequate extreme pressure additives are not present, scuffing and scoring can result, and eventually seizure may occur. Adhesive wear can also occur with extreme pressure lubricants when the reaction kinetics of the additives with the surfaces are such that metal-to-metal contact is not fully controlled.

A number of machines, such as the Almen, Alpha LFW-1, Falex, Four-Ball, SAE, Timken, and Optimol SRV, are used to determine the loading conditions under which

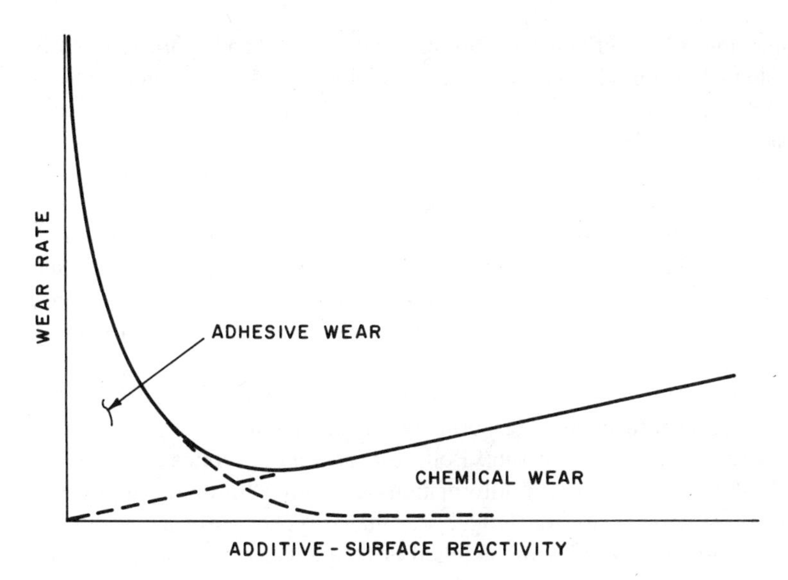

Figure 3.9 Balance of adhesive and chemical wear.

seizure, welding, or drastic surface damage to test specimens would be permitted. Of these, the Alpha LFW-1, Four-Ball, Falex, Timken, and Optimol SRV are also used to measure wear at loads below the failure load. The results obtained with these machines do not necessarily correlate with field performance, but in some cases, the results from certain machines have been reported to provide useful information for specific applications. As shown in Figure 3.9 as additive reactivity increases, adhesive wear decreases and chemical wear increases. To some extent, the machines used to judge additive surface reactivity also can be used in selecting additives for optimum effectiveness in balancing adhesive and chemical wear for certain metal combinations.

Either actual machines or scale model machines are used in testing the antiwear properties of lubricants. Antiwear-type hydraulic fluids are tested for antiwear properties in pump test rigs using commercial hydraulic pumps of the piston-and-vane type. Frequently, industrial gear lubricants are tested in the FZG Spur Gear Tester, a scale model machine that reasonably approximates commercial practice with respect to gear tooth loading and sliding speeds. Correlation of the FZG test with field performance has been good.

4. Fatigue Wear

Fatigue wear occurs under certain conditions when the lubricating oil film is intact and metallic contact of opposing asperities is either nil or relatively small. Cyclic stressing of the surfaces causes fatigue cracks to form in the metal, leading to fatigue spalling or pitting. Fatigue wear occurs in rolling element bearings and gears when there is a high degree of rolling, and adhesive wear, associated with sliding, is negligible. A variation of the fatigue pitting wear is the micropitting wear mechanism that results in small pits in the surfaces of some gears and bearings. Asperity interaction and high loads as well as metallurgy are factors influencing micropitting. An upgraded version of the standard FZG Spur Gear Tester is used to evaluate gear oil micropitting performance according to a German Research Institute test method called FVA (Forschungsvereinigung Antriebstechnik) Method 54.

The understanding of the influence of lubricant composition and characteristics is an evolving technology. Some products are currently available that address fatigue-induced wear in antifriction bearings and gears caused by the presence of small amounts of water in the hertzian load zones. In these instances, the fatigue is called *water-induced fatigue,* and oil chemistry can reduce the negative effects of water in the load zones of antifriction bearings and gears. Two tests have been introduced by Institute of Petroleum for studies of this type. In IP 300/75, Pitting Failure Tests for Oils in a Modified Four-Ball Machine, oils are tested under conditions that cause cyclic stressing of steel bearing balls. To shorten the test time, the load and stress are set higher than is usual for the ball bearings. In another British test, IP 305/74, The Assessment of Lubricants by Measurement of Their Effect on the Rolling Fatigue Resistance of Bearing Steel Using the Unisteel Machine, a ball thrust bearing with a flat thrust ring is used. The flat thrust ring is the test specimen. With a flat thrust ring, stresses are higher than would be normally encountered if a raceway providing better conformity were machined in the ring. Both tests are run in replicate. Results to date indicate that both tests provide useful information on the effect of lubricant physical properties and chemical composition on fatigue life under cyclic stressing conditions. Pitting of gears can be assessed by using the standard FZG Spur Gear Test and the method developed by FVA.

F. Emulsion and Demulsibility Tests

In some cases, lubricating oils are expected to mix with water to form emulsions. Products such as marine steam engine oils, rock drill oils, soluble cutting oils, and certain types of metal rolling lubricant are examples of products that should form stable emulsions with water. Moisture is usually present in the lubricated areas served by marine steam engine oils and rock drill oils, and emulsification with this water is necessary to form adherent lubricating films that protect the metal surfaces against wear and corrosion. Soluble cutting oils and some types of rolling oil are designed to be mixed with water to provide the high cooling capability of water combined with some lubricating properties.

Various proprietary tests for emulsion stability are in use. In addition, the recommendations of ASTM D 1479 Emulsion Stability of Soluble Cutting Oils, and ASTM D 3342, Emulsion Stability of New (Unused) Rolling Oil Dispersion in Water, are used to some extent. Both tests require agitation of the oil sample with water under prescribed conditions. In ASTM D 1479, the emulsion is then allowed to settle for 24 h, a bottom portion of it is broken with acid and centrifuged, and the oil content is compared with the oil content of a freshly prepared emulsion. The difference is reported as the oil depletion of the stored sample. In ASTM D 3324, bottom samples are withdrawn periodically, broken with acid, centrifuged, and analyzed to determine the oil content. A logarithmic plot of the oil content versus time is then prepared to determine an average rate of separation of the emulsion.

Lubricating oils used in circulating systems should separate readily from water that may enter the system as a result of condensation, leakage, or splash of process water. If the water separates easily, it will settle to the bottom of the reservoir, where it can be periodically drawn off. Steam turbine oils, hydraulic fluids, and industrial gear oils are examples of products for which it is particularly important to have good water separation properties. These properties are usually described as the emulsion, demulsibility, or demulsification characteristics.

To determine the emulsion and demulsibility characteristics of lubricating oils, three tests are widely used: ASTM D 1401, ASTM D 2711, and IP 19.

ASTM D 1401, Emulsion Characteristics of Petroleum Oils and Synthetic Fluids, was developed specifically for steam turbine oils with viscosities of 150–450 SUS at 100°F (32–97 cSt at 38°C), but can be used for testing lower and higher viscosity oils and synthetic fluids. The normal test temperature is 130°F (54°C), but for oils more viscous than 450 SUS at 100°F, it is recommended that the test temperature be increased to 180°F (82°C). In the test, equal parts of oil (or fluid) and water are stirred together for 5 min. The time for separation of the emulsion is recorded. If complete separation does not occur after the mixture has been allowed to stand for an hour, volumes of oil (or fluid), water, and emulsion remaining are reported.

ASTM D 2711, Demulsibility Characteristics of Lubricating Oils, is intended for use in testing medium and high viscosity oils such as paper machine oil applications that are prone to water contamination and may encounter the turbulence of pumping and circulation capable of producing water-in-oil emulsions. A modification of the test is suitable for evaluation of oils containing extreme pressure additives. In the test, water and the oil under test are stirred together for 5 min at 180°F (82°C). After the mixture has been allowed to settle for 5 h, the percentage of water in the oil and the percentages of water and emulsion separated for the oil are measured and recorded.

IP 19, Demulsification Number, is intended for testing new inhibited and straight mineral turbine oils. A sample of the oil held at about 90°C is emulsified with dry steam. The emulsion is then held in a bath at 94°C and the time in seconds for the oil to separate is recorded and reported as the demulsification number. If the complete separation has not occurred in 20 min, the demulsification number is reported as 1200+.

Properly used, these three tests have value in describing the emulsion or demulsibility characteristics of new lubricating oils. The IP 19 procedure is sometimes used in describing products that emulsify readily, as well as products that easily separate from water. A demulsification number of 1200+ is sometimes specified for such products as rock drill oils where the formation of lubricating emulsions is desired.

IV. ENGINE TESTS FOR OIL PERFORMANCE

A large number of engine tests are currently used to qualify oils for specific categories of classification. This number continues to expand as new or additional engine and oil characteristics are needed to meet not only the requirements of the engines but also to satisfy environmental regulations. Most of the standardized tests are designed to evaluate oils intended for automotive type engines: that is, relatively small, high speed diesel engines and four-stroke-cycle gasoline engines. Considerable effort has been devoted to standardizing the tests in the United States, Europe, and Asia. The effects of this standardization can be seen in the API licenses granted in a recent 12-month period. During this period, approximately 50% of the API licenses were granted to companies outside the United States, making the API classification system for automotive engine oils an international system. The API went from an Engine Service Classification System (ESCS) to an Engine Oil Licensing and Classification System (EOLCS) in 1994.

The ASTM establishes test procedures and limits and has published the Multicylinder Engine Test Sequences for Evaluation of Automotive Engine Oils (Sequences IID, IIIE, VE, and VIB) in Special Technical Publication STP 315, and the Single Cylinder Engine Tests for Evaluating Crankcase Oils (L-38, 1-G2, 1-H, 1-N, 1-K and 1-P) in STP 509. In addition, the major diesel engine manufacturers have developed engine tests that are part of the API classification system. These tests were initially developed for qualifying oils

Table 3.4 Gasoline and Diesel Engine Oil Tests

Engine test	Principle parameter(s) evaluated
L-38	Bearing corrosion
Sequence IID	Rust and corrosion
Sequence IIIE	Oxidation, varnish, cam wear
Sequence VE	Sludge, cam wear
KA24E	Cam wear
Sequence VIB	Fuel economy
Mack T-6	Oil consumption, oxidation, piston cleanliness
Mack T-7	Oil thickening
Mack T-8 (T-8E)	Oxidation, soot, filterability
Mack T-9	Ring-liner wear, bearing wear, extended drain capability
Cat 1-K	Piston cleanliness, oil consumption, ring and liner scuffing
Cat 1M-PC	Piston cleanliness, varnish, ring and liner wear
Cat 1-N	Piston cleanliness, ring and liner wear, oil consumption
Cat 1-P	Deposit control, oil consumption
DD 6V-92TA	Wrist pin bushing wear, liner scuffing, port deposits
Cummins M-11	Bearing corrosion, sliding wear, filterability, sludge
GM 6.2, 6.5 liter	Roller cam follower wear, ring sticking

for their specific engines. Examples are the Mack EO-L, EO-L Plus, and EO-M using the Mack T-6, T-8, and T-9 engine tests. Cummins Diesel uses the NTC 400 and the M-11 tests. All these engine tests are very expensive to develop and run, significantly increasing the cost of engine oil development and testing. All these tests can be grouped into eight categories of performance measurement:

1. Tests for oxidation stability and bearing corrosion protection
2. Single-cylinder high temperature tests
3. Multicylinder high temperature tests
4. Multicylinder low temperature tests
5. Rust and corrosion protection
6. Oil consumption rates and volatility
7. Emissions and protection of emission control systems
8. Fuel economy

A summary of the common engine oil tests used in the U.S. is shown in Table 3.4. Some additional new engine tests or extensions of existing engine tests are used to evaluate the ability of oils to provide extended drain capability. The ability to extend drain intervals is important to builders because of the pressure from users to reduce costs associated with maintenance and also to conserve nonrenewable resources. Extended drain capability is perceived by users to mean higher quality engines and oils.

A. Oxidation Stability and Bearing Corrosion Protection

Several engine tests are used to evaluate the ability of an oil to resist oxidation and oil thickening under high temperature conditions as well as to protect sensitive bearing materials from corrosion. The tests are operated under conditions that promote oil oxidation and the formation of oxyacids that cause bearing corrosion. Copper-lead inserts are used for connecting rod and main bearings in many gasoline and diesel engines. Even the few engines that use aluminum rod and main bearings will use a lead-tin flashing for break-in purposes. Although the aluminum bearings are more resistant to oxyacid corrosion, they can still experience some effects of acids in the oil. After the tests have been completed, the bearing inserts are examined for surface condition and weight loss to determine the protection afforded by the oil. The engine is also rated for varnish and sludge deposits. The most commonly used bearing corrosion test is the CRC L-38 (CEC L-02-A-78), but the Mack T-9 engine test also evaluates bearing protection capabilities of the oil.

B. Single-Cylinder High Temperature Tests

Single-cylinder engines are designed and operated to duplicate longer term operating conditions in a laboratory. They are specifically designed for oil test purposes. The tests are used primarily for the evaluation of detergency and dispersancy: that is, the ability of the oil to control piston deposits and ring sticking under operating conditions or with fuels that tend to promote the formation of piston deposits. These tests also evaluate oil consumption rates. All the current single-cylinder tests are of the diesel engine design. After completion of the test, the engine is disassembled and rated for piston deposits, top ring groove filling, and wear. The operating conditions for several of these tests are shown in Table 3.5.

C. Multicylinder High Temperature Engine Tests

While the single-cylinder engine tests just discussed provide useful information and are extremely valuable in the development of improved oil formulations, there is a trend toward use of full-scale commercial engines for oil testing and development. This trend has been precipitated by the different needs of the various engine designs as well by the requirement to satisfy demands for reduced emissions and improved fuel economy. The full-scale engine tests are used to evaluate several of the oil's performance characteristics under the different conditions subjected by the various designs. Oxidation stability, deposit control, wear and scuffing, and valve train wear are a few of the oil characteristics evaluated

Table 3.5 Single-Cylinder Engine Oil Test Conditions

Test	1-K	1-N	1-P	MWM-B	Petter AV-B
Area of use	U.S.A.	U.S.A.	U.S.A.	Europe	Europe
Fuel injection	DI	DI	DI	IDI	IDI
Aspiration	SC	SC	SC	NA	SC
Displacement	2.4 liters	2.4 liters	2.2 liters	0.85 liters	0.55 liters
Load	52 kW	52 kW	55 kW	10.5 kW	11 kW
Rpm	2100	2100	1800	2200	2250
Oil capacity	6.0 liters	6.0 liters	6.8 liters	3.2 liters	3.0 liters
Test duration	252 h	252 h	360 h	50 h	50 h
Oil temperature	107°C	107°C	130°C	110°C	90°C

Table 3.6 Multicylinder Test Conditions

Test	Sequence IIIE	Sequence VE	Sequence VIA	Mack T-8
Engine	Buick V-6	Ford OHC-4	Ford V-8	Mack E.7
Fuel	Gasoline	Gasoline	Gasoline	Diesel
Displacement	3.8 liters	2.3 liters	4.6 liters	12.0 liters
Oil capacity	5.3 liters	3.67 liters	5.7 liters	45.4 liters
Test duration	64 h	288 h	50 h	250 h
Rpm	3000	I 2500 II 2500 III 750	800–1500	1800
Load	50.6 kW	25/25/0.75	kW 2.18–15.39 kW	258 kW
Oil temperature C	149°C	68/99/46°C	45–105°C	100–107°C

by these tests. Some of the full-scale engine tests used to evaluate the oil's performance characteristics are the ASTM sequence IIIE, Mack T-6 and T-7, the XUD 11ATE (CEC L-56-T-95), the OM 602A (CEC L-51-T-95), and the Toyota 1G-FE (JASO M 333-93). Several of the multicylinder test conditions are shown in Table 3.6.

D. Multicylinder Low Temperature Tests

Since many engines can idle for long periods of time, particularly in diesel engines in colder weather or in driving conditions characterized by frequent starting and stopping, it is important that the oil have enough detergency and dispersancy to satisfactorily control soot (unburnt fuel) and sludge in engines. Both sludges and soot will increase an oil's viscosity in addition to reducing antiwear protection, filterability, and fuel economy (higher viscosity reduces fuel economy). The oil's detergency and dispersancy characteristics must be balanced with the other performance requirements to handle the negative effects of soot and reduce the buildup of sludge in the engines that can often occur in low temperature operations. Several multicylinder tests are designed to predict the oil's ability to handle the soot and sludging. The Mack T-8, the M111 (CEC L-53-T-95), and the Sequence VE tests are used to evaluate low temperature sludge and soot handling capabilities of oils.

E. Rust and Corrosion Protection Tests

The ability to protect metal parts from rust and corrosion has received more attention in the United States than in many other countries, probably because of the high proportion of stop-and-go driving in combination with severe winter weather. These conditions tend to promote condensation and accumulation of partially burnt fuel in the crankcase oil, both of which promote rust and corrosion in addition to the soot and sludge problems already discussed. Two engine tests are currently used to evaluate rust and corrosion protection properties of oils. These are the Mack T-9 and the sequence IID. At the end of the tests, valve train components, oil pump relief valves, and bearings are rated for rust and any sticking of lifters and relief valves is noted. The Sequence IID test will be gradually replaced by the *ball rust test* (BRT).

F. Oil Consumption Rates and Volatility

There is a direct association with an oil's volatility characteristics and oil consumption rates. Although volatility is not the sole reason for oil consumption in any given engine, it provides a measure of the oil's ability to resist vaporization at high temperatures. Typically, distillations were run to determine volatility characteristics of base stocks used to formulate engine oils. This is still true today. The objective of further defining an oil's volatility led to the introduction of several new *nonengine* bench tests. The most common of these tests is the NOACK Volatility, Simulated Distillation, or ''sim-dis'' (ASTM D 2887), and the GCD Volatility (ASTM D 5480). All these tests measure the amount of oil in percent that is lost upon exposure to high temperatures and therefore, serve as a measure of the oil's relative potential for increased or decreased oil consumption during severe service. Reduced volatility limits are placing more restraints on base stock processing and selection and the additive levels to achieve the required volatility levels. As discussed in Chapter 2 (Refining Processes and Lubricant Base Stocks), more and more of the base stocks used to formulate engine oils will be from API groups II, III and IV, group IV being PAOs. In addition to the bench tests already discussed, many of the actual engine tests monitor and report oil consumption rates as part of the test criteria. Caterpillar, Cummins, and Mack are all concerned about controlling oil consumption for extended service conditions.

G. Emissions and Protection of Emission Control Systems

Control of engine emissions is becoming increasingly important because of health aspects as well as the long-term effects in the earth's atmosphere (greenhouse effects). As a result, pressure is being placed on engine builders and users to reduce engine emissions. These emissions include sulfur dioxide, oxides of nitrogen, carbon monoxide, hydrocarbons, and particulates. An oil can contribute to an engine's emissions in several ways. The most common is by providing a seal between the pressure in the combustion chamber at the rings and pistons. Wear or deposits in this area will reduce combustion efficiency and lead to greater emissions. Control of piston land and groove deposits is crucial to maintaining low oil consumption rates and long-term engine performance. Oil can also contribute to increased emissions by blocking or poisoning catalysts on the engines equipped with catalytic converters. Blocking of catalyst reactions can occur through excessive oil consumption, and the poisoning effects result from chemical components either in the fuel or the oil's additive package. A common additive used in engine oil formulations is phosphorus, an element known to poison catalysts; yet phosphorus is a key element for protecting the long-term performance of engines. Since phosphorus cannot currently be effectively eliminated from engine oil formulations, the oils are formulated to keep oil consumption rates low, which minimizes the effects of phosphorus that gets into the exhaust gases.

H. Fuel Economy

Related to emissions, the trend is to increase the fuel efficiency of both gasoline and diesel engines. The first engine test developed to measure fuel economy was the Sequence VI, which was developed in the mid-1980s with a Buick V-6. This test was well correlated to the now obsolete ASTM five-car test method. The Sequence VIA replaced the Sequence VI in the mid-1990s, substituting a 1993 Ford 4.6-liter V-8. A new test, the Sequence VIB, is replacing Sequence VIA in 2001. All the fuel economy tests indicate the importance of oil viscosity in achieving mandated CAFE (corporate average fuel economy) requirements. The lower viscosity oils such as the 10W-30s and 5W-30s, and now 0W-20s which are the principal recommendations of the automobile manufacturers, provide measurable economy benefits relative to heavier viscosity grades. Although the foregoing tests are used to measure an oil's contribution to fuel economy, the official federal test uses a carbon balance of the tailpipe emissions, and it is this test that is used to establish compliance to CAFE requirements.

V. AUTOMOTIVE GEAR LUBRICANTS

The automotive vehicles manufactured today and in the past have all required gearing of some sort to allow transfer of the engine's power to the driving wheels. This gearing is composed of a range of gear design encompassing spur, helical, herringbone, and/or hypoid gears. All these gears require lubrication. Just as there is a wide range of gearing and application requirements, so is there a range of performance levels to meet mild to severe operating and application conditions.

Finished gear lubricants typically are composed of high quality base stocks (mineral and/or synthetics) and between 5 and 20% additive, depending on desired performance characteristics. Up to as many as 10 different additive materials could be used to formulate these oils and, based on the increasing requirements of extended service intervals and environmental concerns, more may be needed. These additives include antiwear compounds, extreme pressure agents, oxidation stabilizers, metal deactivators, foam suppressors, corrosion inhibitors, pour point depressants, dispersants, and viscosity index improvers. As with the other high performance lubricants, these additives compete with each other to perform their functions and must be balanced to provide the required performance requirements.

Three primary technical societies composed of and working in conjunction with equipment builders, lubricant formulators, additive suppliers, and the users of the equipment have combined efforts to define automotive gear lubricant requirements. These three technical societies are SAE, ASTM, and API. SAE has established the viscosity classification system (SAE J306) for automotive gear lubricants shown earlier (Table 3.2). ASTM establishes test methods and criteria for judging performance levels and defining test limits. API defines performance category language. In addition to SAE, ASTM, and API, the U.S. military has established a widely used specification for automotive gear lubricants: MIL-PRF-2105E.

Unlike automotive engine oils, there are no current licensing requirements for gear oils by API. Some major OEMs, however, offer licenses to use their designations for transmission and axle lubricants.

The API performance categories are as follows:

API GL-1 Lubricants for manual transmissions operating under mild service conditions. These oils do not contain antiwear, extreme pressure, or friction modifier additives. They do contain corrosion inhibitors, oxidation inhibitors, pour point depressants, and antifoam agents.

API GL-4 Lubricants for differentials containing spiral bevel or hypoid gearing operating under moderate to severe conditions. These oils may be used in some manual transmissions and transaxles where EP oils are acceptable.

API GL-5 Lubricants for differentials containing hypoid gears operating under severe conditions of torque and occasional shock loading. These oils generally contain high levels of antiwear and extreme pressure additives.

Table 3.7 Gear Lubricant Testing

Characteristic	API GL-5	API MT-1
Corrosion resistance	CRC L-33	ASTM D130
Load-carrying capability	CRC L-37	ASTM D5182
Scoring resistance	CRC L-42	ASTM D5182
Oxidation stability	CRC L-60	ASTM D5704
Elastomer compatibility	—	ASTM D5662
Cyclic durability	—	ASTM D5579
Foam resistance	ASST. D892	ASTM D892
Gear oil compatibility	FTM 3430/3440	FTM 3430/3440

API MT-1 Lubricants for manual transmissions that do not contain synchronizers. These oils are formulated to provide higher levels of oxidation and thermal stability when compared to API GL-1, GL-4 and GL-5 products.

The military specification MIL-PRF-2105E combines the performance levels of both the API GL-5 and MT-1 (Table 3.7).

In addition to the automotive gear lubricant tests, various car and other automotive axle and transmission manufacturers have gear tests, many of which are conducted in cars or over-the-road vehicles, either on dynamometers or in actual road tests. These tests generally represent special requirements such as the ability of lubricants to provide satisfactory performance in limited slip axles. Generally, most laboratory and bench testing has shown good correlation to field performance.

VI. AUTOMATIC TRANSMISSION FLUIDS

Automatic transmission fluids are among the most complex lubricants now available. In the converter section, these fluids are the power transmission and heat transfer medium; the gearbox, they lubricate the gears and bearings and control the frictional characteristics of the clutches and bands; and in control circuits, the act as hydraulic fluids. All these functions must be performed satisfactorily over temperatures ranging from the lowest expected ambient temperatures to operating temperatures on the order of 300°F (149°C) or higher, and for extended periods of service. Obviously, very careful evaluation is required before a fluid can be considered acceptable for such service.

The major U.S. automotive companies (General Motors, Ford, and DaimlerChrysler) continue to strive for improved automatic transmission fluids (ATF's). These improvements are aimed at fill-for-life applications (100,000–150,000 miles), which means that improvements are needed in oxidation stability, antiwear retention, shear stability, low temperature fluidity, material compatibility, and fluid friction stability. Ford Motor Company is looking at additional improvements to their Mercon V; GM will update Dexron III to Dexron IV; and DaimlerChrysler has improved its MS 7176D specification to MS 9602.

BIBLIOGRAPHY

Mobil Technical Bulletins

Engine Oils Specifications and Tests—Significance and Limitations
Additives for Petroleum Oils
Extreme Pressure Lubricant Test Machines

4

Lubricating Greases

The American Society for Testing and Materials defines a lubricating grease as follows: ''A solid to semifluid product of dispersion of a thickening agent in liquid lubricant. Other ingredients imparting special properties may be included'' (ASTM D 288, Standard Definitions of Terms Relating to Petroleum). This definition indicates that a grease is a liquid lubricant thickened to some extent in order to provide properties not available in the liquid lubricant alone.

I. WHY GREASES ARE USED

The reasons for the use of greases in preference to fluid lubricants are well stated by the Society of Automotive Engineers in SAE Information Report J310, Automotive Lubricating Grease. This report states:

> Greases are most often used instead of fluids where a lubricant is required to maintain its original position in a mechanism, especially where opportunities for frequent relubrication may be limited or economically unjustifiable. This requirement may be due to the physical configuration of the mechanism, the type of motion, the type of sealing, or to the need for the lubricant to perform all or part of any sealing function in the prevention of lubricant loss or the entrance of contaminants. Because of their essentially solid nature, greases do not perform the cooling and cleaning functions associated with the use of a fluid lubricant. With these exceptions, greases are expected to accomplish all other functions of fluid lubricants.

A satisfactory grease for a given application is expected to:

1. Provide adequate lubrication to reduce friction and to prevent harmful wear of components
2. Protect against rust and corrosion
3. Act as a seal to prevent entry of dirt and water
4. Resist leakage, dripping, or undesirable throw-off from the lubricated surfaces

5. Retain apparent viscosity or relationship between viscosity, shear, and temperature over useful life of the grease in a mechanical component that subjects the grease to shear forces
6. Not stiffen excessively to cause undue resistance to motion in cold environments
7. Have suitable physical characteristics for the method of application
8. Be compatible with elastomer seals and other materials of construction in the lubricated portion of the mechanism
9. Tolerate some degree of contamination, such as moisture, without loss of significant characteristics

While the SAE statement is concerned primarily with the use of lubricating greases in automotive equipment, the same considerations and performance requirements apply to the use of greases in other applications.

II. COMPOSITION OF GREASE

In the definition of a lubricating grease given here, the liquid portion of the grease may be a mineral or synthetic oil or any fluid that has lubricating properties. The thickener may be any material that, in combination with the selected fluid, will produce the solid to semifluid structure. The other ingredients are additives or modifiers that are used to impart special properties or modify existing ones. As shown in Figure 4.1, greases are made by combining three components: oil, thickener, and additives.

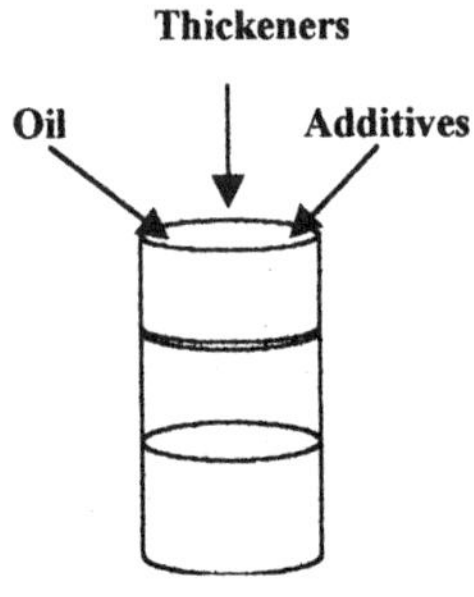

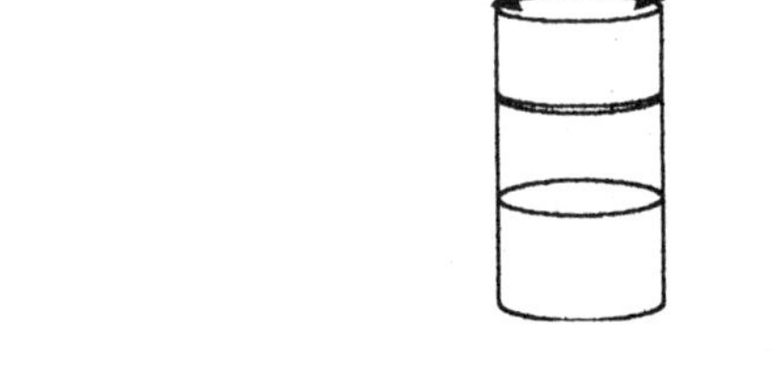

Oil

* Mineral
* Synthetic

Thickener

*Soap
*Non-Soap

Additives

* Oxidation Inhibitors
* Antiwear Agents
* Extreme Pressure Additives
* Rust and Corrosion Inhibitors
* Pour Point Depressants
* Friction Modifiers
* Dyes
* Adhesive (Tackiness) Agents
* Odorants (Perfumes)

Figure 4.1 Grease components.

A. Fluid Components

Most of the greases produced today have mineral oils as their fluid components. These oils may range in viscosity from as light as mineral seal oil up to the heaviest cylinder stocks. In the case of some specialty greases, products such as waxes, petrolatums, or asphalts may be used. Although perhaps these latter materials are not precisely describable as "liquid lubricants," they perform the same function as the fluid components in conventional greases.

Greases made with mineral oils generally provide satisfactory performance in most automotive and industrial applications. In very low or high temperature applications or in applications where temperature may vary over a wide range, greases made with synthetic fluids generally are now used. For a detailed discussion on synthetics, see Chapter 5.

B. Thickeners

The principal thickeners used in greases are metallic soaps. The earliest greases were made with calcium soaps, then greases made with sodium soaps were introduced. Later, soaps such as aluminum, lithium, clay, and polyurea came into use. Some greases made with mixtures of soaps, such as sodium and calcium, are usually referred to as *mixed-base* greases. Soaps made with other metals have been used but have not received commercial acceptance, either because of cost, health, and safety issues, environmental concerns, or performance problems.

The earlier forms of greases were *hydrated* metallic soaps, which were made by combining steric acid with a soap. These low cost greases provided good water resistance, fair low temperature properties, and fair shear stability, but limited temperature performance. Improvements to hydrated greases were necessary to provide higher temperature capability. These improvements were made by use of *12-hydroxysteric acid* with the metallic soaps to produce the next class of greases, *anhydrous* metallic soaps. This change increased dropping points above 290°F but the products were also more costly to make the earlier than hydrated metallic soap greases.

Modifications of metallic soap greases, called *complex* greases, are continuing to gain popularity. These complex greases are made by using a combination of a conventional metallic soap forming material with a complexing agent. The complexing agent may be either organic or inorganic and may or may not involve another metallic constituent. Among the most successful of the complex greases are the lithium complex greases. These are made with a combination of conventional lithium soap forming materials and a low molecular weight organic acid as the complexing agent. Greases of this type are characterized by very high dropping points, usually above 500°F (250°C), and may also have excellent load-carrying properties. Other complex greases—aluminum and calcium—are also manufactured for certain applications.

A number of nonsoap thickeners are in use, primarily for special applications. Modified bentonite (clay) and silica aerogel are used to manufacture nonmelting greases for high temperature applications. Since oxidation can still cause the oil component of these greases to deteriorate, regular relubrication is required. Thickeners such as polyurea, pigments, dyes, and various other synthetic materials are used to some extent. However, since they are generally more costly, their use is somewhat restricted to applications where specific performance requirements are desired. Lithium and lithium complex greases are

Table 4.1 Typical Lubricating Grease Characteristics by Thickener Type

			Calcium	
Properties	Aluminum	Sodium	Conventional	Anhydrous
Dropping point (°F)	230	325–350	205–220	275–290
Dropping point (°C)	110	163–177	96–104	135–143
Maximum usable temperature (°F)	175	250	200	230
Maximum usable temperature (°C)	79	121	93	110
Water resistance	Good to excellent	Poor to fair	Good to excellent	Excellent
Work stability	Poor	Fair	Fair to good	Good to excellent
Oxidation stability	Excellent	Poor to good	Poor to excellent	Fair to excellent
Protection against rust	Good to excellent	Good to excellent	Poor to excellent	Poor to excellent
Pumpability (in centralized systems)	Poor	Poor to fair	Good to excellent	Fair to excellent
Oil separation	Good	Fair to good	Poor to good	Good
Appearance	Smooth and clear	Smooth to fibrous	Smooth and buttery	Smooth and buttery
Other properties		Adhesive and cohesive	EP grades available	EP grades available
Production volume and Trend[a]	No change	Declining	Declining	No change
Principal uses[b]	Thread lubricants	Rolling contact bearings	General uses for economy	Military multiservice

[a] Lithium grease over 50% of production and all others below 10%.
[b] Multiservice includes rolling contact bearings, plain bearings, and others.
Source: Courtesy of NLGI.

the most widely used greases today. Table 4.1 outlines lubricating grease characteristics as determined by thickener type for various major grease soaps.

C. Additives and Modifiers

Additives and modifiers commonly used in lubricating greases are oxidation or rust inhibitors, pour point depressants, extreme pressure additives, antiwear agents, lubricity- or friction-reducing agents, and dyes or pigments. Most of these materials have much the same function as similar materials added to lubricating oils.

Lithium	Aluminum complex	Calcium complex	Lithium complex	Polyurea	Organo clay
350–400	500+	500+	500+	470	500+
177–204	260+	260+	260+	243	260+
275	350	350	350	350	350
135	177	177	177	177	177
Good	Good to excellent	Fair to excellent	Good to excellent	Good to excellent	Fair to excellent
Good to excellent	Good to excellent	Fair to good	Good to excellent	Poor to good	Fair to good
Fair to excellent	Fair to excellent	Poor to good	Fair to excellent	Good to excellent	Good
Poor to excellent	Good to excellent	Fair to excellent	Fair to excellent	Fair to excellent	Poor to excellent
Fair to excellent	Fair to good	Poor to fair	Good to excellent	Good to excellent	Good
Good to excellent	Good to excellent	Good to excellent	Good to excellent	Good to excellent	Good to excellent
Smooth and buttery	Smooth and buttery	Smooth and buttery	Smooth and buttery	Smooth and buttery	Smooth and buttery
EP grades available, reversible	EP grades available, reversible	EP and antiwear inherent	EP grades available	EP grades available	
The leader	Increasing	Declining	Increasing	No change	Declining
Multiservice and industrial	Multiservice industrial	Multiservice automotive and industrial	Multiservice automotive and industrial	Multiservice automotive and industrial	High temperature (frequent relube)

In addition to these additives or modifiers, boundary lubricants such as molybdenum disulfide or graphite may be added to greases to enhance specific performance characteristics such as load-carrying ability. An EP agent reacts with the lubricated surface to form a chemical film. Molybdenum disulfide is used in many greases for applications in which loads are heavy, surface speeds are low, and restricted or oscillating motion is involved. In these applications, the use of "molysulfide," (or "moly" as it is sometimes called) reduces friction and wear without adverse chemical reactions with the metal surfaces. Polyethylene and modified tetrafluoroethane (Teflon) may also be used for applications of this type.

III. MANUFACTURE OF GREASE

The manufacture of a grease, whether by a batch or continuous process, involves the dispersion of the thickener in the fluid and the incorporation of additives or modifiers. This is accomplished in a number of ways. In some cases, the thickener is purchased by the grease manufacturer in a finished state and then mixed with oil until the desired grease structure is obtained. In most cases with metallic soap thickeners, the thickener is produced, through reaction, during the manufacture of the grease.

In the manufacture of a lithium soap grease, for example, hydrogenated castor oil, fatty acids, and/or glycerides are dissolved in a portion of the oil and then saponified with an aqueous solution of lithium hydroxide. This produces a wet lithium soap that is partially dispersed in the mineral oil and is then dehydrated by heating. After drying, the mixture is cut back with additional oil and additives to produce the desired consistency and formulation characteristics intended of the finished grease. In this case, the dehydrated soap–oil mixture would be a plastic mass with a grainy structure. During or following the cutback operation, the grease might be further processed by kettle milling or homogenization to modify this structure. Once the proper structure and consistency have been obtained, the grease is ready for finishing and packaging.

As noted in the preceding discussion, manufacture of one of the basic greases involves all or some of the following five steps:

1. Saponification
2. Dehydration
3. Cutback
4. Milling
5. Deaeration

These basic processing steps are used in the manufacture of most soap-thickened greases. In certain manufacturing environments some of these steps may be accomplished simultaneously. Two primary items of equipment are required for the conventional production of soap thickened lubricating greases: a heated vessel, usually capable of containing pressure, in which the saponification product, is made, and a vessel, usually an open mixing kettle, in which the saponification product can be heated and cooled and mixed with oil and additives to reach the intended grease formulation. Hot oil or steam usually heats the pressure vessel. In the mixing kettle, counterrotating mixing paddles move the grease to first dehydrate it in the heated kettle and then to aid with mixing of additional oil and additives. Circulating an appropriate hot or cold fluid through the jacketed portion of the mixing kettle carries out heating and cooling of the grease. Adding oil to the grease also cools the grease to a temperature appropriate for both including additives and packaging. A typical batch manufacturing process is illustrated in Figure 4.2.

As mentioned earlier, the structure may be modified by milling. This milling may be continuous in the kettle during the cooling period or it may be accomplished in a separate operation. If milling is done in a separate operation, a high shear rate pump, homogenizer, or colloid mill may be used. Usually, the purpose of milling is to break a fibrous structure or to improve the dispersion of the soap in the lubricating fluid. Kettle milling will break a fibrous structure, but milling in a homogenizer or other milling equipment is required to improve dispersion.

During processing, grease may become aerated. Generally, aeration does not detract from the performance of a grease as a lubricant, but it does affect the appearance and the

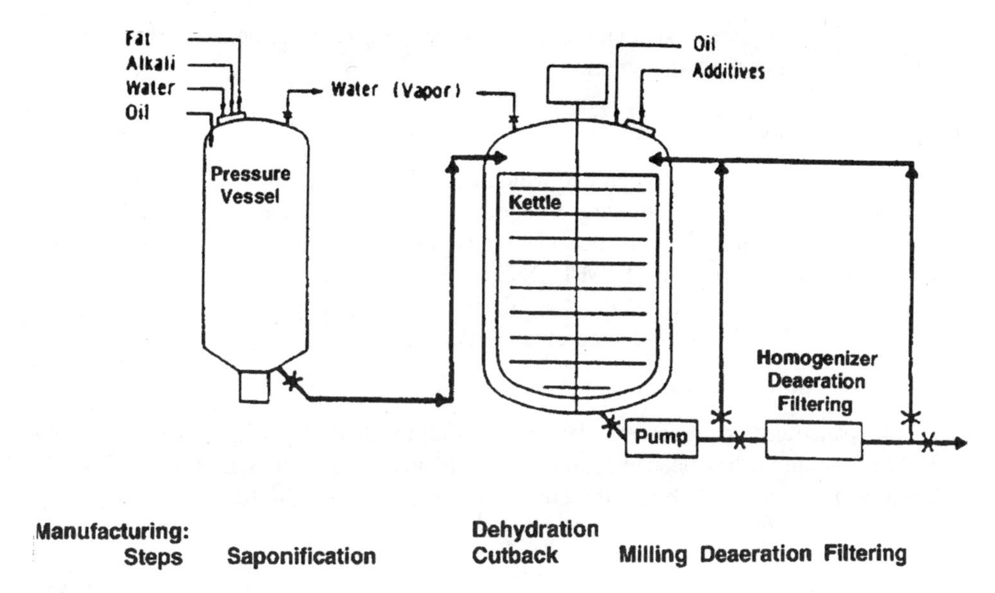

Figure 4.2 Typical batch manufacturing of lubricating grease.

volume-to-weight ratio. To improve customer appeal, some modern greases are deaerated. Various types of equipment are used for this purpose, but basically they all expose a thin film of grease to a vacuum. The vacuum draws off the entrained air, giving a much brighter appearance to the grease.

Considerable work has been done on the development of in-line or continuous manufacturing processes for greases. In many ways, in-line grease manufacturing can be thought of as an automation of the batch manufacturing process. Advantages of the in-line process include less labor and a more uniform final product.

IV. GREASE CHARACTERISTICS

The general description of a grease is in terms of the materials used in its formulation and physical properties, some of which are visual observations. The type and amount of thickener and the viscosity of the fluid lubricant are formulation properties. Color and texture, or structure, are observed visually. There is some correlation between these descriptive items and and performance. For example:

1. Certain types of thickener usually impart specific properties to a finished grease.
2. The viscosity of the fluid lubricant is very important in selecting greases for some applications.
3. Light-colored or white greases may be desirable in certain applications (e.g., in the textile and paper industries, where staining is a consideration).

This description normally is supplemented by tests for the consistency and dropping point of the grease (Section V.C), and sometimes by data on the apparent viscosity of the grease. Most of the other tests that are used to describe greases come under the category of evaluation and performance tests.

A. Consistency

Consistency is defined as the degree to which a plastic material resists deformation under the application of a force. In the case of lubricating greases, it is a measure of the relative hardness or softness and may indicate something of flow and dispensing properties. Consistency is reported in terms of ASTM D 217, Cone Penetration of Lubricating Grease, or National Lubricating Grease Institute (NLGI) grade. Consistency is measured at a specific temperature, 77°F (25°C) and degree of shear (working).

1. Cone Penetration

The cone penetration of greases is determined with the ASTM penetrometer, see Figure 4.3. After a sample has been prepared in accordance with ASTM D 217, the cone is released and allowed to sink into the grease, under its own weight, for 5 s. The depth the

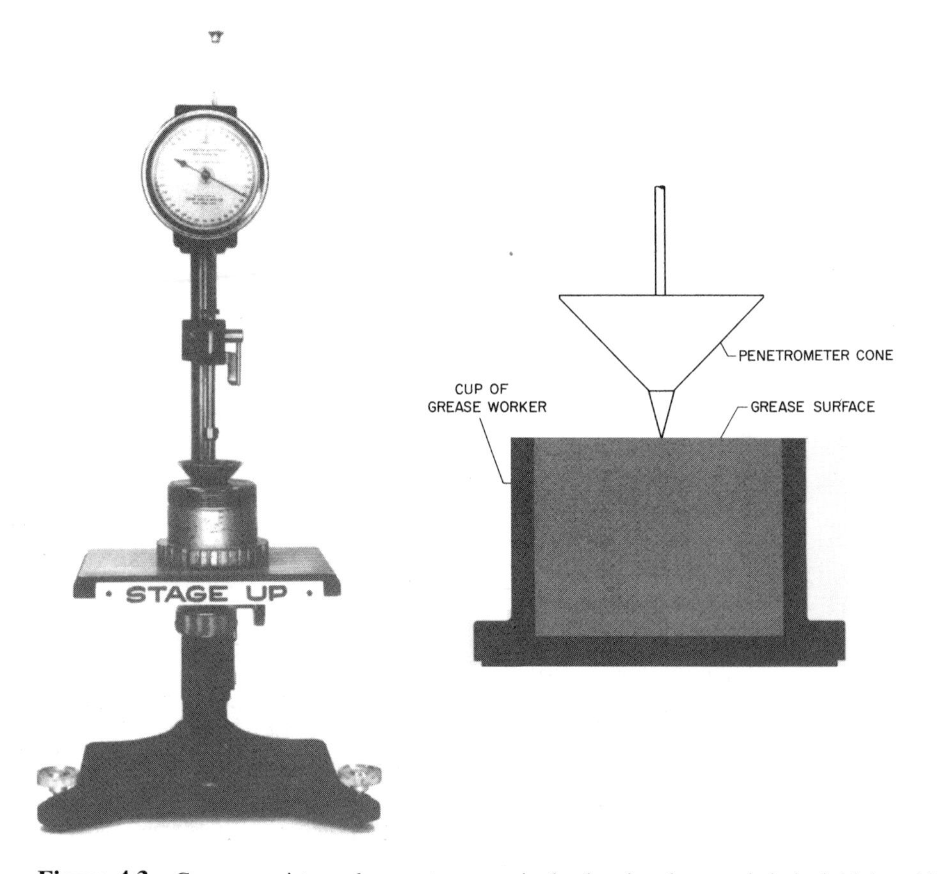

Figure 4.3 Grease consistency by penetrometer: in the drawing the cone is in its initial position, just touching the surface of the grease in the cup; in the photograph, the cone has penetrated into grease, and the amount of penetration is recorded on the dial.

cone has penetrated is then read, in tenths of a millimeter, and reported as the penetration of the grease. Since the cone will sink farther into softer greases, higher penetrations indicate softer greases. ASTM penetrations are measured at 77°F (25°C).

In addition to the standard equipment (ASTM D 217) shown in Figure 4.3, quarter- and half-scale equipment (ASTM D 1403) is available for determining the penetrations of small samples. An equation is used to convert the penetrations obtained by ASTM D 1403 to equivalent penetrations for the full-scale test.

Penetrations are reported as undisturbed penetrations, unworked penetrations, worked penetrations, or prolonged worked penetrations. *Undisturbed penetrations* are measured in the original container, without disturbance, to determine hardening or softening in storage. *Unworked penetrations* are measured on samples transferred to the grease cup with minimum disturbance. This value may have some significance with regard to transferring greases from the original containers to application equipment. The value normally reported is the worked penetration, measured after the sample has been worked 60 strokes in the ASTM grease worker (see Figure 4.4). It is considered to be the most reliable test, since the amount of disturbance of the sample is controlled and repeatable. *Prolonged worked penetrations* are discussed in Section V.A, Mechanical or Structural Stability Tests.

2. NLGI Grease Grade Numbers

On the basis of ASTM worked penetrations, the NLGI has standardized a numerical scale for classifying the consistency of greases. The NLGI grades and corresponding penetration ranges, in order of increasing hardness, are shown in Table 4.2. This system has been well accepted by both manufacturers and consumers. It has proved adequate for specifying the preferred consistency of greases for most applications.

B. Apparent Viscosity

Newtonian fluids, such as normal lubricating oils, are defined as materials for which the shear rate (or flow rate) is proportional to the applied shear stress (or pressure) at any given temperature. That is, the viscosity, which is defined as the ratio of shear stress to shear rate, is constant at a given temperature. Grease is a non-Newtonian material that does not begin to flow until a shear stress exceeding a yield point is applied. If the shear stress is then increased further, the flow rate increases more proportionally, and the viscosity, as measured by the ratio of shear stress to shear rate, decreases. The observed viscosity of a non-Newtonian material such as grease is called its *apparent viscosity*. Apparent viscosity varies with both temperature and shear rate; thus, it must always be reported at a specific temperature and flow rate.

Apparent viscosities of greases are determined in accordance with ASTM D 1092. In this test, samples of a grease are forced through a set of capillary tubes, at predetermined flow rates. From the dimensions of the capillaries, the known flow rates, and the pressure required to force the grease through the capillaries at those flow rates, the apparent viscosity of the grease, in poise, can be calculated. Results usually are reported graphically as apparent viscosity versus shear rate at a constant temperature, or as apparent viscosity versus temperature at a constant shear rate.

Apparent viscosity is used to predict the handling and dispensing properties of a grease. In addition, it can be related to starting and running torque, in grease lubricated mechanisms, and is useful in predicting leakage tendencies.

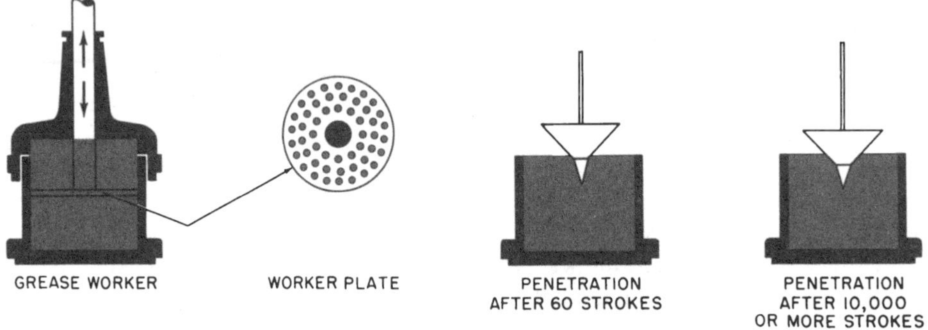

Figure 4.4 Worked penetration equipment.

C. Dropping Point

The dropping point of a grease is the temperature at which a drop of material falls from the orifice of a test cup under prescribed test conditions (Figure 4.5). Two procedures are used (ASTM D 566 and ASTM D 2265) that differ in the type of heating units and, therefore, the upper temperature limits. An oil bath is used for ASTM D 566 with a measurable dropping point limit of 500°F (260°C); ASTM D 2265 uses an aluminum block oven with a dropping point limit of 625°F (330°C). Greases thickened with or-

Table 4.2 NLGI Grease Classification

NLGI grade	ASTM worked penetration[a]
000	445–475
00	400–430
0	355–385
1	310–340
2	265–295
3	220–250
4	175–205
5	130–160
6	85–115

[a] Ranges are the penetration in tenths of a millimeter after 5 s at 77°F (25°C).

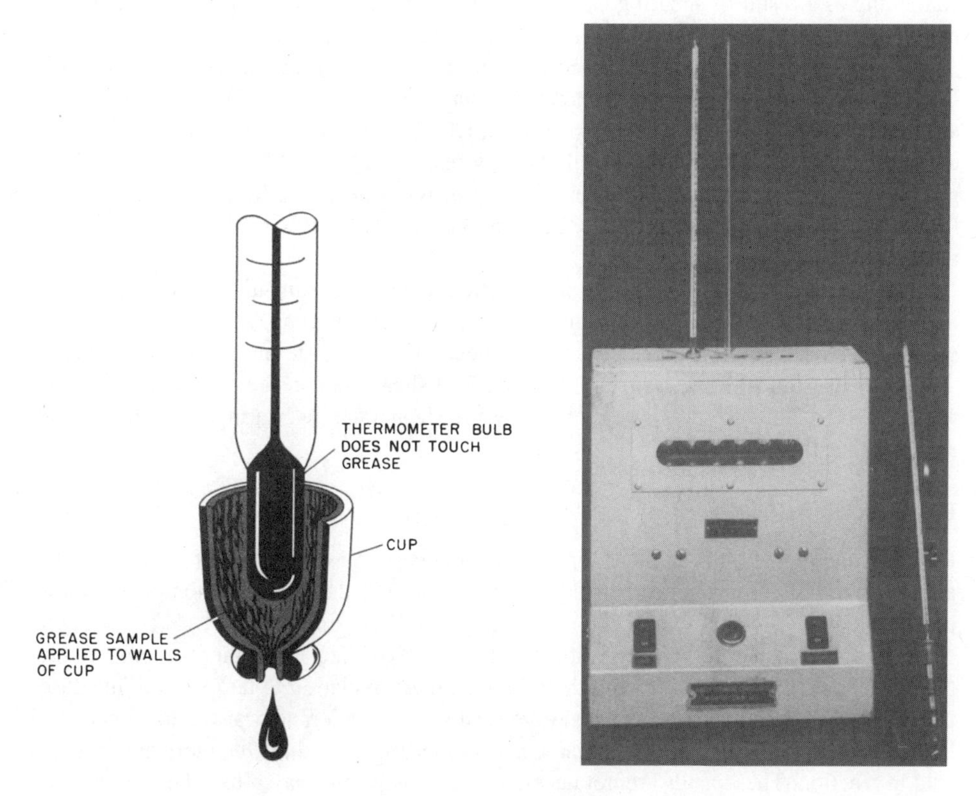

Figure 4.5 Dropping point test: photo shows complete apparatus with viewing window. Assembled grease cup and thermometer at lower right with assembly rig. Sketch at left gives enlarged view of cup and thermometer.

ganoclay soaps do not have a true melting point, instead, they have a melting range during which they become progressively softer. Some other types of grease may, without change in state, separate oil. In either case, only an arbitrary, controlled test procedure can provide a temperature that can be established as a characteristic of the grease.

The dropping point of a grease is only loosely related to the upper operating temperature to which a grease can successfully provide adequate lubrication. Additional factors must be taken into account in high temperature lubrication with grease. It is useful for characterization, and also as a quality control during grease manufacture.

V. EVALUATION AND PERFORMANCE TESTS

The tests described in Section IV characterize greases. Most of the other tests for lubricating greases are designed to be useful in predicting performance under certain conditions.

A. Mechanical or Structural Stability Tests

The ability of a grease to resist changes in consistency during mechanical working is termed its mechanical or structural stability. This is important in most applications because a grease that softens excessively as a result of the mechanical shearing encountered during service, may begin to leak. Such loss of lubricant may in turn cause equipment failure. Hardening as a result of shearing can be equally harmful in that it can prevent the grease from feeding oil properly to the equipment and can also result in its failure.

Generally, two methods are used for determining the structural stability of greases. Determinations of prolonged worked penetration (see above: Section IV.A, Consistency) are made after a grease has been worked 10,000, 50,000, or 100,000 double strokes in the ASTM grease worker. In the roll stability test (ASTM D 1831), a small sample of grease is milled in a cylindrical chamber by a heavy roller for 2 h at room temperature (Figure 4.6). The penetration after milling is then determined with the 1/2 or 1/4 scale cone equipment.

In both these tests, the change in consistency with mechanical working is reported as either the absolute change in penetration or the percent change in penetration. The significance of the tests is somewhat limited because of the differences in test shear rates and the actual rates of shearing in a bearing. The shear rates in the tests range between 10^2 and 10^3 reciprocal seconds (s^{-1}), while the shear rates in bearings may be as high as, or higher than, 10^6 s^{-1}.

B. Oxidation Test

Resistance to oxidation is an important characteristic of greases intended for use in rolling element bearings. Improvement in this property through the use of oxidation inhibitors has enabled the development of the ''packed for life'' bearings.

Both the oil and the fatty constituents in a grease oxidize; the higher the temperature, the faster the rate of oxidation, and the relationship of oxidation to temperature increases is exponential. When grease oxidizes, it generally acquires a rancid or oxidized odor and darkens in color. Simultaneously, organic acids usually develop, and the lubricant becomes acid in reaction. These acids are not necessarily corrosive but may affect the grease structures causing hardening or softening.

Laboratory tests have been developed to evaluate oxidation stability under both static and dynamic conditions. In the static test ASTM D 942, Oxidation Stability of Lubricating

Figure 4.6 Roll stability test: the heavy cylindrical roller (standing at left) rolls freely inside the tubular chamber, which is driven by the motor and gears at the right. This machine can run two rollers simultaneously.

Greases by the Oxygen Bomb Method, the grease is placed in a set of five dishes. The dishes are placed first in a pressure vessel, or bomb, which is pressurized with oxygen to 110 psi (758 kPa) and then in a bath held at 210°F (99°C), where the test materials are allowed to remain for a period of time, usually 100, 200, or 500 h. At the end of this time, the pressure is recorded and the amount of pressure drop reported. For specification purposes, pressure drops of 5–25 psi (34–172 kPa) are usually referenced, depending on the test time and the intended use of the grease.

The results of this test are probably most indicative of the stability of thin films of a grease in extended storage, as on prelubricated bearings. They are not intended to predict the stability of a grease under dynamic conditions or in bulk storage in the original containers.

A number of tests are used to evaluate oxidation stability under dynamic conditions. Two tests that were formerly used, and are still used to some extent, are ASTM D 1741, Functional Life of Ball Bearing Greases, and Method 333 of Federal Test Method (FTM) Standard 791b. A more recent test, ASTM D 3336, Performance Characteristics of Lubricating Greases in Ball Bearings at Elevated Temperatures, is designed to replace both these earlier tests. All the tests are run in ball bearing test rigs with the bearings loaded and heated. The tests differ principally in the maximum operating temperature: ASTM D 1741 provides for operation at temperatures up to 125°C (247°F); Method 333 FTM 791b, to 450°F (232°C); and ASTM D 3336, to 371°C (700°F). The tests are run until the bearing fails or for a specific number of hours if failure has not occurred. All the tests are considered to be useful screening methods for determining projected service life of ball bearing greases operating at elevated temperatures.

C. Oil Separation Tests

The resistance of a grease to separation of oil from the thickener involves certain compromises. When greases are used to lubricate rolling element bearings, a certain amount of bleeding of the oil is necessary to perform the lubrication function. On the other hand, if the oil separates too readily from a grease in application devices, a hard, concentrated soap residue may build up, which will clog the devices and prevent or retard the flow of grease to the bearings. In bearings, excessive oil separation may lead to the buildup of a hard soap in bearing recesses, which in time could be troublesome. Further, leakage of separated oil from bearings can damage materials in production or equipment components such as electric motor windings.

In application devices, such as central lubrication systems and spring-loaded cups where pressure is applied to the grease on a more of less continuous basis, oil can be separated from greases by a form of pressure filtration. The pressure forces the oil through the clearance spaces around plungers, pistons, or spool valves; but since the soap cannot pass through the small clearances, it is left behind. This may result in blockage of the devices and lubricant application failure.

Some oil release resulting in free oil on the surface of the grease in containers in storage is normal. However, excessive separation is indicative of off-specification product, which should be discussed with the supplier and possibly returned.

Generally, there is no accepted method for evaluating the oil separation properties of a grease in service. Trials in typical dispensing equipment may be conducted, and some of the dispensing equipment brands such as Trabon, Alemite, and Lincoln have tests in their specific equipment to try to identify oil separation characteristics. Some useful information may also be obtained from tests such as ASTM D 1741 and D 3336, and Method 333 of FTM 791b (see above, Section V.A: Oxidation Test). The tendency of a grease to separate oil during storage can be evaluated by means of ASTM D 1742, Oil Separation from Lubricating Grease During Storage. In this test, air pressurized to 0.25 psi (1.72 kPa) is applied to a sample of grease held on a 75 μm (No. 200) mesh screen. After 24 h at 77°F (25°C), the amount of oil separated is determined and reported. The test correlates directly with oil separation in containers of other sizes.

The tendency of a grease to separate oil at elevated temperatures under static conditions can be evaluated by Method 321.2 of FTM 791b. In this test, a sample of grease is held in a wire mesh cone suspended in a beaker. The beaker is placed in an oven, approximately at 212°F (100°C), for the desired time, usually 30 h. After the test, the oil collected in the beaker is weighed and calculated as a percentage of the original sample. Sometimes the test is used for specification purposes.

D. Water Resistance Tests

The ability of a grease to resist washout under conditions where water may splash or impinge directly on a bearing is an important property in such applications as paper

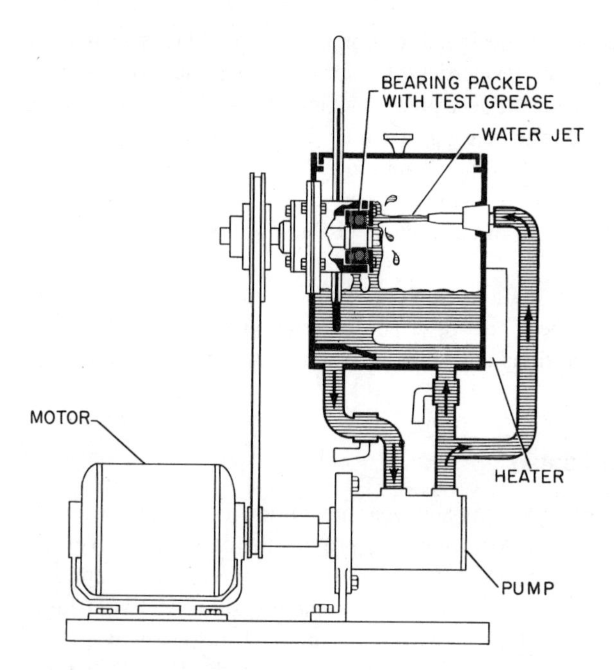

Figure 4.7 Water washout test.

machines and automobile front wheel bearings. Comparative results between different greases can be obtained with ASTM D 1264, Water Washout Characteristics of Lubricating Greases.

In this test, a ball bearing with increased clearance shields is rotated with a jet of water impinging on it (Figure 4.7). Resistance to washout is measured by the amount of grease lost from the bearing during the test. This test is considered to be a useful screening test for greases that are to be used wherever water washing may occur.

In many cases, direct impingement of water may not be a problem, but a moist atmosphere or water leakage may expose a grease to water contamination. One method of evaluating a grease for use under such conditions is to homogenize water into it. The grease may then be reported on the basis of the amount of water it will absorb without loss of grease structure, or on the amount of hardening or softening resulting from the admixture of a specific proportion of water. Table 4.1 contains some data on the water resistance characteristics of greases, based on thickener type.

E. Rust Protection Tests

In many applications, greases are not only expected to provide lubrication, but are also expected to provide protection against rust and corrosion. Some types of grease have inherent rust-protective properties, while others do not. Rust inhibitors can be incorporated in greases to improve rust-protective properties.

Both static and dynamic tests are used to evaluate the rust-protective properties of greases. Often, the test specimen is a rolling element bearing lubricated with the grease under test and then exposed to conditions designed to promote rusting. One typical static test is ASTM D 1743, Rust Preventive Properties of Lubricating Greases. In this test, tapered roller bearings are packed with the test grease, which is distributed by rotating the bearings for 60 s under light load. The bearings are then dipped in distilled water and stored for 48 h at 125°F (52°C) and 100% relative humidity. After storage, the bearings are cleaned and examined for rusting or corrosion. A bearing that shows no corrosion is rated 1. Incipient corrosion (no more than three spots of visible size) is rated 2; anything more is rated 3.

This test was developed some years ago as a cooperative project to correlate with difficulties experienced in aircraft wheel bearings. The correlation with service performance, particularly under static conditions and without water washing, is considered to be quite good.

F. EP and Wear Prevention Tests

While the results of laboratory extreme pressure and wear prevention tests do not necessarily correlate with service performance, the tests presently provide the only means to evaluate these properties at a reasonable cost. ASTM has standardized the test procedures to determine EP properties of greases using the Four-Ball Extreme Pressure (ASTM D 2596) and the Timken (ASTM D 2509) machines (Figures 4.8 and 4.9, respectively). Also, ASTM has standardized tests for wear prevention properties using the Four-Ball Wear Tester Machine (ASTM D 2266).

The two extreme pressure tests (ASTM D 2509 and D 2596) are considered to be capable of differentiating between greases having low, medium, and high levels of extreme

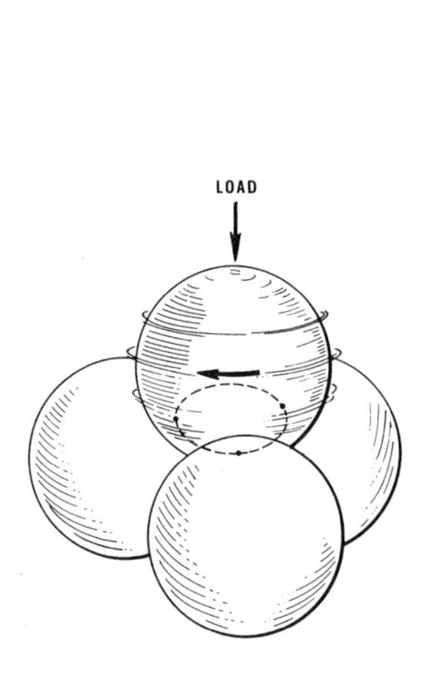

Figure 4.8 Four-ball EP test.

Figure 4.9 Timken load tester.

pressure properties. The wear prevention test (ASTM D 2266) is intended to compare only the relative wear-preventive characteristics of greases in sliding steel-on-steel applications. It is not intended to predict wear characteristics with other metals.

G. Compatibility

Greases are available with many thickener types, additives, and base oils. As a result, mixing of different greases could result in altering performance or physical properties (incompatibility), which could lead to a grease (mixture) that exhibits characteristics inferior to those of either grease before mixing. The mixing of incompatible greases will alter properties such as consistency, pumpability, shear stability, oil separation, and oxidation stability. Generally, when two incompatible greases are mixed, the result is a softening, which can lead to increased leakage as well as loss of other performance features.

Equipment performance problems as a result of mixing incompatible greases could manifest themselves after relatively a short period of operation but usually occur over longer time periods, sometimes making it difficult to trace the source of the problem back to mixing of incompatible greases. When it becomes necessary to use different greases, it is best to mix greases of the same thickener type, but in all cases compatibility charts

Table 4.3 Grease Compatibility Guidelines[a]

	Aluminum complex	Barium	Calcium	Calcium 12-hydroxy-steric acid	Calcium complex	Clay	Lithium	Lithium 12-hydroxy-steric acid	Lithium complex	Polyurea
Aluminum complex	X	I	I	C	I	I	I	I	C	I
Barium	I	X	I	C	I	I	I	I	I	I
Calcium	I	I	X	C	I	C	C	B	C	I
Calcium 12-hydroxy-steric acid	C	C	C	X	B	C	C	C	C	I
Calcium complex	I	I	I	B	X	I	I	I	C	C
Clay	I	I	C	C	I	X	I	I	I	I
Lithium	I	I	C	C	I	I	X	C	C	I
Lithium 12-hydroxy-steric acid	I	I	B	C	I	I	C	X	C	I
Lithium complex	C	I	C	C	C	I	C	C	X	I
Polyurea	I	I	I	I	C	I	I	I	I	X

[a] B, borderline compatibility; C, compatible; I, incompatible; X, same grease.
Source: NLGI Spokesman.

(Table 4.3) and the grease suppliers should be consulted. However, the safest practice is to avoid mixing of greases.

BIBLIOGRAPHY

Mobil Technical Bulletins

Extreme Pressure Lubricant Test Machines
Lubricating Grease Tests—Significance and Applicability

5 Synthetic Lubricants

Considerable attention has been focused on synthetic lubricants since the introduction into the retail market of synthetic-based or synthesized automotive engine oils. The use of synthetic-based lubricants in aviation and industrial applications extends back over many years. Past interests in synthetic lubricants were due to their ability to resist burning to a greater degree than mineral oils and to provide equipment protection advantages under extremes of operating conditions. These conditions included very low or very high temperatures. More recent interest still entails taking advantage of these performance capabilities, but researchers also want to see how synthetic lubricants may be able to minimize direct or indirect effect on the environment. More details on synthetic lubricants and environmental characteristics are covered in Chapter 6.

The terms ''synthetic'' and ''synthesized'' are both used to describe the base fluids used in these lubricants. A synthesized material is one that is produced by combining or building individual units into a unified entity. The production of synthetic lubricants starts with synthetic base stocks that are often manufactured from petroleum. The base fluids are made by chemically combining (synthesizing) low molecular weight compounds that have adequate viscosity for use as lubricants. Unlike mineral oils, which are a complex mixture of naturally occurring hydrocarbons, synthetic base fluids are man-made and tailored to have a controlled molecular structure with predictable properties.

As developed in Chapter 3, the properties of a mineral lubricating oil result from the selection of crude oil compounds that have the best properties for the intended application. This is accomplished through fractionation, solvent refining, hydrogen processing, solvent dewaxing, and filtration. However, even with extensive treatment, the finished product is a mixture of many compounds. There is no way to select from this mixture only the materials with the best properties, and if there were, the yield would be so low that the process would be uneconomical. Thus, the mineral oils produced have properties that are the average of the mixture, including both the most and the least suited components. This is not to say that lubricants made from mineral-based oils are unsatisfactory. On the contrary, lubricants properly formulated from

Table 5.1 Synthetic Lubricants Applications and Recommendations

Equipment: lubricated unit	Operating conditions	Advantages of synthetic oils
Industrial		
Calenders—rubber, plastics, board, tile	High temperature	Extended service; reduced deposits, oxidation, and thermal cracking
Paper machines—dryers, calenders, drive gear units	High temperatures, water contamination	Extended service; reduced deposits, oxidation, and thermal cracking
Nuclear power plants—vertical coolant motors, 6000–9000 hp	Annual oil change 8000 hours min	Extended service, fewer deposits
Gas engines	Low temperature start-up	Extended drains, cold temperature starting, fuel economy, lower rates of degradation
Gas turbines	High temperature and low ambients	Extended service, broader temperature application range, fewer deposits
Steam turbines—electro hydraulic control, throttle/governor	Near-superheated steam lines	Fire resistance
Tenter frame and high temperature conveyor chains	150–260°C (302–500°F)	Reduced deposits and improved wear protection, extended service, less consumption, little or no smoke
Hydraulic systems	−40°C to 93°C (−40°F to 200°F)	Better low temperature pumpability and high temperature stability
Enclosed gears—Parallel, right angle, worm	Moderate to heavy-duty, shock-loaded, severe service	Extended service, better oxidation resistance at elevated temperatures, improved efficiency
Refrigeration compressors—SRM license screw and reciprocating compressors	Severe service	Improved refrigerant solubility, compatibility with HFC refrigerants
Machine tool spindles, freezer plants—motors, conveyors, bearings	High speeds, low temperatures	Extended service, minimized thermal distortion, low temperature start-up
Metal diecasting hydraulic systems	Molten metal, source of ignition	Fire resistance
Mining—continuous miners and associated equipment	Fire hazards exist	Fire resistance
Primary metals—slab, continuous casters, rolling mills, shears, ladles, furnace controls	Fire hazards exist	Fire resistance
Air compressors—reciprocating	Severe service	Extended service, fewer deposits
Air compressors—rotary screw	Severe service	Extended service, fewer deposits
Greased/lubricated bearings	Low to high speeds	Wide temperature service range, longer relubrication intervals

(continued)

Table 5.1 Continued

Equipment: lubricated unit	Operating conditions	Advantages of synthetic oils
Industrial *(continued)*		
Greased/lubricated bearings *(continued)*	High to very high speeds	Improved low-temperature starting, lower stabilized temperature in service
Grease/lubricated gears	Heavy-duty severe service	Extended service, low temperature starting
Automotive		
Passenger car gasoline engines and front wheel drive manual transmissions	All	Improved fuel economy, low temperature starting, oil economy, wear protection
On- and off-highway gas and diesel engines	All	Improved low-temperature starting and operation, longer drain intervals, fuel and oil economy
Truck and rear wheel drive cars, drive axles: hypoid spiral bevel, and spur gears	All	Improved low-temperature starting and operation, wear protection
Passenger car automatic transmissions—Ford, GM	All	Longer life, improved high and low temperature operation, greatly improved wear protection
Commercial manual transmissions	All	Improved low temperature starting and operation, longer drain intervals, fuel and oil economy
Wheel and clutch bearings	All	Longer relubrication intervals, improved low-temperature starting, reduced leakage vs. oils
Aviation—military and commercial		
Commercial turbine engines—Pratt & Whitney; Allison; GE; SNECMA, Rolls-Royce Avon, IAE, MIL-L-23699D approved	Temperatures to 220°C (428°F)	Wide-temperature service range, high temperature stability
Military turbine engines—MIL-L-7808J approved	Temperatures to 190°C (374°F)	High-temperature stability
Aircraft all—wheel bearings, wing flap-screws—MIL-G-81322D approved	Temperature −55°C to 180°C (−67°F to 351°F)	Wide temperature service range, high, temperature stability
Marine		
High to medium speed marine diesel engines (1.5% sulfur fuel)	Very severe	Improved fuel economy, low temperature starting, oil economy and wear protection, high operating temperature stability

Source: Mobil Oil Corporation, 1994.

mineral-based oils will provide good performance in the majority of applications. Synthetics, on the other hand, possess additional performance advantages.

With synthetic lubricant base stocks, on the other hand, the process of combining individual units can be controlled so that a large proportion of the finished base fluid is either one or only a few compounds. Depending on the starting materials and the combining process, the compound (or compounds) can have the properties of the best compounds in a mineral base oil. It can also have unique properties, such as miscibility with water or nonflammability, that are not found in any mineral oil.

In either case, the special properties of the finished synthetic lubricants prepared from the compound may justify the additional cost in applications where mineral oil lubricants do not provide adequate performance. Some of the primary applications for synthetic lubricants are listed in Table 5.1.

The primary performance advantage of synthetic lubricants is the extended service life capability and handling a wider range of application temperatures. Their outstanding flow characteristics at low temperatures and their stability at high temperatures mark the preferred use of these lubricants. Figure 5.1 compares the operating temperature limits of mineral oil and synthetic lubricants. Other advantages, as well as limiting properties, are outlined in Table 5.2.

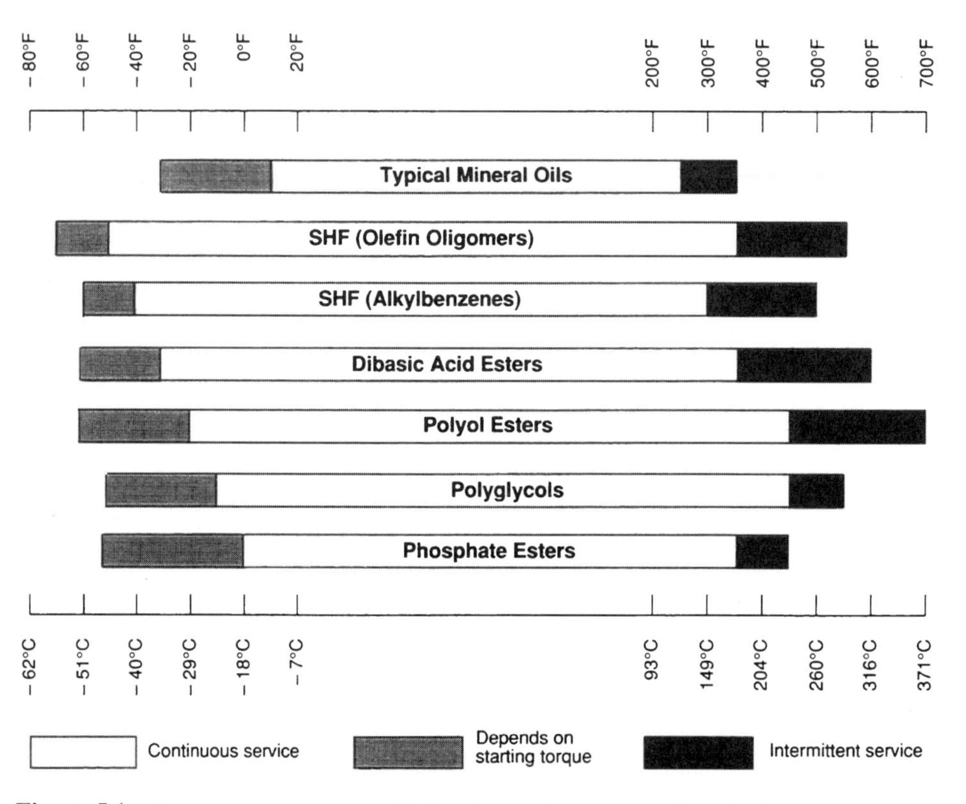

Figure 5.1 Comparative temperature limits of mineral oil and synthetic lubricants.

Table 5.2 Advantages and Limiting Properties of Synthetic Base Stocks

Synthetic	Advantages vs. mineral oil	Limiting properties
SHF	High temperature stability Long life Low temperature fluidity High viscosity index Improved wear protection Low volatility, oil economy Compatibility with mineral oils and paints No wax	Solvency/detergency[a] Seal compatibility[a]
Organic Esters	High temperature stability Long life Low temperature fluidity Solvency/detergency	Seal compatibility Mineral oil compatibility Antirust[a] Antiwear and extreme pressure[a] Hydrolytic stability Paint compatibility
Phosphate Esters	Fire resistancy Lubricating ability	Seal compatibility Low viscosity index Paint compatibility Metal corrosion[a] Hydrolytic stability
Polyglycols	Water versatility High viscosity index Low temperature fluidity Antirust No wax	Mineral oil compatibility Paint compatibility Oxidation stability[a]

[a] Limiting property of synthetic base fluids that can be overcome by formulation chemistry.

Various ways have been used to classify synthetic base fluids. Some of these classification approaches neglect similarities between certain types of material, while others may lead to confusion by grouping materials that have similarities in chemical structure but are totally dissimilar from a performance or application standpoint. For the purposes of this discussion, the materials are classified as follows:

Synthesized hydrocarbon fluids (SHFs)
Organic esters
Polyglycols
Phosphate esters
Other synthetic lubricating fluids

The first four classes account for over 90% of the volume of synthetic lubricant bases now used. The fifth class includes a number of materials, some of which are quite costly, that are generally used in somewhat lower volumes in specialized applications. Table 5.3 shows a further breakdown of synthetic lubricant classes and types.

Table 5.3 Synthetic Lubricants

Class	Type
Synthesized hydrocarbon	Polyalphaolefins
Fluids	Alkylated aromatics
	Polybutenes
	Cycloaliphatics
Organic esters	Dibasic acid ester
	Polyol ester
Polyglycols	Polyalkylene
	Polyoxyaklylene
	Polyethers
	Glycol
	polyglycol esters
	polyalkylene glycol ester
	polyethylene glycol
Phosphate esters (phosphoric acid esters)	Triaryl phosphate ester
	Triakyl phosphate ester
	Mixed alkylaryl phosphate esters
Other	Silicones
	Silicate esters
	Fluorocarbons
	Polyphenyl ethers

I. SYNTHESIZED HYDROCARBON FLUIDS

The SHFs comprise the fastest growing type of synthetic lubricant base stock. They are pure hydrocarbons, manufactured from raw materials derived from crude oil. Three types are used in considerable volume: olefin oligomers, alklated aromatics, and polybutenes. A fourth type, cycloaliphatics, is used in small volumes in specialized applications.

A. Olefin Oligomers (Polyalphaolefins)

Olefin oligomers, also called polyalphaolefins (PAOs), are formed by combining a low molecular weight material, usually ethylene, into a specific olefin. This olefin is oligomerized into a lubricating oil type material, and then hydrogen-stabilized. In this sense, oligomerization is a polymerization process in which a few, usually three to ten, of the basic building block molecules are combined to form the finished material. Therefore, the product may be formed with varying molecular weights and attendant viscosities to meet a broad range of requirements. Polymerization in the usual sense involves condensation of the same basic building blocks (monomers). The structure of a typical olefin oligomer base oil molecule is shown in Figure 5.2.

Olefin oligomers can be considered to be a special type of paraffinic mineral oil, comparable in properties to the best components found in petroleum derived base oils. They have high viscosity indexes (usually > 135 VI), excellent low temperature fluidity, and very low pour points. Their shear stability is excellent, as is their hydrolytic stability. As a result of the saturated nature of the hydrocarbons, both their oxidation and thermal stability are good. Volatility is lower than with comparably viscous mineral

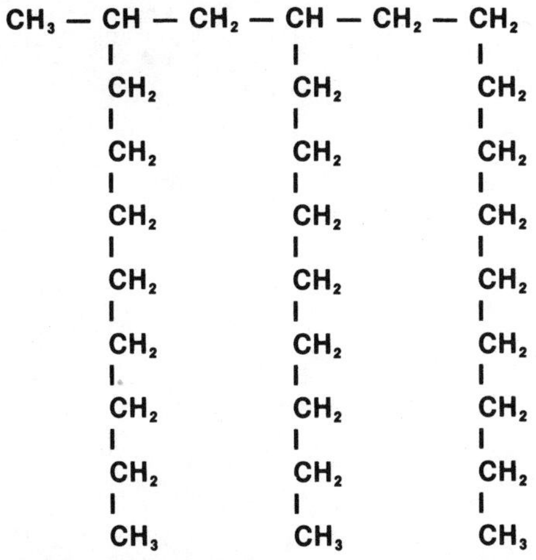

$$
\begin{array}{llllll}
CH_3 - & CH - & CH_2 - & CH - & CH_2 - & CH_2 \\
 & | & & | & & | \\
 & CH_2 & & CH_2 & & CH_2 \\
 & | & & | & & | \\
 & CH_2 & & CH_2 & & CH_2 \\
 & | & & | & & | \\
 & CH_2 & & CH_2 & & CH_2 \\
 & | & & | & & | \\
 & CH_2 & & CH_2 & & CH_2 \\
 & | & & | & & | \\
 & CH_2 & & CH_2 & & CH_2 \\
 & | & & | & & | \\
 & CH_2 & & CH_2 & & CH_2 \\
 & | & & | & & | \\
 & CH_2 & & CH_2 & & CH_2 \\
 & | & & | & & | \\
 & CH_3 & & CH_3 & & CH_3
\end{array}
$$

Figure 5.2 Olefin oligomer. This compound, an oligomer of 1-decene, is a low viscosity oil suitable for blending low viscosity engine oils, as well as a variety of other products.

oils; thus, evaporation loss at elevated temperatures is lower. In many applications, it is important that olefin oligomers be similar in composition to mineral oils and that they be compatible with mineral oils, additive systems developed for use in mineral oils, and machines designed to operate on mineral oils. The olefin oligomers do not cause any significant softening or swelling of typical seal materials, which may be a disadvantage when slight swelling of the seals is desirable to keep a system tight and pliable to prevent leakage. However, proper formulation of finished lubricants can overcome this problem.

Olefin oligomers (PAOs) are used widely as automotive and industrial lubricants. They are often combined with one of the organic esters or other synthetic compounds as the base fluid in engine oils, gear oils, circulating oils, and hydraulic fluids.

In industrial applications, olefin oligomers may be combined with an organic ester or other synthetic compounds as the base fluid in high temperature gear and bearing oils and as lubricants for land-based gas turbines or rotary air compressors. SHFs are widely used in gear applications because they possess two unique advantages over mineral oils: lower traction coefficients and higher oxidation stability. Lower traction coefficients translate directly into greater energy efficiency (up to 8%), and their higher oxidation stability enables longer drain intervals. Properly formulated SHF lubricants can yield six to eight times the drain intervals of comparable mineral oils barring any significant contamination of the fluid. They are also formulated as wide temperature range hydraulic fluids, power transmission fluids, and heat transfer fluids. Wide temperature range greases made from an olefin oligomer combined with an inorganic thickener are finding increasing acceptance as long-life rolling element bearing greases for severe duty applications.

In commercial aviation applications, greases formulated from olefin oligomers and inorganic thickener are widely used as general-purpose aircraft greases. These greases are also used for military aircraft applications, as are less flammable hydraulic fluids formulated with olefin oligomer base fluids.

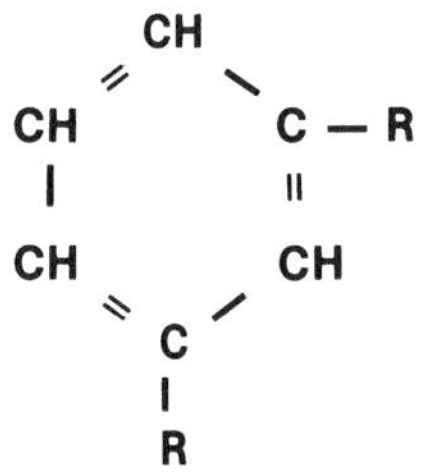

```
        CH
     //    \
CH           C — R
 |           ||
CH           CH
    \\    /
       C
       |
       R
```

Figure 5.3 Alkylated aromatic. This structure represents dialkylated benzene; R is an alkyl group that usually contains 10–14 carbon atoms for products in the lubricating oil range.

B. Alkylated Aromatics

The alkylation of an aromatic compound, usually benzene or naphthalene, is another widely used method of synthesizing hydrocarbons. The alkylation process involves the joining to the aromatic molecule of substituent alkyl groups. Generally, the alkyl groups used contain from 10 to 14 carbon atoms and are of normal paraffinic configuration. The properties of the final product can be altered by changing the structure and position of the alkyl groups. Dialkylated benzene, a typical alkylated aromatic in the lubricating oil range, is shown in Figure 5.3.

Alkylated aromatics have excellent low temperature fluidity and low pour points, and they exhibit good solubility for additives and degradation materials. Their viscosity indexes are about the same as, or slightly higher than, those of high VI mineral oils. They are less volatile than comparable viscous mineral oils, and more stable to oxidation, high temperatures, and hydrolysis. However, it is more difficult to incorporate inhibitors, and the lubrication properties of specific structures may be poor. As with the olefin oligomers, the alkylated aromatics are compatible with mineral oils and systems designed for mineral oils.

Alkylated aromatics are used as the base fluid in some engine oils, gear oils, hydraulic fluids, and greases in subzero applications. They may be used as the base fluid in power transmission fluids and in gas turbine, air compressor, and refrigeration compressor lubricants. They are also used as additives or supplements in the formulation of some specialty lubricants.

C. Polybutenes

Polybutenes are not always considered in discussions of synthesized lubricant bases, although there are a number of lubricant-related applications for the materials. Polybutenes are produced by controlled polymerization of butenes and isobutene (isobutylene). A typical structure is shown in Figure 5.4. The lower molecular weight materials produced by this process have lubricating properties, while higher molecular weight materials, usually referred to as polyisobutylenes (PIBs), are often used as VI improvers and thickeners.

Polybutenes in the lubricating oil range have viscosity indexes between 70 and 110 and fair lubricating properties; they can be manufactured to have excellent dielectric properties. An important characteristic in a number of applications is that above their decomposition temperature, which is approximately 550°F (288°C), the products decompose completely to gaseous materials.

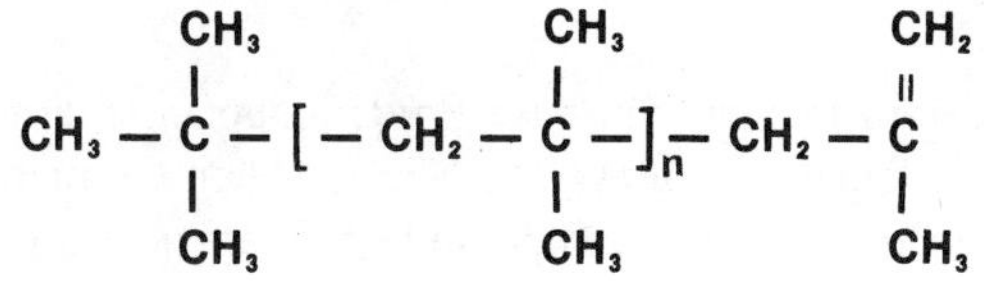

```
         CH3              CH3             CH2
         |                |               ||
CH3 — C — [— CH2 — C —]n — CH2 — C
         |                |               |
         CH3              CH3             CH3
```

Figure 5.4 Typical molecular structure found in polybutenes. Viscosity is a function of the factor *n*, higher values of *n* giving higher viscosities.

The major use of polybutenes in the lubricant-related field is as electrical insulating oils. They are used as cable oils in high voltage underground cables, as impregnants for insulating paper for cables, as liquid dielectrics, and as impregnants for capacitors. Significant volumes have been used as lubricants for rolling, drawing, and extrusion of aluminum when the aluminum is to be annealed afterward. Other applications include gas compressor lubrication, open gear applications, and as carriers for solid lubricants.

D. Cycloaliphatics

Cycloaliphatics are a class of synthesized hydrocarbons now used in small quantities because of certain special properties they possess. One typical structure is shown in Figure 5.5. Cycloaliphatics are sometimes referred to as traction fluids. Under high stresses they develop a glasslike structure and can transmit shear forces; that is, they have high traction coefficients. At the same time, they perform somewhat of a lubricating function in that they prevent welding and metal transfer from one surface to the other. Their stability is excellent under these conditions.

The main application for the cycloaliphatics at present is in stepless, variable speed drives in which the torque is transmitted from the driving member to the driven member by the resistance to shear of the lubricating fluid. The high traction coefficients of the cycloaliphatics permit higher power ratings than do conventional lubricants.

Also, cycloaliphatics have also found some application in rolling element bearings where speed and load conditions may cause skidding of the rolling elements when conventional lubricants are used.

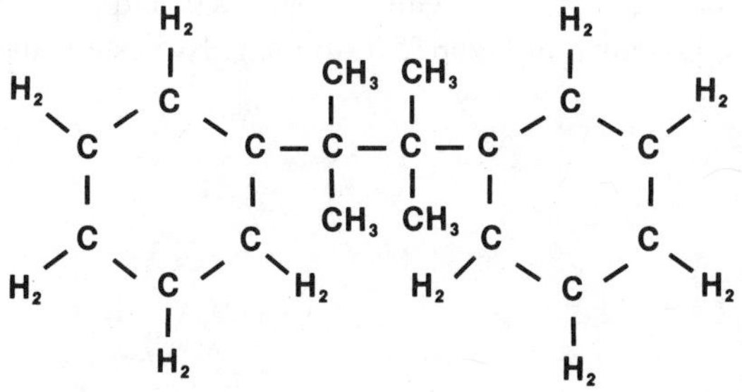

```
        H2                                H2
        |                                 |
H2      C        CH3  CH3                 C       H2
  \   /   \      |    |                 /   \   /
    C       C — C — C — C                   C
    |       |    |    |    |                |
    C       C    CH3  CH3  C                C
  /   \   /   \          /   \            /   \
H2      C       H2     H2      C        /       H2
        |                      |
        H2                     H2
```

Figure 5.5 Typical cycloaliphatic structure.

II. ORGANIC ESTERS

Organic esters have been an important class of synthesized base fluids longer than any of the other materials now in use. Their use dates back to World War II, when German chemists used them in mineral oil blends to improve low temperature properties and to supplement scarce supplies of mineral oils. They were first used as aircraft jet engine lubricants in the 1950s and are now used as the base fluid for essentially all aircraft jet engine lubricants. They are also used as the base fluid in many wide temperature range aircraft greases.

Organic esters are oxygen-containing compounds that result from the reaction of an alcohol with an organic acid. The two commonly use classes of organic esters are dibasic acid esters and polyol esters.

A. Dibasic Acid Esters

The various diesters differ in their acid and alcohol components. For all diesters, the acid and alcohol are reacted either thermally or in the presence of a catalyst in an esterification reactor. After the ester has been formed, the water by-product is distilled off and unreacted dibasic acid is neutralized and removed by filtration. The base stock is then suitable for final product blending. As shown in Figure 5.6, the backbone of the structure is formed by the acid, with the alcohol radicals joined to its ends.

Dibasic acid esters exhibit good metal-wetting ability, high film strength, high oxidation and thermal stability, and good shear stability. Diesters will dissolve system deposits and keep metal surfaces clean. This property could be a disadvantage in dirty systems. The hydrolytic stability and antirust properties of diesters are fair. In changing over to diesters from another family of products, care should be taken to assure a thorough cleaning and flushing of the system prior to their installation. Also the compatibility with elastomers and paints used in the system should be reviewed. Diesters are compatible with mineral oils.

Dibasic acid esters have been used as the base fluid for older type I jet engine oils. Generally the use of these oils has been restricted mainly to older military jet engines, and some very limited use in jet engines for industrial service. Diesters are used as the base oils, or components of the base oil, for automotive engine oils and air compressor lubricants. They are also used as the base fluid in some aircraft greases.

These products were developed originally for use as jet engine oils, but now have been replaced, to a large extent, by polyol esters.

B. Polyol Esters

Polyol esters are made by reacting a polyhydric alcohol with a monobasic acid to give the desired ester. In contrast to diesters, as shown in Figure 5.7, in the polyol esters, the

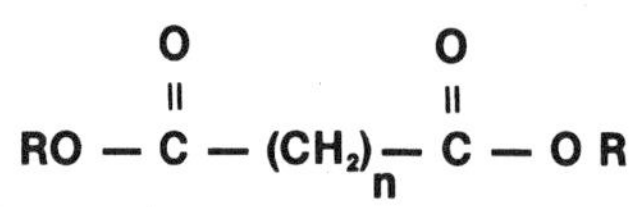

$$\mathrm{RO - \overset{\overset{\displaystyle O}{\|}}{C} - (CH_2)_n - \overset{\overset{\displaystyle O}{\|}}{C} - O\,R}$$

Figure 5.6 Dibasic acid ester: Commonly used acids are adipic ($n = 4$), azelaic ($n = 7$), and sebacic ($n = 8$).

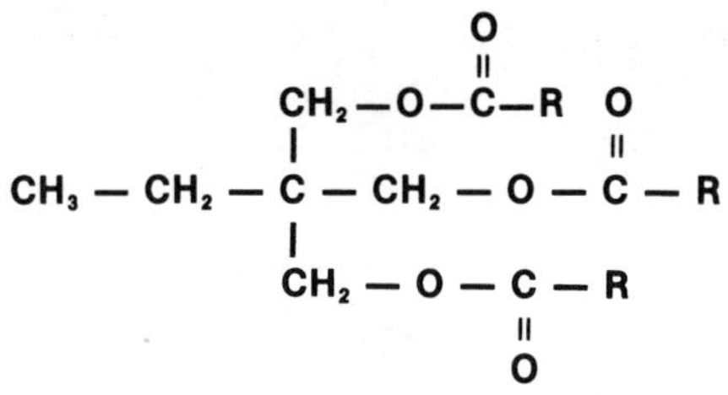

```
                          O
                          ||
                 CH2—O—C—R   O
                  |           ||
CH3 — CH2 — C — CH2 — O — C — R
                  |
                 CH2 — O — C — R
                            ||
                            O
```

Trimethylol Propane Ester

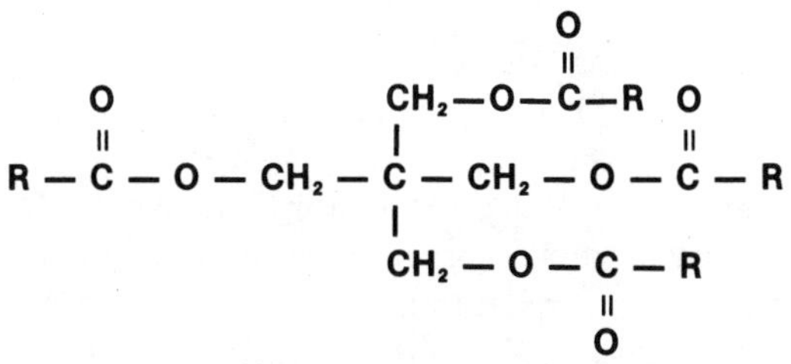

```
                               O
                               ||
     O               CH2—O—C—R   O
     ||               |            ||
R — C — O — CH2 — C — CH2 — O — C — R
                      |
                     CH2 — O — C — R
                                ||
                                O
```

Pentaerithritol Ester

Figure 5.7 Typical polyol esters: acid radicals R typically used contain 6–10 carbons, with those containing an odd number of carbon atoms being generally favored.

polyol forms the backbone of the structure with the acid radical attached to it. As with diesters, the physical properties of polyol esters can be varied by using different polyps or acids. Trimethylol propane and pentaerithritol are two of the polyols that are commonly used. Usually, the acids are obtained from animal or vegetable oils.

Polyol esters have better high temperature stability than diesters. Their low temperature properties and hydrolytic stability are about the same, but their VIs may be lower. Their volatility is equal or lower. The polyol esters also may have more effect on paints and cause more swelling of elastomers.

The primary use of polyol diesters is in type II jet engine oils. They also are used in air compressor oils and as components in some synthesized hydrocarbon blends. More recently, polyol esters have become widely used as refrigeration lubricants, to take advantage of their miscibility with hydrofluorocarbon (HFC) refrigerants. Polyol esters are also the lubricant of choice in refrigeration systems where nonchlorine HFC refrigerants such as R134a are used. They are also used in environmentally sensitive products such as hydraulic fluids and other lubricants for which biodegradable structures are required. Additional information on lubrication of refrigeration compressors can be found in Chapter 17, Compressors.

III. POLYGLYCOLS

Polyglycols cover a wide range of products and properties. Presently, they are the largest single class of synthetic lubricant bases.

Polyglycols are variously described as polyalkylene glycols, polyethers, polyglycol ethers, and polyalkylene glycol ethers. The latter term is the most complete and accurate

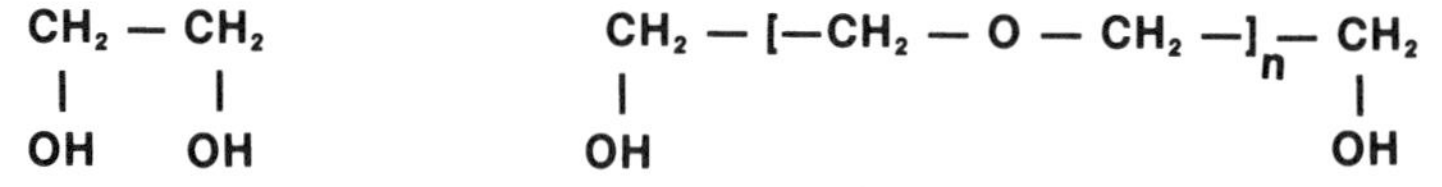

$$\begin{array}{cc} CH_2 - CH_2 \\ | \quad\quad | \\ OH \quad OH \end{array}$$

Ethylene Glycol

$$\begin{array}{l} CH_2 - [-CH_2 - O - CH_2 -]_n - CH_2 \\ | \qquad\qquad\qquad\qquad\qquad\qquad | \\ OH \qquad\qquad\qquad\qquad\qquad\quad OH \end{array}$$

Polyethylene Glycol Ether

Figure 5.8 Glycols. Ethylene glycol is the simplest glycol; when polymerized, the oxygen bond is formed and it becomes a polyglycol ether.

for the bulk of the materials used in lubricants. Small quantities of simple glycols, such as ethylene and polyethylene glycol, are also used as hydraulic brake fluids. Typical structures for the two types are shown in Figure 5.8.

Polyglycols are polymers made from ethylene oxide (EO), propylene oxide (PO), or their derivatives. The primary raw materials are ethylene or propylene, oxidized to form cyclic ethers (alkylene oxides). Combining ethers derived from ethylene oxide (EO) to propylene oxide (PO) has a profound effect on the solubility of the product in other fluids:

EO:PO = 4:1 Water-soluble, not soluble in hydrocarbons
EO:PO = 1:1 Soluble in cold water, soluble in alcohol and glycol ethers, not soluble in hydrocarbons
EO:PO = 0:1 Not soluble in water, conditionally soluble in hydrocarbons

These comparisons help explain the differences between the various types of, and uses for, polyglycol: automotive antifreeze, brake fluid, water-based hydraulic fluids, hydrocarbon gas compressors, and high temperature bearing lubricants. In addition to ethylene and propylene oxides, butylene oxide is used to provide some polyglycols with specific properties and is oil soluble. Polyglycols made with butylene oxide are more expensive and do not exhibit traction coefficients equal to combinations of EO and PO.

One of the major advantages of polyglycols is that they decompose completely to volatile compounds under high temperature oxidizing conditions. This results in low sludge buildup under moderate to high operating temperatures, or complete decomposition without leaving deposits in certain extremely hot applications. Polyglycols have good viscosity–temperature characteristics, although at low temperatures they tend to become somewhat more viscous than some of the other synthesized bases. Pour points are relatively low. High temperature stability ranges from fair to good and may be improved with additives. Thermal conductivity is high. Not generally compatible with mineral oils or additives developed for use in mineral oils, polyglycols may have considerable effect on paints and finishes. They have low solubility for hydrocarbon gases and some refrigerants. Seal swelling is low, but with the water-soluble types some care must be exercised in seal selection to be sure that the seals are compatible with water. Even if the glycol fluid does not initially contain any water, it has a tendency to absorb moisture from the atmosphere.

The applications for the polyglycols are divided into those for the water-soluble types and those for the water-insoluble types.

The largest volume application of water-soluble polyglycols is in hydraulic brake fluids. Other major applications are in metalworking lubricants, where they can be removed by water flushing or being burned off, and in fire-resistant hydraulic fluids. In the latter application, the polyglycol is mixed with water, which provides the fire resistance. Water-soluble polyglycols are also used in the preparation of water-diluted lubricants for rubber bearings and joints.

Water-insoluble polyglycols are used as heat transfer fluids and as the base fluid in industrial hydraulic fluids of certain types, as well as high temperature gear and bearing oils. They are also finding application as lubricants for screw-type refrigeration compressors operating on R12 and hydrocarbon gases, and compressors handling hydrocarbon gases. (The use of R12, a potential ozone layer depleting substance, is being phased out; see Chapter 17 for additional information on refrigeration compressors and refrigerants.) In these latter applications, the low solubility of the gases in the polyglycol minimizes the dilution effect, contributing to better high temperature lubrication.

IV. PHOSPHATE ESTERS

Phosphate esters are one of the other commonly used classes of synthetic base fluids. A typical phosphate ester structure is shown in Figure 5.9.

One of the major features of the phosphate esters is their fire resistance, which is superior to mineral oils. Their lubricating properties are also generally good. The high temperature stability of phosphate esters is only fair, and the decomposition products can be corrosive. Generally, they have poor viscosity–temperature characteristics (low VI), although pour points are reasonably low and volatility is quite low. Phosphate esters have considerable effect on paints and finishes, and they may cause swelling of many seal materials. Their compatibility with mineral oils ranges from poor to good, depending on the ester. Their hydrolytic stability is only fair. They have specific gravities greater than 1, which means that water contamination tends to float rather than settle to the bottom, and pumping losses are higher than is the case with products lower in specific gravity. Their costs are generally high, and they are limited in viscosity.

The major application of phosphate esters is in fire-resistant fluids of various types. Hydraulic fluids for commercial aircraft are phosphate ester based, as are many industrial fire-resistant hydraulic fluids. These latter fluids are used in applications such as the electrohydraulic control systems of steam turbines and industrial hydraulic systems, where hydraulic fluid leakage might contact a source of ignition. In some cases they may also be used in the turbine bearing lubrication system.

Phosphate esters are also used as lubricants for compressors where discharge temperatures are high, to prevent receiver fires and explosions that might occur with conventional lubricants. Some quantities of phosphate esters are also used in greases and mineral oil blends as wear and friction reducing additives.

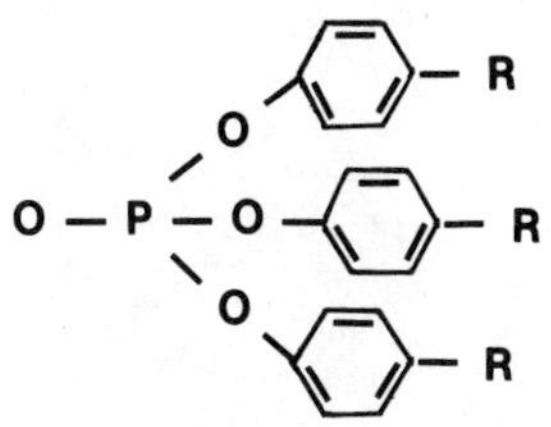

Figure 5.9 Phosphate ester. The R group can be either an aryl or an alkyl type. If, for example, the methyl group (CH_3) is used, the ester is tricresyl phosphate.

V. OTHER SYNTHETIC LUBRICATING FLUIDS

Brief descriptions and principal applications of some of the other synthetic base fluids are given in Sections V.A–V.D.

A. Silicones

Silicones are among the older types of synthetic fluid. As shown in Figure 5.10, their structure is a polymer type with the carbons in the backbone replaced by silicon.

Silicones have high viscosity indexes, some on the order of 300 or more. Their pour points are low and their low temperature fluidity is good. They are chemically inert, nontoxic, fire resistant, and water repellent, and they have low volatility. Seal swelling is low. Compressibility is considerably higher than for mineral oils. Thermal and oxidation stability of silicones are good up to quite high temperatures. If oxidation does occur, oxidation products include silicon oxides, which can be abrasive. A major disadvantage of the common silicones is that their low surface tension permits extensive spreading on metal surfaces, especially steel. As a result, effective adherent lubricating films are not formed. Unfortunately, the silicones that exhibit this characteristic also show poor response to additives aimed at reducing wear and friction. Some newer silicones show promise of overcoming these deficiencies.

Silicones are used as the base fluid in both wide temperature range and high temperature greases. They are also used in specialty greases designed to lubricate elastomeric materials that would be adversely affected by lubricants of other types. Silicones are also used in specialty hydraulic fluids for such applications as liquid springs and torsion dampers, where their high compressibility and minimal change in viscosity with temperature are beneficial. They are also being used as hydraulic brake fluids and as antifoam agents in lubricants. Some newer silicones are also offered as compressor lubricants.

B. Silicate Esters

Silicate esters have excellent thermal stability and, with proper inhibitors, show good oxidation stability (see Figure 5.11 for their chemical structure). They have excellent viscosity–temperature characteristics, and their pour points are low. Their volatility is low, and they have fair lubricating properties. A major factor for their limited use is their poor resistance to hydrolysis.

Small quantities of silicate esters are used as heat transfer fluids and dielectric coolants. Some specialty hydraulic fluids are formulated with silicate esters.

C. Polyphenyl Ethers

Polyphenyl esters are organic materials that have excellent high temperature properties and outstanding radiation resistance. They are thermally stable to above 800°F (450°C)

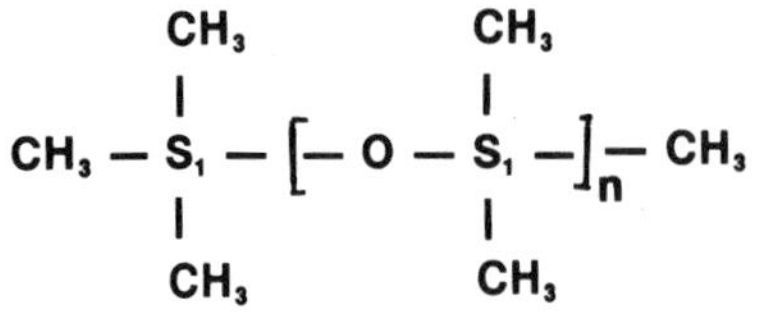

$$CH_3 - \underset{\displaystyle CH_3}{\overset{\displaystyle CH_3}{S_i}} - \left[- O - \underset{\displaystyle CH_3}{\overset{\displaystyle CH_3}{S_i}} - \right]_n - CH_3$$

Figure 5.10 Dimethyl polysiloxane, one of the more widely used silicone polymer fluids.

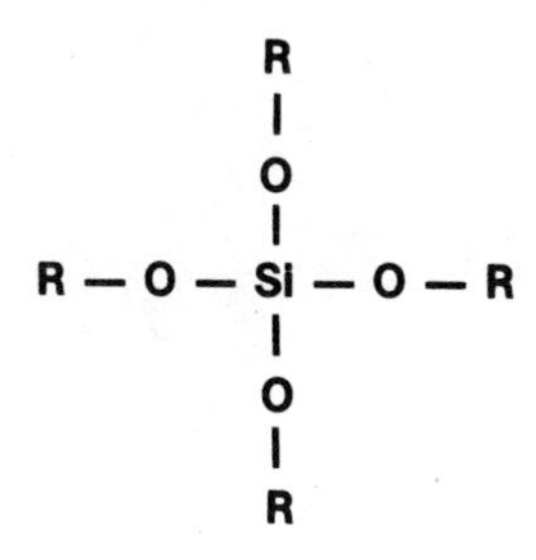

Figure 5.11 Silicate ester: the R groups contain either alkyl or aryl groups. The physical properties are dependent on the nature of these groups.

and have excellent resistance to oxidation at elevated temperatures. High viscosities at normal ambient temperatures, however, tend to restrict their use.

Small quantities of polyphenyl ethers are used as heat transfer fluids, as lubricants for high vacuum pumps, and as the fluid component of radiation-resistant greases.

D. Halogenated Fluids

Chlorine or fluorine or combinations of the two are used to replace part (or all) of the hydrogen in hydrocarbon or other organic structures to form lubricating fluids. Generally, these fluids are chemically inert and essentially nonflammable, often showing excellent resistance to solvents. Some have outstanding thermal and oxidation stability, being completely unreactive even in liquid oxygen. Also, their volatility may be extremely low.

Some of the lower cost halogenated hydrocarbons are used alone or in blends with phosphate esters as fire-resistant hydraulic fluids. Other halogenated fluids are used for such applications as oxygen compressor lubricants, lubricants for vacuum pumps handling corrosive materials, solvent-resistant lubricants, and other lubricant applications entailing highly corrosive or reactive materials.

6

Environmental Lubricants

With concerns for protecting and preserving the environment in all aspects of our daily lives, it is only natural that as lubricant technology has advanced, so has the development of oils and greases that would be less detrimental to the environment if inadvertently spilled or leaked. Accelerating research and development in this area has also been driven by public demand, industry concerns, and governmental agencies to find better ways to protect the biological balance in nature, or at least to reduce the negative impact of spills or leakage of lubricants that do occur. Many names are used for the class of lubricants to address these concerns: environmentally friendly, environmentally acceptable, biodegradable, nontoxic, and others. For purposes of discussions in this chapter, we shall refer to this class of lubricants as environmentally aware (EA).

Of primary interest in selection and use of this class of EA lubricants is defining and measuring the product attributes that could affect the environment. In addition, the lubricants must provide performance in key areas such as oxidation stability, viscosity–temperature properties, wear protection, friction reduction, rust and corrosion protection, and hydrolytic stability where water or moisture may be present. In other words, the EA products must perform at levels equivalent to those achieved by conventional mineral- or synthetic-based lubricants in the equipment, while providing characteristics that reduce the negative impact in the event of inadvertent introduction into the environment.

I. ENVIRONMENTAL CONSIDERATIONS

Environmental acceptability of lubricants is not well defined and can encompass a broad range of potential environmental benefits: use of renewable resources, resource conservation, pollutant source reduction, recycling, reclamation, disposability, degradability, and so on. Therefore, any claim of environmental acceptability must be supported by appropriate technical documentation. Most petroleum-based lubricants can be considered to be environmentally acceptable by various standards. For example, long-life synthetics (discussed in Chapter 5) and other lubricants that provide extended oil drain capability might be classed as EA materials because they conserve resources and aid in potential pollutant source reduction (since quantities for disposal will be lower). Many oils can be reclaimed, recycled, or burned for their heat energy value, again resulting in conservation of resources. All these efforts to help reduce the environmental impact of lubricants have positive effects and should be an integral part of the planning to establish an environmental program. The remainder of this chapter is devoted to discussions of a class of lubricants exhibiting specific characteristics such as biodegradability and low toxicity.

II. DEFINITIONS AND TEST PROCEDURES

The two environmental characteristics most desirable in EA lubricants are speed at which the products will biodegrade if introduced into nature and toxicity characteristics that might affect bacteria or aquatic life. Most lubricants are inherently biodegradable, which means that given enough time, they will biodegrade by natural processes. They will not persist in nature. In certain applications, however, much faster rates of biodegradation are desired. These are referred to as *readily biodegradable* products. All lubricants range in toxicity from low (sometimes called nontoxic) to relatively high. Toxicity has a direct effect on naturally occurring bacteria and aquatic life and therefore needs to be an important part of the development of EA lubricants.

Unlike traditional lubricant development, where the predominant focus is on product performance in equipment, a major part of developing EA lubricants involves understanding and defining environmental test criteria and developing ways to assess the effects of new and used lubricants in actual applications where environmental sensitivity is an issue. Since both base fluids and additive systems impact the environmental characteristics, these tests must evaluate the ecotoxicity of base fluids, additives, and finished lubricants.

A. Toxicity

The impact of lubricants on the aquatic environment is evaluated by conducting acute aquatic toxicity studies with rainbow trout (a freshwater fish that is sensitive to environmental changes) or other aquatic life-forms that are sensitive to changes in their environment. Since oil is insoluble in water, the aquatic specimens are exposed under oil–water dispersion (mechanical dispersion) conditions to increasing concentrations of test materials up to a maximum concentration of 5000 ppm. This oil–water dispersion technique follows a modification of the procedure used by the British Ministry of Agriculture, Fisheries and Food (MAFF). In the oil–water dispersion procedure, the test materials are added to aquaria equipped with a central cylinder-housed propeller system that provides mechanical agitation to continuously disperse the test material as fine droplets in the water column. The propeller is rotated to produce flow in the cylinder by drawing small quantities of water and test material from the surface into the top of the cylinder and expelling a suspension of oil droplets in water through apertures near the bottom of the cylinder. This procedure, which simulates physical dispersion by wave and current action, is used to evaluate the relative toxicity of lighter-than-water materials. The aquatic specimens are exposed to five concentrations of test material and a control (without test material) during each study. Toxicity is expressed as the concentration of test material in parts per million (wt/vol) required to kill 50% of the aquatic specimens after 96 hours of exposure (LC_{50}).

B. Biodegradability

Two tests are most commonly used to assess the biodegradability of lubricants. The shake flask test* is used to determine ultimate biodegradability (conversion to CO_2) of the test material. The second, the CEC (Coordinating European Council) test, is not as discriminating but is widely used in Europe for assessing the biodegradability of lubricants and was, in fact, specifically designed to evaluate the aerobic aquatic biodegradation potential of two-stroke-cycle engine oils. Both tests use a mineral salts mix for the growth medium, with the carbon substrate being supplied only by the test material. Both the shake flask and CEC tests use unacclimated sewage inoculum, which is typically obtained from a municipal wastewater treatment plant that has no industrial inputs. The shake flask test, in addition, utilizes a soil inoculum.

The shake test flasks, closed with neoprene stoppers from which are suspended alkali traps, are placed on a rotary shaker and heavily shaken at 25°C. Periodically, over a 28-day period, the flasks are removed and titrated to quantify the trapped CO_2. The medium is then sparged with air to maintain aerobic conditions, and the fresh traps are placed back in the flasks. Blank controls, which are run alongside the flasks containing the test material, have all components present in the test flasks except the test material. At each time point, the quantity of CO_2 evolved from the blanks is subtracted from CO_2 values in the test material flasks. A positive control containing a readily biodegradable material is also run to ensure inoculum viability.

ASTM has issued a test for biodegradability to standardize testing for biodegradability of environmental type products. This test (ASTM D 5846) is a modified Sturm test and very similar to the shake flask test just described. It also measures CO_2 evolution as the bacteria metabolizes the test material.

The CEC test utilizes cotton-stoppered flasks and, as with the shake flask test, the flasks are placed on a rotary shaker table and heavily shaken at 25°C. At 0, 7, and 21 days, flasks are extracted with Freon 113 and the quantity of test material in each of the extracts is determined by infrared (IR) analysis at 2930 cm^{-1} (C—H stretch). The percent of material biodegraded after 7 and 21 days is determined by comparing the intensity of the IR absorbance in the test flask extracts, after each period of time, against zero time values and against values in the abiotic controls ($HgCl_2$-poisoned).

C. Environmental Criteria

At the present time, there are no generally accepted worldwide regulations to define criteria for lubricants used in environmentally sensitive areas. There are products with limited applications such as those receiving the German Blue Angel Label for lubricants. A lubricant can carry a Blue Angel label if all major components meet OECD ready biodegradability criteria and all minor components are inherently biodegradable. Secondary criteria include a ban on specific hazardous materials, and lubricants must meet aquatic toxicity limits. Based on an evaluation of current legislation for new product registration by the European Inventory of Existing Commercial Chemical Substances (EINECS) and on marine transport requirements by Marpol, the International Maritime Organization (IMO), as well as a review of proposed labeling schemes, there is some consensus in industry for

* ''Shake flask test'' refers to either the U.S. Environmental Protection Agency test described in EPA 560/6-82-003 or the Organization for Economic Development and Co-operation tests described in OECD 301.

biodegradation and aquatic toxicity criteria for lubricants that will be used in environmentally sensitive areas. A product may be considered acceptable if it meets the following criteria:

Aquatic toxicity >1000 ppm (50% min survival of rainbow trout)
Ready biodegradability > 60% conversion of test material carbon to CO_2 in 28 days, using unacclimated inoculum in the shake flask or ASTM D 5846 test

Aquatic toxicity and ready biodegradation studies were conducted on products formulated with mineral oils and non–mineral oil base stocks (Table 6.1). In general, the base stocks that comprise the major component of most lubricant formulations are nontoxic. The aquatic toxicity observed following exposure to the formulated products in Figure 6.1 is caused by one or more of the additives.

Vegetable oils, such as Mobil EAL 224H, and a number of synthetic esters easily met the ready biodegradation criterion (> 60% conversion to CO_2 in 28 days) and always had CEC test results exceeding 90% conversion after 21 days. None of the formulations tested containing mineral oil base stocks were able to meet the ready biodegradation criterion, although 42–49% of these materials were converted to CO_2 in 28 days (Figure 6.1). This does not appear to be a significant difference from the 60% criterion, but in actual field conditions, it is a major difference.

The polyglycol-based materials, although soluble in water, failed to meet the ready biodegradability criterion, with only 6–38% of the test material converted to CO_2 in 28 days. The biodegradation of polyglycols is determined by the ratio of propylene oxide to ethylene oxide, with polyethylene glycols being more biodegradable. The average molecular weight of the material is also critical, with material under a molecular weight of 1000 being rapidly biodegraded. The rate and extent of biodegradation diminishes with increasing molecular weight. Some additional studies of the polyglycol materials is needed to further quantify biodegradation rates of these materials.

Evaluations of the impact of base stocks used to formulate hydraulic oils, formulated conventional hydraulic oils and EA hydraulic fluids have been conducted to determine the various levels of aquatic toxicity that these materials may exhibit. The toxicity testing was done using the EPA 560/7-82-002 (for all intents and purposes, this is the same test as the OECD 203:1–12). The results of this study of base stocks and fully formulated hydraulic fluids, given in Figure 6.2, indicate that the toxicity of most lubricants is due to one or more of the additives in the formulation.

Table 6.1 Ecotoxicology Data for Select Hydraulic Fluids

		Biodegradability (%)	
Product base stock	Trout LC_{50} (ppm)	Shake flask	CEC test
Mineral oil	389 to >5000	42–48	(Not tested)
Vegetable oil	633 to >5000	72–80	>90
Synthetic ester	>5000	55–84	>90
Polyglycol	80 to >5000	6–38	(Not tested)

Sources: Shake Flask Test Measures Carbon Dioxide Evolution, EPA Method 560/6-82-003; CEC Method-CEC-L-33-T-82.

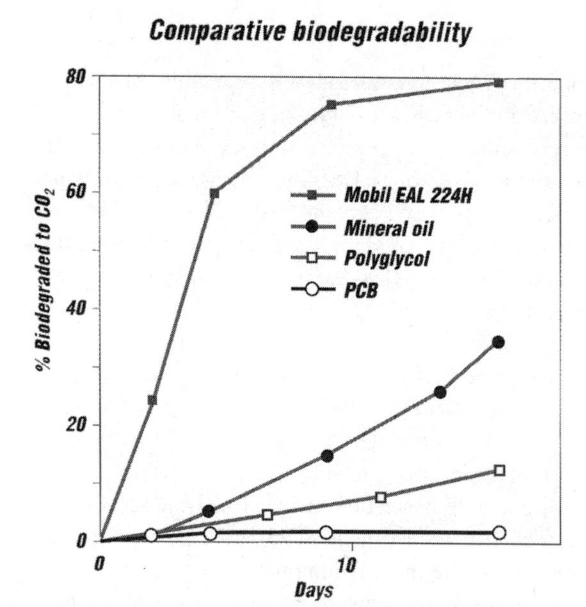

Figure 6.1 Determining comparative biodegradability with the EPA 560/6-82-003 test method.

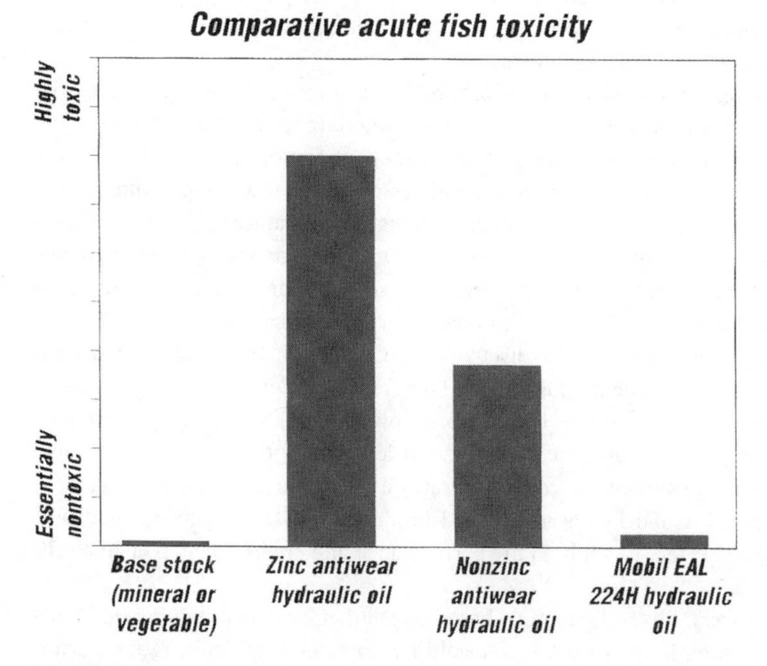

Figure 6.2 Comparison of toxicity characteristics of various lubricants.

III. BASE MATERIALS

One of the primary choices of base oils for EA lubricants today is vegetable oils. This is due to their good natural biodegradability and very low toxicity in combination with very good lubricity characteristics. These renewable resources also provide a cost advantage over other EA base materials such as synthetic base stocks. But, there are some performance limitations of the vegetable-based lubricants that have been and continue to be addressed. These characteristics are mainly high temperature oxidation stability, low temperature performance, viscosity limitations, and cost. Although less expensive than synthetic alternatives, vegetable-based products can cost several times as much as conventional mineral-based lubricants. Genetic engineering will provide improved performance in the future in areas of oxidation stability and low temperature performance by increasing the high oleic acid content as well as other by means of genetic alterations (branching). In most applications, the vegetable-based EA lubricants can be formulated to perform in all but the most severe equipment. It is important to note that not all vegetable-based EA lubricants will provide the same levels of performance. Vegetable-based oils derived from rapeseed plants, cotton seeds, soybean oil, sunflower seed oil, corn oil, palm oil, and peanut oils are frequently used materials, with rapeseed being the most common.

Synthetic-based materials such as polyglycols (discussed earlier), polyol esters, pentaerithritol esters, and certain PAOs (see Chapter 5) are used to formulate the synthetic EA lubricants. Their advantages over vegetable-based EA lubricants are wider temperature range application, longer drain capability (oxidation stability), and excellent performance in systems with close-tolerance servo valves.

Some of the more general performance characteristics of the various base materials can be discussed in the following categories.

1. *Vegetable oils.* The choices of correct processes to refine, bleach, and deodorize vegetable-based oils can yield very satisfactory base materials for the formulation of finished lubricants. This renewable resource provides excellent natural lubricity, low volatility, and good environmental characteristics. Weaknesses are in low temperature performance, hydrolytic stability, and oxidation stability in high temperature applications. These products are also currently limited to low viscosity (ISO 32–68) materials. Properly manufactured and formulated vegetable-based lubricants can equal conventional mineral oil based lubricants in performance in all but the most severe applications.

2. *Polyalphaolefins (PAOs).* As discussed in Chapter 5, PAOs provide a good option for formulating environmental lubricants. Their ready biodegradability in the lower viscosity range is good. They also provide excellent low and high temperature (oxidation stability) performance, good hydrolytic stability, and low volatility. Their disadvantages are in costs and lower rates of biodegradability rates as viscosities increase. To achieve the good characteristics of the PAOs in finished products, they are often blended with biodegradable synthetic esters to get both the performance and environmental characteristics desired.

3. *Synthetic esters.* Several materials based on synthetic esters exhibit good biodegradability as well as high levels of oxidation stability, low and high temperature performance, and good hydrolytic stability and seal swell performance. The synthetic esters will allow formulation of higher viscosity lubricants typically used in circulating systems and some gear oils.

Table 6.2 Comparison of Fully Formulated EA Lubricants with Various Bases[a]

Properties	Mineral	Vegetable	PAOs	Diesters	Polyol esters	PAGs
Viscosity temperature characteristics	Fair	Good	Good	Fair	VG	VG
Low temperature properties	Poor	Poor	VG	Good	Good	Good
Oxidation stability	Fair	Poor	VG	Good	Good	Good
Compatibility with mineral oils	Exc	Exc	Exc	Good	Fair	Poor
Low volatility	Fair	Good	Exc	Exc	Exc	Good
Varnish and paint compatibility	Exc	VG	Exc	Poor	Poor	Poor
Seal swell (NBR)	Exc	Exc	VG	Fair	Fair	Good
Lubricating properties	Good	VG	Good	VG	VG	Good
Hydrolytic stability	Exc	Poor	Exc	Fair	Fair	VG
Thermal stability	Fair	Fair	Fair	Good	Good	Good
Additive solubility	Exc	Exc	Fair	VG	VG	Fair

[a] These ratings are generalizations. Specific manufacturers of products should be consulted for current data. Exc, excellent; VG, very good.

Table 6.2 shows a general comparison, against mineral oils, of some of the more common performance characteristics of fully formulated EA lubricants using the various base materials. Actual finished product performance could vary from these ratings as a result of technological advancements in such areas as additive technology, use of blends of base materials, and manufacturing processes.

As a result of the higher costs, EA lubricants will typically be used in areas where environmental sensitivity is an issue. In many instances of spillage or leakage that are reportable to governmental agencies such as the U.S. National Response Center, the added costs of EA lubricants may be offset by the potential for lower fines and remediation costs.

EA lubricants are not meant for use in all applications but only when their use can be economically justified or the environmental sensitivity issues are of prime importance. In many cases, economic justification of the EA lubricants based solely on equipment performance is not sufficient to merit their use. The economics must be derived from reducing costs of remediation in the event of spills or leakage. Also, in some localities, limited legislation or regulations promote or require the use of such products. Environmental sensitivity issues prevail in the following specific areas.

Dredging operations for waterways
Operation of equipment for dams and locks
Offshore drilling
Marine equipment
Recreation and parks
Construction sites on or near water or groundwater systems
Agricultural operations
Forestry and logging
Mining
Automotive service lifts
Hydraulic elevators

A. Product Availability and Performance

Because of the large volumes of hydraulic fluids used around the world and the tendency of these products to leak under conditions of relatively high pressures and severity of some applications, the first category of EA fluids to be developed and widely marketed were hydraulic oils. Once readily biodegradable base oils and low toxicity additive systems had been identified, the next hurdle was to provide fully formulated oils that exhibited required equipment performance as established by the builders of equipment as well as the users. Equipment builders have received many requests to approve the use of EA lubricants by their customers and need to be assured that the EA products will perform satisfactorily in their equipment and meet the service life requirements of the customers.

Certain EA lubricants are formulated not only to meet the environmental criteria but also to provide performance equal to that of conventional mineral based lubricants. Much of the performance in hydraulic systems is determined by industry standard pump tests. We list three common tests.

1. *Vickers V-104C Pump Test (ASTM D 2882).* A rapeseed-based EA antiwear hydraulic fluid provided little or no pump wear in the ASTM D 2882 pump test. In addition to the standard 100 h dry test, a more severe 200 h test was undertaken in which 1% water was added at 0 and 100 hs. Because water contamination affects some EA fluids with poor hydrolytic stability, this 200 h test simulates wet systems and evaluates the oil as it degrades at accelerated rates or loses pump wear protection in the presence of water. Test results (Table 6.3) indicate low cumulative wear as well as good viscosity control and low total acid number (TAN) increase. Fluid characteristics did not change appreciably and wear protection was excellent, even in the presence of high moisture levels.

2. *Vickers 35VQ pump test.* This severe industry-accepted antiwear vane pump test is based on the Vickers 35VQ25 vane pump run at 3000 psi and 200°F. Standard procedures require that the same fluid be subjected to three successive 50 h test runs, and total ring and vane wear be less than 90 mg for each run. The tests (Table 6.4) showed low wear and good pump component appearance at the end of five successive 50 h pump test inspections. While fluid color darkened rapidly, increases in viscosity and TAN were small.

Table 6.3 Extended "Wet" Vickers 104C Vane Pump Test Results with Mobil EAL 224H Vegetable-Based Oil

		Test hours	
Properties	New oil	100	200
Viscosity, cSt at 40°C (ASTM D 445-3)	35.4	35.1	35.6
Total acid number (ASTM D 664)	0.92	0.97	1.14
Water addition	1%	1%	
Cumulative wear (vanes and ring)	—	16 mg	25 mg
Test conditions			
Duration:	200; 1% water added at 0 and 100 test h; wear measured at 100 and 200 h		
Temperature:	150°F		
Pressure:	2000 psi		

Table 6.4 Vickers 35VQ Vane Pump Test Results with Mobil EAL 224H

	Normal test				Extended test	
Properties	New oil	First run	Second run	Third run	Fourth run	Fifth run
Viscosity cSt at 40°C (ASTM D 445-3)	35.0	35.3	35.6	35.8	36.2	36.6
Total acid number (ASTM D 664)	1.03	1.27	1.27	1.29	1.18	1.32
Color (ASTM D 1500)	<1.5	<3.5	<4.5	6.0	<7.5	7.5
Total wear (vanes and ring)	—	10 mg	3 mg	56 mg	3 mg	32 mg
Test conditions						
Duration:	50 h and a new cartridge per run					
Temperature:	200°F					
Pressure:	3000 psi					

3. *Denison hydraulics HF-0 test.* Another industry-accepted pump test is the Denison HF-0 axial piston pump test, which evaluates multimetal compatibility of a fluid and corrosiveness to soft metals under severe operating conditions. The 100 h test uses a 46 series piston pump and is run at 160°F (71°C) for the first 60 h; the temperature is raised to 210°F (99°C) for the remaining 40 h. The 100 h test is run at a pressure of 5060 psi (340 bar). The pass/fail criterion for this pump test is based on physical inspection and measurement for distress, wear, and chemical etching on both the bronze bearing plate and the piston shoes. In addition to the axial piston pump test, part of meeting the HF-0 requires passing a 600 h (300 h dry/300 h wet) vane pump test (TP-30283A). Wear measurements and surface examination showed critical parts to be in good condition, and the pump manufacturer confirmed a passing result for the EA fluid.

The results of these three relatively severe pump tests indicate that EA antiwear hydraulic fluids not only can be formulated to meet the biodegradability and toxicity requirements but can provide performance levels at least equal to conventional mineral-based antiwear hydraulic fluids in these tests. These test results have also been verified by years of field application experience. Again, some EA lubricants may or may not be equal in performance to conventional hydraulic fluids, and suppliers should be consulted for specific test results and examples of field performance.

In addition to EA antiwear hydraulic fluids, several gear oils, some greases, engine oils, circulating oils, metalworking fluids, dust control oils, transmission fluids, and chain saw oils are available in readily biodegradable and low toxicity formulations. Table 6.5 shows one manufacturer's partial line of EA lubricants, along with application information.

B. Vegetable Oil Based EA Lubricant Performance Concerns

As mentioned earlier, the potential performance concerns in vegetable-based lubricants relative to comparable premium mineral oil based lubricants are oxidation stability, low temperature performance, and hydrolytic stability. These are not significant as long as they are recognized and proper care is taken during selection, application, and operation.

Table 6.5 EA Lubricants and Application Data

Product	Description	Application	Benefits
Vegetable-based hydraulic and circulating oil	An ISO viscosity grade 32/46 vegetable oil-based antiwear hydraulic oil that is readily biodegradable and virtually nontoxic	Primary application is for industrial and mobile equipment hydraulic systems operating at temperatures of 0°F to 180°F. Meets requirements of major hydraulic pump manufacturers	Provides good system performance while substantially reducing the negative impact on the environment when inadvertently leaked or spilled.
Synthetic-based hydraulic and circulating oil	Ester-based, readily biodegradable, and virtually nontoxic antiwear hydraulic and circulating oil. Product available in four ISO viscosity grades: 32, 46, 68, and 100, all with excellent oxidation stability and wide temperature range application capability	Primary application is for hydraulic and circulation systems operating in moderate to severe applications such as mobile equipment hydraulic systems where low and high temperatures exceed the limitations of vegetable-based EA oils, Recommended for a temperature range from −20° to 200°F. Owing to their high degree of antiwear and high FZG (12+ stages), they can also be used in gear units not requiring EP additives.	Provides excellent performance over a wide temperature range while demonstrating excellent biodegradation and virtual nontoxicity. If inadvertently spilled or leaked, they significantly reduce the environmental impact relative to mineral oils and can help reduce or elimiante fines and the costs of remediation.
Multipurpose grease	Synthetic-based EP NLGI #1 and #2 grade greases	Designed for multipurpose outdoor applications where grease leakage or run out could contaminate soil, groundwater, or surface water systems. They can be used for indoor applications where grease leakage could enter plant water systems. They are recommended for plain and rolling element bearings and couplings that operate at temperatures from −15° to 250°F.	Provides excellent lubrication characteristics over a wide temperature range while reducing the potential for negative impact on the environment. They are compatible with most other greases.

In most of the other performance areas, the EA lubricants can be formulated to provide performance at least equal to that of the premium mineral oil based lubricants.

1. Oxidation Stability

The conventional laboratory oxidation tests designed for turbine oils but often used for other oils [ASTM D2272 Rotary Bomb Oxidation Test (RBOT) and ASTM D943 Oxidation Characteristics of Turbine Oils (TOST)] yield poorer results for vegetable-based lubricants. These tests were designed for evaluating the oxidation characteristics of highly refined mineral oil based products with higher levels of oxidation inhibitors but overall low levels of other additives that could interfere with achieving high RBOT and TOST results. For example, there is poor correlation between high RBOT and TOST values and long-term performance of antiwear mineral oil based hydraulic oils. In fact, some premium quality hydraulic oils with lower RBOT and TOST values perform much better and provide longer service life than those with higher values. It is well recognized that vegetable oil base stocks do exhibit poorer oxidation stability. With proper processing and use of correct additives, however, the finished formulation can provide satisfactory performance in the majority of applications. Most manufacturers and suppliers of EA lubricants will provide application guidelines for application such as those shown in Table 6.5.

2. Low Temperature Performance

The low temperature performance of vegetable oil based products will naturally be poorer than those that of lubricants based on highly refined mineral oil based products. Without the use of VI improvers and pour point depressants, paraffinic mineral oil based products also exhibit poor low temperature performance. Vegetable oils respond to VI improvers and pour point depressants by exhibiting substantially lower finished product pour points and low temperature fluidity. Pour points are less meaningful for vegetable oil based products than for lubricants based on mineral oils and should not be used as an indication of the lower use temperature for application. A more meaningful indication of low temperature performance of EA lubricants consists of the solidification point and low temperature pumpability. These better represent how the product performs under longer cold soak conditions. This information is important for outdoor applications such as mobile equipment, where fluids may be subjected to subzero temperatures for extended periods of time. If the application involves low temperatures, data on low temperature performance should be obtained from the supplier.

3. Hydrolytic Stability

It is almost impossible to keep moisture out of most lubrication systems, and water can be detrimental to lubricant performance regardless of the base material used to formulate the lubricant. Vegetable oils, as well as all natural and synthetic esters, have poorer hydrolytic stability than comparable mineral oil based products. The proper formulation of EA products is the key to minimizing the potential for negative hydrolysis effects. Current studies of a specific EA fluid indicate that only severe water contamination ($> 0.1\%$ allowable limit for both conventional and EA fluids in critical systems) will adversely affect a lubricant's performance because of hydrolysis or additive depletion.

IV. PRODUCT SELECTION PROCESS

Because of the many choices of available EA lubricants and the lack of clear and universally accepted guidelines to define environmental criteria, the selection process is more difficult than that used for selecting conventional non-EA products. Even so, many of the product performance aspects in actual equipment applications are essentially the same. The selection process needs to include the following aspects of product performance:

Environmental acceptability
Product physical specifications
Equipment builder approvals
Evidence of proven field performance

Credibility of the supplier
Operating and maintenance considerations

The major difference lies in the first and last items in this list: environmental acceptability and operating and maintenance considerations.

A. Environmental Acceptability

The area of environmental acceptability itself encompasses the greatest difference in selecting EA lubricants versus selecting conventional mineral-based lubricants. Although most conventional lubricants are *inherently* biodegradable and can be low in toxicity, EA products need to meet more stringent requirements, as was discussed earlier in this chapter (see Section II: Definitions and Test Procedures). Again, some industry consensus has been established for environmental acceptability, but at present there are no universal industry/regulatory agency agreements on definitions and test procedures. This should not be a deterrent to the use of EA products, particularly in areas that are sensitive to spills or leakage of conventional lubricants. Where such spills have inadvertently occurred, EA lubricants have clearly demonstrated much less negative impact on the environment than would have been expected from the older formulations. Environmental performance requirements also must include the specification that a product not lose its environmental characteristics during its projected service life.

B. Specifications

A product that is going to be relied on in a given piece of equipment must exhibit certain physical and chemical characteristics that have been shown to be important to the performance of that equipment. The primary concern is to have adequate viscosity characteristics. Too low a viscosity could result in metal-to-metal contact and consequent wear. Too high a viscosity can result in improper flow or excessive internal shear, resulting in excessive heating and energy losses. Other characteristics would include compatibility, antiwear, oxidation stability (long life), rust and corrosion protection, filterability, and demulsibility.

C. Equipment Builder Approvals

Much of the initial thrust to develop an EA class of lubricants came from equipment builders whose customers were requesting guidance on such products. Chances are that if the equipment builders have approved the use of these products, they have satisfactorily tested the products or have test data showing that desired performance requirements are met. Also, builders have followed field applications to assure that the product not only meets laboratory test requirements but works in the equipment in question under field service conditions. In some cases, builders will grant conditional approvals only if they can limit temperatures and pressures and may in some cases derate equipment for which EA products must be used. This generally means lower service pressures, speeds, and/or temperatures. If the equipment is under warranty, the builders should be consulted before its use to ensure that the requirements of the warranty are met.

D. Proven Field Performance

Similar to selecting conventional lubricants, testimonials of customers who have used the products in equipment that represents a certain application, indicate that the products are likely to provide satisfactory performance. Since most major lubricant manufacturers conduct extensive laboratory testing as well (up to several years of field testing) even with the so-called new products, field application data should be available.

E. Supplier Credibility

A reputable supplier is important with conventional established lubricants but even more important when new technology is introduced. It is very important that suppliers support their products and recommendations. This provides an increased level of confidence that the new products will work, or at least that a new customer will not be handling problems alone. Reputable suppliers have generally performed adequate laboratory and field evaluations of their products prior to introduction so that chances of success are high.

F. Operating and Maintenance Conditions

Equipment reliability and service life are strongly influenced by operating conditions and maintenance practices regardless of the lubricant. This is especially true where EA lubricants are used. Three of the more important areas of operating conditions and maintenance practices involve control of temperatures, elimination of contamination, and good system maintenance. For example, one high quality vegetable-based hydraulic oil is recommended for temperatures higher than 0°F at start-up and less than 180°F for operations. Contamination should also be minimized to avoid loss of the environmental performance capability as well to assure long performance life. Maintenance practices should include regular inspections of the systems to correct any unsatisfactory conditions involving, for example, filtration, breathers, temperature control, pressure control, correct makeup oil, regular oil analysis, and oil/filter changes as required.

V. CONVERTING TO EA LUBRICANTS

In most cases, the EA lubricants will be compatible with the mineral oils, elastomers, and paints used in industrial and hydraulic systems designed for mineral oils. Some products such as polyglycols may not be compatible with mineral oils or paints. A primary requirement with any conversion is to determine the compatibility of the EA lubricant with all system components before making a conversion. Table 6.2 showed limited general compatibility information with mineral oils and selected elastomers, but additional data should be obtained from the manufacturer or supplier of the product. Equipment builders should also be asked to provide input on component materials if it is suspected that compatibility concerns may develop.

Once compatibility issues are understood, it is recommended that all systems being converted to EA lubricants be drained, cleaned, and thoroughly flushed before the final charge of EA lubricant is added. This will help ensure that the environmental and performance characteristics are not jeopardized by excessive contamination with previously used non-EA lubricants or other contaminants. Even with new systems, it is important to clean and flush to remove materials such as rust preventives, assembly lubes, and sealants that may be in the system. These materials can negatively affect both the environmental and the physical performance characteristics of the EA lubricants. For example, preservative oils and assembly lubes can affect demulsibility, air separation, oxidation stability, and antiwear performance.

The extent of cleaning and flushing is dependent on the system condition as well as the current product in service. If the system is dirty or contains deposits that have built up over time, it may be necessary to flush with a noncorrosive solvent mixed with a conventional mineral oil to remove the deposits. If this is required, care should be taken to install jumper lines across critical tolerance components such as electrohydraulic servovalves. This will reduce the potential for operational problems after start-up due to loosened debris getting into the close clearances. Where high detergent oils (engine oils) have been in service, more attention to flushing is recommended, particularly if vegetable-based EA lubricants will be used. This is due to potential reactions between the highly additized oils with the additive packages used to formulate the EA oils.

As a general rule of thumb, no more than 3% of the previously used non-EA fluid should remain in the system when the final charge of the EA fluid is installed. There are relatively simple laboratory checks to determine whether this has been accomplished. One method is by determining the additive metals such as zinc, calcium, or phosphorus in the previous fluid and comparing these amounts to the additive metals in the final charge. Another way involves analyzing for aromatic content of the fluid except where severely hydrotreated or hydrocracked base stocks were used. Again, the supplier of the EA lubricant will be able to provide assistance in determining the success of the flushing and the degree of contamination in the final system fluid.

7
Hydraulics

Hydraulics covers a wide spectrum of industrial and commercial applications in which it is necessary or advantageous to be able to convert mechanical energy to fluid energy and then to mechanical work. Because this type of system allows a very wide range of flexibility in both control and force, hydraulic systems are almost universally applicable to all areas of our daily lives. From the use of a simple hydraulic jack to lift our vehicles to hydrostatics that operate recreational amusement park rides to very complex and very precise machines such as robotics capable of producing extremely close tolerance parts to exploration of space, the principles of hydraulics are applied because of the versatility and dependability of such systems to meet the requirements of many applications. As a result, hydraulics represents one of world's largest areas of oil (hydraulic fluid) usage. Some of the reasons for the universal application of hydraulics are as follows:

Large forces can be transferred over distances with relatively small space requirements.
Operations can commence from rest while the system is at full load.
Speed can be easily controlled (by controlling oil flow).
Smooth adjustments to speed, torque, and force are possible.
Protection against system overloading is simple to implement.
Hydraulic systems are suitable for both quick and slow controlled sequences of movements.
Simple centralized drive systems are available (several machines can be operated off one central hydraulic system).

Hydraulic systems use a fluid medium under pressure to create motion in machine components. The fluid can be steam, gases (including air), water, oil or other media. In some cases, even grease is used, although its fluidity characteristics are much less desirable for transferring energy over distances. Most of the discussion in this chapter is centered around the use of lubricating oil as the hydraulic medium, but the same principles apply to other fluid media. In service, properly selected hydraulic fluids not only are suitable for pressure transmission and controlled flow, but also minimize wear, reduce friction, provide cooling, prevent rust and corrosion, and help keep the system components free of deposits. High quality hydraulic oils are able to maintain their initial characteristics and provide long satisfactory service for long periods of time—often years in hydraulic systems that are well-designed and maintained.

I. PRINCIPLES

The whole art of hydraulic actuation rests on Pascal's discovery in the seventeenth century that pressure developed in a fluid acts throughout the fluid and equally (neglecting static pressure due to gravity) in all directions (Figure 7.1). The amount of pressure p in the fluid is equal to force F divided by the area A to which that force is applied. For example, in a stationary fluid, if the force is generated by a 100 lb weight and the area is 10 in.^{2}, the pressure is 10 psi (pounds per square inch). So $p = F/A$ (in this example it is 100 lb divided by 10 in.^{2} = 10 psi). The pressure always acts perpendicular to the surfaces of the container. This led to what might be thought of as a hydraulic lever (or force multiplier). Figure 7.2 shows that a 10 lb force acting on a 1 in.^{2} piston develops the same pressure as a 100 lb weight acting on a 10 in.^{2} piston. The 10 psi pressure is easily transmitted around corners, and the cylinders and pistons can be separated by considerable distances.

The system shown in Figure 7.2 is in static equilibrium: no movement of either piston occurs. When the forces acting on the pistons are unbalanced, there will be motion as the system attempts to get back to equilibrium conditions. For example, slightly more weight (>10 lb) applied to the 1 in.^{2} piston will force it downward, pushing fluid into the larger chamber, in turn pushing the 10 in.^{2} piston supporting the 100 lb weight upward. The small piston will have to move downward 10 in. to cause the large piston to move upward 1 in. This is necessary because the volume of fluid (noncompressible) leaving one cylinder must equal the volume of fluid entering the other cylinder. The motion continues as long as the forces (pressure) are unbalanced or stops are applied within the system.

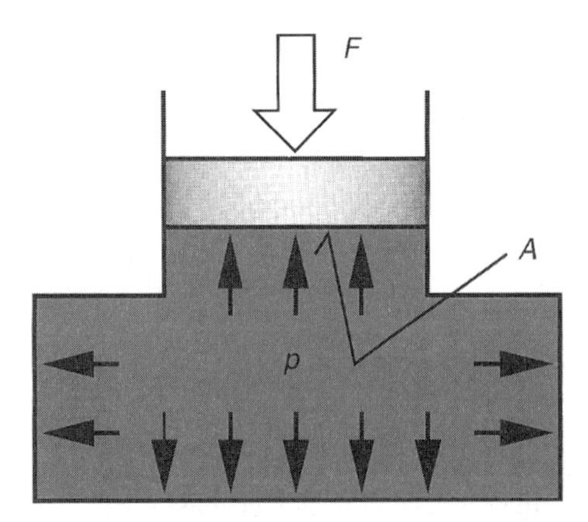

Figure 7.1 Pascal's law-force applied to a stationary liquid acts in all directions within that liquid.

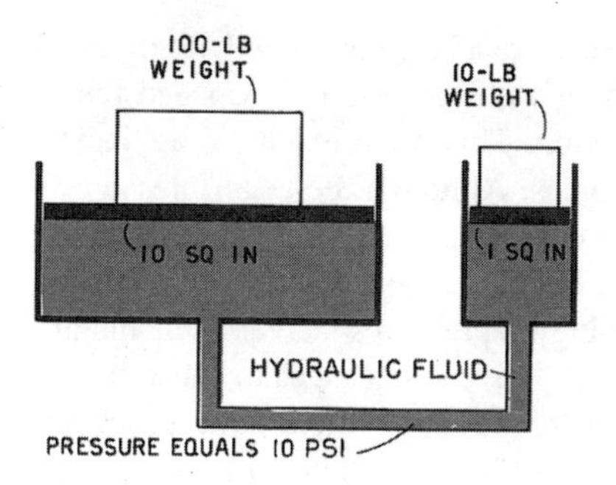

Figure 7.2 Basis for hydraulic actuation.

A. Hydromechanics

It is important to recognize that hydraulic systems are subject to the laws of hydromechanics. This means that balances within the hydraulic system are subject to both *hydrostatic* and *hydrodynamic* forces. By some definitions, hydrostatics are included in hydrodynamics but for our purposes, we separate static forces from the mechanics of fluids in motion.

Hydrostatic principles include the pressure that exists in system (at rest or in motion) that is caused by a combination of gravity, the height of the fluid in the container (reservoir), and the density of that fluid. The shape of the container is irrelevant. Only the height (or head) of the fluid in a container and its density determine the static pressure at any point within a given system. The static pressure at the bottom of the container will always be higher than the pressure at the top of the fluid level. This is still true in a closed system, where external pressure is exerted on the fluid. Volume has no affect on static pressure. For example, a column of water (density of 62.36 lb/ft^3 at 60°F) will exert a pressure of 1 psi for each 2.304 ft of height of that column. A 100 ft high water column will exert a static pressure of 43.4 psi at the bottom of that column. This pressure is the *same* whether the column is a fraction of an inch in diameter or thousands of feet in diameter, and it is not affected by the physical shape of the container. This is more easily understood if you consider the pressures to which divers are subjected at various depths of water bodies such as oceans or lakes. The static pressure (or head) is not affected by surface area or total volume; rather, it is determined by height and density of the fluid. If we substituted mineral oil for the water in this example, the pressure would be slightly less, since mineral oils have lower densities than water.

Hydrostatic considerations are important in design and application of hydraulic systems in such areas as needing to recognize that higher density fluids require more energy to pump, location of pump suctions relative to fluid levels, and distances and heights over which fluids will need to perform work. Pumps designed with a positive static head at their suctions are much less susceptible to cavitation than pumps with suctions located above fluid levels. Where actuators must be located at a considerable height above pump discharge, system design must compensate for pressures (and flows) needed to provide the expected performance.

Hydrodynamics, sometimes referred to as hydrokinetics, involves the energy of fluids in motion. Although hydrodynamics is not generally considered part of fluid power, it is important in certain specific areas. Fluids in motion contain an energy component above that indicated by pressure and flow. This energy level is related to mass and acceleration (velocity). Suppose, for example, that you are running water from a fully opened faucet at home and shut it off completely in one swift motion. This often results in a ''bang'' (shock wave) in the system due to hydrodynamic forces. Although the water pressure in your supply system may only be 40–50 psi, the instantaneous pressure rise generated by the sudden stoppage of motion can be very high. In hydraulic systems, the pressures are considerably higher, and sudden stoppage of flow at higher pressures increases the magnitude of these shock waves—making them high enough in some instances to cause failure of system components. Where these conditions exist, accumulators (hydraulic shock absorbers) should be installed at appropriate locations within the system to minimize the negative affects of these shock waves. Accumulators are discussed in Section III.F.

B. Fundamental Hydraulic Circuits

Although water hydraulic circuits have been used for many centuries, the basic concepts were first put into practice in hydraulic presses that came into use in last part of the eighteenth century—after Pascal's time. Water was used as the hydraulic medium in these early presses. Instead of having a small piston move through a relatively great distance, as in Figure 7.2, a hand-operated pump requiring many short strokes was used. A schematic illustration of such a press is shown in Figure 7.3. The small piston *A* had to be given

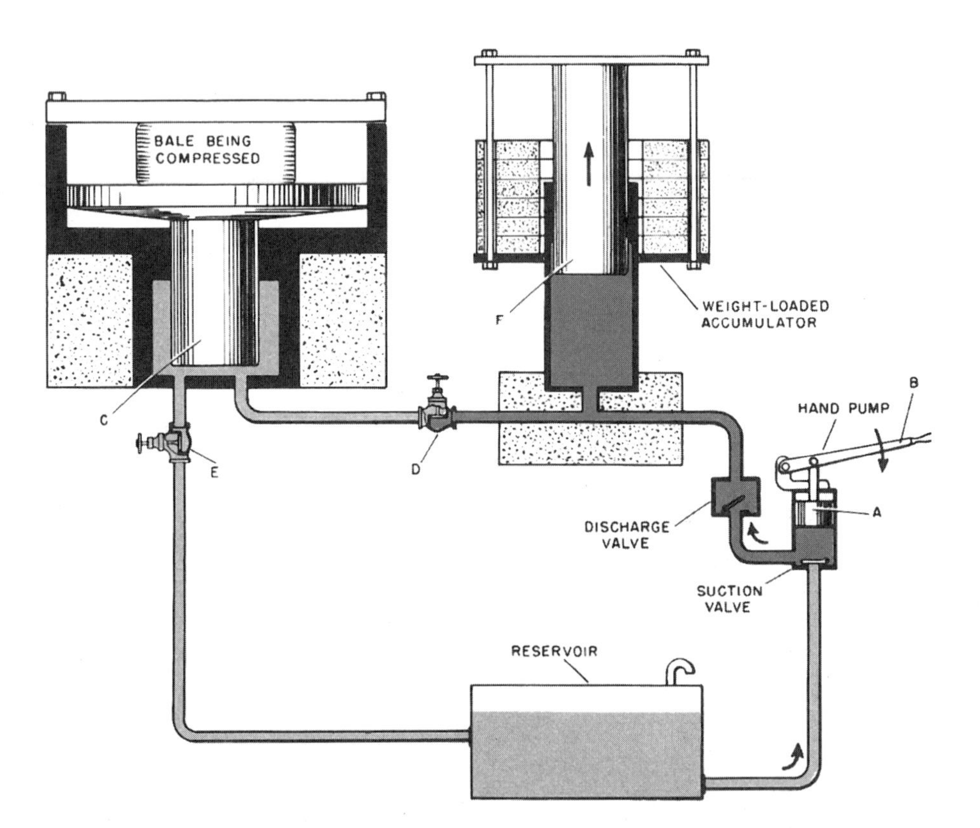

Figure 7.3 Eighteenth-century hydraulic press circuit.

many strokes by means of hand lever *B,* to supply the large volume of fluid required for the upward movement of large piston *C.* The weight-loaded accumulator was added so that energy could be stored during unloading and loading of the press, then released during the work stroke of the press. In operation, with valves *D* and *E* closed and press ram *C* at the bottom of its stroke, the accumulator ram *F* was pumped up to the top of its stroke, the hydraulic pressure being determined by the weight on the accumulator platform. After

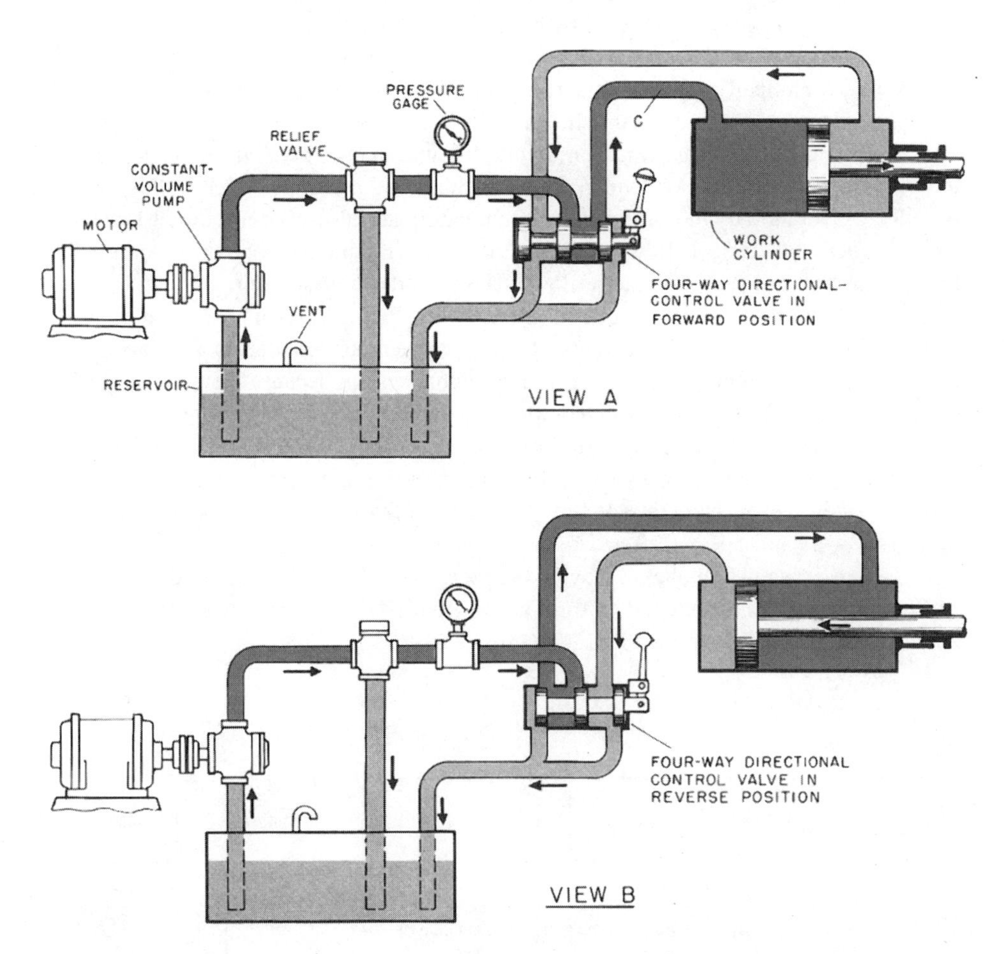

Figure 7.4 Constant volume hydraulic circuit elements common to many hydraulic systems: reservoir, pump, piping, control valves, and hydraulic cylinder (or motor). The hydraulic fluid shown in red is the ''life blood'' of such systems. In view A, the four-way directional control valve is in its forward position so that hydraulic fluid under pressure is delivered by the pump to the left of the piston, causing it to move to the right on its work stroke. Hydraulic fluid from the right-hand side of the piston flows through the four-way valve to the reservoir. A relief valve controls pressure by shunting excess hydraulic fluid to the reservoir. By doing this, the relief valve also acts as a safety valve. In view B, the four-way valve is in its reverse position, and hydraulic fluid under pressure is delivered by the pump to the right of the piston, causing it to move left on its retraction stroke. Hydraulic fluid from the left-hand side of the piston flows through the four-way valve to the reservoir.

the work had been loaded on the press, valve *D* was opened (with valve *E* still closed), and fluid under pressure flowed from the accumulator into the press cylinder, causing ram *C* to move upward. At the end of the work stroke, valve *D* was closed and valve *E* opened to let ram *C* down. The hydraulic fluid drained back to the reservoir to be used again.

The weight-loaded accumulator (Figure 7.3), a device for storing energy in hydraulic systems, was invented around 1850 and since then, much development work has been done with respect to hydraulically actuated machines. The accumulator is still an important part of hydraulic systems. Hydraulic presses have been used for forging operations since about 1860, and an adjustable-speed hydraulic transmission was effectively used in 1906. Some consider this last date as the beginning of modern hydraulics, but larger scale manufacture of hydraulically actuated machines did not occur until the 1920s.

The elementary press circuit shown in Figure 7.3 has several parts common to all hydraulic systems. These are reservoir, hydraulic fluid, pump, pump drive (hand-operated in this illustration), piping or tubing, control valves, and actuator (which in this example is called a hydraulic cylinder or ram). Accumulators are also used in current systems to reduce shock and dampen flow pulsations resulting in smoother operation, particularly for systems that are subject to large flow changes under high pressure.

Other features of hydraulic circuits are shown in Figure 7.4, which is a simple system that could be used to reciprocate a work table of a machine tool such as a surface grinder or milling machine. This is called a constant volume system because the pump is positive constant displacement type, in which output flow remains constant regardless of the speed of the work performed by the system. Pressure is controlled by a relief valve that shunts excess oil back to the reservoir. By preventing excessive pressures, the relief valve also acts as a safety valve. Direction of flow to and from the hydraulic cylinder is controlled by means of a hand-operated four-way spool valve.

Figure 7.5 shows an overall view of the energy transfer process of a hydraulic circuit. Mechanical energy is converted to fluid energy, which is converted back to mechanical energy. This principle applies to all hydraulic circuits.

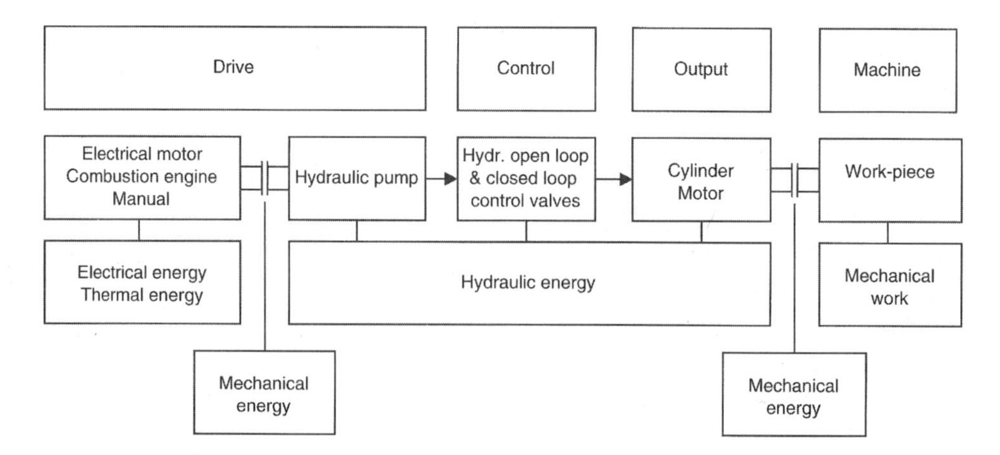

Figure 7.5 Transfer of energy in a hydraulic circuit. (Courtesy of Rexroth.)

II. SYSTEM COMPONENTS

A. Hydraulic Pumps

All hydraulic systems use some sort of mechanism to create system flow. Resistance or stoppage of that flow creates pressure. Flow generates motion, and pressure generates force. A combination of flow and pressure in a hydraulic system is the basis for work. In most applications, pumps are used for this purpose and their function is to convert mechanical energy to fluid energy. The pumps can be hand-operated or driven by electric motors, internal combustion engines, turbines, or other types of prime mover that provide rotative or reciprocating motion. The various pump designs provide flows from less than 1 gallon per minute to well over 600 gpm and are capable of generating pressures from less than 100 psi to over 15,000 psi. Most of the hydraulic systems encountered in industrial and commercial applications will have flows from a few gallons per minute to over 100 gpm and pressures that range from 750 psi to about 5000 psi.

Although centrifugal pumps are used in some hydraulic systems, this chapter emphasizes positive displacement type hydraulic pumps. These positive displacement pumps are classified into two categories: constant volume (fixed displacement) and variable volume (variable displacement). The three basic pump designs in these two classifications are the *gear, vane,* and *piston* types. *Screw*-type pumps are a variation of gear pumps that use worm-type enveloping gears similar to screw-type air compressors. *Plunger* pumps (reciprocating) are a variation of piston pumps. All these designs and variations of these designs are used in constant volume systems. If variable volume systems are required, only vane and piston type pumps will be used.

1. Gear Pumps

The four types of gear pump design: external gear, internal gear, ring gear, and screw are shown in Figure 7.6. Gear pumps are generally simple in design, lower in cost than other types, and compact. They can operate over a wide range of speeds, provide smooth (nonpulsating) flow, and use oils over a wide range of viscosities. They are available with pressure ranges up to about 4000 psi, but in most applications are used in systems with pressures less than 1500 psi.

Table 7.1 General Pump Selection Criteria

Pump type	Speed range	Pressure range	Viscosity range	Noise level[a]	Service life	Relative costs
Internal gear	Good	Good	Good	Low	Good	Moderate
External gear	Very good	Good	Very good	High	Moderate	Low
Ring gear	Good	Moderate	Moderate	Moderate	Good	Moderate
Screw	Good	Moderate	Very good	Low	Good	Moderate
Vane	Moderate	Moderate	Moderate	Moderate	Good	Moderate
Piston	Good	High	Very good	Moderate	Good	High

[a] Some of the newer design pumps may contain features to reduce noise levels and therefore not fit these general criteria.

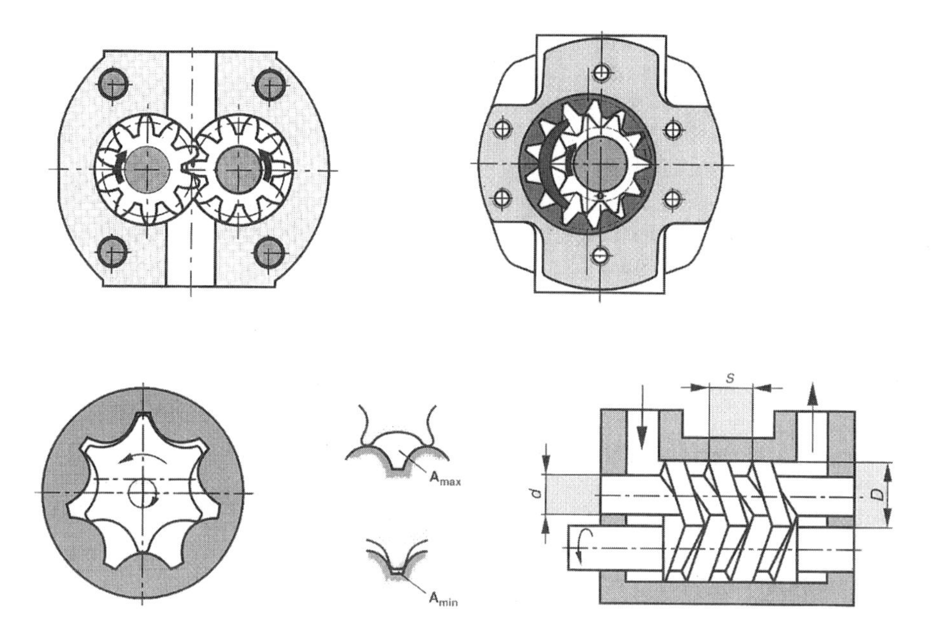

Figure 7.6 Four common gear pump designs: internal, external, ring, and screw.

Gear pumps are constant discharge types. Their fluid pumping chambers are formed between the gears, the housing, and side plates. As the gears rotate, they create a low pressure area when the teeth separate from mesh on the suction side, allowing the hydraulic fluid to fill the pumping chambers. The fluid is then trapped between the gear teeth and the housing and carried toward the discharge side. As the gears enter mesh on the discharge side, the fluid is forced from between the teeth and out the discharge. To minimize fluid loss from the higher pressure discharge side back to the lower pressure suction side, the close clearances must be maintained, particularly as discharge pressures increase. The hydraulic fluid serves to help seal these close clearances as well as to seal and lubricate the meshing gears. As can be seen, small amounts of bearing wear could result in contact of the gears with the housings or side plates reducing the efficiency of the pump. Minimizing wear is critical to optimizing pump performance.

A common design external gear pump (Figure 7.7) consists of two meshing gears. One of these gears is driven and the other is an idler. This is the simplest and lowest cost type of gear pump, and it produces the highest noise levels. Internal gear designs are generally much quieter. Screw pumps are also relatively quiet. Ring gear pumps exhibit noise levels between external and internal gear designs. Much attention is being devoted to development of *''quiet pumps''* of all designs to aid in overall noise reduction programs in industry.

2. Vane Pumps

Vane pumps can be of fixed displacement or variable displacement design. Vanes are designed to slide in and out of close tolerance slots, machined into the rotor, by contact with the cam ring. The rotor–vane assembly is fit into a stationary eccentric cam ring (or machined housing). At start-up and under low pressure conditions, centrifugal forces keep the vanes in contact with the cam ring. As the pressure increases, centrifugal forces in

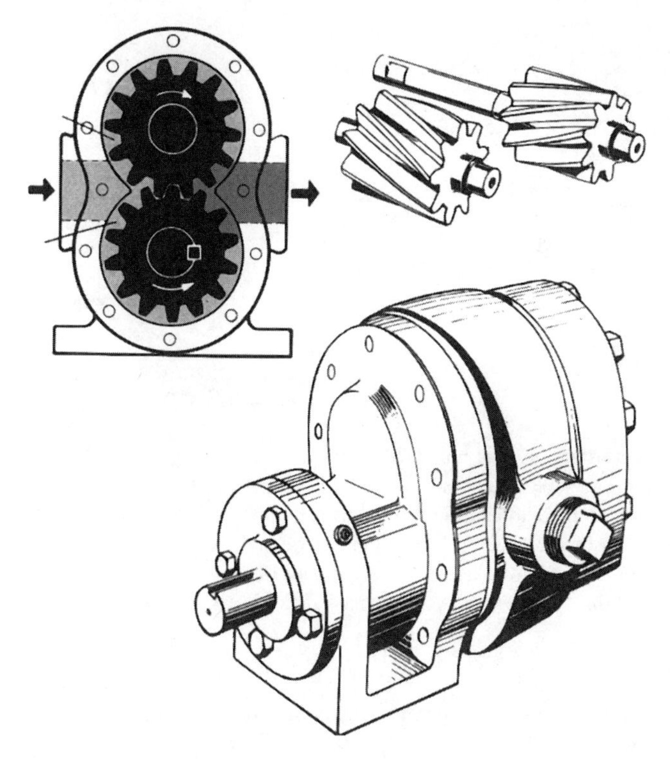

Figure 7.7 External gear pump: Oil is trapped in the pockets (top left) between the case and the teeth of the revolving gears and is carried around the periphery of both gears from the suction side to the discharge side. Gear pumps like this can use gears of spur, helical, or herringbone type. Special gears, however, are employed to make pumps, especially internal gear types.

combination with internal oil pressure acting on the bottom of the vanes helps keep them in contact with the cam ring. The close-clearance machined surfaces are further sealed and separated from contacting each other by the hydraulic fluid. In addition to controlling wear, the sealing effects help minimize leakage of fluid from the discharge side back to the suction side as well as through side plate clearance spaces.

As the vanes slide out of the rotor slots on the suction side, they form expanding volume pockets with the cam ring, creating an area of reduced pressure, drawing the fluid into the pockets. When the vanes pass the rotational point of maximum volume, the confined fluid is forced out the discharge by the increased pressure caused by the decreasing volume of the pockets.

Vane pumps of the balanced design use an elliptical cam ring and operate with two suction and two discharge ports 180 degrees apart. This design (i.e., two pumping chambers) provides an offsetting pressure balance so that there are minimal thrust forces acting in one direction; that is, the high pressure in the discharge side is pushing the rotor toward the lower pressure suction side. Figure 7.8 shows a balanced design, constant volume vane pump in which each half rotation creates a suction and discharge cycle.

Variable volume vane pumps operate with adjustable cam rings (relative to stationary rotors), allowing variation of the pocket volumes to produce discharge volumes that range

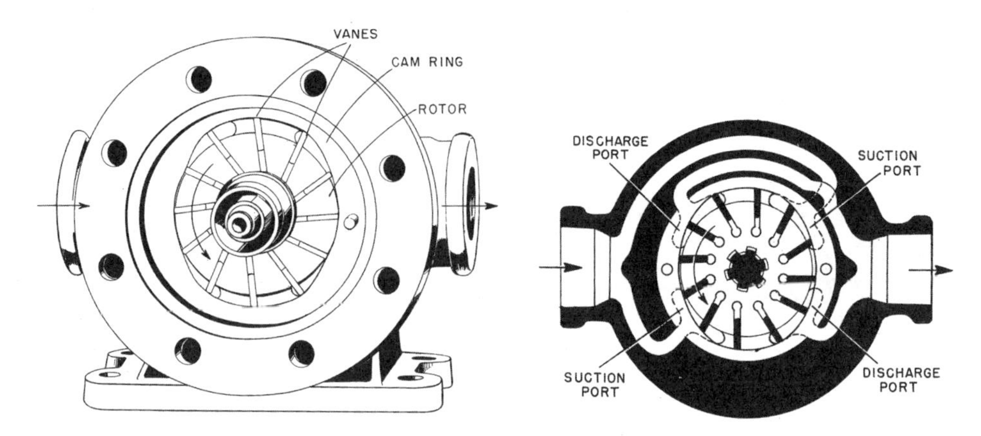

Figure 7.8 Vane pump: In this constant volume pump, the vanes are forced by contact with the cam ring to slide in and out of the rotor slots. The vanes are held in contact with the cam ring by centrifugal force and by oil pressure on their inner edges. Suction ports are located where the wall of the cam ring recedes from the hub, since the increasing volume between the vanes at these points results in reduced pressure. Discharge ports are placed where the wall of the ring approaches the hub. At these points, decreasing volume between the vanes causes pressure to increase.

from "0" volume to rated maximum volume. Control of the cam ring position is generally based on system pressure, but some pumps are equipped with manual controls. Also, some pumps that need to provide very precise flows may contain devices for temperature compensation to adjust for changes in fluid viscosity with changes in temperature.

3. Piston Pumps

Radial and axial piston pumps are positive displacement pumps used when operating pressures are high. Piston pumps can be either fixed displacement or variable displacement. Some piston pumps are designed for pressures exceeding 15,000 psi, although typically they are used for applications in which the pressures range from 2500 psi to about 6000 psi. In industrial service, piston pumps will be the primary choice when pressure requirements are consistently above 3000 psi and very accurate control of discharge flow is desirable. Although pumps of the gear and vane types can attain these pressures, piston pumps are cost-effective in severe service because their life expectancy is greater. Another advantage of variable displacement piston pumps is that in addition to infinitely adjustable flow (up to maximum volume), they are reversible. This means that the pump can act as a pump or a hydraulic motor if the direction of flow is changed. Circuits using these pumps can be simplified, since reversing valves are not needed. With reversing pumps, rapidly moving masses can be smoothly decelerated, reversed, and accelerated without system shocks. For example, some presses use a circuit similar to that shown in Figure 7.9, with a reversible piston pump in a closed-loop system instead of a volume and directional control valve arrangement. The pumping direction is controlled by means of mechanical linkage connected to the press platen or by electronic sensors that send signals back to the pump control unit.

4. Radial Piston Pumps

Radial piston pumps encompass two designs: in one, a rotating eccentric shaft causes the pistons to reciprocate within their cylinders; the second uses a rotating cylinder block

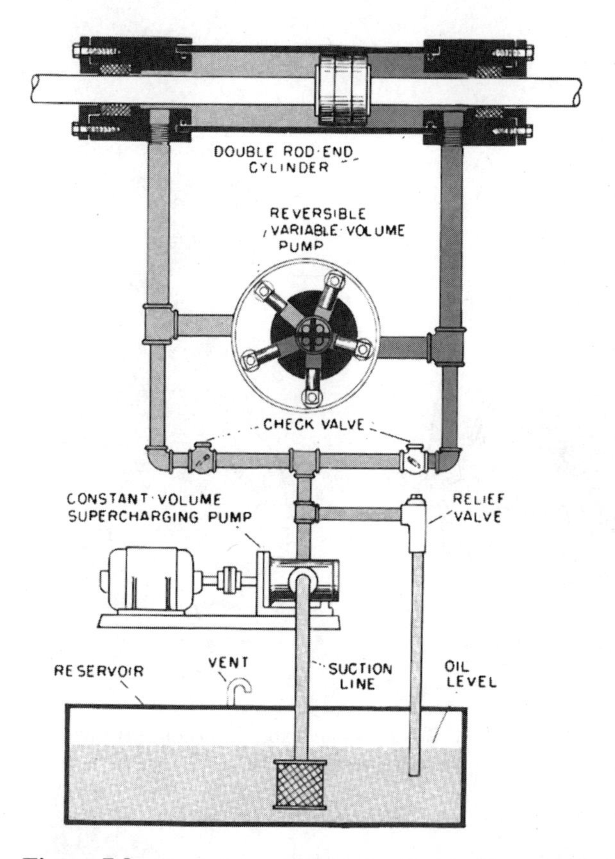

Figure 7.9 Reversing pump circuit with a double-rod-end cylinder, which has the same displacement on both ends. The reversing pump circuit can theoretically consist of nothing but the pump, cylinders, and suitable pipe connections. Speed is adjusted by adjusting the pump stroke, and no reversing valve is needed. A small fixed volume pump, usually a gear pump, is required to make up main pump leakage losses and keep the entire system under positive pressure so that no air can get in.

mounted eccentrically with the pistons and block rotating within a rigid external ring. In both designs, the reciprocating movement is perpendicular to shaft rotation (Figure 7.10). This reciprocating radial motion of the pistons within their respective cylinders creates a reduced pressure at the suction, drawing fluid into the cylinder on one stroke and expelling the fluid into the discharge on the other stroke. Figure 7.11 shows a variable volume radial piston pump of the eccentric cylinder block design.

5. Axial Piston Pumps

Both the swashplate (or wobble plate) and the bent axis designs can be used in axial piston pumps. Both designs are positive displacement and are available in fixed and variable volume outputs. Axial piston pumps operate on the same design principles as radial piston pumps except that the pump pistons reciprocate parallel to the rotor axis as shown in Figure 7.12. The angle between the cylinder block and the driving flange determines the

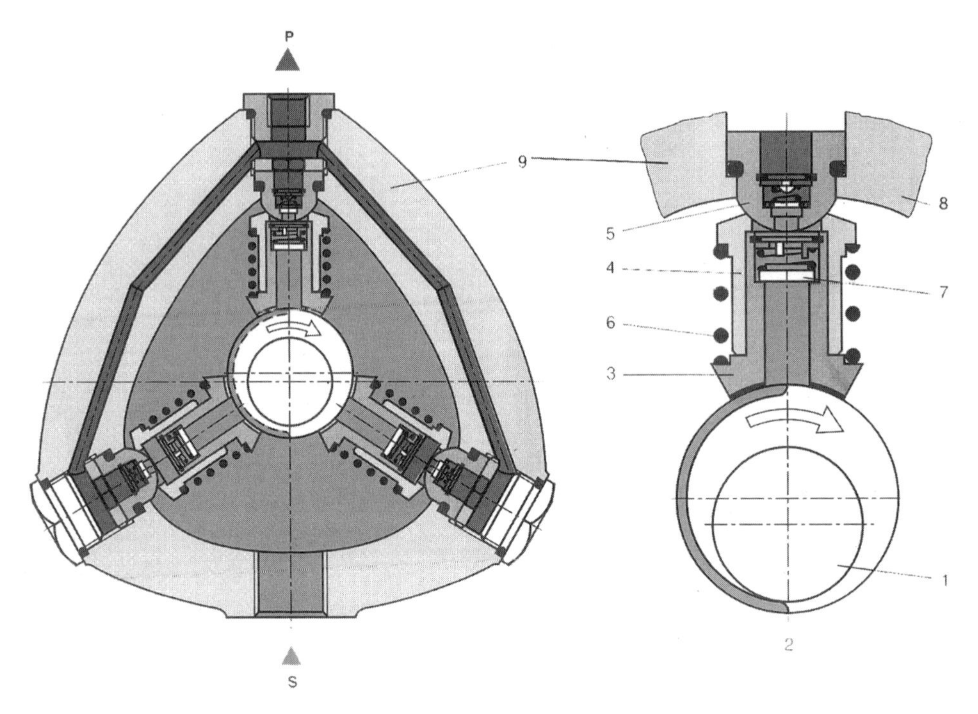

Figure 7.10 Radial piston pump: (1) drive shaft, (2) eccentric shaft, (3) piston, (4) pivot, (5) cylinder sleeve, (6) compression spring, (7) suction valve, (8) pressure control valve, and (9) housing.

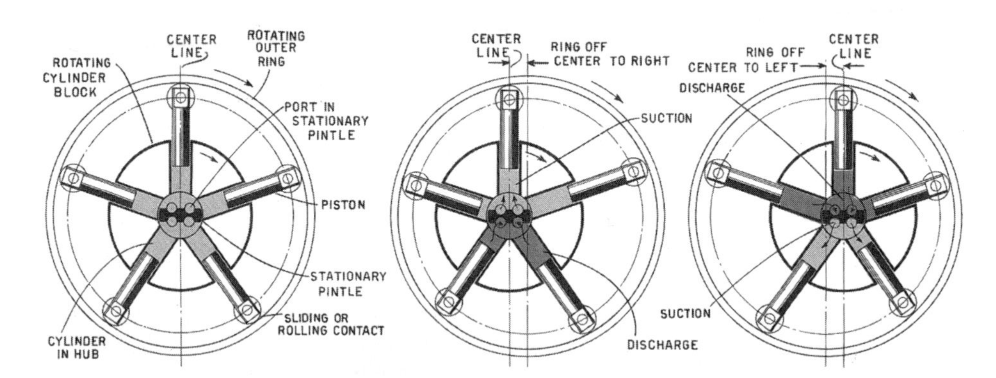

Figure 7.11 Principle of the variable displacement radial piston pump. This type of variable stroke pump has a cylinder block that rotates about a stationary spindle, or pintle. Several closely fitted pistons are carried around the block, their outer ends held in contact with a rotating outer ring by centrifugal force and by oil pressure. Suction and discharge ports are in the pintle and communicate with the open ends of the cylinders. When the ring and the pintle are concentric, rotation occurs without any reciprocating movement of the pistons and therefore output is ''0.'' If the ring is moved off-center, rotation causes the pistons to reciprocate, drawing oil from the suction ports on the outward stroke and delivering it through the discharge ports on the inward stroke. Moreover, by adjusting the eccentricity of the outer ring to the right or to the left (as shown), it is possible to reverse the direction of oil flow in either direction without stopping the pump or changing its direction of rotation.

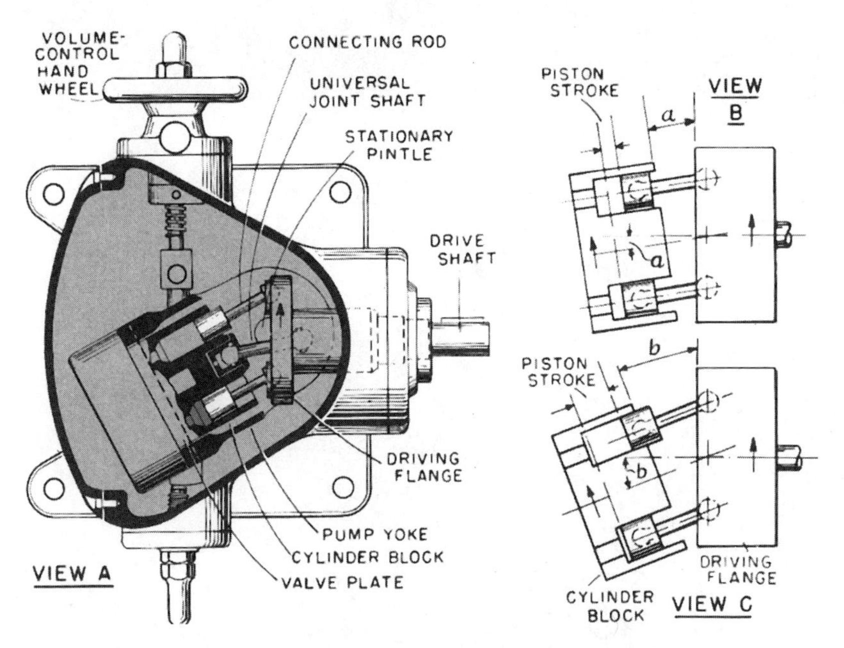

Figure 7.12 Variable volume axial piston pump. As in the case of the radial piston pump in Figure 7.11, variable volume (displacement) is accomplished by varying the piston strokes. The cylinder block is carried in a yoke that can be set at various angles with the drive shaft. The block with its several pistons is rotated by the drive shaft through an intermediate shaft having a universal joint at each end. The drive shaft and cylinder block are supported by antifriction bearings. All working parts are submerged in oil that is continuously recirculated. In view B, with the cylinder block set at angle *a,* rotation of the structure through 180° produces the relatively short stroke of the piston as shown. In view C, with the block set at the larger angle *b,* the stroke (and volume of oil pumped) is seen to be increased. With the block set at ''0'' angle (in line with drive shaft), there is no movement of the pistons in the cylinder block and no oil is pumped. Finally, when the cylinder block is set at an angle above the drive shaft in the views shown, the direction of oil flow is reversed. In other words, flow can be adjusted in small steps from zero to maximum in either direction during operation. Suction and discharge takes place through sausage shaped ports in the valve plate (view A). Passages in the pump yoke lead from the valve plate to the stationary pintles (on which the yoke pivots) and through these to external suction and discharge connections. In addition to manual control (as shown), volume output can be adjusted by means of a hydraulic cylinder or electric motor. Servo and automatic (pressure-compensated) controls are also used. There are several makes of variable volume axial piston pumps and the mechanism for varying piston strokes are quite different from one design to another.

volume output (along with speed and cylinder diameter). In variable displacement pumps, when this angle is ''0'' (in line with driving shaft), there is no reciprocating motion and the volume output is ''0.'' As the angle increases, the displacement increases, creating greater volume output. In this type of pump, the discharge flow is infinitely variable between ''0'' and maximum capacity. In a constant volume pump, the angle is fixed and the displacement remains constant.

Piston pumps, whether radial or axial, are designed with an odd number of pistons. The odd number of pistons reduces the flow and pressure pulsations in the discharge when

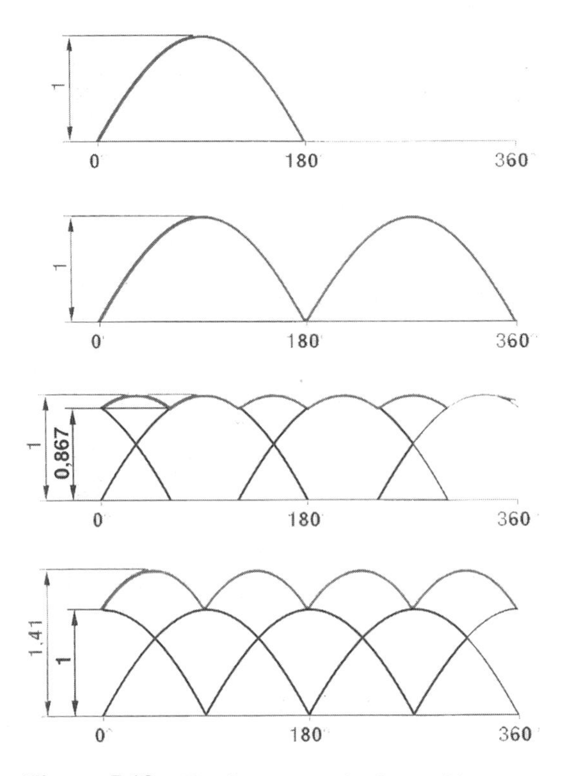

Figure 7.13 Flow/pressure pulsations with one, two, three, and four pistons. (Courtesy of Rexroth.)

the individual piston flows are added together. Figure 7.13 shows the additive flow/pressure pulsations as influenced by the number of pistons through one 360-degree cycle of rotation.

III. CONTROLLING PRESSURE AND FLOW

The pumps create flow, and generally valves are used to control and direct flow as well as maintain system pressures. Valves are essential components of all hydraulic systems. They range from a simple check valve, which permits flow only in one direction (Figure 7.14) to complex electrohydraulic control valves (Figure 7.15), which may be required to

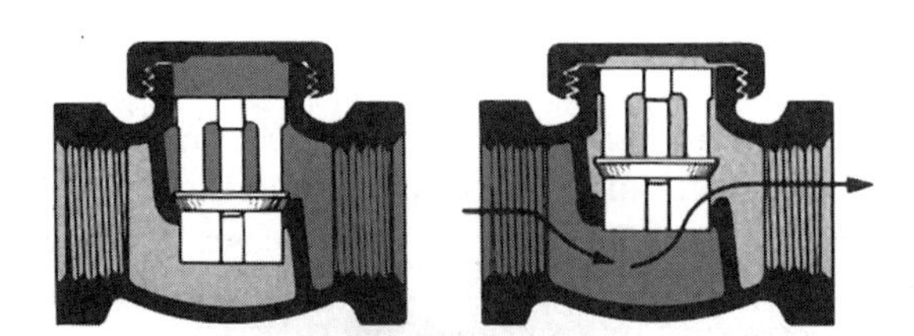

Figure 7.14 A check valve permits free flow in one direction but prevents flow in the other. This check valve is one of several types used in hydraulic circuits such as that shown in Figure 7.21.

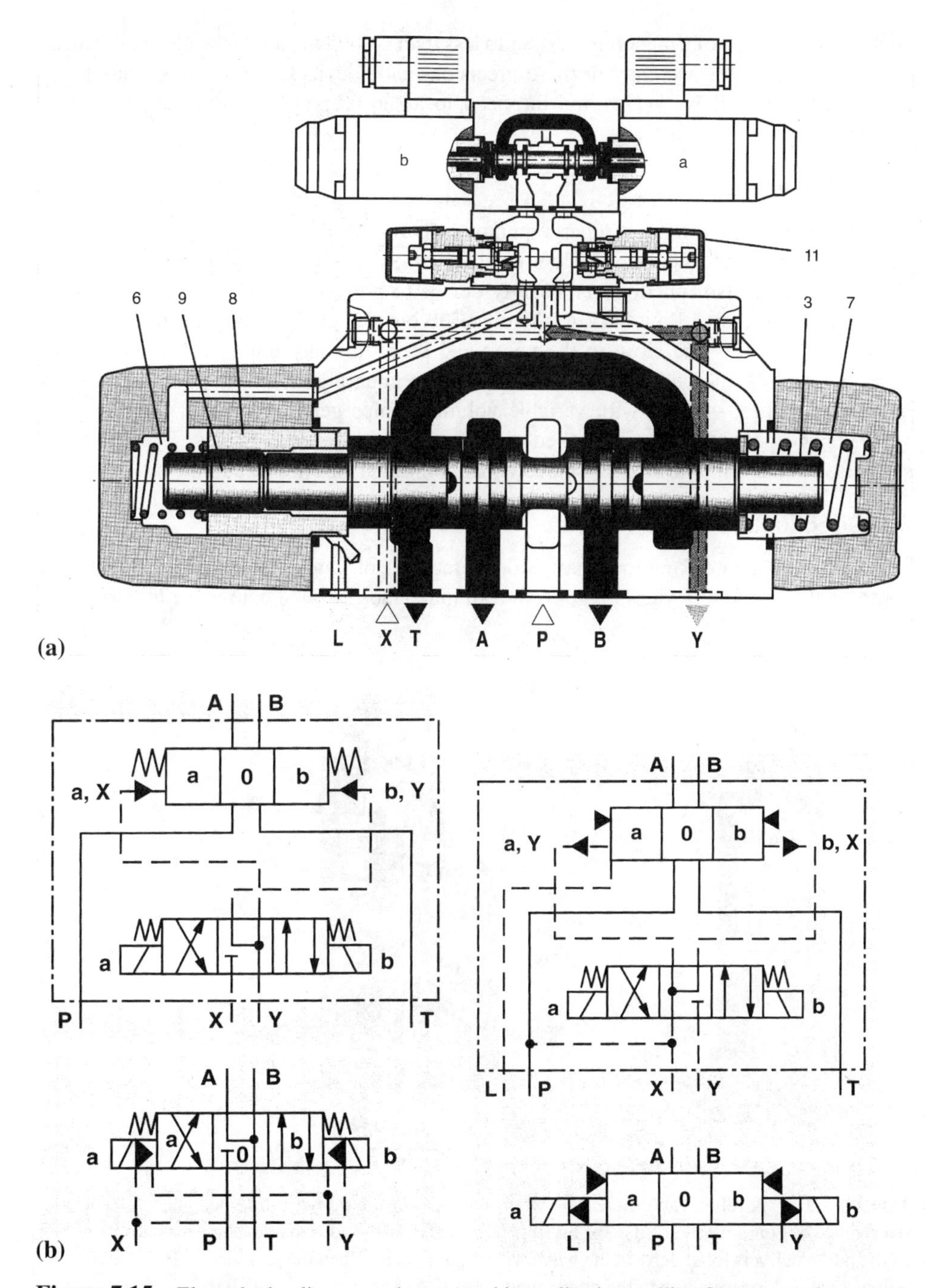

Figure 7.15 Electrohydraulic servovalve, as used in applications calling for very precise control of a machine. Servovalves are used to control speeds of turbines and critical machine tools where close tolerances are required in finished parts. (a) Electrohydraulically operated directional spool valve, pressure-centered, for sandwich plate. (b) Symbol for electrohydraulically operated directional valve, pressure-centered: (top) detailed; (bottom) simplified.

control the accuracy of a machining center to less than 0.0005 in. Given the close tolerances and requirements for accuracy in these precision-made devices, it is easy to understand the importance of these valves and the need to assure very good fluid conditions and maintenance to the system.

A. Relief Valves

The use of constant volume discharge pumps calls for a relief valve (Figure 7.16) in the pump discharge line to control system pressure below a predetermined maximum level. This relief valve also functions as a safety device to prevent overpressure in the system by returning excessive flow back to the oil reservoir. Relief valves should be located in the pump discharge line close to the pump and preceding any other valves and system components. In some pumps, the relief valve is an integral part of the pump design. As mentioned earlier, systems with variable volume pumps generally do not require relief valves, since pump flow is controlled by system pressure requirements. Even in these systems, a relief valve may be installed as an additional safety precaution.

B. Directional Control Valves

All valves that control motion (start, stop, direction) of actuators are called directional control valves. These devices range from a simple hand-operated gate valve to very com-

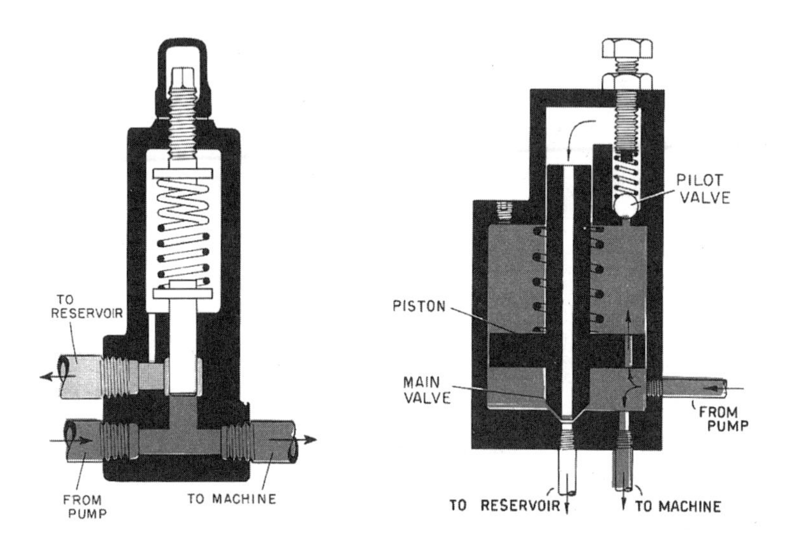

Figure 7.16 A relief valve functions when pressure in the delivery line becomes sufficient to overcome the compression of the spring (left). The valve lifts and bypasses the excess oil. As long as the pump delivers too much oil, the valve will remain open. When the pressure in the system drops below a value corresponding to the adjustment spring, the valve closes, and all the oil discharged by the pump is delivered to the system. The *simple* relief valve (left) may permit a rather wide variation of pressure and may tend to chatter. A *compound* relief valve (right) provides closer control of pressure and has less tendency to chatter. Under normal conditions (as shown), the pressure on one side of the piston of the compound relief valve is balanced with the pressure on the other side, and the spring holds the main valve closed. When the pressure builds up sufficiently to raise the pilot valve, the pressure on the upper side of the piston is relieved, the main valve is lifted, and the excess oil is bypassed to the reservoir.

plex servovalves. Directional control valves are among the most common of hydraulic system control components in that they activate and deactivate hydraulic mechanisms such as cylinders and fluid motors. The operation of directional control valves was briefly discussed in connection with Figure 7.4.

Directional control valves such as the four-way spool valve shown in Figure 7.17 can be operated in several ways. They can be operated by hand levers such as those found on most home-use hydraulic log splitters; by mechanical linkage, cams, or stops on moving parts such as a ram or workpiece; or by means of air or oil pressure supplied by a pilot valve. Alternatively, they can be electronically activated by position sensors, solenoids, or various other electronic signals generated by pressure, flow, and/or temperature. A typical four-way valve has three control positions: foreward, neutral (or stop), and reverse. For a four-way valve connected to a hydraulic cylinder, this corresponds to work stroke, neutral, and return stroke.

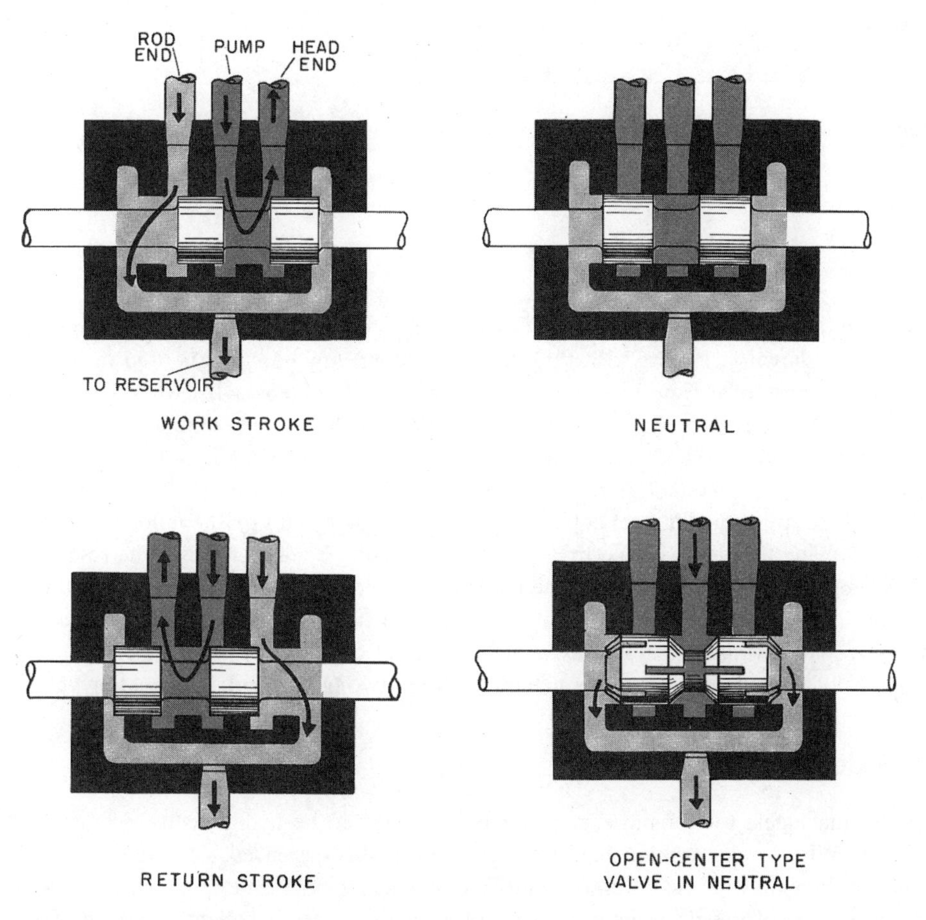

Figure 7.17 Four-way spool-type directional control valve: a closed-center valve in work stroke, neutral, and return stroke positions. With the closed-center type of valve, there is very little leakage between ports that are blocked off by the valve spools. In the neutral position, there is intended to be no flow in any of the four connections. When the valve spool is shifted from one position to another, flow tends to start or stop immediately in the various connecting lines.

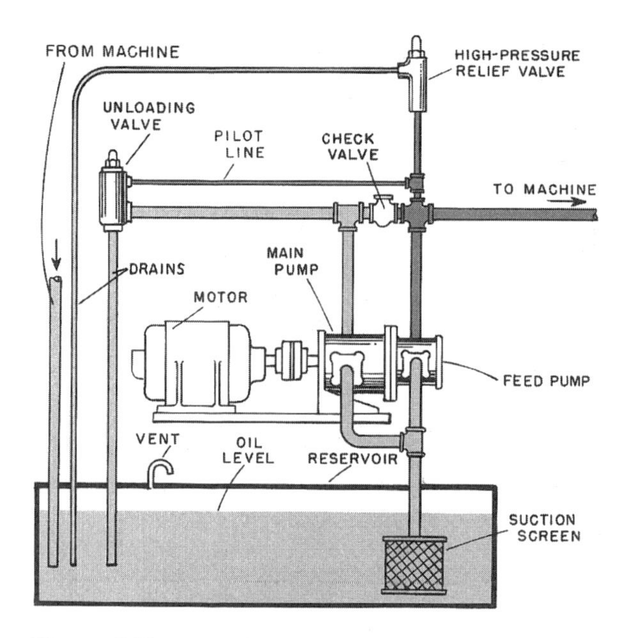

Figure 7.18 Two-pump circuit with unloading valve.

C. Unloading Valves

In some circuits, two pumps may be used to meet substantially variable flow requirements. In the case, shown in Figure 7.18, flow from the larger volume pump is used to ensure the rapid advance of the machine tool to a certain point but then, as the work activity is performed, the volume of the flow decreases markedly. During the rapid advance cycle, the discharge volume of both pumps is required but while actual work is being performed, only the small pump volume is required, causing a rise in system pressure. As the pilot pressure rises (pilot pressure set below relief valve pressure setting), the unloading valve opens, allowing the flow volume from the larger pump to be discharged to the reservoir at low pressure. If one constant volume pump were used, most of the oil pumped during the work cycle would be discharged at full system pressure through the relief valve. Its energy would be wasted and excessive heating could occur. Use of a variable volume pump could be an alternative to the method featuring two pumps and an unloading valve.

D. Sequence Control Valve

In some machines, two or more movements may need to be hydraulically operated in sequence. When one movement must not begin until another has ended, a pressure-operated *sequence* valve may be used. Referring to the circuit shown in Figure 7.19, when oil flow stops at the end of the clamping-cylinder stroke, pressure rises sufficiently to open valve *A*. Full line pressure is then available to activate the feed cylinder. Sequence valves can also be activated by pressure-sensing pilot valves or electronically, by position or other pressure-sensing devices.

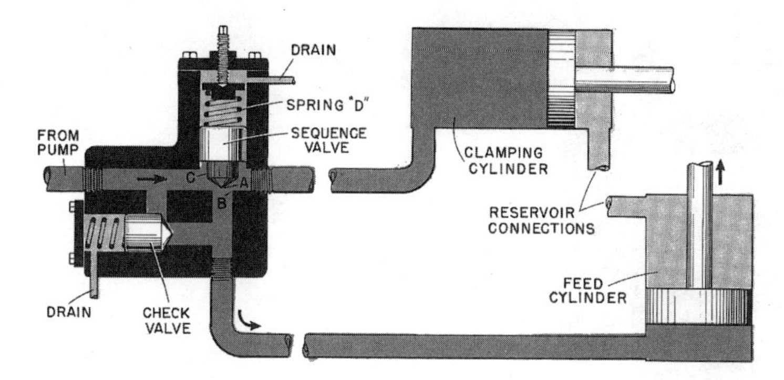

Figure 7.19 Sequence valve circuit.

E. Flow Control Valves

Valves that control the speed of actuators by regulating the volume of oil to the actuators are called *flow control valves.* In some applications, it is desired to have slow work stroke and a rapid return stroke. One simple way of accomplishing this in a nonprecise application is to use an orifice (flow restrictor) in the supply line to the work stroke as shown in Figure 7.20 with a check valve to allow unrestricted flow on the return stroke. With this simple metering method, any variation in pressure in the head end of the work cylinder due to varying load will cause oil flow and work speed to vary. To avoid this, a pressure-compensated flow control valve (Figure 7.21) can be used. When this valve is used, the circuit is called a *meter-in* circuit. Sometimes the flow control/check valve combination is placed in the oil supply line from the rod end of the work cylinder. In this circuit, called a *meter-out* circuit, adequate pressure is maintained in the rod end to prevent lunging if the workload drops suddenly. In Figure 7.21, the pressure at *A* acts through passages *A* and tends to push the spool to the left. The pressure at *B,* on the other side of the metering orifice, acts through passage *B* and together with the spring, tends to push the spool to

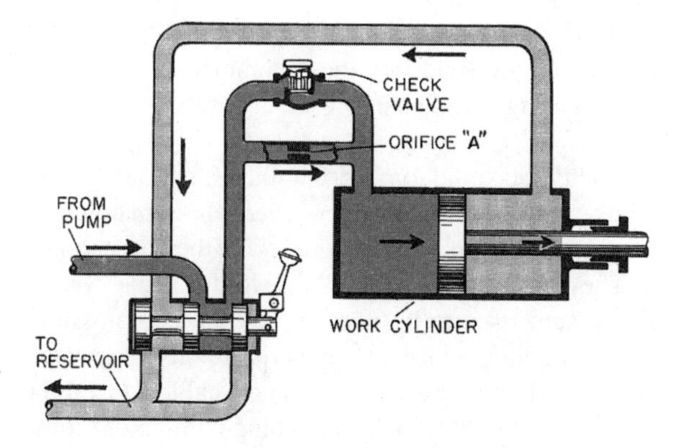

Figure 7.20 Metering circuit: orifice ''A'' restricts the oil flow to the head end of the cylinder during the work stroke, but on the return stroke, the check valve opens and permits unrestricted flow.

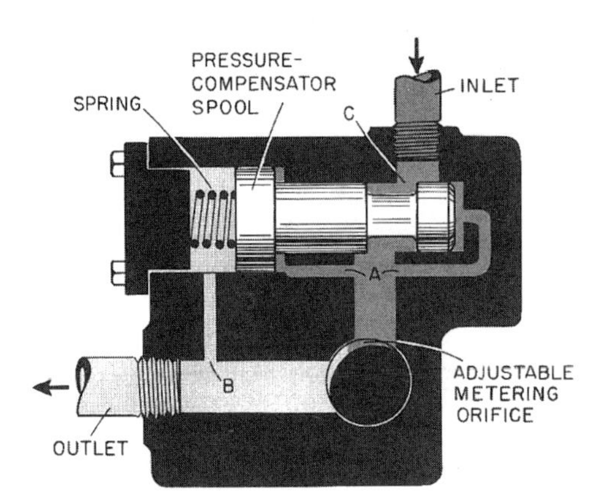

Figure 7.21 Flow control valve: the orifice shown in the meter-in circuit in Figure 7.20 usually will not maintain constant speed throughout the work stroke. The workload will usually vary, causing the pressure drop across the orifice and flow to vary. To maintain constant pressure drop and flow, a *pressure compensator* such as shown may be used.

the right. Since the two pressures act on equal areas, the pressure at *A* will be greater than the pressure at *B* by a constant amount determined by the spring tension. Any imbalance will cause the spool to move in a way that opens or throttles the inlet, at *C,* and restores the balance. With constant pressure drop across the orifice, the flow remains constant.

F. Accumulators

Although not control valves, *accumulators* can act as flow control mechanisms. As we briefly discussed in connection with Figure 7.3, accumulators can be used to store energy of an incompressible fluid. Some presses and other machines require large volumes of oil under pressure for short duration cycles with relatively long periods of time between cycles. A pump and motor large enough to generate the necessary flow and pressure for such an application could be very costly. Instead, accumulators (Figure 7.22), in conjunction with small pumps and motors, can often be used as shown in Figure 7.23. Energy is stored in the accumulator by the pump during the long periods between high energy requirements. This is done by pumping hydraulic fluid into the accumulator and raising a weight, compressing a spring, or compressing a gas charge. The energy is returned to the system when required.

In addition to energy storage capacity, accumulators serve other functions. Many hydraulic systems are subject to rapid or sudden flow changes where the dynamics of fluid (incompressible) in motion can create high levels of system *shock.* In these situations, accumulators act as shock absorbers. They are able to reduce the severity of the system shock by allowing the instantaneous pressure rises to be taken up by the compressible mechanisms within the accumulators. This helps reduce the potential for line breakage and component failures in those high flow, high pressure systems that are subject to abrupt changes in flow. They can also be used to smooth out flow by absorbing pump pulsations, maintain constant pressures for long periods of time, such as in clamping operations, and make up for system internal leakage. Most hydraulic applications use gas-pressurized accumulators.

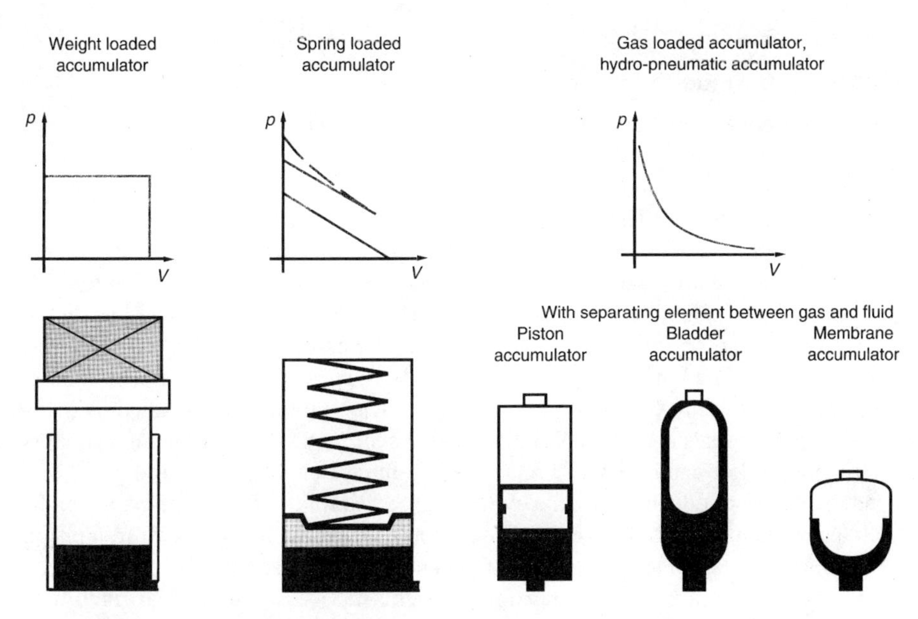

Figure 7.22 Various types of accumulator. (Courtesy of Rexroth.)

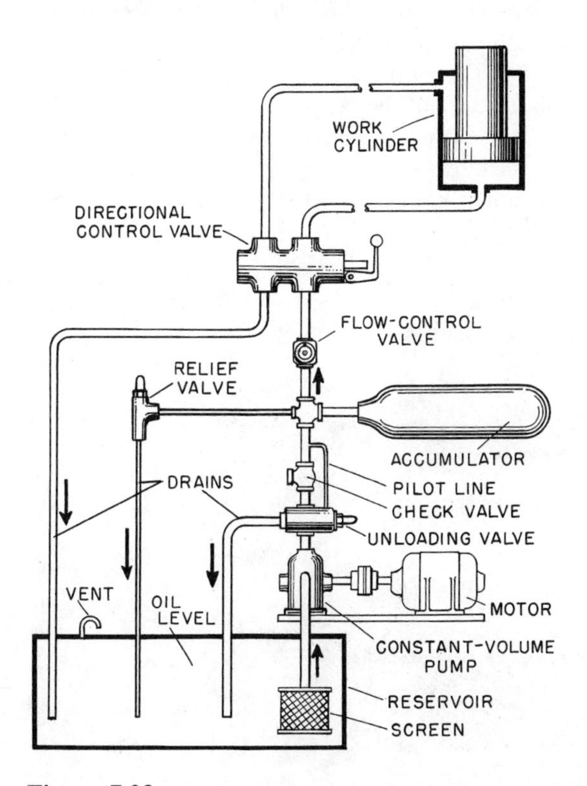

Figure 7.23 An accumulator circuit: In this press circuit, the pump is automatically unloaded (unloading valve circuit shown in Figure 7.18) when the accumulator pressure reaches a preset value. A flow control valve determines the work stroke speed, and a check valve prevents loss of high pressure hydraulic fluid.

IV. ACTUATORS

A. Hydraulic Cylinders

At the point of use, hydraulic energy must be converted to mechanical energy and motion. The device most frequently used is the *hydraulic cylinder* (Figure 7.24), sometimes called a reciprocating motor. Hydraulic cylinders may be single-acting (pressure applied in one direction) or double-acting. They are made in sizes ranging from less than an inch to several feet in diameter. Working strokes of up to several feet and pressure ratings of up to several thousand psi are available. Single-rod-end cylinders, as shown in Figure 7.24, are most common, but double-rod-end cylinders (rod extends through both heads, as shown earlier: Figure 7.9) can be obtained. By means of flanges, extended tie rods, and adapters, cylinders can be mounted in many positions. Cylinders are usually made of steel tubing, bored and honed. Rods are often hard chrome-plated steel, ground and polished. All surfaces in contact with seals and gaskets are finely finished. Typical double-acting cylinders are illustrated. The left-hand view of Figure 7.24 features a cartridge-type rod gland that can be removed without dismantling the cylinder. Accessibility of the rod gland is important from the maintenance point of view. The piston is sealed by means of cast iron piston rings, which are very satisfactory where a small amount of leakage can be permitted. O-ring static seals, which are very popular for hydraulic service, are used. The right-hand view of Figure 7.24 is cushioned at both ends. As the piston nears the end of its stroke, the cushion plunger enters the head counterbore, shutting off oil flow. The stroke can then continue only as fast as oil can flow through passage *A* and past the cushioned valve. At the beginning of the next stroke, oil flows past the spring-loaded ball-type cushion check valve and builds up pressure against the full face of the piston. Unless other means are provided to decelerate the piston, cushioning becomes more and more necessary at higher piston speeds. The piston in this case is sealed by means of cup-type seals, which permit little or no leakage. V-ring packings are similar in this respect.

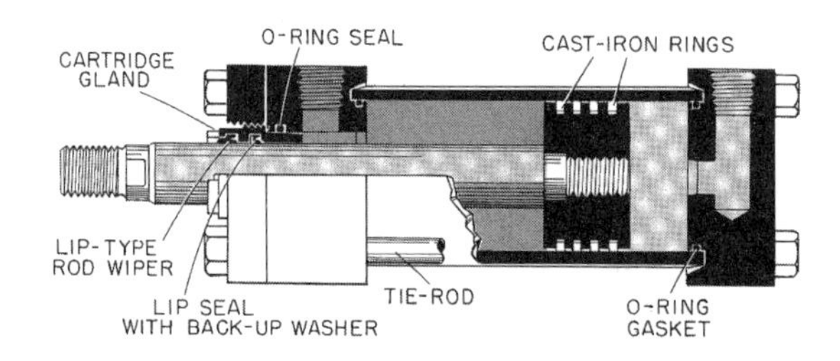

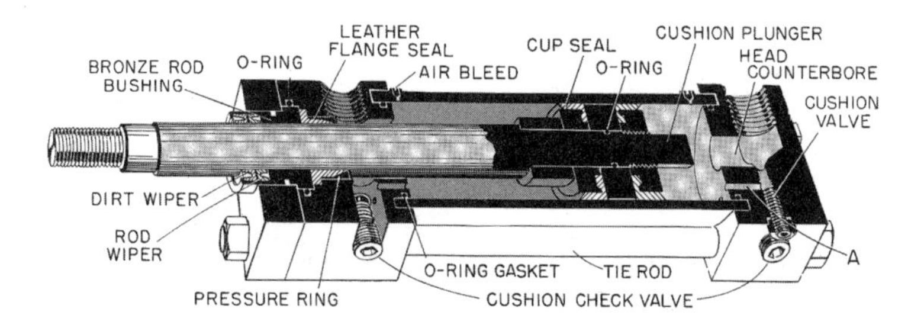

Figure 7.24 Hydraulic cylinders.

In addition to single and double-acting cylinders described earlier, and the ram and double-rod cylinders also discussed earlier in this chapter, several other types are commonly used in industry. These are *spring return, telescoping,* and *tandem* types. Spring return cylinders are components of the single-rod type with pressure to activate the work stroke, and the spring in the nonpressurized end returns the piston to its original position when pressure is released. Telescoping cylinders are used where long strokes are required but space limits the length of the cylinder. These applications also have lower load requirements as the cylinder rod extends. These are made of multitubular rod segments fit into each other to provide the telescoping action. Tandem cylinders are used when high force is required and length is not limited but larger diameter cylinders cannot fit in the available space. These are mounted in-line with a common piston rod.

B. Rotary Fluid Motors

Rotary fluid motors are used instead of hydraulic cylinders to convert fluid energy to mechanical motion, especially when rotary motion is required or continuous or long movements in one direction are involved. They compete with electric motors under the following conditions: (1) when a variable speed transmission having a wide range of closely controlled speeds and torques is required, (2) when space limitations demand a very compact power source, or (3) when torques or loading might occasionally be severe enough to overload an electric motor. Rotary fluid motors can be of gear, vane, or piston (radial or axial) design. They are similar to their counterparts in pumps but differ in certain details that affect efficiency. In fact, some radial and axial piston pumps are designed to act as both a pump and a motor, as described in connection with Figure 7.11, but in these designs, they do lose some efficiencies. Rotary fluid motors are supplied with oil under pressure and rotate at a speed and torque dependent on the available volume of oil that flows through the motor.

Radial and axial piston motors are usually of the constant displacement design but are available in variable displacement forms. In a constant displacement axial piston motor (Figure 7.25), the motor shaft is supported by ball bearings *A* and *B*. The cylinder barrel

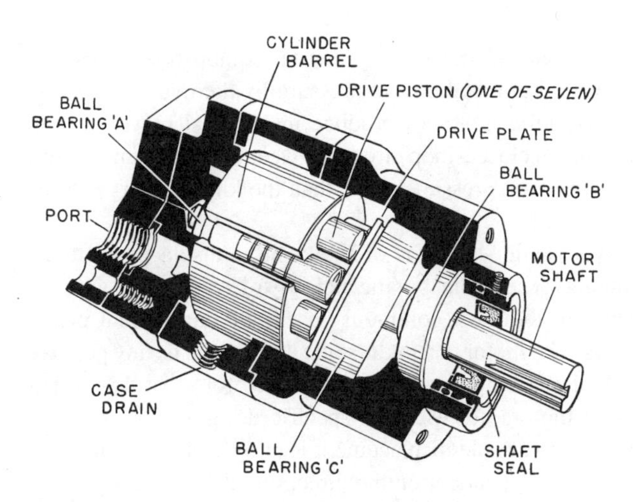

Figure 7.25 Axial piston fluid motor.

is supported by the motor shaft and is keyed to by means of a splined joint. The drive plate is supported at an angle by ball bearing *C*. Ports are arranged so that oil under pressure starts to enter a cylinder when its piston is at the end of its inward stroke. The pressure causes the piston to push against the drive plate. Part of this force is perpendicular to the drive plate and part is tangential. The tangential force causes the drive plate, cylinder barrel, pistons, and motor shaft to rotate. The torque developed depends on oil pressure and the motor dimensions. The speed depends on the rate of oil flow.

Axial piston motors exhibit high volumetric efficiencies and excellent operation over a wide range of speeds. They can be used where torque requirements are up to more than 17,000 ft-lb or speeds up to 4000 rpm. Radial piston motors can attain several hundred thousand foot-pounds of torque or speeds up to 2000 rpm. Gear-type motors will provide up to 6000 ft-lb of torque or up to 3000 rpm. It is important to recognize that since the torque of hydraulic motors is inversely proportional to rotational speed, their highest torques will occur at low speeds.

V. HYDRAULIC DRIVES

Hydraulic drives are classified as *hydrostatic* or *hydrodynamic*. These can be designed to produce power in three ways: variable power and torque, constant power and variable torque, and variable power and constant torque. Hydrostatic drives use oil under pressure to transmit force while hydrodynamic drives use the effects of high velocity fluid to transmit force. Engine-driven transmissions widely used in main and auxiliary drives of mobile construction and farming equipment is an example of hydrostatic drives. Torque converters (sometimes referred to as *hydrokinetic* drives) are commonly found in automotive applications but are finding increased use in industrial applications. Another form of a hydrokinetic drive is the *hydroviscous* drive. This form uses the viscosity characteristics of the fluid rather than the energy from fluid in motion to develop the drive torque. Hydrodynamic drives include hydrokinetic and hydroviscous drives.

A. Hydrostatic Drives

In a hydrostatic system, power from an electric motor, internal combustion engine, or other form of prime mover is converted into static fluid pressure by the hydraulic pump. This static pressure acts on the hydraulic motor to produce mechanical power output. While the fluid actually moves through a closed-loop circuit between the pump and motor, energy is transferred primarily by the static pressure rather than the kinetic energy of the moving fluid.

The hydraulic pump in a hydrostatic system is of the positive displacement type. Either fixed or variable displacement is acceptable, but the majority of systems use variable displacement. Axial piston pumps are the most commonly used, although radial piston pumps are used in some applications. The motor in a hydrostatic system can be any positive displacement hydraulic motor. Axial piston motors are usually used for most drives, but both gear motors and radial piston motors are used for specific designs. The motor is usually of fixed displacement type, but variable displacement is acceptable. The motor is reversible, with the direction of rotation dependent on the direction of flow in the closed-loop circuit from the pump. Figure 7.26 shows a diagram of a typical hydrostatic drive.

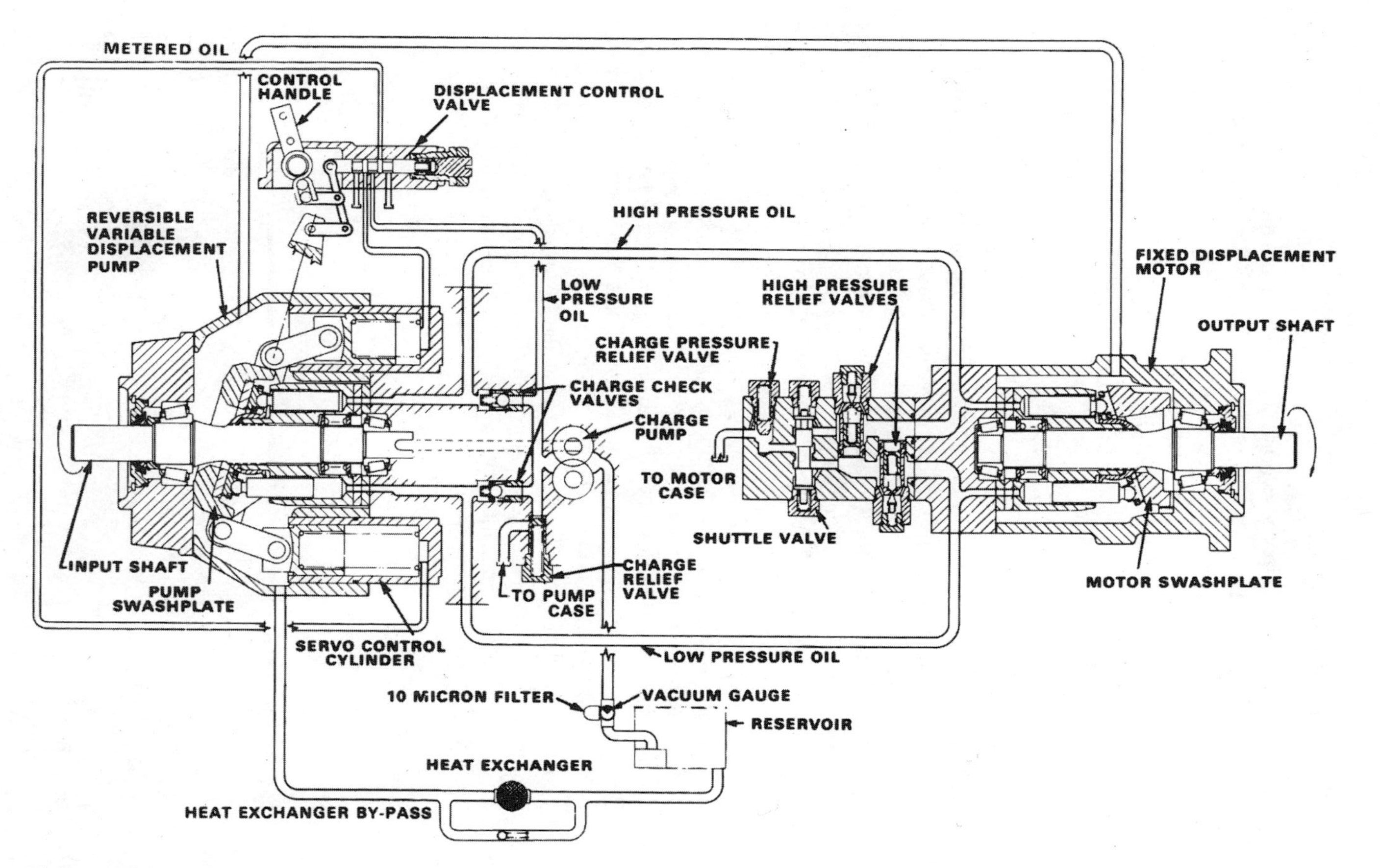

Figure 7.26 Diagram of typical hydrostatic drive.

VI. OIL RESERVOIRS

The oil reservoir is also a very important component of the hydraulic system. It contains the oil supply, provides radiant and convection cooling, allows solid contamination and water to drop out, and helps reduce entrained air from circulating to the critical control components. In addition, in relation to oil levels and pump location, the oil reservoir facilitates easy flow of the oil to the pump suction, reducing the potential for cavitation or starvation conditions. Proper design and care of oil reservoirs will help assure satisfactory component service life and long oil life.

A. Reservoir Design

When possible, the oil capacity of a hydraulic reservoir should be at least 2.5 times the rate of oil circulation (pump capacity) at full operating conditions. This rule does not apply to closed-loop systems, such as the hydrostatic drives, in which the oil flows back and forth between the pump and motor without returning to the reservoir. In these systems, the reservoir is used only for oil makeup to the system necessitated by internal leakage (or other system leakage). Figure 7.27 shows a typical reservoir configuration for non-closed-loop systems. The proportions shown are suitable. If the reservoir is too shallow, there may not be enough sidewall area in contact with the returning oil for effective cooling and conditions that allow air to enter the pump suction (vortexing) may be promoted. On the other hand, if the reservoir is too deep and narrow, there may not be enough surface area for separation of the air in the oil. The baffle aids cooling and contamination separation by promoting flow along the sidewalls. The baffle also helps to prevent short-circuiting of hot oil from the return on one side to the pump suction line on the other. The bottom of the reservoir is dished, and the drain is located at the lowest point. This design not only permits complete drainage at oil change time, but allows occasional removal of water and other heavier-than-oil contaminants that separate in the reservoir. Space below the

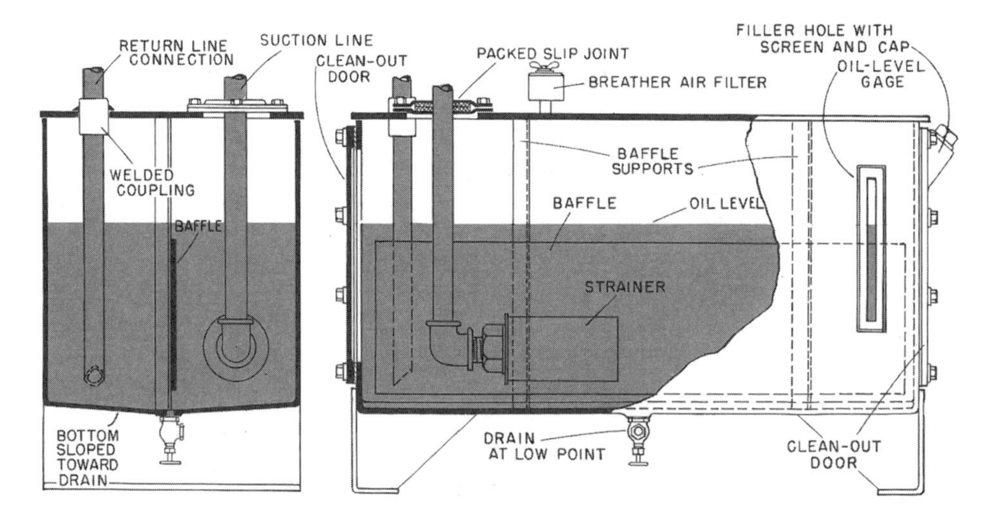

Figure 7.27 Oil reservoir design.

reservoir permits the use of hoses, pans, and so on, to permit oil to be drained without spillage. The reservoir cover is welded on in this example, but bolted and gasketed covers are often used. Large clean-out doors are provided at one or both ends. Other design features include a filler hole with cap and screen, a level gage, a breather with a filter, and gasketed seals for clearance holes where pipes pass through the cover. Return lines should be extended well below the oil level to reduce misting and aeration of the oil.

The reservoir design discussed is a separate component of most typical hydraulic systems and is exposed to atmospheric pressure. In a pressurized reservoir, sometimes used to positively charge the pump suction, the level of pressure must be considered in the design. The pressurized reservoir reduces the potential for atmospheric contamination such as moist air or other airborne contamination to enter the reservoir. In addition to these designs, a space-saving integral design can be used. In these, the reservoir is built into the machine as, for example, a fluid-tight base or hollow member of the support structure that can hold oil without requiring additional space for a separate oil reservoir. The same rules and precautions apply to these alternate designs. The needs remain for the oil to be cooled, contamination removed, and adequate pump suction supply provided.

VII. OIL QUALITIES REQUIRED BY HYDRAULIC SYSTEMS

As the discussion of hydraulic system components has indicated, the hydraulic fluid is a very important component that is often casually considered. Most often, satisfying the requirements of only the pump seems to be the primary consideration for fluid selection. Although the costs of hydraulic pump failures are generally one of the more costly occurrences within hydraulic systems, erratic operation of valves and actuators due to inadequate oil performance characteristics such as oil degradation (oxidation) that causes deposits to form in critical clearance areas, often leads to costly production losses. With the close clearances, different metallurgies, various elastomers, and high pressures and temperatures, service life and performance of all the system components depend on proper selection and maintenance of the hydraulic fluids.

Hydraulic fluids perform many functions in addition to transmitting pressure and energy. These include minimizing friction and wear, sealing close-clearance parts from leakage, removing heat, minimizing system deposits, flushing away wear particles and contamination, and protecting surfaces from rust and corrosion. The important characteristics of a hydraulic fluid vary by the components used and the severity of service. Chapter 3 dealt with product characteristics and testing in some detail. A number of the physical characteristics and performance qualities of hydraulic fluids commonly required by most hydraulic systems are listed as follows:

Viscosity
Viscosity index (VI)
Wear protection capability
Oxidation stability
Antifoam and air separation characteristics
Demulsibility (water-separating characteristics)
Rust protection
Compatibility

Some specific applications may require the following:

Soluble oils
High water content fluids
Fire-resistant fluids
Environmental performance

A. Viscosity

The single most important physical characteristic of a hydraulic fluid is its viscosity. Viscosity is a measure of the oil's resistance to flow, so in hydraulic systems that are dependent on flow, viscosity is important with respect to both lubrication and energy transmission. Although viscosity requirements are to some extent dictated by the components (pumps, valves, motors, etc.) and by system manufacturers, certain effects of improper viscosity selection need to be recognized. Too low a viscosity can lead to excessive metal-to-metal contact of moving parts, as well as to wear and leakage. Too high a viscosity can result in excessive heating, sluggish operation (particularly at start-up), higher energy consumption, lower mechanical efficiencies, and increased pressure drops in transmission lines and across filters. Since viscosity decreases as temperature increase, viscosity requirements are generally specified *at operating temperature.* If temperatures are higher than those specified for *normal operation,* a higher viscosity oil may be required to provide long service life of the components. If start-up and operating temperatures are lower than those specified, then a lower viscosity oil may prove to be better for overall system performance. Systems operating over a wide range of temperatures may require oil that exhibit high VIs. Hydraulic systems normally use oil with a viscosity range of 32–68 cSt at 40°C (150–315 SUS at 100°F). To ensure flow to the pump, most hydraulic equipment builders require that the viscosity at start-up temperatures not exceed 1515 cSt (7000 SUS). Some builders, however, limit the start-up viscosity to 866 cSt (4000 SUS).

B. Viscosity Index (VI)

The viscosity of all oils varies substantially with changes in temperature. In some hydraulic systems, subjected to wide variations in start-up and operating temperatures, it is desirable to use an oil that changes relatively little in viscosity for a given temperature range. An oil that does this is said to have a *high viscosity index.* High viscosity indexes can be achieved by using mineral oil base stocks that have been refined through a severe hydroprocessing technique (Chapter 2) in conjunction with the use of long chain polymers called *viscosity index improvers* (VIIs). Mineral oil base stocks refined through conventional methods can also achieve wide temperature range performance by the addition of viscosity index improvers. Synthetic hydraulic fluids with naturally high VIs are an alternative for severe application temperatures—both high and low.

Hydraulic fluids are subjected to high shear rate conditions, particularly as pressures and speeds rise. Viscosity index improvers are generally long chain polymers and, depending on the type of polymer used, are subject to shearing over time. This shear can result in loss of viscosity and, if severe enough, in poorer hydraulic system performance owing to increased potential for metal-to-metal contact, increased leakage, and loss of some of the friction-reducing qualities. In the selection of a mineral oil based hydraulic fluid with a high viscosity index obtained by the use of viscosity index improvers, good shear stability performance should be required.

C. Antiwear (Wear Protection)

To assure satisfactory hydraulic component life, the hydraulic fluid must minimize wear. Wear results in loss of mechanical efficiency as well as higher costs due to shorter component life. In some hydraulic systems, such as low pressure, low temperature systems with gear pumps, antiwear additives are not necessary. Other high pressure, high temperature systems using vane pumps do require antiwear additives in the hydraulic fluid. Piston pumps may or may not require antiwear additives depending on the metallurgy and design. A properly refined petroleum oil has naturally good wear protection (without the use of antiwear additives). This quality, sometimes referred to as lubricity or film strength, is present to a greater degree in oils of higher viscosity than those of lower viscosity. However, certain additive materials are used to improve antiwear performance in hydraulic oils. These additives work by chemically reacting with the metal surfaces forming a strong film preventing metal-to-metal contact under boundary lubrication conditions. The use of effective antiwear additives may allow the use of lower viscosity fluids without sacrificing potential wear.

Antiwear fluids are generally required in gear and vane pumps operating at pressures above 1000 psi and over 1200 rpm. Piston pumps may or may not require antiwear additives depending on the specific manufacturer and the metallurgy used. For example, Denison Hydraulics Inc. prefers R & O oils for their piston pumps, whereas antiwear additives are required for Vickers piston pumps. Denison Hydraulics typically uses bronze piston shoes against a steel swashplate, while Vickers may use steel on steel. Steel against steel at high pressure will always require antiwear formulations. Premium quality antiwear hydraulic fluids are formulated to provide performance in all pumps and hydraulic systems. Because a given industrial plant will use many different pumps and other component, it may be advisable to consolidate the number of hydraulic fluids by using antiwear formulations that meet all the requirements.

The antiwear performance of hydraulic fluids is evaluated in several standard industry-recognized tests. The major ones include the ASTM D 2882 (Vickers V-104C Pump), the Vickers M-2952-S (Vickers 35VQ25 Pump), and the Denison Hydraulics HF-0 (combination of the P-46 piston pump test and the TP-30283A Vane Pump Test). These tests were discussed in some detail in Chapter 6 in connection with environmental hydraulic fluids.

D. Oxidation Stability

Oil is circulated over and over during long periods in hydraulic systems. It is heated by the churning and shearing action in pumps, valves, tubing, and actuators. Also, the energy released as the oil goes from high pressure to low pressure in a relief valve is converted to heat, which raises its temperature. Oil can be further increased by convection or conduction heating while performing its work in applications such as the hot molds in plastic injection molding operations and continuous caster hydraulics in steel mills.

The oil is in contact with warm air in the reservoir. Air is also dissolved or entrained in the oil. Because of this contact with air, oxygen is intimately mixed with the oil. Under these conditions (exposure to temperature and oxygen), the oil tends to chemically combine with the oxygen, creating oxidation products. The tendency to oxidize is greatly increased as temperatures increase, as agitation or splashing becomes excessive, and by exposure to certain materials that catalyze the oxidation reactions. Catalysts such as iron, copper, rust, and other metallic materials are commonly present in hydraulic systems.

Slight oxidation is not harmful, but if the oil has poor resistance to this chemical change, oxidation may become excessive. If this occurs, substantial amounts of both soluble and insoluble oxidation products are formed, and the oil gradually increases in viscosity. Some variation of viscosity within a range that has proved satisfactory in service is not necessarily harmful. However, viscosity higher than necessary is accompanied by higher fluid friction and more heating. With many of today's critical systems using electrohydraulic servovalves with extremely close clearances, slight oxidation could result in the formation of deposits on the servovalve spools, restricting their movement and resulting in production problems. Low quality oils have poor resistance to oxidation under severe conditions. With such oils, troubles of the kind just described often occur. In high quality hydraulic oils, the natural ability of well-refined, carefully selected base oils to resist oxidation is greatly improved by the use of additives that retard the oxidation process.

E. Antifoam/Air Separation Characteristics

The positive and accurate motion of actuators within a hydraulic system is dependent on the virtual incompressibility of the hydraulic fluid. Under high pressure, mineral oils can see a very slight reduction in volume (4.0% at 10,000 psi and 140°F) and a corresponding increase in density. For purposes of the vast majority of hydraulic systems, this is considered to be insignificant. Introduction of air into the fluid can substantially change the compressibility. Air causes *spongy* or *erratic* motion, which will result in poor system performance, particularly during the production of close-tolerance parts. Antifoaming and air separation characteristics are two different concepts, although somewhat connected. "Air separation" means that the entrained air is released from the oil, while "antifoam" means that the air bubbles getting to the surface of the oil are readily dissipated. Both aspects are important to the performance of a hydraulic oil. Contamination can alter both these characteristics so it is not only important to select an oil that will provide good antifoam and air separation performance but it is necessary to minimize contamination in order to maintain this good performance.

F. Demulsibility (Water-Separating Ability)

Water contamination is sometimes a problem in hydraulic systems. It may be present as a result of water leaks in heat exchangers or washdown procedures, but more commonly it accumulates because of condensation of atmospheric moisture. Most condensation occurs above the oil level in reservoirs as machines cool during idle or shutdown periods. A clean hydraulic oil of suitable type will have little tendency to mix with water; and in a still reservoir, the water will tend to settle at a low point. During operation, the water may be picked up by oil circulation, broken up into droplets, and mixed with the oil, forming an emulsion. The water and oil in such an emulsion should separate quickly in the reservoir, but when solid contaminants or oil oxidation products are present, emulsions tend to persist and to join with other deposit-forming materials present to form sludge. The emulsion may be drawn into the pump and made more permanent by the churning action of the pump and the mixing effect of flow at high velocities through control devices.

To prevent such contamination from occurring, it is essential that a hydraulic oil have the ability to separate quickly and completely from water. Properly refined oils have this ability when new, but only oils having exceptionally high oxidation stability are able to retain good water-separating ability over long service periods. In addition to using such an oil, every effort should be made to keep systems free of water, dirt, and other contaminants.

G. Rust Protection

Water and oxygen can cause rusting of ferrous surfaces in hydraulic systems. In as much as air is always present (except in specialized nitrogen blanket systems), oxygen is available, and some water is often present. The possibility of rusting is greatest during shutdown, when surfaces that are normally covered with oil may be unprotected and subject to gathering condensation as they cool. This is particularly important for operations that experience high humidity conditions and temperature changes within the reservoirs.

Rusting results in surface destruction, and rust particles may be carried into the system where they will contribute to wear and the formation of sludge like deposits certain to interfere with the operation of pumps, actuators, and control mechanisms. Rusting of piston rods or rams causes rapid seal or packing wear, resulting in increased leakage and system contamination; moreover, external contaminants can enter the hydraulic fluid through worn seals or packing.

High quality hydraulic fluids contain an additive material, called a rust inhibitor, which has an affinity for metal surfaces. It plates out on the surfaces, forming a barrier film that resists displacement by water and, therefore, protects the surfaces from contact with water. The rust inhibitor must be carefully selected to provide adequate protection without reducing other desirable properties, especially water-separating ability.

H. Compatibility

An often overlooked characteristic of the hydraulic fluid during the selection process is compatibility. Use of oils that exhibit undesirable reactions with system components such as elastomers, paints, or gasket materials can result in leakage or contamination within the system that can reduce overall performance. The metallurgy of system components must also be considered when selecting an oil in order to avoid metallic corrosion or staining due to additive reactions in the presence of heat and contaminants. In changing from one oil to another (e.g., when switching from a conventional mineral oil to a fire-resistant fluid or a synthetic hydraulic fluid) attention to compatibility issues is particularly important.

VIII. SPECIAL CHARACTERISTICS IN HYDRAULIC FLUIDS

A. Soluble Oils

Water serves as a hydraulic fluid in some very large noncritical systems. Large forging and extrusion presses with vertical in-line pumps or special axial piston pumps, for example, operate at 2000–3000 psi (136–204 bar). Because water has little ability to lubricate, seal, or prevent rust, it can be mixed with 2–5% soluble oil to form an oil-in-water emulsion for these systems. Operating temperatures are limited to 60°C (140°F) to prevent excess evaporation.

B. High Water Content Fluids (HWCF)

High water content fluids, sometimes referred to as 95/5 fluids, contain 2–5% water-soluble chemicals that impart some lubricity, rust protection, and wear protection of specially designed pumps, valves, packing glands, and cylinders. They work well at pressures as high as 10,000 psi (680 bar) in reciprocating plunger pumps, or in axial piston pumps at pressures in the range of 1000 psi (68 bar). Because their antiwear protection is limited, they are used in vane pumps up to 1000 psi (68 bar) and are not recommended for systems with gear pumps. Again, operating temperatures must be limited to avoid excessive loss of water by evaporation.

C. Fire-Resistant Fluids

When the possibility exists that the hydraulic fluid may come in contact with a source of ignition, fire-resistant fluids may be used. The use of any fire-resistant fluid will mean additional costs, but these may be offset by safety factors as well as high costs of equipment in the event of fires. The potential for fires exists in applications such as die-casting operations, and continuous casting hydraulics in steel mills or presses that are operated near furnaces or ovens. In these and many other types of service, hydraulic line breakage could result in serious fires.

Three approaches to this problem follow. One or a combination of them may be adopted, depending on cost for any particular application: (1) redesign or relocate machines to remove or isolate sources of ignition; (2) install fire control measures such as fire suppression devices to assure the safety of the personnel and prevent or localize property damage; and (3) use a *fire-resistant* hydraulic fluid.

In addition to the use of soluble oils and HWCFs just discussed, there are three basic types of general-use fire-resistant fluid: synthetic, emulsion, and water glycol. The term ''fire-resistant'' does not mean that the fluid will not burn. For example, emulsions and water glycol fluids depend on the water content for their fire-resistant characteristics. Once the water has evaporated, these materials will burn when subjected to a source of ignition. Even the synthetics will burn when subjected to high enough temperatures.

It should be noted that pumps and other system components must be specially designed or operating conditions derated for use of certain fire-resistant fluids. In most cases, pumps are designed with special materials to provide satisfactory service life in equipment using fluids with high water content. In addition, the pumps have specified maximum operating parameters placing limits, for example, on pressures, temperatures, and speed. Component as well as system manufacturers should be consulted before designing systems or using fire-resistant fluids.

1. Synthetics

Synthetic fire-resistant fluids (typically phosphate ester materials) are the most expensive choice. They provide good overall performance compared to the other fluids, and lower maintenance. Their densities are high (heavier than water) and the VIs are low. They have potential problems with compatibility with paints, elastomers (seals and hoses), and other system materials. They are not compatible with emulsions or water glycol fluids, and special precautions are necessary when converting to these from one of the other fire-resistant fluids. A synthetic fluid may be partially or fully compatible with conventional mineral oils. Because the densities of such fluids are relatively high, additional pumping requirements may be necessary. Polyol esters are also used as fire-resistant fluids and provide better compatibility but may not provide the same level of fire resistance.

2. Emulsions

Emulsions are classified in two categories: conventional (such as the soluble and HWCF types already discussed) and invert emulsions. Invert emulsions are more commonly used in hydraulic systems as fire-resistant fluids because they are water-in-oil emulsions having oil as the continuous phase. Conventional emulsions contain about 95% water, and the inverts contain between 40 and 45% water. Fire resistance characteristics are provided by the water, and maintenance of the correct water content is necessary to assure fire resistance. Invert emulsions provide good lubrication characteristics, and various products range in viscosity from 65 to 129 cSt at 40°C (300–650 SUS at 100°F). The viscosity of both invert and conventional emulsions is affected by water content. Invert emulsions will increase in viscosity as the water content is increased, while conventional emulsions will decrease in viscosity as water content increases. To maintain proper viscosity and fire resistance, operating temperatures should be limited to 60°C (140°F) and preferably should remain below 49°C (120°F) to prevent evaporation of the water. In properly designed and maintained systems, invert emulsions will provide a reasonable cost alternative where fire-resistant fluids are used.

3. Water Glycols

Water glycols are true solutions that contain about 40–50% water and glycol (typically diethylene glycol), and additives to impart specific performance levels. Because of the water content, they contain liquid and vapor phase rust inhibitors. Since glycols do not provide good natural antiwear protection without the use of antiwear additives, these are used in most formulations for hydraulic applications. Water content controls the viscosity of the fluid as well as the fire-resistant characteristics. Maximum operating temperatures should be limited to 60°C (140°F).

D. Environmental Hydraulic Fluids

Chapter 6 provided a detailed discussion of environmental hydraulic fluids. The performance characteristics of fluids of these types are the same as those of conventional hydraulic fluids, but, in addition, high levels of biodegradability and low levels of toxicity are required.

E. Changing or Converting to Fire-Resistant Fluids

Changing to a fire-resistant fluid generally requires draining of existing fluid and thorough flushing and cleaning of the system to assure minimal contamination. If the system contains painted surfaces, compatibility of the fluid with the paint should be verified, as well as compatibility with seals, hoses, and other system components. In some instances, it may be necessary to change suction pipe size (to allow adequate flow of fire-resistant fluid to pump suction) and filters to accommodate the fluid being used. Specific manufacturers of both the fluid and the system components should be consulted for any additional precautions.

IX. HYDRAULIC SYSTEM MAINTENANCE

The degree of system maintenance is based on specific performance expectations, the fluid used, and the system operating parameters. The various hydraulic fluids, ranging from mineral oil based to synthetic to water-containing fire-resistant fluids, demand various levels of maintenance to assure performance. Water-containing fluids require higher levels of maintenance to assure not only that the fire protection properties are retained but that the fluid will provide proper lubrication characteristics while in service. This topic was discussed earlier in this chapter, but selecting the proper, fluid matched to system needs and understanding the limitations of that fluid comprise the basic starting point for an effective maintenance program.

Once the proper fluid has been selected, the equipment and operating conditions will dictate the degree of maintenance required to keep that fluid in service for long periods of time while retaining its lubrication characteristics. These maintenance procedure objectives should include the following:

Keeping fluid clean/controlling contamination
Maintaining proper temperatures
Maintaining proper oil levels
Periodic oil analysis
Routine inspections
- Noise levels
- Vibration
- Pressures
- Shock loads
- Leakage
- Fluid odor and color
- Filtration
- Temperatures
- Foaming

A. Keeping Fluid Clean/Controlling Contamination

The first maintenance objective starts at the time the oil is received and stored and continues through the period of time it is in service in the system. Contamination, such as moisture, can enter ''sealed'' containers while in storage, through normal expansion and contraction of the fluid due to temperature changes. Moisture thus allowed to enter then condenses inside the containers. Contamination can also result while the oil is being transferred from storage (or containers) to the system. Dirty transfer containers might be at fault, or equipment that has been sitting in the open or has been used for other materials such as gear oils, engine oils, and coolants. Dirty reservoirs and debris around the fill location are also sources of contamination during filling or adding makeup oil to the system. In critical systems, sometimes quick-disconnect fittings are installed on reservoirs, or portable filter carts are used to facilitate adding oil to clean oil to these systems. Such measures minimize the potential for contamination.

The fluid in service must be clean, and the level of cleanliness depends on the system. Numerically controlled (NC) machine tools, for example, require high levels of cleanliness to accommodate the close-tolerance servovalves, whereas the hydraulics used to operate hydraulic lifts in automotive repair shops can run satisfactorily with minimal filtration. It should be noted that conventional filters will not remove water- or oil-soluble contaminants. Special coalescing-type filters are available for the removal of limited amounts of water.

B. Filtration

Full flow filtration is the most common type used on hydraulic systems to control the levels of solid contaminants. These filters are generally installed in the supply (pressure) line but can also be installed in return (low pressure) lines to the reservoir. Full flow filter

Table 7.2 Typical Critical Clearance for Fluid System Components

Component	Clearance	
	micrometers	inches
Antifriction bearings	0.5	0.000019
Vane pump: tip to vane	0.5–1.0	0.000019–0.000039
Piston pump: valve plate to cylinder	0.5–5.0	0.000019–0.000197
Gear pump: gear to side plate	0.5–5.0	0.000019–0.000197
Gear pump: gear tip to case	0.5–5.0	0.000019–0.000197
Servovalve spool (radial)	1.4	0.000055
Control valve spool (radial)	1.0–23.0	0.000039–0.000904
Hydrostatic bearings	1.0–25.0	0.000039–0.000984
Vane pump sides of vanes	5.0–13.0	0.000197–0.000511
Piston pump: piston to bore	5.0–40.0	0.000197–0.001575
Servovalves: flapper wall	18.0–63.0	0.000708–0.002363
Actuators	50.0–250.0	0.001969–0.009843

housings are generally equipped with bypass valves that open when the pressure drop across the filter exceeds a predetermined level. This assures that components will receive oil in the event of filter plugging or restriction of oil flow through the filter due to start-up or cold oil where viscosities are high. When filters go on bypass, unfiltered oil is supplied to components. Some filters are equipped with condition indicators or differential pressure gages to warn of restrictions or plugging.

Selection of appropriate filtration levels (fluid cleanliness) will be based on specific system components and operation. Table 7.2 shows some of the typical clearances in hydraulic system components. The vast majority of hydraulic systems function properly on 10 μm filtration with filter efficiencies of 98.7% or greater. Filter efficiencies are often referred to as *beta ratios* (β). A beta ratio of 75 for a 10 μm filter would mean that 98.7% of the particles in the 10 μm and larger range will be removed. Tables 7.3 and 7.4 give a brief explanation of beta ratios and filter efficiencies.

Bypass filters, sometimes referred to as *polishing filters,* generally are installed in an independent system where from 5–15% of the system's oil capacity (in gpm) is filtered to a finer degree. An auxiliary pump is used to take the oil from a low point in the reservoir, whereupon it is filtered and returned to the reservoir. With this type of system, oil purification can be continued whether the hydraulic system is in operation or shut down. An alternative to the independent system is to use a continuous bypass mode, in which a percentage of the oil flow from the pressure or return line is passed through suitable purification equipment and returned to the reservoir. Bypass purification equipment can be relatively small because only a portion of the total oil capacity is handled.

Portable filters, sometimes called *filter carts or buggies,* are also used to supplement permanently installed system filters. These units consist of a motor-driven pump and filter arrangement, which circulates fluid from the reservoir, through a fine filter, and back to the reservoir. The suction and return hoses should be connected to opposite ends of the reservoir with quick-disconnect fittings. Generally, portable filters will operate for at least 24 h on each system to ensure that the full oil charge is filtered effectively. Portable filter units can be used in place of bypass filters if periodic or as-needed filtration is sufficient

Table 7.3 Filter Efficiency and Beta Ratios: The filtration ratio or beta is calculated by dividing the number of particles entering the filter by the number of particles exiting the filter. β_5 represents the filtration ratio at 5 μm or the ratio of the upstream to the downstream particles larger than 5 μm.

In Out $\beta = \frac{\text{In}}{\text{Out}}$

Filter	Particles > 5 μm		β_5
	In	Out	
Filter A	10,000	5,000	2
Filter B	10,000	100	100
Filter C	10,000	1	10,000

to maintain the desired levels of fluid cleanliness. Portable filtration units can be a simple arrangement, as just discussed, or may be *reclamation units* consisting of motor-driven pumps, oil heating elements, vacuum chambers, and fine filters. The advantage of reclamation units is that they can remove water and some volatile contaminants (such as some solvents) in addition to removal of particulates.

Batch filtration may be used where the fluid volume is very large or is heavily contaminated. The large volume of oil may be removed and reclaimed by using batch-through settling processes, filtration, centrifuging, and/or reclamation units. The disadvantage of this process is that the machine must be shut down for removal of the fluid charge.

Table 7.4 Significance of Beta (β) Values in Particle Count Control: Beta values can be directly related to efficiency. To determine the relative performance of two filters with different beta ratios, the downstream particle count from each filter can be calculated by using the beta ratio and an assumed upstream particle count (10^6 in this example). The filter with the highest beta value will have the lowest downstream particle count.

β Value at X μm (β_x)	Removal efficiency (%): particles > X μm	Downstream count: > X μm when filter is challenged upstream with 10^6 particles > X μm
1.0	0	1,000,000
1.5	33	670,000
2.0	50	500,000
20	95	50,000
50	98.0	20,000
75	98.7	13,000
100	99.0	10,000
200	99.5	5,000
750	99.87	1,333
1,000	99.90	1,000
10,000	99.99	100

C. Controlling Temperatures

Excessive temperatures, in addition to possibly reducing oil viscosities so greatly that metal-to-metal contact results, will oxidize the oil and lead to varnish and sludge in the system. These deposits plug or restrict the motion of valves, plug suction screens, and cause shortened filter life.

Heat develops as the fluid is forced through the pumps, motors, tubing, and relief valves. Temperatures should be maintained between 49 and 65°C (120–149°F) in conventional hydraulic systems. Some variable volume pump systems, closed-loop hydraulic systems, and hydraulic transmissions can operate up to 120°C (248°F), temperatures at which premium fluids must be used or drain intervals shortened. Systems operating on water-based fluids should be kept below 60°C (140°F) to prevent the water from evaporating.

To allow heat to radiate from the system, keep the outside of the reservoir clean and the surrounding area clear of obstructions. Make sure the oil cooler is functioning properly and keep air-cooled radiators free of dirt and debris. Keep the reservoir filled to the proper level to allow enough fluid residence time for the heat to dissipate.

Oil degradation is even more critical in numerically controlled machine tools with electrohydraulic servovalves. Because of space constraints, these systems typically are designed with small reservoirs, which results in short residence times for the hydraulic fluid. With minimal rest times and high system pressures, entrained air bubbles can cause extremely high localized temperatures; this is because of the adiabatic compression of the air bubbles as they pass from low suction pressure to the high discharge pressure. Nitrogen fixation then results and, when combined with oil oxidation, can lead to the formation of deposits that plug oil filters and cause servovalve sticking.

D. Maintaining Proper Reservoir Oil Levels

Systems are designed to provide a certain amount of oil residence time in the reservoir. This allows the fluid time for separation of air and water and solid contaminants to settle. It also provides for a degree of cooling. Operating with low oil levels reduces the effectiveness of these processes. In addition, if levels are low enough, air may be pulled into the pump suction, resulting in air entrainment. This could lead to excessive *foaming* and *cavitation.* Foaming and air entrainment can also result from air leaks in the suction, low fluid temperatures, or use of a fluid too viscous to release air bubbles. Oxidation and contamination increase the fluid's tendency to foam and retain air. Entrained air is a major cause of destructive pump cavitation. The intense pressures and temperatures created by the collapse of the bubbles erode metal parts of pumps and valves, resulting in excessive wear. Pump or valve cavitation may cause irregular operation or ''spongy'' response of the system. In addition to air leaks, a restricted suction can cause cavitation. The cause of this restriction could be plugged strainers, oil viscosity too high, or inadequate suction design conditions. Cavitation is apparent by a high-pitched whine or scream in the pump. And, the pump may sound as if there are marbles trapped in the housing.

E. Periodic Oil Analysis

Laboratory oil analysis can be set up as a routine or determined by visual inspection of the fluid. A visual inspection may reveal the type and degree of contamination. Take a sample of the fluid and allow it to settle overnight in a clear container. Inspect the sample for color, appearance, and odor. If there is no evidence of water, corrosion, or excessive accumulation of deposits, sediment, or sludge, and if the fluid has the color and odor of new fluid, generally laboratory analysis is not necessary. A slight ''burnt odor'' is common in conventional hydraulic systems that use petroleum oils. However, a burnt oil odor in an NC machine oil sample may be a cause for concern. In some cases, even conventional laboratory oil analysis cannot determine the low levels of contamination, such as the nitrogen fixation discussed earlier (Section IX.C: Controlling Temperatures), which will cause system operational problems. These systems generally require high quality oils as well as a specialized oil analysis program.

If visual inspections cannot identify specific contaminants and potential sources, take a sample for laboratory analysis. It is a good practice to establish a program to periodically submit oil samples for laboratory analysis. In critical high temperature machines, this may be as often as every 3 months. Noncritical machines will require laboratory oil analysis only every 6–12 months, but specific schedules should be based on operation, equipment, and oil supplier's recommendations.

F. Routine Inspections

Routine inspections of operating systems can provide insight into potential problem areas and suggest ways to initiate corrective actions. These corrective actions will result in longer equipment life and lower cost operation of the hydraulic system. Commonly included in routine inspection programs are the following:

Noise levels. Increases in noise levels may signal problems with cavitation. Noise levels may also increase when excessive temperatures allow the oil's viscosity to become too low, resulting in metal-to-metal contact.

Vibration. Loose mounting or misalignment of components will cause vibration, resulting in accelerated wear or failure.

Shock loads. System components, such as hoses and fittings, subjected to shock loads due to abrupt changes in flows and pressures, can lead to leakage and failures. Where shock loads are experienced, accumulators should be installed in the system.

Leakage. Oil leakage can result in low reservoir oil levels, leading to poor system performance. In addition, leakage can be costly and can create safety problems.

Temperatures. As discussed earlier, controlling temperatures in the appropriate range is important both from the oil life and system performance standpoints. Excessive temperatures may be caused by plugged or dirty heat exchangers, excessive pressures, high rates of internal leakage, or low oil levels. Use of fluids with too high (excessive shearing) or too low (inadequate films) a viscosity will also result in higher temperatures.

Filtration. The condition of fill screens, breathers, and filters (indicators or differential pressure gages) should indicate the need for cleaning or replacement.

Foaming. A little foam on top of the oil in the reservoir is normal. Excessive foam may indicate air leaks in the suction line, unsatisfactory contamination levels, or inadequate antifoam characteristics of the oil.

Fluid odor and color. Although odor and color are not characteristics used to judge an oil's ability to provide proper lubrication, changes in these physical properties may indicate contamination (solvents, wrong oils added, degradation, etc.) or that the oil has reached the useful end of its service life. If there is doubt about the causes of the color and odor changes, a sample should be submitted for laboratory oil analysis. If these changes are accompanied by undesirable machine operating characteristics, change the oil but still submit a sample for laboratory oil analysis.

8

Lubricating Films and Machine Elements: Bearings, Slides, Ways, Gears, Couplings, Chains, Wire Rope

The elements of machines that require lubrication are bearings—plain, rolling element, slides, guides, and ways; gears; cylinders; flexible couplings, chains, cams, and cam followers; and wire ropes. These elements have fitted or formed surfaces that move with respect to each other by sliding, rolling, and advancing and receding, or by combinations of these motions. If actual contact between surfaces occurs, high frictional forces leading to high temperatures, and possibly wear or failure, will result. Therefore, the elements are lubricated to prevent or reduce the actual contact between surfaces.

Without lubrication, most machines would run for only a short time. With inadequate lubrication, excessive wear is usually the most serious consequence, since a point will be reached, usually after a short period of operation, when the machine elements cannot function and the machine must be taken out of service and repaired. Repair costs—material and labor—may be high, but lost production or lost availability of the machine may exact by far the greatest cost. With inadequate lubrication, even before failure of elements occurs, frictional forces between surfaces may be so great that drive motors will be overloaded or frictional power losses excessive. Finally, inadequately lubricated machines will not run efficiently, smoothly, or quietly.

Machine elements are lubricated by interposing and maintaining fluid films between moving surfaces. These films minimize actual contact between the surfaces, and because shearing occurs readily, the frictional force opposing motion of the surfaces is low.

I. TYPES OF LUBRICATING FILM

Lubrication films may be classified as follows:

1. *Fluid films* are thick enough that during normal operation they completely separate surfaces moving relative to each other. Fluid film lubrication includes hydrodynamic, elastohydrodynamic, hydrostatic, and squeeze types.

2. *Thin films* are not thick enough to maintain complete separation of the surfaces all the time. Thin film lubrication includes mixed film and boundary lubrication.
3. *Solid films* are more or less permanently bonded onto the moving surfaces.

A. Fluid Films

Fluid film lubrication is the most desirable form of lubrication because during normal operation, the films are thick enough to completely separate the load-carrying surfaces. Thus friction is at a practical minimum, being due to shearing of the lubricant films alone, and wear does not occur because there is essentially no mechanical contact. Fluid films are formed in three ways:

1. Hydrodynamic and elastohydrodynamic films are formed by motion of lubricated surfaces through a convergent zone such that sufficient pressure is developed in the film to maintain separation of the surfaces.
2. Hydrostatic film is formed by pumping fluid under pressure between surfaces that may or may not be moving with respect to each other.
3. Squeeze films are formed by movement of lubricated surfaces toward each other.

Two types of hydrodynamic film lubrication are now recognized. In plain journal bearings and tilting pad or tapered land thrust bearings operating under correct lubrication conditions, thick hydrodynamic films are formed. These films are usually in excess of 0.001 in. (25 μm) thick. In hydrodynamic applications, the loads are low, and the areas over which the loads are distributed are relatively large; thus the amount of deformation of the load-carrying area is not great enough to significantly alter that area. Pressures in full hydrodynamic lubrication films range from a few psi to several thousand psi. Typical pressures found are in the range of 50–300 psi. Load-carrying surfaces of this type are often referred to as ''conforming,'' although it is obvious that in the case of tapered land thrust bearings, for example, the surfaces do not conform in the normal concept of the word. However, the term is a convenient opposite for the term ''nonconforming,'' which quite accurately describes surfaces of the types at which elastohydrodynamic films are formed.

The surfaces of the balls in a ball bearing theoretically make contact with the raceways at points; rollers of roller bearing make contact with the raceways along lines; and meshing gear teeth bearing also make contact along lines. These types of surface are nonconforming. Under the pressure (30,000–400,000 psi) applied to these elements by load conditions through the lubricating film, the metals deform elastically, expanding the theoretical points or lines of contact into discrete areas. The oil is able to generate films at these extremely high pressures because of the characteristic of viscosity doubling for each 5000 psi increase in applied pressure. This is discussed later in this chapter.

Since a convergent zone exists immediately before these areas of contact, a lubricant will be drawn into the contact area and can form a hydrodynamic film. This type of film is referred to as an elastohydrodynamic film, the ''elasto'' part of the term referring to the requirement that elastic deformation of the surfaces occur before the film can be formed. This type of lubrication is called elastohydrodynamic lubrication (EHL: the acronym EHD is also used). EHL films are very thin, in the order of 10–50 μin. (0.25–1.25 μm) thick. However, even with these thin films, complete separation of the contacting surfaces can be obtained.

Any material that will flow at the shear stresses available in the system may be used for fluid film lubrication. In most applications, petroleum-derived lubricating oils or synthetic fluids are used. There are some applications for greases. Some materials not usually considered to be lubricants, such as liquid metals, water, and gases, are also used in some applications. The following discussions are concerned mainly with fluid film lubrication with oils and greases, since some rather complex special considerations are involved when other materials are used.

1. Thick Hydrodynamic Films

The formation of a thick hydrodynamic fluid film that will separate two surfaces and support a load can be described with reference to Figure 8.1. The two surfaces are submerged in a lubricating fluid. As the upper surface moves, internal friction in the fluid causes it to be drawn into the space between the surfaces. The force drawing the fluid into space *A* is equal to the force tending to draw it out; but since the cross-sectional area at the outlet section is smaller than the inlet, the flow of fluid is restricted at the outlet. The moving surface tries to ''compress'' the fluid to force it through this restricted section, with the result that the pressure in the fluid rises. This pressure rise tends to do the following:

Retard the flow of fluid into space *A*
Increase the flow at the restricted outlet section
Cause side leakage in the direction normal to the direction of motion

The most important effect, however, is that the pressure in this wedge-shaped film enables it to support a load without contact occurring between the two surfaces.

For the elements of Figure 8.1 to work as a bearing, the moving surface would have to be of infinite length. A more practical approach is to distribute a series of wedge-shaped sections around the circumference of a disk as shown in the inset. The moving surface

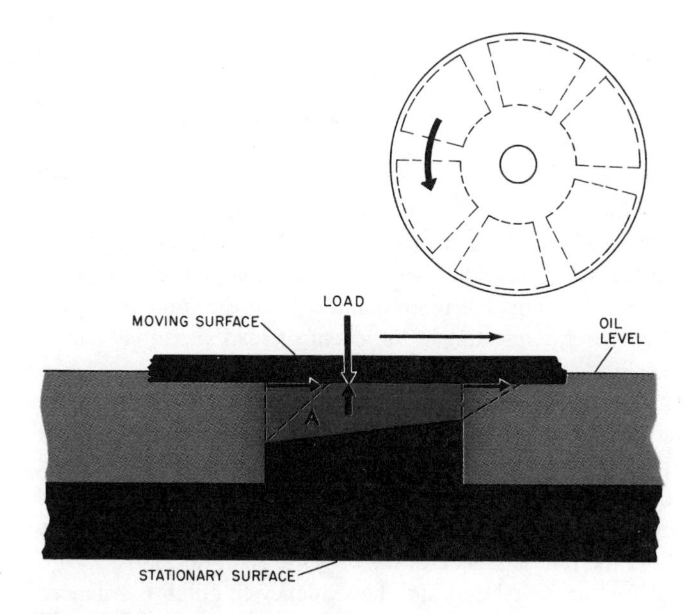

Figure 8.1 Converging wedge.

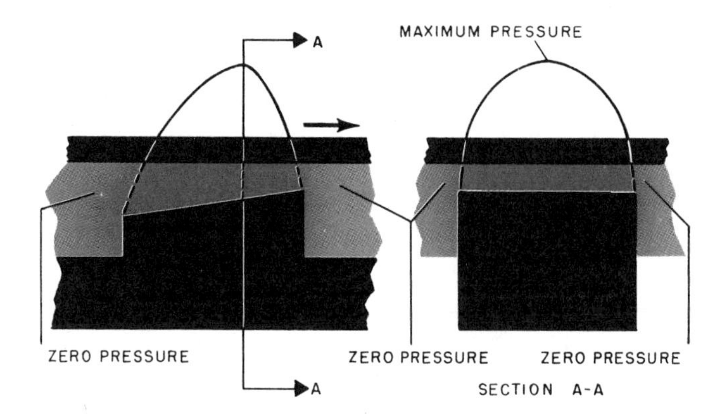

Figure 8.2 Pressure distribution in the wedge film: left, pressure distribution in the direction of motion; right, pressure distribution in the direction normal to motion.

then becomes a disk to which load can be applied, and the complete assembly can be used as a thrust bearing.

The pressure distribution in the wedge-shaped fluid film of Figure 8.1 is shown in Figure 8.2. In the direction of motion (left view), the pressure gradually rises more or less to a peak just ahead of the outlet section and then falls rapidly. In the direction normal to motion (right view), the pressure is at its maximum in the middle of the shoe, but because of side leakage, falls off rapidly toward each side.

In Figure 8.2, the taper of the shoe and the thickness of the fluid film are greatly exaggerated for illustrative purposes. In a practical bearing of this type, the amount of taper might be on the order of 0.002 in. (50 μm) in a length of 6 in. (150 mm), and the film thickness at the outlet would be on the order of 0.001 in. (25 μm). With proper design and fluid viscosity, there would be no contact between the disk and shoes when the speed is above a certain minimum during normal operation. The bearing would not be frictionless, however, since a force would have to be applied to the disk to overcome the resistance to shear of the fluid films.

The same principles outlined in Figure 8.2 can be used to generate a fluid film, which will lift and support a shaft or journal in a bearing. If the moving surface and tapered shoe of Figure 8.2 are "rolled up," the journal and bearing of Figure 8.3 will result. A journal turning at a suitable speed will then draw fluid into the space between the journal and the bearing and develop a load-supporting film. This bearing is called a partial bearing and is suitable for some applications in which the load on the journal is always in one direction. However, bearings are more commonly of the full 360-degree type as shown in Figure 8.4 to provide restraint in the event of load variations and to permit easier enclosure and sealing.

The stages in the formation of a hydrodynamic film in a full journal bearing are illustrated in Figure 8.4. First the machine is at rest with the oil shut off, and the oil has leaked from the clearance space. Metal-to-metal contact exists between the journal and bearing. When the machine has been started (left) and the oil supply turned on, filling the clearance space, the shaft begins to rotate counterclockwise and friction is momentarily high, so the shaft tends to climb the left-hand side of the bearing. As it does this, it rolls onto a thicker oil film so that friction is reduced and the tendency to climb is balanced by the tendency to slip back. As the journal gains speed (center), drawing more oil through

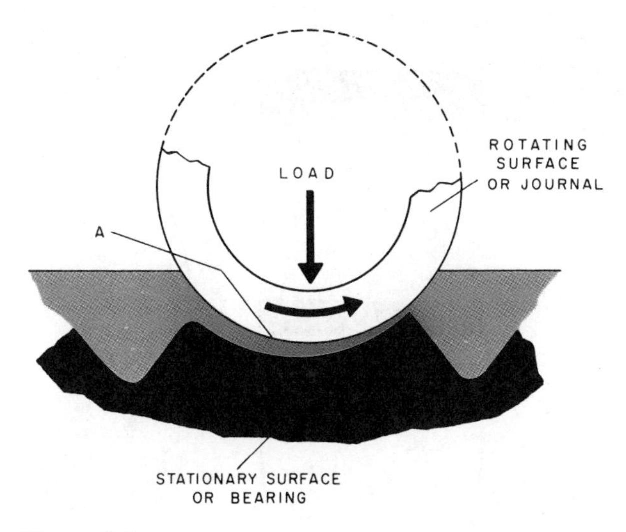

Figure 8.3 Partial journal bearing.

the wedge-shaped space between it and the bearing, pressure is developed in the fluid in the lower left portion of the bearing that lifts the journal and pushes it to the right. At full speed (right), the converging wedge has moved under the journal, which is supported on a relatively thick oil film with its minimum thickness at the lower right, on the opposite side to the starting position shown (left). Under steady conditions, the upward force developed in the oil film just equals the total downward load, and the journal is supported in the slightly eccentric position shown. Variations in speed, load, and oil viscosity will cause changes in this eccentricity, and also in the minimum film thickness. The pressures developed in the oil film are illustrated in Figure 8.5.

Thick hydrodynamic film lubrication is used in journal and thrust bearings for many applications. Characteristics and lubricant requirements of various applications are discussed more fully in later sections.

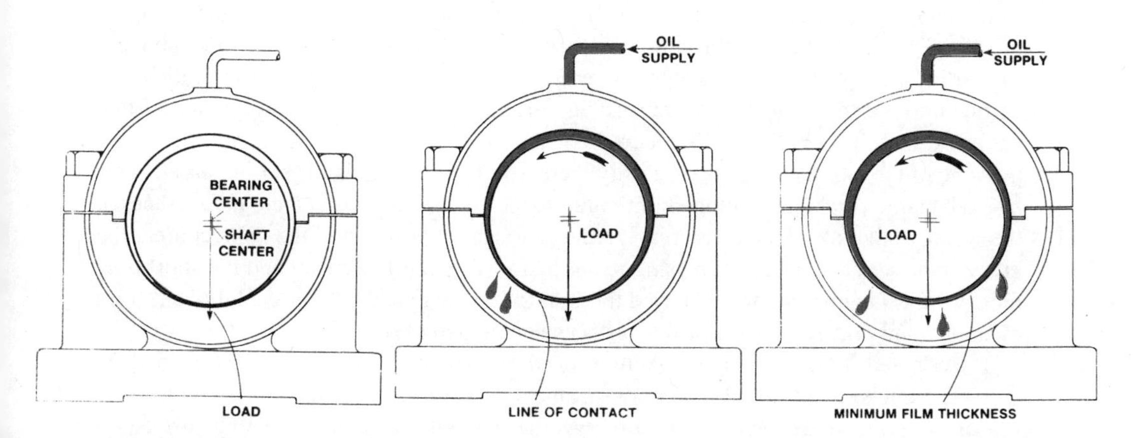

Figure 8.4 Development of hydrodynamic film in a full journal bearing with downward load.

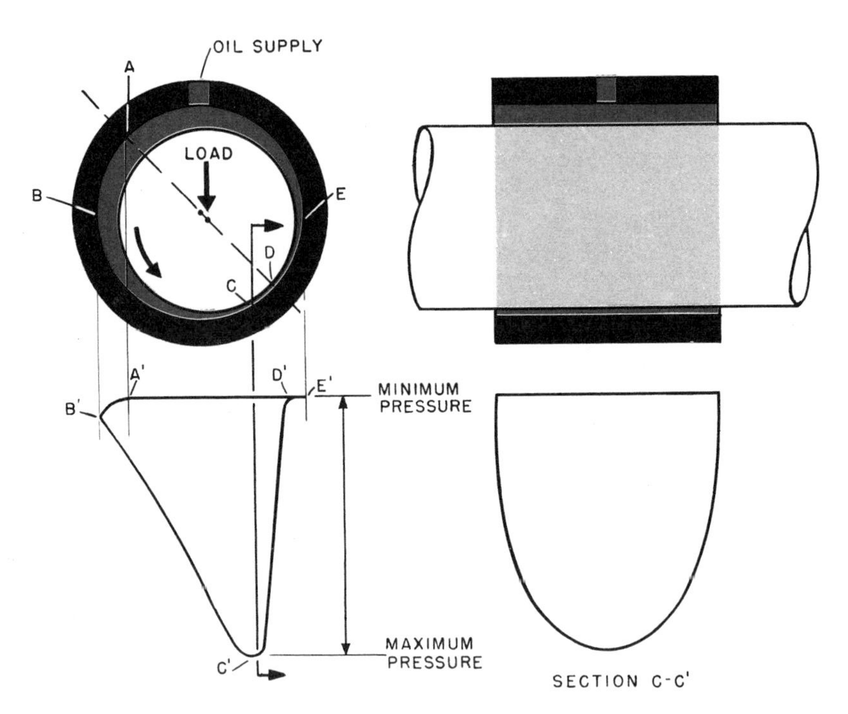

Figure 8.5 Pressure distribution in a full journal bearing. In the left view from *C*, the point of minimum film thickness, to *D*, the pressure drops rapidly to the minimum, which may be atmospheric, slightly above atmospheric, or slightly below atmospheric depending on such factors as bearing dimensions, speed, load, oil viscosity, and supply pressure. From *A* to *C* the pressure increases as the clearance decreases. Pressure distribution along the line of minimum film thickness is shown at the right. From a maximum value near the center, it drops rapidly to zero at the ends because of end leakage.

2. Elastohydrodynamic Lubricating (EHL) Films

In the lubrication of EHL contacting surfaces, such as antifriction bearings, the viscosity of the oil must increase as the pressure on the oil increases (Figure 8.6) to establish a supporting film at the very high pressures in the contact area. Thus, as the lubricant is carried into the convergent zone approaching the contact area, elastic deformation of two surfaces, such as a ball and its raceway or a roller and its raceway, results from the pressure of the lubricant. As the viscosity increases, the pressure is further increased. This hydrodynamic pressure developed in the lubricant is sufficient to separate the surfaces at the leading edge of the contact area. As the lubricant is drawn into the contact area, the pressure on it rises further, increasing its viscosity. This high viscosity and the short time required for the lubricant to be carried through the contact area prevent the lubricant from escaping, and thus separation of the surfaces can be achieved.

As noted, EHL films are very thin, on the order of about 10–50 μin. (0.25–1.25 μm). The thickness of the film increases as the speed is increased, the lubricant viscosity is increased, the load is decreased, or the geometric conformity of the mating surfaces is improved. Load has comparatively little effect on the film thickness because at the pres-

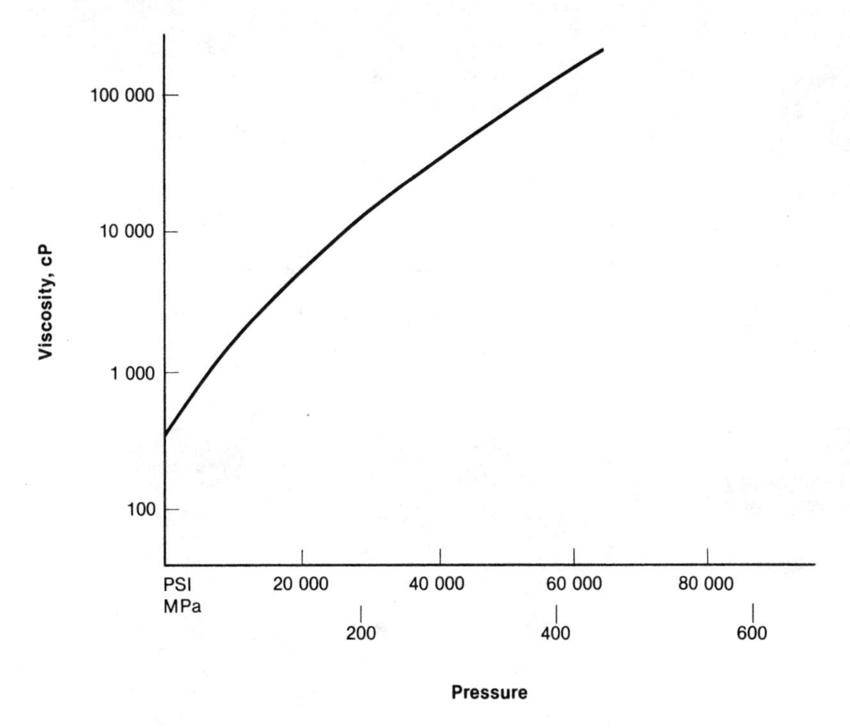

Figure 8.6 Effect of pressure on viscosity.

sures in the contact area, the oil film is actually more rigid than the metal surfaces. Thus an increase in load, for example, mainly has the effect of deforming the metal surfaces more and increasing the area of contact rather than decreasing the film thickness.

If the surfaces shown in Figure 8.7 were rougher, that is, if the asperities were higher, the opposing surface asperities would be more likely to make contact through the film. Thus the relationship between the roughness of the surfaces and the film thickness is important in the consideration of elastohydrodynamic lubrication. This has led to the introduction of the quantity specific film thickness λ in EHL considerations. The specific film thickness is the ratio of the film thickness h to the composite roughness σ of the two surfaces:

$$\lambda = h/\sigma$$

The composite roughness of the two surfaces can be calculated as the square root of the sum of the squares of the individual surface roughness:

$$\sigma = \sqrt{\sigma_1^2 + \sigma_2^2}$$

where σ_1 and σ_2 are the root-mean-square (rms) roughness values of the two surfaces.*

* A number of calculations are given in the literature for the composite roughness of using centerline average (CLA) values. The relationship between rms and CLA values is $\sigma_{RMS} = 1.3\sigma_{CLA}$.

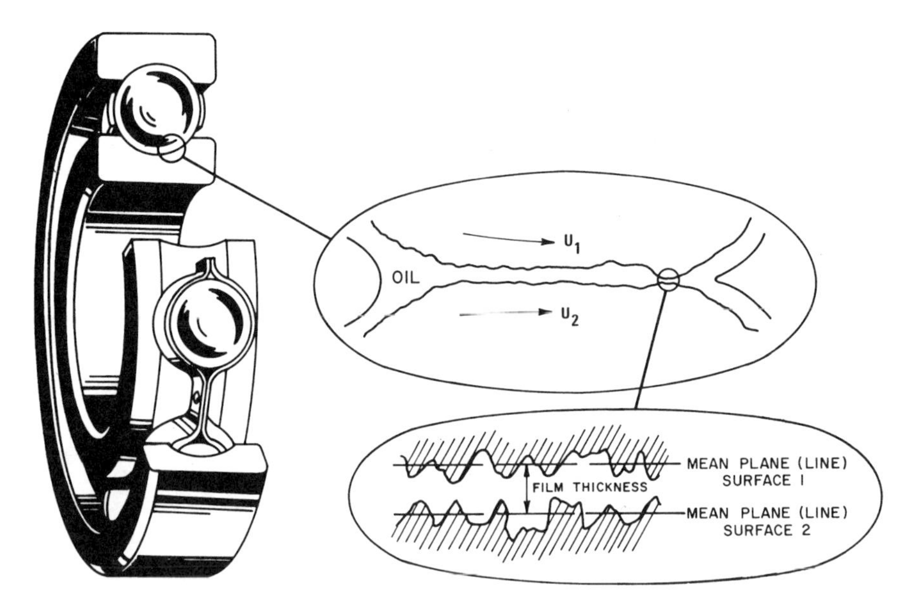

Figure 8.7 Schematic of contact area between ball and inner race.

The specific film thickness is related to the ability of the lubricating film to prevent or minimize wear or scuffing. Figure 8.8 shows the effect of λ on the L_{10} fatigue life† of a series of cylindrical roller bearings. For values of λ of approximately 3–4, the L_{10} fatigue life is relatively constant, so is essentially independent of the oil film thickness. This region is labeled ''full EHL'' and corresponds to full fluid film conditions that exist in thick hydrodynamic film lubrication. In the region labeled ''partial EHL,'' the film becomes progressively thinner, and more and more surface asperities penetrate the film.

Fatigue life is decreased accordingly. This region is one of mixed film lubrication. When λ is below about 1, boundary lubrication prevails and the surfaces are in contact nearly all the time. Fatigue failure is generally rapid unless lubricants with effective anti-wear additives are used. For λ values above 4, full film conditions exist; however, the thicker oil films can result in higher bearing temperatures and lower overall efficiency.

These ranges for λ are considered to be applicable to rolling element bearings. As will be discussed later, somewhat lower values of the calculated safe minimum film thickness may be applicable to gears.

Increased understanding of EHL film characteristics and the use of specific film thickness has led to improved understanding of the lubricant function in many applications where unit loads are high, as well as to improved methods of selecting lubricants for such applications.

3. Hydrostatic Films

In hydrostatic film lubrication, the pressure in the fluid film that lifts and supports the load is provided from an external source. Thus, relative motion between opposing surfaces

† The L_{10} fatigue life is the average time to failure of 10% of a group of identical bearings operating under the same conditions.

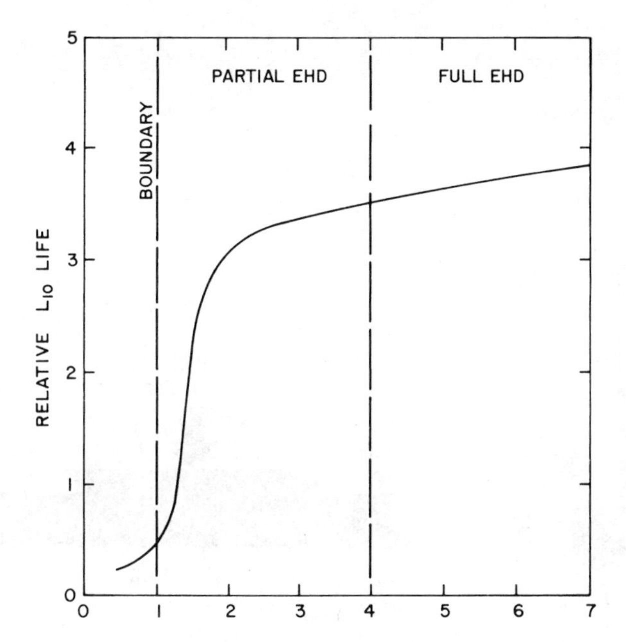

Figure 8.8 L_{10} fatigue life relative to book value as a function of specific film thickness; the data are for a series of cylindrical roller bearings.

is not required to create and maintain the fluid film. The principle is used in plain and flat bearings of various types, and it offers such advantages as low friction at very low speeds or, when there is no relative motion, more accurate centering of a journal in its bearing, and freedom from stick–slip effects.

The simplest type of hydrostatic bearing is illustrated in Figure 8.9. Oil under pressure is supplied to the recess or pocket. If the supply pressure is sufficient, the load will be lifted and floated on a fluid film. At equilibrium conditions, the pressure across the pad will vary approximately as shown. The total force developed by the pressure in the pocket and across the lands will be such that the total upward force is equal to the applied load.

The clearance space and the oil film thickness will be such that all the oil supplied to the bearing can flow through the clearance spaces under the pressure conditions prevailing.

4. Squeeze Films

As discussed in connection with EHL films (Section 1.A.2), as the applied pressure on an oil increases, the viscosity of the oil increases. This fact contributed to the formation of what are called squeeze films.

The principle of the squeeze film is shown in Figure 8.10, where application of a load causes plate *A* to move toward stationary plate *B*. As pressure develops in the oil layer, the oil starts to flow away from the area. However, the increase of pressure also causes an increase in the oil viscosity, so the oil cannot escape as rapidly and a heavy load can be supported for a short time. Sooner or later, if load continues to be applied, most of the oil will flow or be forced from between the surfaces, and metal-to-metal

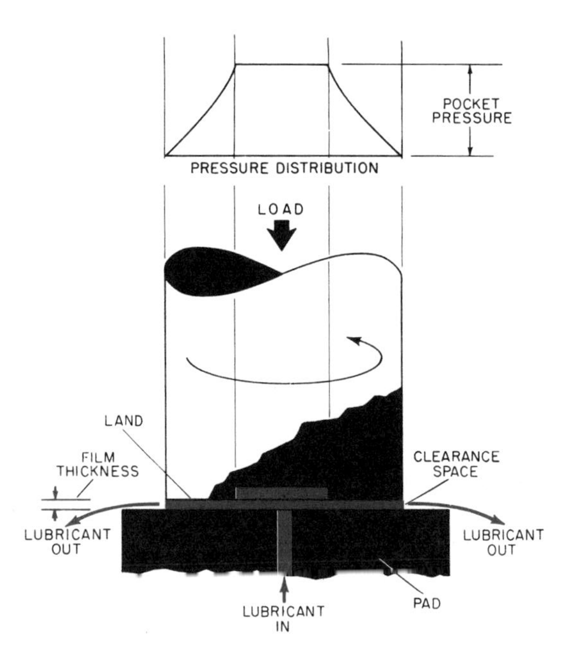

Figure 8.9 Simple hydrostatic thrust bearing.

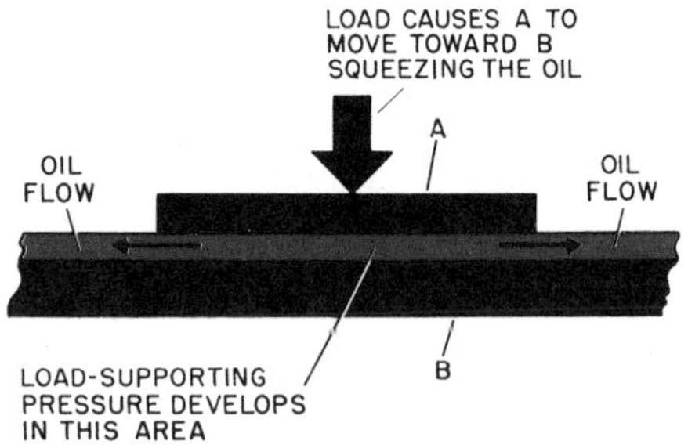

Figure 8.10 Squeeze film principle.

contact will occur. For short periods, however, such a lubricating film can support very heavy loads.

One application in which squeeze films are formed is found in piston pin bushings. At the left in Figure 8.11, the load is downward on the pin and the squeeze film develops at the bottom. Before the film is squeezed so thin that contact can occur, the load reverses (right view) and the squeeze film develops at the top. The bearing oscillates with respect to the pin, but this motion probably does not contribute much to film formation by hydrodynamic action. Nevertheless, bearings of this type have high load-carrying capacity.

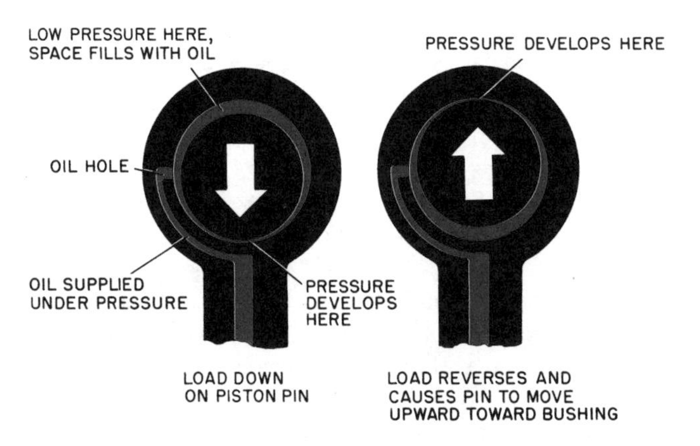

Figure 8.11 Squeeze film in piston pin bushing.

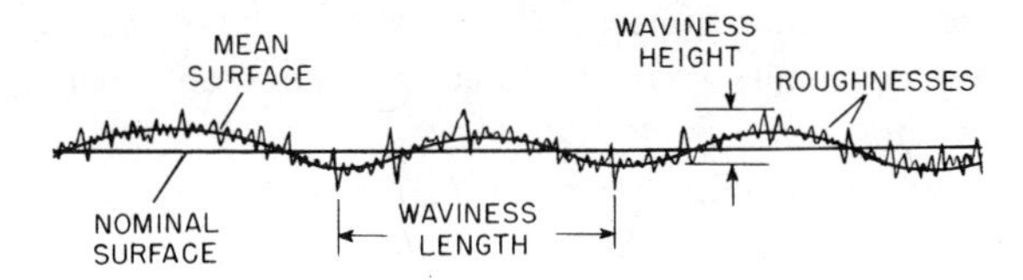

Figure 8.12 Magnified surface profile.

B. Thin Films

A copious, continuous supply of lubricant is necessary to maintain fluid films. In many cases, it is not practical or possible to provide such an amount of lubricant to machine elements. In other cases (e.g., during starting of a hydrodynamic film bearing), loads and speeds are such that fluid film cannot be maintained. Under these conditions, lubrication is by the so-called thin films.

When surfaces run together under thin film conditions, enough oil is often present to permit part of the load to be carried by fluid films and part by contact between surfaces. This condition is often called *mixed film lubrication.* With less oil present, or with higher loads, a point is reached at which fluid oil plays less of a role in separating the metal surfaces. This condition is often called *boundary lubrication.*

1. Nature of Surfaces

Machined surfaces are not smooth but are wavy because of inherent characteristics of the machine tools used. Minute roughnesses—''hills and valleys''—are superimposed on these waves, somewhat as shown in Figure 8.12. The surfaces are not clean even when freshly machined, but are coated with incidental films (moisture, oxides, cutting fluids, rust preventives) and, in service, with lubricating films. These films must be pierced or rubbed away before contact between clean surfaces can occur.

2. Surface Contact

When rubbing contact is made between the surface peaks, known as asperities, a number of actions take place, as shown in Figure 8.13, which represents a highly magnified contact area of a bearing and journal. These actions are as follows.

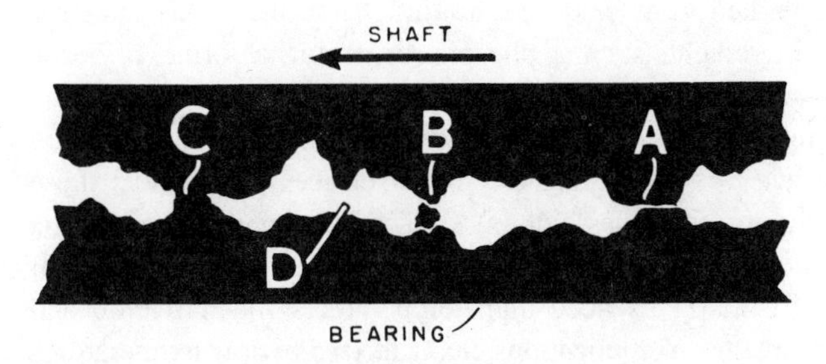

Figure 8.13 Actions involved in surface contacts.

1. There is a heavy rubbing as at *A;* surface films are sheared and elastic or plastic deformation occurs. Real (vs. apparent) areas of contact are extremely small, and unit stresses will be very high.
2. The harder shaft material plows through the softer bearing material *B,* breaking off wear particles and creating new roughness.
3. Some areas *C* are rubbed or sheared clean, and the clean surfaces weld together. The minute welds break immediately as motion continues, but depending on the strength of the welds, the break may occur at another section, with the result that metal is transferred from one member to the other. New roughnesses are formed, some to be plowed off to form wear particles.

These actions account for friction and wear under boundary lubrication conditions. Under mixed film conditions, the force necessary to shear the oil films *D* also is part of the frictional force. However, the presence of partial oil films reduces friction and wear that result from the other actions.

The actions involved in rubbing contact result in the development of very high temperatures in the minute areas of contact. These local temperatures, which are of short duration, are known as flash temperatures. With light loads and low speeds, the small total amount of heat is conducted away through the surfaces without excessive rise in the general temperatures of the surfaces. However, with higher loads or speeds, or both, the number of contacts in a given area may increase to the point at which the heat cannot be dissipated fast enough, and the surfaces then may run hot.

Under *mixed film* conditions, the moving surfaces are kept apart by very thin films of lubricant. These thin films are subjected to high shear conditions, which can lead to increased temperatures, thereby reducing the effective viscosity of the oil. Quite often, owing to the increase in temperature, excessive loads, shock loads, slower speeds, stop and start operation or break-in of new components, the fluid films collapse and contact of the surfaces occurs. The collapse of the *mixed film* conditions result in *boundary lubrication.*

To help reduce the friction and wear under boundary conditions, it may be necessary to increase the viscosity or to use an oil with additives that provide special properties. These additives include film strength agents and extreme pressure (EP) additives. The film strength agents, sometimes referred to as oiliness or lubricity additives, attach themselves to the metal surfaces by polar attraction. This results in tenacious films that help separate the metal surfaces under moderate boundary conditions. Mild antiwear additives also help under moderate conditions by chemically attaching themselves to the metal surfaces.

As boundary lubrication conditions become more severe, increased additive activity may be needed to prevent accelerated wear or destruction of the surfaces. EP additives composed of chemically active ingredients such as phosphorus, sulfur, chlorine, or boron are needed for the more severe conditions. The EP additives work by chemically reacting with the rolling and sliding surfaces to form a strong surface film. The surfaces in motion then tend to wear away the surface film rather than the metal surfaces. Even under these conditions, shearing of the high spots (asperities) coming into contact may still occur. The EP additives also reduce the potential for welding or fusing of these sheared surfaces. It is important to recognize that EP additives need high temperatures for activation and therefore do not provide any advantages in applications characterized by low temperatures or surfaces separated by full fluid films.

C. Solid or Dry Films

In many applications, oils or greases cannot be used because of difficulties in applying them, sealing problems, or other factors such as environmental conditions. A number of more or less permanently bonded lubricating films have been developed to reduce friction and wear in applications of this type. The solid or dry film lubricants reduce the effective surface roughness. They attach themselves to the metal surfaces through rubbing action or by chemical reaction. This effect is dependent on load, temperature, and types of material.

The simplest type of solid lubricating film is formed when a low friction, solid lubricant such as molybdenum disulfide is suspended in a carrier and applied more or less in the manner of a normal lubricant. The carrier may be a volatile solvent, grease, or any of several other types of material. When the carrier is squeezed or evaporates from the surfaces, a layer of molybdenum disulfide remains to provide lubrication.

Solid lubricants are also bonded to rubbing surfaces with resins of various types, which cure to form tightly adhering coatings with good frictional properties. In the case of some plastic bearings, the solid lubricant is sometimes incorporated in the plastic. This also occurs with some sintered metal bearings. During operation, some of the solid lubricant may then be transferred to form a lubricating coating on the mating surface.

In addition to molybdenum disulfide, polytetrafluoroethylene (PTFE), graphite, polyethylene, and a number of other materials are used to form solid films. In some cases, combinations of several materials, each contributing some characteristics to the film, are used.

II. PLAIN BEARINGS

The simplest types of plain bearing are shown in Figure 8.14. The bearing may be only a hole in a block (left); it may be split to facilitate assembly (center); or in some cases, where the load to be carried is always in one direction, the bearing may consist of only a segment of a block (right). The part of the shaft within a bearing is called the journal, and plain bearings are often called journal bearings.

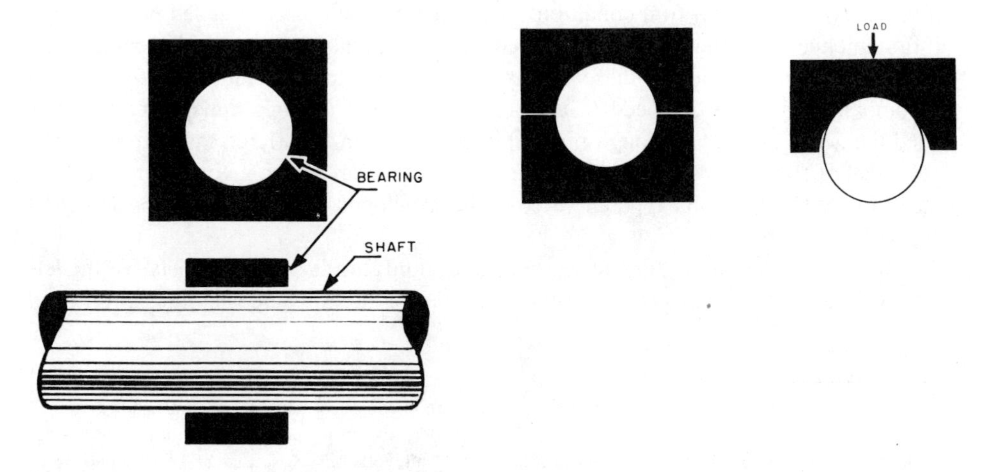

Figure 8.14 Elementary plain bearings.

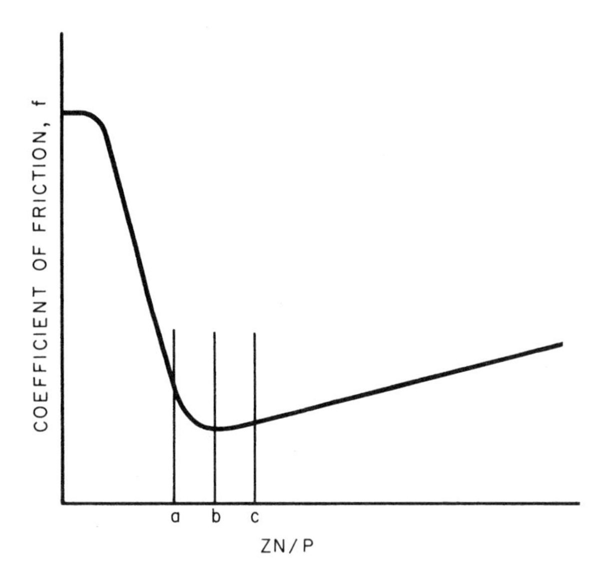

Figure 8.15 Effect of viscosity, speed, and load on bearing friction. For each hydrodynamic film bearing, there is a characteristic relation, such as that shown, between the coefficient of friction and the factor combining viscosity (Z), speed (N), and load (P).

Plain bearings are designed for either fluid film lubrication or thin film lubrication. Most fluid film bearings are designed for hydrodynamic lubrication, but increasing numbers of bearings for special applications are being designed for hydrostatic lubrication.

A. Hydrodynamic Lubrication

The primary requirement for hydrodynamic lubrication is that oil of correct viscosity and sufficient quantity be present at all times to flood the clearance spaces.

The oil wedge formed in a hydrodynamic bearing is a function of speed, load, and oil viscosity. Under fluid film conditions, an increase in viscosity or speed increases the oil film thickness and the coefficient of friction, while an increase in load decreases them. The separate consideration of these effects presents a complex picture that is simplified by combining viscosity Z, speed N, and unit load P, into a single dimensionless factor called the ZN/P factor.* Although no simple equation can be offered that expresses the coefficient of friction in terms of ZN/P, the relationship can be shown by a curve such as that in Fig 8-15. A similar type curve could be developed experimentally for any fluid film bearing.

In Figure 8.15, in the zone to the right of c, fluid film lubrication exists. To the left of a, boundary lubrication exists. In this latter zone, conditions are such that a full fluid

* The expression ZN/P is dimensionless when all quantities are in consistent units: for example, Z in poises, N in revolutions per second, and P in dynes per square centimeter; or Z in pascal-seconds, N in revolutions per second (reciprocal seconds), and P in pascals; or Z in reyns, N in revolutions per second, and P in pounds-force per square inch.

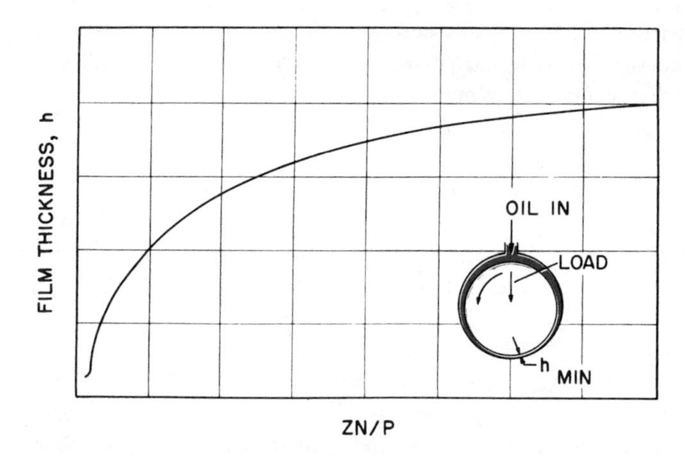

Figure 8.16 Effect of viscosity, speed, and load on film thickness.

film cannot be formed, some metallic friction and wear commonly occur, and very high coefficients of friction may be reached.

The portion of the curve between points *a* and *c* is a mixed film zone including the minimum value of *f* corresponding to the *ZN/P* value indicated by *b*. From the point of view of low friction, it would be desirable to operate with *ZN/P* between *b* and *c,* but in this zone any slight disturbance such as a momentary shock load or reduction in speed might result in film rupture. Consequently, good practice is to design with a reasonable factor of safety so that the operating value of *ZN/P* is in the zone to the right of *c*.* The ratio of the operating *ZN/P* to the value of *ZN/P* for the minimum coefficient of friction (point *b*) is called the bearing safety factor. Common practice is to use a bearing safety factor on the order of 5.

In an operating bearing, if it becomes necessary to increase the speed, *ZN/P* will increase and it may be necessary to decrease the oil viscosity to keep *ZN/P* and the coefficient of friction in the design range. An increase in load will result in a decrease in *ZN/P,* and it may be necessary to increase the oil viscosity to keep *ZN/P* and the coefficient of friction in the design range.

Film thickness can be related to *ZN/P* in the manner shown in Figure 8.16. The curve is typical of large, uniformly loaded, medium speed bearings such as are used in steam turbines. In general, film thickness increases if *ZN/P* is increased—for example, if the load is reduced while the oil viscosity and journal speed remain constant. With a proper bearing safety factor, the film thickness will be such that normal variation in speed, load, and oil viscosity will not result in the reduction of film thickness to the point at which metal-to-metal contact will occur.

---

* Equations, procedures, and data for plain bearing design and performance calculations are available in many technical papers and books. Among the latter are the following: *Bearing Design and Application,* Wilcock and Booser, McGraw-Hill, *Theory and Practice of Lubrication for Engineers,* Fuller, John Wiley & Sons; *Analysis and Lubrication of Bearings,* Shaw and Mack, McGraw-Hill.

The work done against fluid friction results in power loss, and the energy involved is converted to heat. Most of the heat is usually carried away by the lubricating oil, but some of it is dissipated by radiation or conduction from the bearing or journal. The normalized operating temperature is the result of a balance between the heat generated, overcoming fluid friction, and the total heat removal. Certain oils, such as some synthetics, have naturally lower frictional characteristics, which can reduce power requirements.

The effect of increasing temperature is to decrease oil viscosity. The reduction in viscosity results in a lower *ZN/P* and coefficient of friction (provided boundary or mixed film lubrication conditions do not exist). Also, less work is required to overcome fluid friction, less heat is developed, and the temperature tends to decrease. This has a stabilizing influence on bearing temperatures.

In general, if excessive temperatures develop even though load, speed, and oil viscosity are within the correct range, it may be that there is insufficient oil flow for proper cooling. It may then be necessary to provide extra grooving or increase the clearance in order to increase the flow of oil through the bearing.

1. Grease Lubrication

While the grease in a rolling element bearing acts as a two-component system in which the soap serves as a sponge reservoir for the fluid lubricant, greases in plain bearings behave like homogeneous mixtures with unique flow properties. These flow properties are described by the apparent viscosity (see Chapter 4), that is, the observed viscosity under each particular set of shear conditions. As the rate of shear is increased, the apparent viscosity decreases and, at high shear rates, it approaches the viscosity of the fluid lubricant used in the formulation. In many plain bearings, the shear rate in the direction of rotation is high enough to cause the apparent viscosity of a grease to be in the same general range as the viscosities of lubricating oils normally used for hydrodynamic lubrication. As a result, fluid film formation can occur with grease, and it is now believed that some grease-lubricated plain bearings operate on fluid films, at least part of the time. In addition, hydrodynamic film bearings designed for grease lubrication are used in some applications.

The pressure distribution in a grease-lubricated hydrodynamic film bearing is similar to that in an oil-lubricated bearing (Figure 8.5). However, toward the ends of the bearing, because of reduced pressure in the film, the shear stress is lower, the apparent viscosity of the grease remains high, and end leakage is lower. As a result, high pressures are maintained farther out toward the ends of the bearing; moreover, the average pressure in the film is higher, and the maximum pressure is correspondingly lower. The minimum film thickness for the same bearing load and speed will be greater. The coefficient of friction may be equal or less than that with an equivalent oil-lubricated bearing, depending on such factors as the type of grease used and the viscosity of the oil component in the grease.

Fluid film bearings lubricated with grease have some advantages compared to those lubricated with oil. As a result of the lower end leakage, the amount of lubricant required to be fed to the bearing is less, so grease-lubricated bearings can be supplied by an all-loss system with either a slow, continuous feed, or a timed, intermittent feed in conjunction with adequate reservoir capacity in the grooves of the bearing.

When a grease-lubricated bearing is shut down for a period of time with the flow of lubricant shut off, the grease usually does not drain or squeeze out completely. Some grease remains on the bearing surfaces, and thus a fluid film can be established almost immediately when the bearing is restarted. Starting torque and wear during starting may be greatly reduced. During shutdown periods, retained grease also acts as a seal to exclude dirt, dust, water, and other environmental contaminants, and to protect bearing surfaces from rust and corrosion. If the grease provides a lower coefficient of friction, power consumption during operation will also be lower.

When grease lubrication is used for fluid film bearings, the cooling is not as efficient as the cooling obtained from oils. This disadvantage may be partially offset if the coefficient of friction is lower with a grease; if speeds or loads are high, however, it may be a limitation.

B. Hydrostatic Lubrication

In a hydrostatic bearing, the oil feed system used must be such that the pressure available, when distributed across the pocket and land surfaces, is sufficient to support the maximum bearing load that may be applied. The system must also be designed to provide an equilibrium condition for loads below the maximum. Three types of lubricant supply are used to accomplish this-constant volume system, constant pressure system with flow restrictor, and constant pressure system with flow control valve.

1. Constant Volume System

In the first type of system, the pump delivers a constant volume of oil at whatever pressure is necessary to force that volume through the system. That is, if the backpressure increases, the pump pressure automatically increases sufficiently to maintain the flow rate. In most cases, the volume delivered by the pump actually decreases somewhat as the pressure increases, but this has relatively little effect on the way the system operates.

A constant volume system must have adequate pressure capability to support any applied load. Referring to Figure 8.9, when the pump is turned on, oil will flow into the pocket and the pressure will increase until the load is lifted sufficiently to establish a clearance space through which the volume of oil flowing in the system will be discharged. The clearance space and oil film thickness will be functions of the volume of flow in the system, the viscosity of the lubricant, and the applied load.

If the load is then increased, the clearance space and film thickness will decrease, and the pump pressure will have to increase to permit the discharge of the same volume of oil through the reduced clearance space. Only small changes in clearance space and film thickness accompany fairly large variations in load, so the bearing is said to be very ''stiff.''

The disadvantage of the constant volume system is that it does not compensate for variations in the point of application of the load in multiple pocket bearings. In the two-pocket bearings of Figure 8.17, using a constant volume system, if the load is shifted to the right, the runner will tend to tilt. This will decrease the clearance at the right-hand land and increase the clearance at the left-hand land. Oil can then flow more freely out of the left pocket, the pressure in the system will decrease, and the load will sink until metallic contact might occur at the right side. This problem can be compensated for with either of the following systems.

2. Constant Pressure System with Flow Restrictor

A constant pressure system requires an accumulator or manifold to maintain the pressure at a relatively constant value. If this constant pressure is applied to the pockets of the bearing (Figure 8.17) through flow restrictors, such as capillaries or orifices, a compensat-

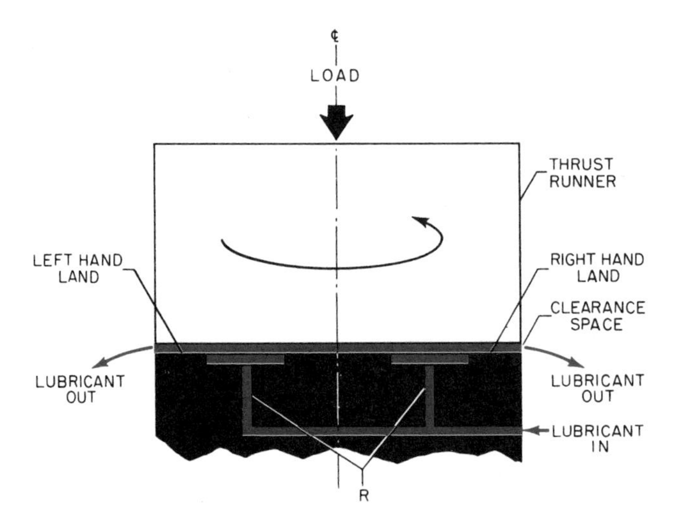

Figure 8.17 Two-pocket hydrostatic thrust bearing. When flow restrictors are used, they are installed at the locations marked *R*.

ing force will be developed when the runner tends to tilt. The pressure drop across a flow restrictor increases as the flow through it increases, and decreases as the flow decreases. Thus, the pressure in the pocket, being the difference between the system pressure and the pressure drop across the restrictor, decreases as flow increases and increases as flow decreases. With this system, if an off-center load causes the runner to tilt toward the right, the clearance at the right-hand land will decrease, the flow to that pocket will decrease, the pressure drop across the restrictor will decrease, and the pocket pressure will increase. The opposite effect will occur in the other pocket. This will tend to lift the load over the right-hand pocket and let it sink over the left-hand pocket, restoring the runner to a more or less coaxial position.

In single-pocket bearings, a constant pressure system with a flow restrictor is less stiff than in a constant volume system. As the load is increased with the constant pressure system, the pocket pressure must increase to support it. This requires that the oil flow decrease to reduce the pressure drop across the flow restrictor. With less flow and higher pressure, a considerably smaller clearance space is required to discharge the oil. However, the compensating feature results in wide use of constant pressure systems in both thrust and journal bearings with multiple pockets.

3. Constant Pressure with Flow Control Valve

If a flow control valve is used as the restrictor in a constant pressure system, the system becomes essentially constant pressure and constant volume. The bearing will have the stiffness of a constant volume system as well as the compensating features of the constant pressure system.

4. Hydrostatic Bearing Applications

One of the more common applications of the hydrostatic principle is in "oil lifts" for starting heavy rotating machines, such as steam turbines, large motors in steel mills, and

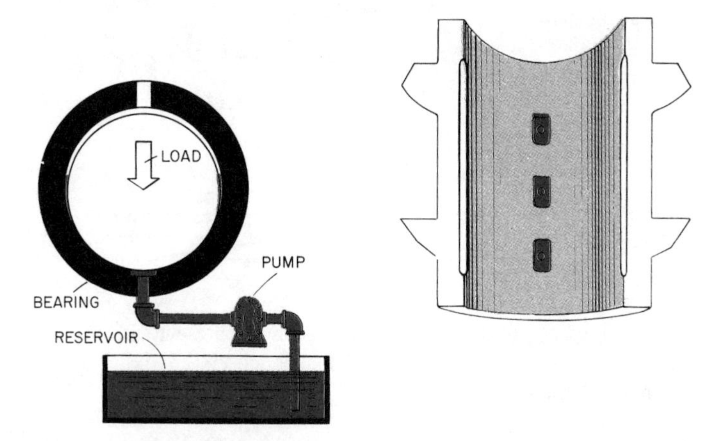

Figure 8.18 Hydrostatic lift; the view at the right shows one type of shallow pocket through which the oil pressure can be applied.

rotary ball and rod mills. Hydrostatic lifts for plain bearings are also used for turning gear operation during start-up and cooldown periods of large steam and gas turbines, where the turbine rotors are rotated at speeds too slow to establish hydrodynamic films. Because metal-to-metal contact exists between the journal and the bearings when the journal is at rest, extremely high torque may be required to start rotation, and damage to the bearings may occur. By feeding oil under pressure into pockets machined in the bottoms of the bearings, the journal can be lifted and floated on fluid films (Figure 8.18). The pockets are generally kept small to prevent serious interference with the hydrodynamic film capacity of the bearings. When the journal reaches a speed sufficient to create hydrodynamic films, the external pressure can be turned off and the bearings will continue to operate in a hydrodynamic manner. The reverse procedure may be used during shutdown.

The low friction characteristics of hydrostatic film bearings at low speeds are being used in a variety of ways. One application is in ''frictionless'' mounts or pivots for dynamometers. Another is in the bearings for tracking telescopes where the relative motion is extremely slow but must be completely free of stick–slip effects. Increasingly, the hydrostatic principle is being applied to the guides and ways of large machine tools, particularly when extremely precise movement and location of the ways is required.

The characteristic of controlled film thickness of hydrostatic film bearings is being used in high speed applications such as machine tool spindles for high precision work. Spindles of this type are equipped with multiple pocket bearings with a constant pressure system and a flow restrictor for each pocket. With this arrangement, any change in the lateral loading on the spindle as a result of a change in the cutting operation is automatically compensated for by changes in the pressures in the individual pockets. Lateral movement of the spindle is thus minimized, and very accurate control of the centering of the spindle in the bearings can be achieved.

C. Thin Film Lubrication

Many bearings are designed to operate on restricted lubricant feeds as the most practical and economic approach. The lubricant supplied to the bearings gradually leaks away and

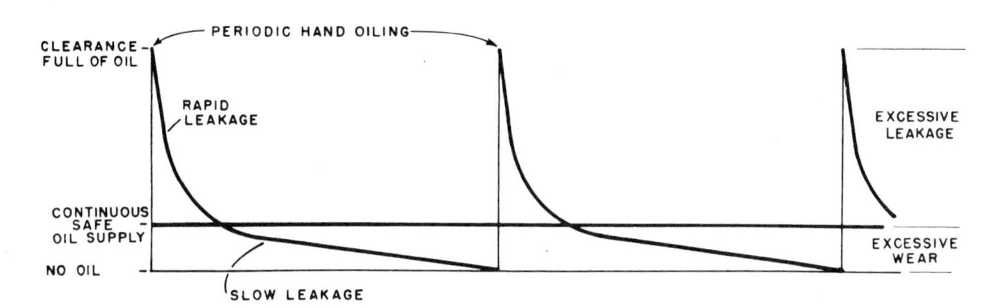

Figure 8.19 Hand oiling: the condition of ''feast or famine'' that is always present with periodic hand oiling is compared with the safe continuous supply of oil that is closely approximated by devices that feed oil frequently in small quantities.

is not reused; thus this type of lubrication is generally referred to as ''all-loss'' lubrication. Because of the restricted supply of lubricant, these bearings operate on thin lubricating films, either of the mixed film or boundary type. The simplest type of all-loss lubrication is hand oiling (see Figure 8.19). Hand oiling results in flooded clearances immediately after lubrication. This condition may permit formation of fluid films for a brief period of time; however, the oil quickly leaks away to an amount less than that considered to be acceptable for safe operation. In short, the bearing passes through the regime of mixed film lubrication and operates much of the time under boundary conditions.

A closer approach to maintaining a safe oil supply may be accomplished with application devices such as wick feed oilers, drop feed cups, waste-packed cups, bottle oilers, and central dispensing systems such as force feed lubricators or oil mist systems. These devices supply oil on either a slow, constant basis or at regular, short intervals. With greases, leakage is not as serious a problem, but the use of centralized lubrication systems will provide a more uniform lubricant supply than grease gun application (see Chapter 8).

Even with regular application of small amounts of lubricant, thin film bearings require proper design and installation, as well as proper lubricant selection to control wear and provide satisfactory service life.

1. Wearing In of Thin Film Bearings

In a new bearing, the journal normally will make contact with the bearing over a fairly narrow area (Figure 8.20, left). Generally lighter loads should be carried by such a bearing under thin film conditions, since unit loads beyond the ability of the oil film to prevent metallic contacts would probably exist. Under favorable conditions, wear will occur, but it will have the effect of widening the contact area (Figure 8.20, right) until the load is distributed over a region so large that wear becomes practically negligible. New plain bearings generally are supplied with a thin ''flashing'' (approximately 0.0005 in.) of a softer material to help facilitate break-in. Under unfavorable conditions, this initial wear may be so rapid that bearing failure occurs.

Large bearings are often fitted prior to operation by hand scraping, or by counterboring the loaded area to the radius of the journal. Fitting of this type can be done only when

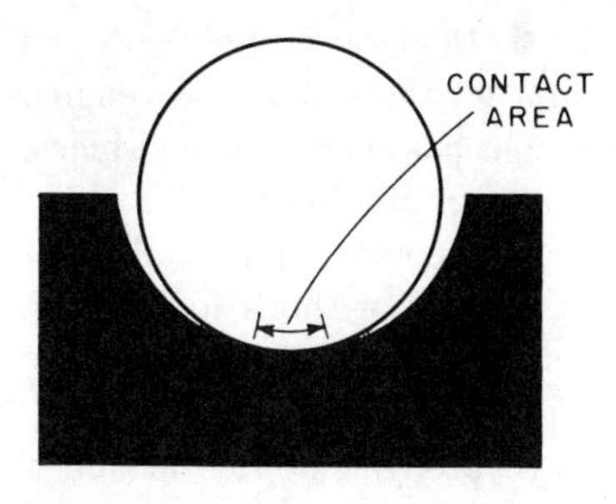

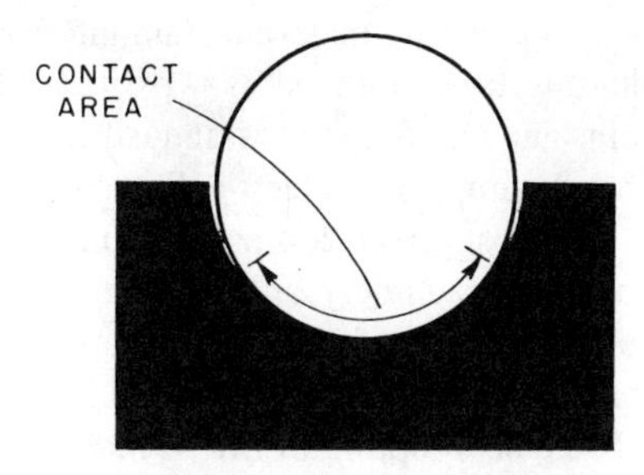

Figure 8.20 Contact area before and after wearing-in of plain thin film bearing.

the direction of loading is known and is constant. After fitting, the area of contact should be in the range of 90–120 degrees.

D. Mechanical Factors

In plain bearings, there are several mechanical factors that affect lubrication. These include the following:

1. Ratio of the length of the bearing to its diameter
2. Projected area of the bearing
3. Clearance between journal and bearing
4. Bearing materials
5. Surface finish
6. Grooving (where used)

1. Length-to-Diameter Ratio

The diameter of a journal is determined by mechanical requirements, which involve the torque transmitted, shaft strength, and shaft rigidity. The bearing length is governed by the load to be carried, the space available, and the operating characteristics of the particular type of machine, especially with regard to retention of bearing alignment and shaft flexing. More and more, space limitations are becoming extremely important in the determination of bearing length.

As a general rule, the shorter the bearing, the higher end leakage will be, and the more difficult it will be to develop load-supporting lubricant films. On the other hand, with too long a bearing, metal-to-metal contact, high friction, high temperatures, and wear may occur at the bearing ends if there is even moderate misalignment or shaft deflection under load. Length-to-diameter (*L*/*D*) ratios of as much as 4 have been used in the past, but the trend for many years has been to lower ratios. Most fluid film bearings now have *L*/*D* values of 1.5 or less; connecting rod bearings for internal combustion engines may have *L*/*D* of 0.75 or less. Thin film bearings have *L*/*D* ratios in about the same range.

2. Projected Area

The axial length of a bearing multiplied by its diameter is called the projected area, and it is common to express unit loads in force per unit of projected area. Established practices pertaining to unit loads on this basis vary considerably for different classes of machinery and different bearing materials, ranging from as low as 15 psi (103 kPa) for lightly loaded line shafting to as high as 5000 psi (24.5 MPa) or more for internal combustion engine crankpins and wristpins. Most industrial bearings carrying constant loads—as in turbines, centrifugal pumps, and electrical machinery—fall in the range of 50–300 psi (345–2700 kPa), with most under 200 psi (1380 kPa). Heavier loads are encountered in bearings of reciprocating machinery and in other bearings subject to varying or shock loads. Peak hydraulic pressures within the oil films (Figure 8.5) are usually three to four times these unit loads based on projected area.

To achieve optimum life in plain bearings, full film (hydrodynamic) lubrication is necessary. Other contributing factors to bearing life are speeds, loads, temperatures, and the compressive strength of the bearing materials. If the compressive strength of the materials used for metallic plain bearings is known, a good rule of thumb to achieve good life is that bearing loads not exceed 33% of the compressive strength of the materials. The limiting load and speed conditions can be expressed as a factor *PV,* with *P* being the pressure on the bearing (psi) multiplied by the surface speed *V* of the shaft (ft/min). The *PV* factor varies by bearing design and materials used. Data on *PV* factors and compressive strengths of materials can be obtained from the bearing manufacturers or, if the materials used in the bearing are known, is readily available technical manuals.

3. Clearance

A full bearing must be slightly larger than its journal to permit assembly, to provide space for a lubricant film, and to accommodate thermal expansion and some degree of misalignment and shaft deflection. This clearance between journal and shaft is specified at room temperature.

One of the principal factors controlling the amount of clearance that must be allowed is the coefficient of thermal expansion of the bearing material. The higher the coefficient of thermal expansion, the more clearance must be allowed to prevent binding as the bearing warms up to operating temperature. Babbitt metals and bearing bronzes have the lowest coefficients of thermal expansion of common bearing materials. Clearances for these materials in general machine practice range from 0.1 to 0.2% of the shaft diameter (0.001–0.002 in. per inch of shaft diameter). Many precision bearings have less clearance than this, while a rough machine bearing may have more. Because of their higher coefficients of thermal expansion, aluminum bearings require somewhat more clearance than babbitt metals or bronzes, and some of the plastic bearing materials require considerably more, in some cases as much as 0.8% of the shaft diameter.

4. Bearing Materials

During normal operation of a fluid film lubricated bearing under constant load, the most important property required in the bearing material is adequate compressive strength for the hydraulic pressures developed in the fluid film. When cyclic loading is involved, as in reciprocating machines, the material should have adequate fatigue strength to operate without developing cracks or surface pits. With shock loading, the material should be of such ductility that neither extrusion nor crumbling occurs. Under boundary lubrication conditions, the material also requires the following:

1. Scoring resistance, requiring appreciable hardness and low shear strength
2. The ability to conform to shaft irregularities and misalignment
3. The ability to embed abrasive particles

If operating temperatures are high, resistance to corrosion and softening may be important.

Although these properties are somewhat conflicting, numerous materials have been developed to obtain satisfactory bearings for the wide range of conditions encountered.

Plain bearing materials most often encountered in industrial machines are bronzes and babbitt metals. Suitable bronzes and babbitt metals are available for practically all conditions of speed, load, and operating temperature encountered in general practice. Steel and cast iron are used for a limited number of purposes, usually involving low speeds or shock loads. There has been considerable growth in the use of plastic and elastomeric materials such as nylon, thermoplastic polyesters, laminated phenolics, polytetrafluoroethylene, and rubber for bearings, particularly in applications where contamination of, or leakage from, oil-lubricated bearings might result in high maintenance costs or short bearing life. Some of these materials can be lubricated with water or water soluble oil emulsions in certain applications. Allowable unit loads for these bearing materials usually are lower, although in a number of cases, filled nylon bearings have been used as direct replacements for bronze bearings.

For internal combustion engines, babbitt metal bearings are made with a very thin layer of babbitt over a backing of copper and/or steel to increase the load carrying capacity. Even then, the loads may be greater than babbitts can handle, so a number of stronger bearing materials have been developed. Aluminum bearings are being used in some diesel and gas engine applications because of their longer potential life and greater resistance to acid attack. Because the aluminum is harder, it will not embed particles as well as the softer bearing materials, and therefore contamination is more critical. Engine bearings are usually fabricated in the form of precision inserts (Figure 8.21), which are interchangeable and require no hand fitting machining at installation.

Precision insert bearings, which are usually constructed of layers of different materials, provide the following:

A thin surface layer (sometimes as little as 0.0003 in., 0.0075 mm) having good surface characteristics—such as low friction, scoring resistance, conformability, and resistance to corrosion

A thicker layer (0.008–0.025 in., 0.2–0.6 mm) of bearing material having adequate compressive strength and hardness, suitable ductility, and good resistance to fatigue

A still thicker (usually 0.05–0.125 in., 1.25–3.2 mm) back or shell of bronze or steel

Some of the more common combinations used with this type of construction are babbitt metal over leaded bronze over steel, lead alloy over copper-lead over steel, silver alloy over lead over steel, and tin over aluminum alloy over steel. These bearings all require smooth hardened journals, rigid shafts and minimum misalignment.

5. Surface Finish

Machined surfaces are never perfectly smooth. The peak-to-valley depth of roughness in machined surfaces ranges from about 160 μin. (4 μm) for carefully turned surfaces to about 60 μin. (1.5 μm) precision-ground surfaces. Finer finishes, approximately 10 μin. (0.25 μm), can be obtained by other commercial methods.

Finely finished surfaces would, in general, be damaged less than rough surfaces by the metal-to-metal contact that occurs under boundary lubrication conditions. However,

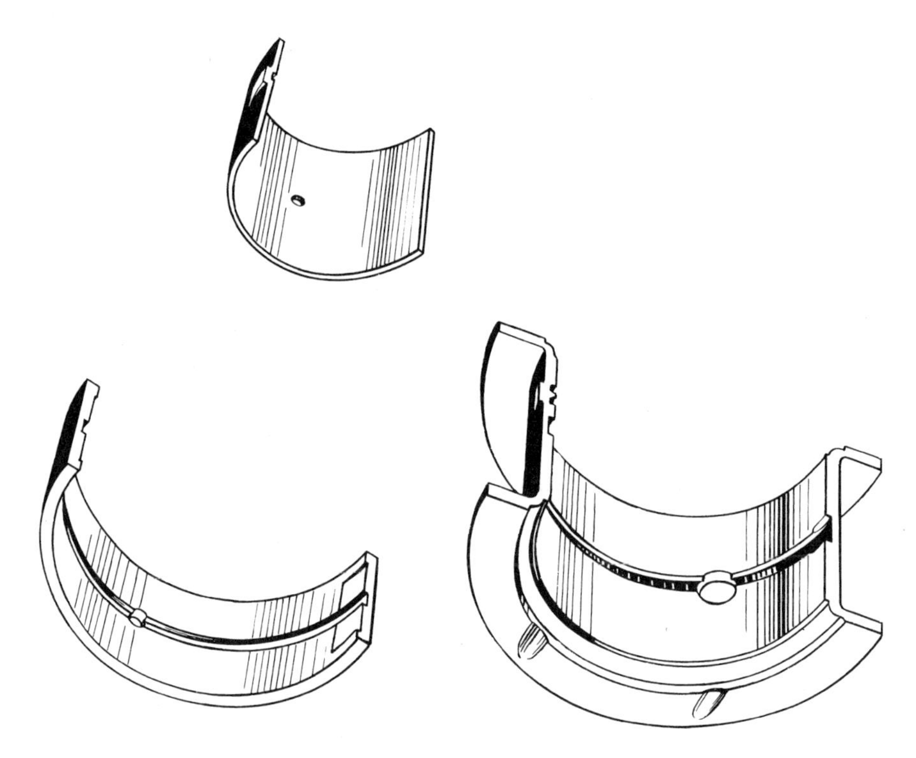

Figure 8.21 Precision insert bearings: typical main and connecting rod insert bearings used in internal combustion engines. Bearing at lower right is designed to carry thrust as well as radial loads.

some degree of "wearing in" of new bearing surfaces always occurs. New surfaces, which may be relatively rough, tend under favorable conditions and careful wearing in to become smoother. In some cases, certain degrees of initial roughness aid this wearing-in process (by holding lubricant in place), as long as loads are minimized for break-in.

Under fluid film conditions, the minimum safe film thickness is a function of the roughness of the surfaces; rougher surfaces requiring thicker films to prevent contact of surface asperities through the film. On the other hand, the finer the surface finish, the lower the minimum safe film thickness, and the less clearance is necessary. Since the film thickness decreases with increases in unit loading, if the minimum safe film thickness is lower as a result of finer surface finishes, the allowable unit loading is higher, all other factors being equal. Conversely, it can be said that bearings designed for high unit loads and small clearances must have finely finished surfaces. Tests show that fluid films may also be formed at a lower speed when starting up a bearing with a smooth finish than when starting one with a rough finish.

6. Grooving

In all plain bearings, some provision must be made to admit the lubricant to the bearing and distribute it over the load-carrying surfaces. Lubricant is generally admitted through an oil port or ports and then distributed by means of grooves cut in the bearing surface. The location of the supply port and the type of grooving used depend on several factors, including the type of supply system, the direction and type of load, and the requirements of the bearings. Certain basic principles apply to all cases.

(a) Grooving for Oil. The distribution of oil pressure in a typical fluid film bearing with steady load is shown in Figure 8.5. Usually, oil should be fed to a bearing of this type at a point in the no-load area where the oil pressure is low. When the shaft is horizontal and the steady load is downward, it is usually convenient to place the supply port at the top of the bearing, as shown.

Generally, grooves should not be extended into the load-carrying area of a fluid film bearing. Grooves in the load-carrying area provide an easy path for oil to flow away from the area. Oil pressure will be relieved and load-carrying capacity will be reduced. This effect for an axial and a circumferential groove is shown in Figures 8.22 and 8.23. However, to provide increased oil flow for better cooling in certain force-feed-lubricated bearings, it is sometimes necessary to extend the grooves through the load-carrying area. With variable load direction, it may also be necessary to extend the grooves through the load-carrying area. This is done in some precision insert bearings for internal combustion engines, mainly to increase cooling and oil distribution.

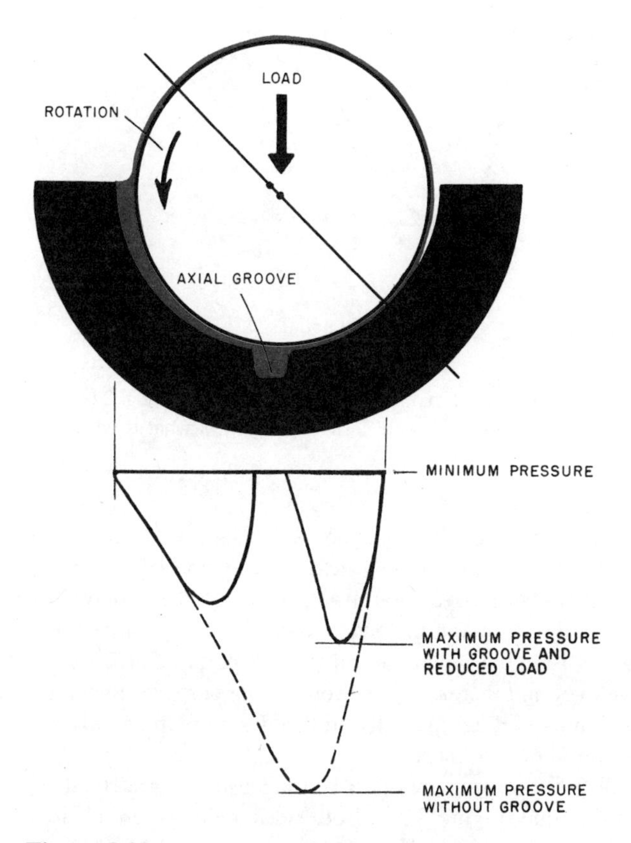

Figure 8.22 Axial groove reduces load-carrying capacity. An axial groove through the pressure area of a fluid film bearing provides an easy path for leakage and relief of oil pressure. Solid lines in the lower sketch represent the approximate pressure distribution when the groove is present; dashed line represents approximately what it would be without the groove.

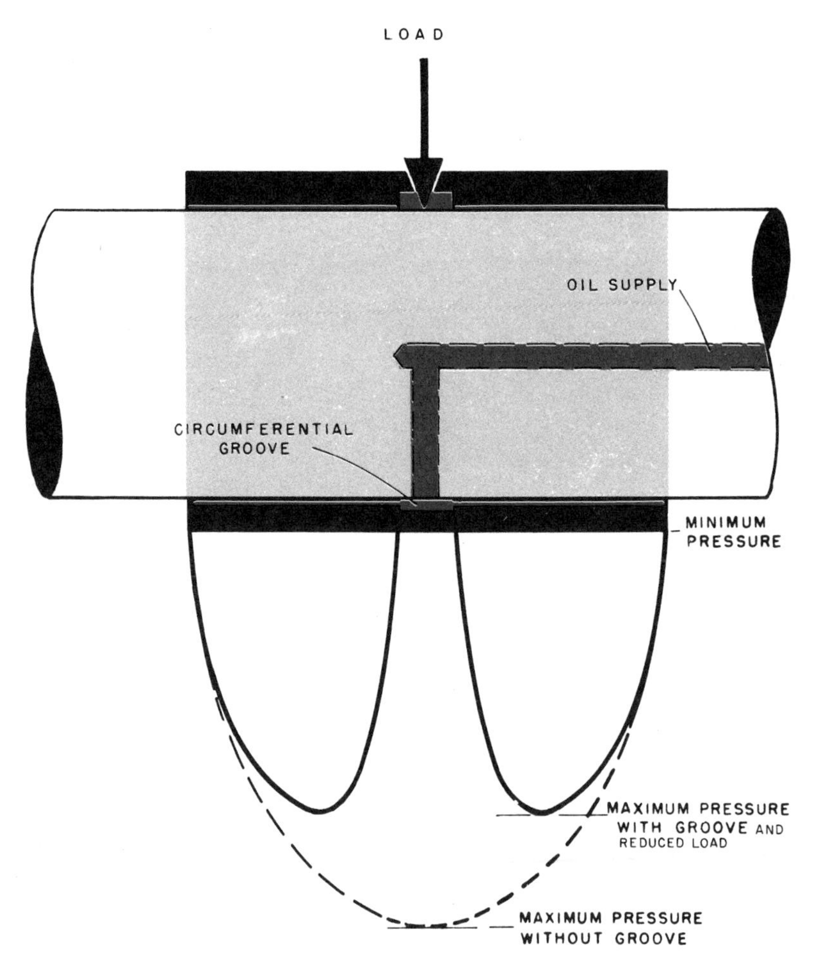

Figure 8.23 Circumferential groove reduces load-carrying capacity. As in Figure 8.22, solid lines show the pressure distribution with the groove present, dashed line shows what it would be without the groove.

With constant load direction, a single oil hole may be sufficient. To increase oil flow or improve distribution, an axial groove cut through the oil supply port (Figure 8.24) often is all that is required. Normally the groove should not extend to the ends of the bearing, since that would allow the oil to flow out the ends rather than being carried into the oil wedge. An exception to this is found when carefully sized ports or orifices are provided at the ends of the groove to permit increased flow for cooling purposes in certain high speed bearings. Also, in medium speed equipment, end bleeder ports frequently are provided to give a continuous flushing of dirt particles.

The grooving needed for distribution in a two-part bearing usually is formed by chamfering both halves at the parting line (Figure 8.25). Both sides are chamfered if it is necessary to provide for rotation in both directions. These chamfers should also stop short of the ends of the bearings. Frequently, oil inlet port and distribution grooves are combined with the chamfer at the split. If a top inlet port is used, an overshot feed groove may be machined in the upper half as shown in Figure 8.26.

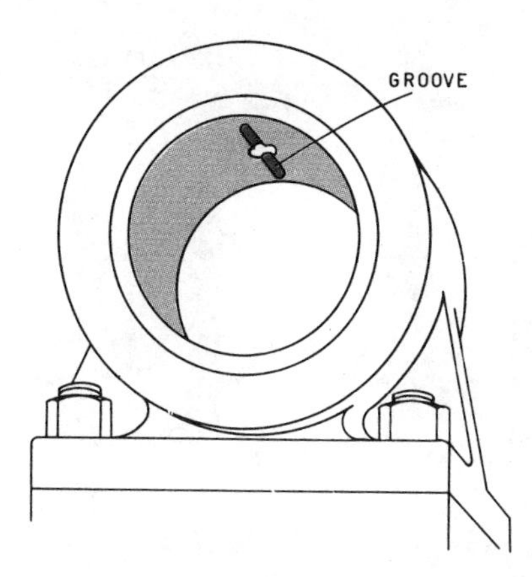

Figure 8.24 Axial distribution groove in one-part bearing.

If a stationary journal and a rotating bearing are used, oil may be fed through a port and axial groove in the journal. Again, the groove should be placed on the no-load side.

Where heavy thrust loads are to be carried, fluid film bearings of the tilting pad or tapered land type are often used. Tilting pad bearings require no grooving, since the oil can readily flow out around the pad mountings. Tapered land bearings require radial grooves located just ahead of the point where the oil wedge is formed. If thrust load is carried by one end face of a journal bearing, the axial groove or chamfers may be extended to the thrust end so that oil will flow directly to the thrust surfaces. The end of the bearing should be rounded or beveled to aid in the flow of oil between the end face and thrust collar or shoulder.

Circumferential grooves are sometimes cut near one or both ends of a bearing to collect end leakage and drain it to the sump or reservoir. This oil might otherwise flow along the shaft and leak through the shaft seals. When collection grooves are used, they mark the effective ends of the bearing.

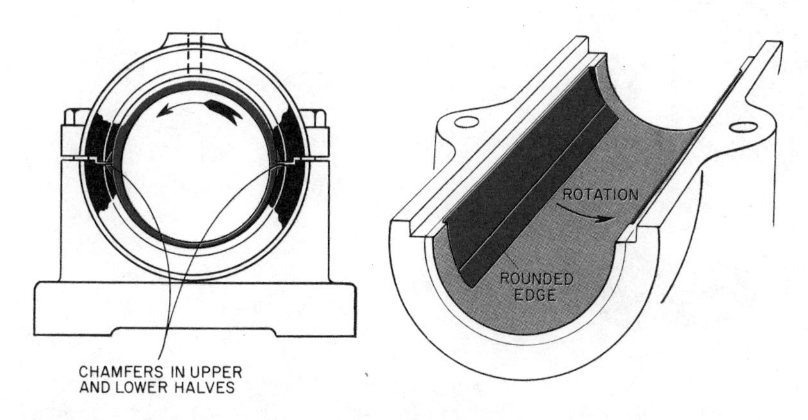

Figure 8.25 Distribution grooves in two-part bearing.

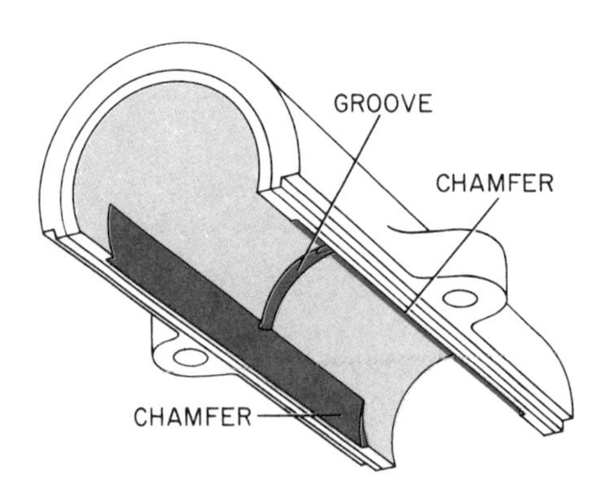

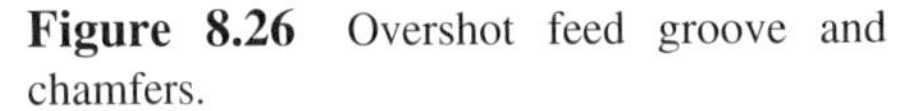
Figure 8.26 Overshot feed groove and chamfers.

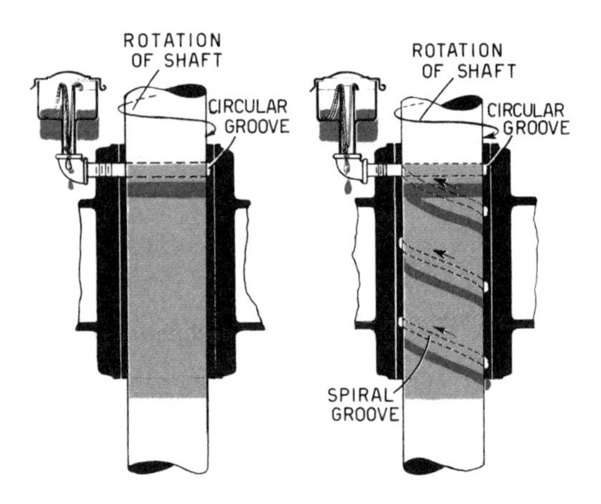

Figure 8.27 Grooving for vertical bearing.

Vertical shaft bearings often require only a single oil port in the upper half of the bearing in the no-load area. In general, the lower the supply pressure, the higher the port should be. Sometimes a circumferential groove may be added near the top of the bearing to improve distribution (Figure 8.27, left). If leakage from the bottom of the bearing is excessive, a spiral groove is sometimes cut in the bearing in the proper direction relative to shaft rotation so that oil will be pumped upward (Figure 8.27, right).

Increased oil flow to cool a hot running bearing can be obtained by simple forms of grooving. An axial groove on the no-load side, for example, will increase oil flow by three to four times compared to a single port alone. Circumferential grooves also increase oil flow, but not as much as an axial groove. They also have the disadvantage of reducing the load-carrying capacity of the bearing. Increased clearance often can be used in lightly loaded, high speed bearings to increase oil flow. When increased clearance might reduce

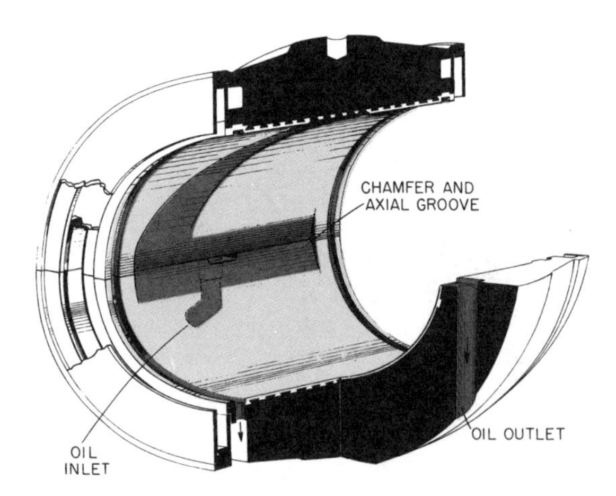

Figure 8.28 Grooving to increase oil flow for cooling: cutaway of a large turbine bearing shows a wide groove cut diagonally in the top (unloaded) half to permit a large flow of oil for cooling purposes. A relatively small part of the oil passing through this bearing would be needed for the fluid film.

load-carrying capacity too much, extra grooving or a clearance relief in the unloaded portion can be used to increase oil flow for cooling (see Figure 8.28).

Where the direction of bearing load changes as in reciprocating machines, it is still essential that oil be fed into an unloaded or lightly loaded area. One way of doing this is with a circumferential groove. While this, in effect, divides the bearing into two shorter bearings of reduced total load-carrying capacity, it may be the most effective alternative. Also, it may be desirable to provide a path for oil flow to other bearings—for example, as in many internal combustion engines (Figure 8.29). An axial groove or chamfer may be used with a circumferential groove to improve oil distribution or to increase oil flow (Figure 8.30).

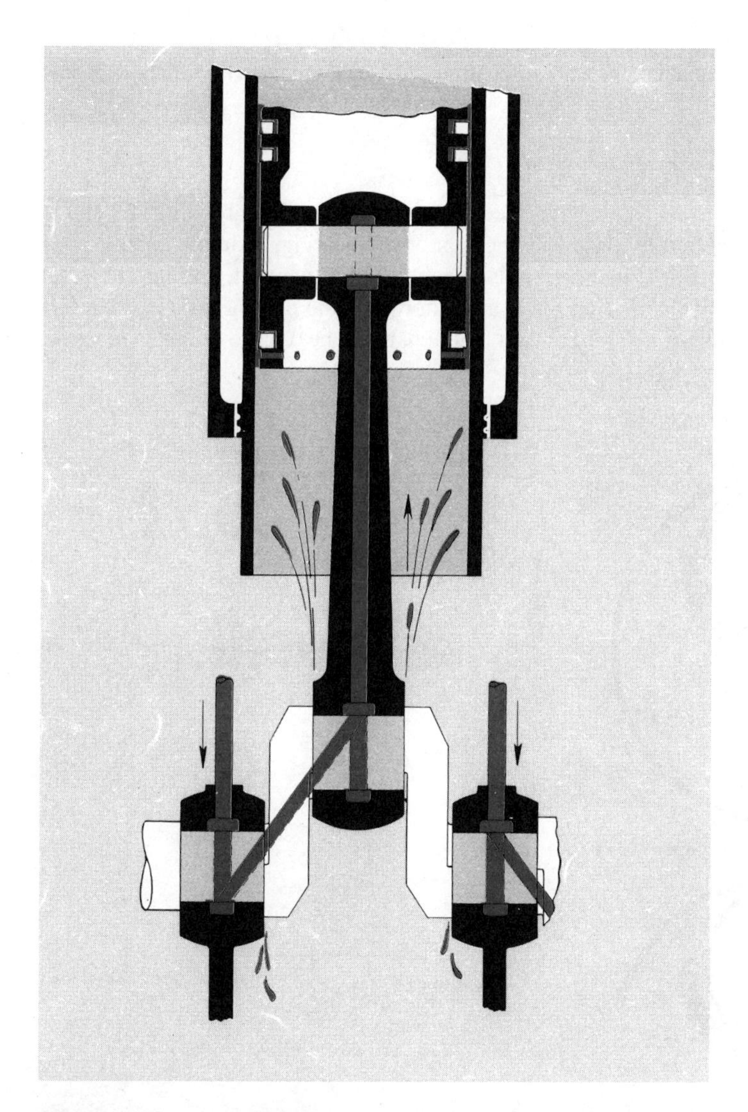

Figure 8.29 Circumferential grooves. In this circulation system an oil pump, driven from the crankshaft, takes oil from the crankcase sump and delivers it under pressure to the crankpin and wrist pin bearings through passages in the crankshaft and connecting rods.

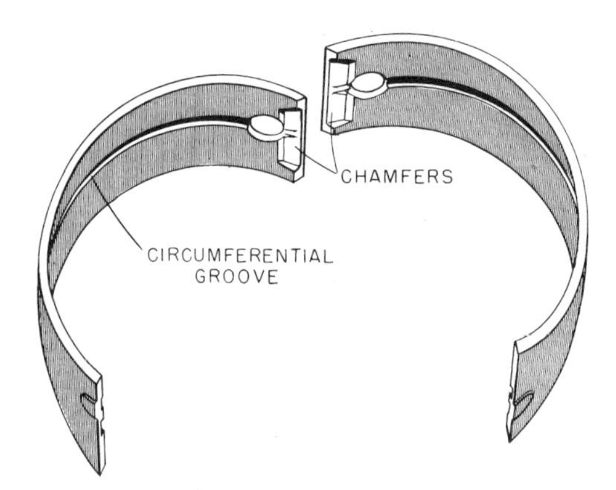

Figure 8.30 Circumferential groove and axial chamfer. An axial groove (also called a spreader groove) is often used with a circumferential groove in precision bearings.

Ring-oiled bearings are a somewhat special case with regard to grooving. When the direction of load is fixed and is toward the bottom of the bearing, a simple axial groove connecting with the ring slot (Figure 8.31) is adequate. In a two-part bearing, the axial groove may be formed by chamfering the bearing halves at the parting line. However, if loads are sideways because of belt pull or gear reaction, this type of groove can be blocked,

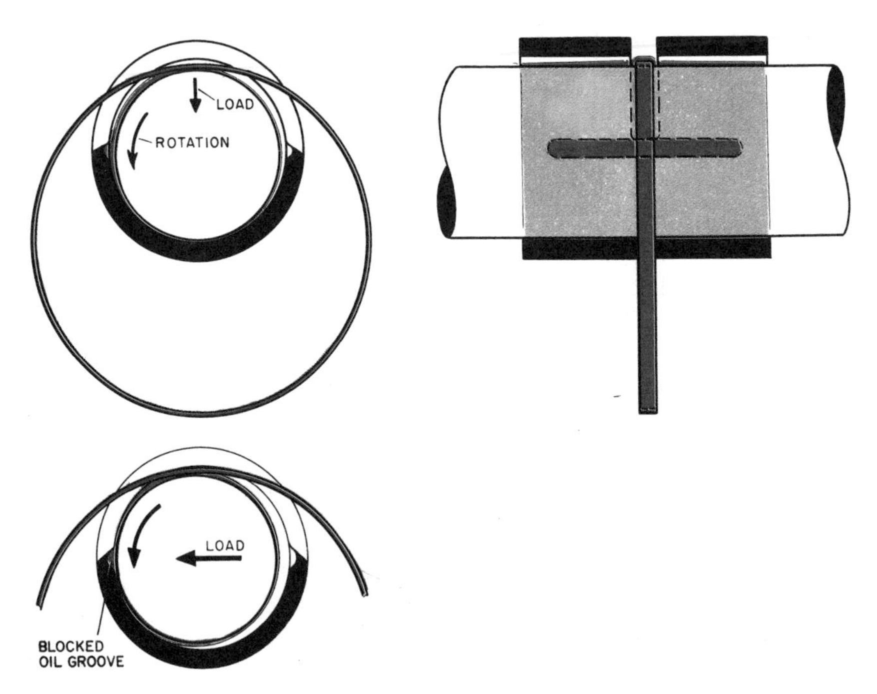

Figure 8.31 Ring-oiled bearing, downward load. A side thrust on the shaft can cause blockage of this type of grooving (lower left).

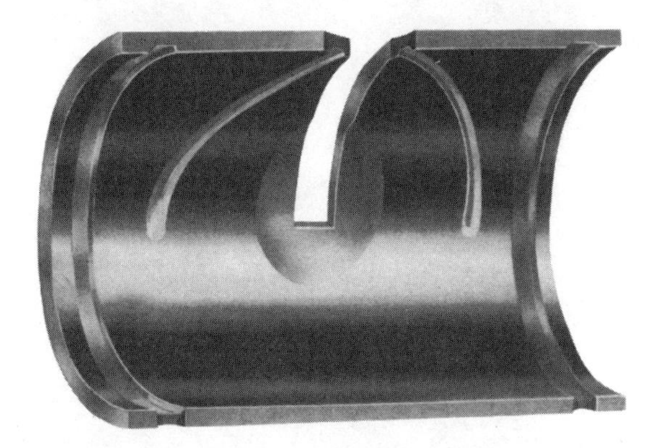

Figure 8.32 Section of ring-oiled bearing with X grooves, which cross in the top of the bearing. The area around the end of the ring slot is relieved to aid in the distribution of oil.

as shown in Figure 8.31. To overcome this problem, grooving such as shown in Figures 8.32 and 8.33 is often used. Grooving of this type is used for electric motors, which may be belted or geared to the load; thus the motor manufacturer does not need to know the contemplated direction of loading and shaft rotation.

Grooving for thin film lubricated bearings generally follows the same pattern seen with for fluid film lubricated bearings. The restricted supply of oil to this type of bearing, however, may necessitate an additional reservoir capacity in the grooves, which in turn would have to be designed to retain and distribute the oil (Figures 8.34–8.36).

(b) Grooving to Grease. Grooving principles for grease are practically the same as for oil. Because grease has high resistance to flow at the low shear rates in the supply system, grooves need to be made wider and deeper than oil grooves. Annular grooves cut near

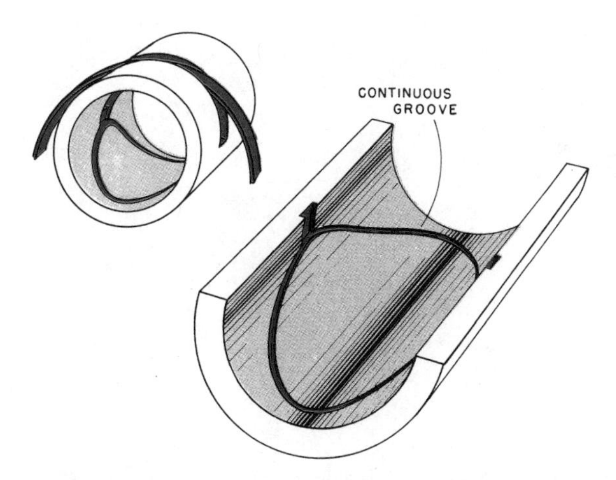

Figure 8.33 Ring-oiled bearing, continuous groove. With this pattern of grooving, oil feed cannot be blocked, and there is little interruption of load-carrying surface along any axial line.

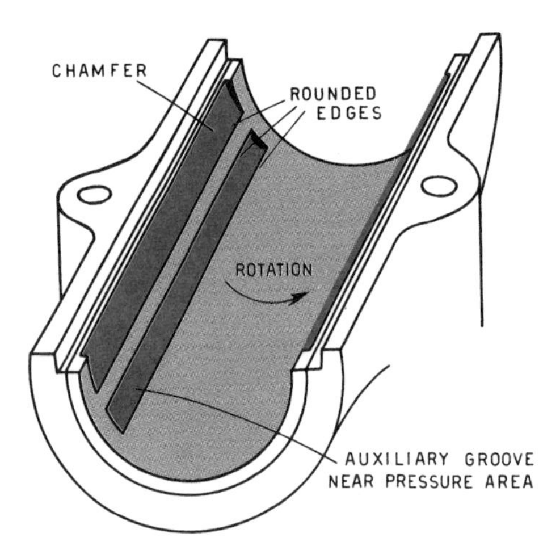

Figure 8.34 Auxiliary groove. When oil application is infrequent, extra storage capacity is sometimes needed in a bearing. This can be provided by means of an auxiliary groove cut just ahead of the load-carrying area.

the ends of the bearing, similar to oil collection grooves but without the drain hole, are often used to reduce grease leakage. Some of the grease forced into these grooves in the loaded zone flows back out in an unloaded area and enters the bearing for reuse.

E. Lubricant Selection

Lubricants for plain bearings must be carefully selected if the bearings are to give long service life with low friction and minimum power losses.

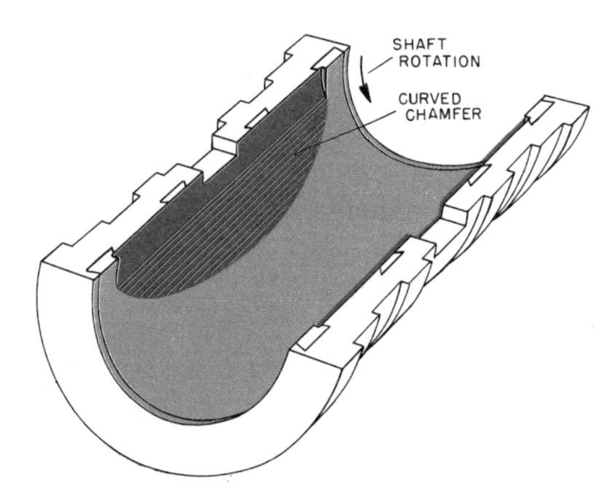

Figure 8.35 Curved chamfer. If clearance at the horizontal centerline is large, end leakage of oil can be reduced by using a curved chamfer, which tends to guide a restricted supply of oil toward the center of the bearing so that it will be drawn into the load-carrying area.

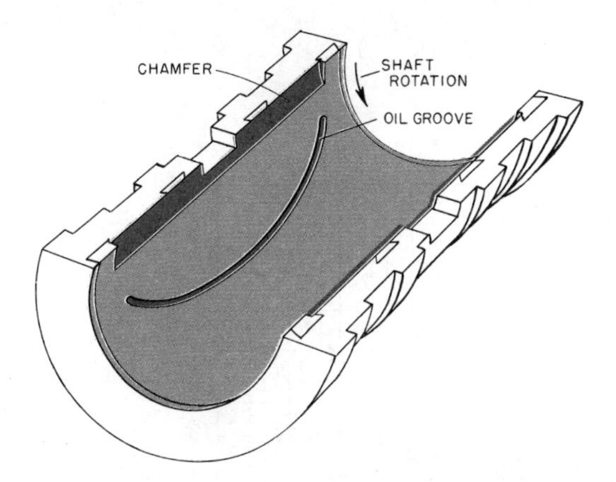

Figure 8.36 Straight chamfer and curved groove. The chamfer provides oil storage capacity and aids distribution. The curved groove catches oil flowing toward the bearing ends and guides it back toward the center of the bearing.

1. Oil Selection

Most of the lubrication systems for fluid film bearings are essentially circulation systems; thus, the oil is reused many times over long periods.

Moisture and other contaminants may enter the system. These conditions require that the oil have the following characteristics:

Good chemical stability to resist oxidation and deposit formation in long-term service
Ability to provide protection against rust and corrosion
Ability to separate readily from water to resist the formation of troublesome emulsions
Ability to resist foaming
Viscosity control in applications where dilution could occur

These properties are generally found in high quality circulation oils.

To select the proper viscosity of oil, three operating factors must be determined: speed, load, and operating temperature (including ambient temperatures at start-up). Speed (rpm) is usually readily determined. Bearing loads may be calculated from the total weight supported by the bearings divided by the total projected area of the bearings. For simplicity, in industrial-type applications, bearing loads of 200 psi (1380 kPa) can often be assumed for bearings operating under light to moderate loads with no shock loads, and 500 psi (3450 kPa) for heavily loaded bearings or where shock loads are present. Operating temperatures can be approximated by using the oil exit temperature of the hottest bearing.

When these three factors are known, the optimum viscosity of oil can be determined from a chart such as that shown in Figure 8.37. Since this chart gives the viscosity at the operating temperature, ASTM viscosity–temperature charts must then be used to determine the desired viscosity at one of the standard measuring temperatures (see discussion of viscosity in Chapter 3).

If bearings are to be started or operated at low temperatures, the pour point and low temperature fluidity of the oil selected must be considered. Synthesized hydrocarbon fluids (SHFs) are often used for low temperature bearing lubrication.

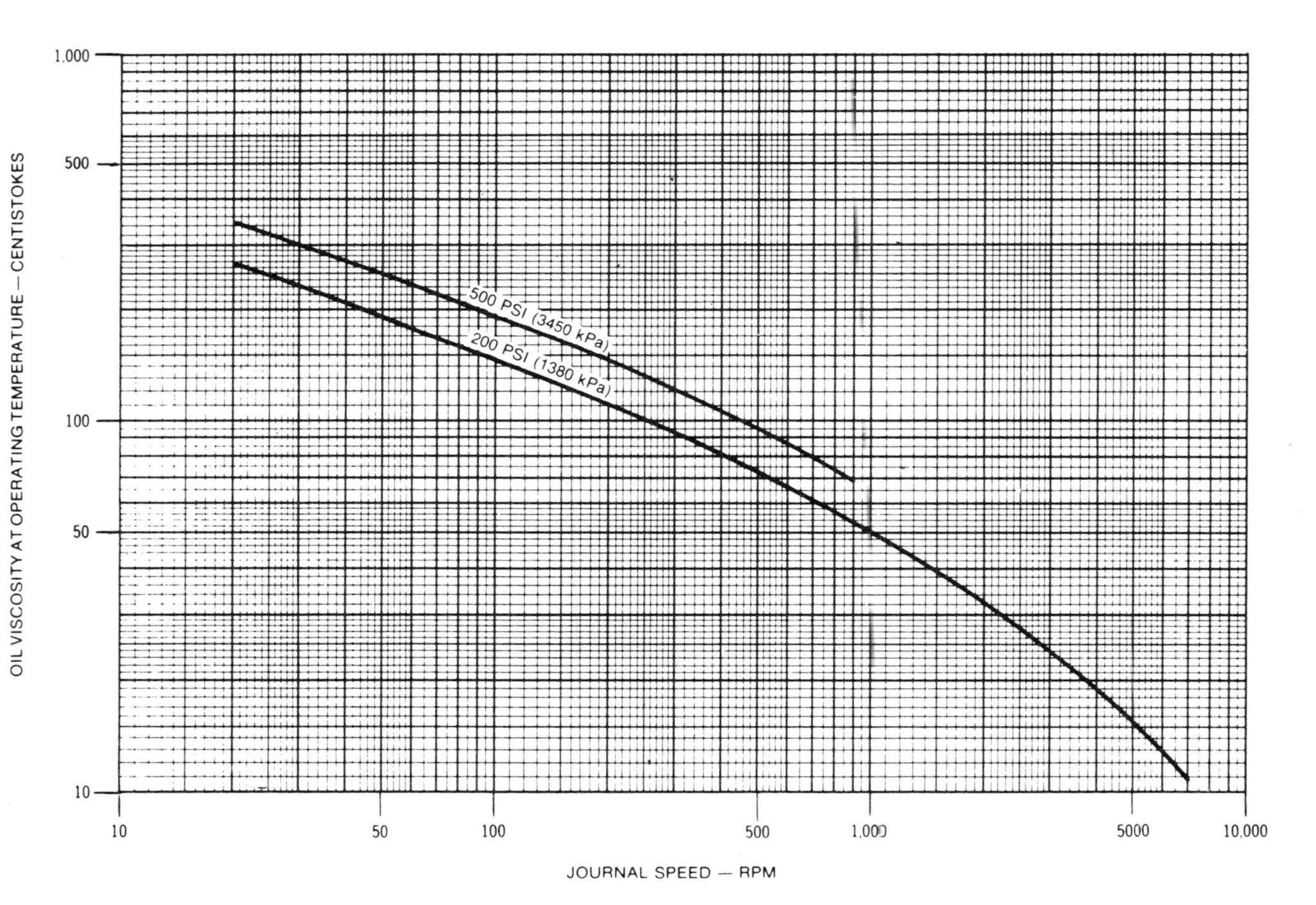

Figure 8.37 Oil viscosity chart for fluid film bearings. To determine the viscosity required, follow up from the appropriate journal speed to either of the lines, then to the left to read the oil viscosity at the operating temperature. The 200 psi line should be used for bearings operating under normal loads without shock loads, and the 500 psi line where loads are heavy or shock loads are present.

In thin film lubricated bearings, the oil is normally not reused, so it is important that the oil selected be readily distributed and form tough, tenacious films on the bearing surfaces. Generally, oils with good film strength and friction reducing properties are preferred.

With thin film lubricated bearings, it may be assumed that loads are high enough for the bearings to operate under boundary conditions at least part of the time. Therefore, only speed and operating temperature must be determined to select the correct viscosity of oil.

Once speed and operating temperature are known, the oil viscosity may be obtained from a chart such as Figure 8.38. Again, the viscosity must be converted to a value at one of the standard reporting temperatures by means of viscosity–temperature charts. Ambient temperatures must also be considered, and so must the type of application device, since some of these devices are limited with respect to the maximum viscosity of oil they will dispense.

2. Grease Selection

The greases used for fluid film and thin film bearings are essentially similar. Enhanced film strength is probably required more frequently in thin film bearings, but where heavy, shock, or vibratory loads are encountered in fluid film bearings, this property is usually required.

Since the methods used to grease lubricate plain bearings are all-loss systems, the grease used is not subjected to long-term service that might cause oxidative breakdown. On the other hand, operating temperatures may be higher in grease-lubricated bearings than in comparable oil-lubricated bearings because of the poorer cooling ability of grease. Thus, there may be exposure to high temperatures while the grease is in the bearings. There is also severe mechanical shearing of the grease, particularly as it passes through the load-carrying zone, which may cause softening and lead to increased end leakage.

The method of application has considerable influence on both the type of grease and the consistency selected for plain bearings. With centralized lubrication systems, the grease must be a type suitable for dispensing through such systems. Where bearings must be lubricated at low ambient temperatures, softer greases or greases with good low temperature properties are usually needed, since they pump more easily at low temperatures.

These requirements generally dictate the use of greases with good mechanical stability, adequate dispensing and pumpability characteristics, corrosion protection properties, adequate film strength, and adequate high temperature characteristics for the operating temperatures. In addition, for fluid film bearings, the apparent viscosity of the grease must be sufficient to permit the formation of fluid films at the shear rates prevailing, but not so high that friction losses will be excessive.

III. ROLLING ELEMENT BEARINGS

The term "rolling element bearings" is used to describe that class of bearings in which the moving surface is separated from the stationary surface by elements such as balls, rollers, or needles that can roll in a controlled manner. These bearings are often referred to as "antifriction" bearings.

The essential parts of a rolling element bearing (Figure 8.39) are a stationary ring (cup or raceway), a rotating ring (cup or raceway), and a number of rolling elements. The inner ring fits the shaft or spindle, and the outer ring fits in a suitable housing. Shaped

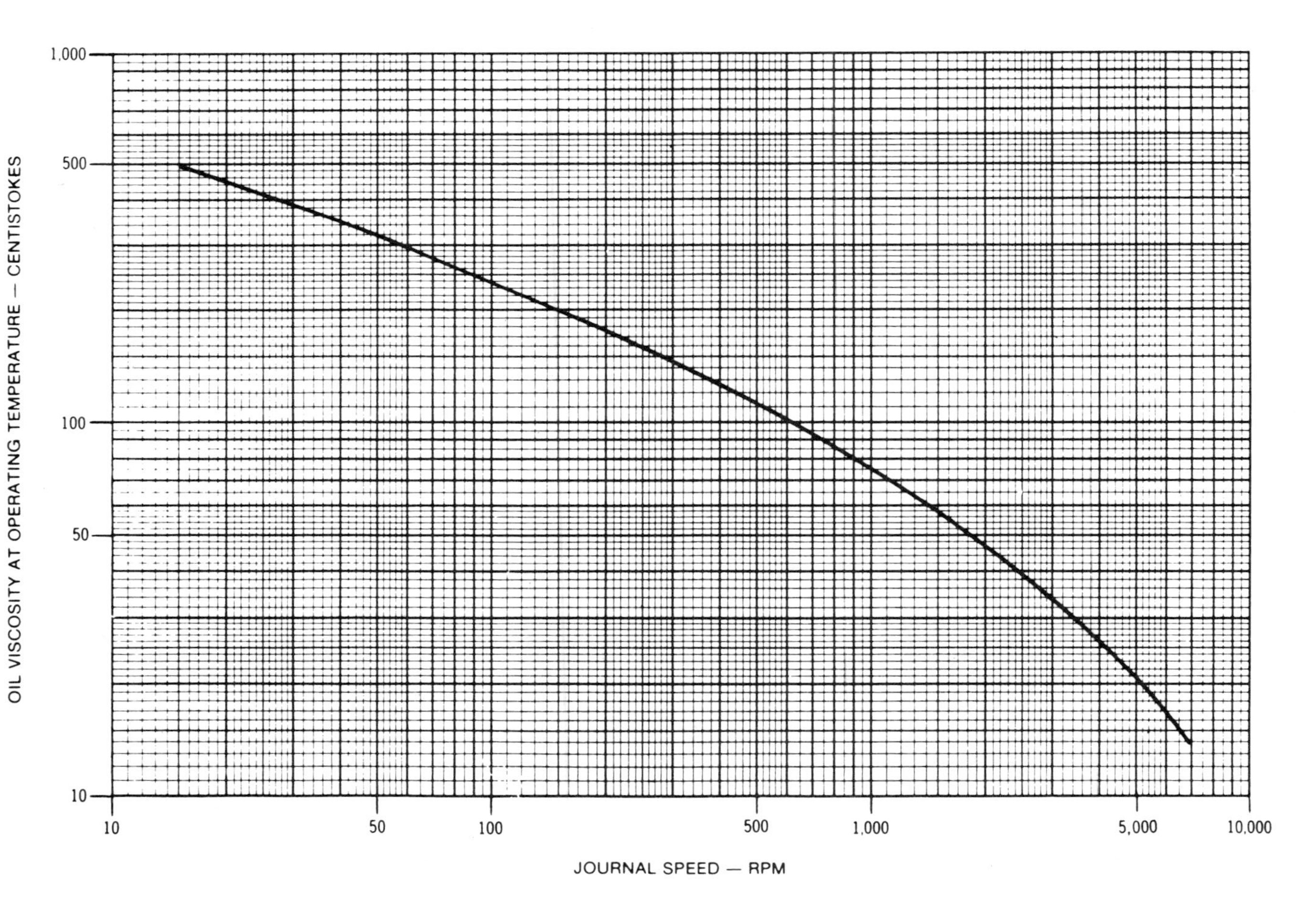

Figure 8.38 Oil viscosity chart for thin film bearings. Since loads can be assumed to be heavy enough for the bearing to operate in a thin film regime, only journal speed and operating temperature affect oil viscosity selection.

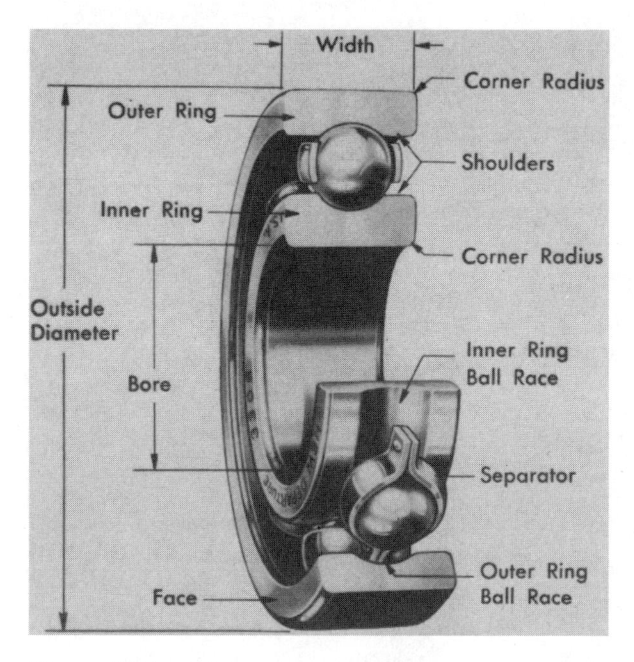

Figure 8.39 Ball bearing nomenclature.

raceways are machined in the rings to confine and guide the rolling elements. These rolling elements are usually held apart from each other, and their relative positions maintained to keep the shaft or spindle centered by the separator, which is also called a cage or retainer. In full-complement bearings, the rolling elements completely fill the space between the rings and no separator is used. Rolling element bearings have been manufactured in sizes ranging from smaller than a pinhead to 18 ft (6 m) or more in outside diameter.

Manufacturers now supply a wide variety of designs of rolling element bearings, of which those in Figures 8.40 and 8.41 represent only some of the more popular types. The bearing in Figure 8.39 is a single-row, deep-groove ball bearing, which is usually the initial choice for any application. One of the other types can be selected when the fatigue life of this design is inadequate, when space is limited, when self-alignment is required, when thrust loads must be carried, or when any of a variety of other conditions must be met. When one of these other types is selected, the desired performance characteristics are very often obtained at the expense of higher cost, a lower speed limit, or more severe lubrication requirements.

When a rolling element bearing is properly lubricated, its load capacity and life are limited primarily by the fatigue strength of the bearing steel. Normal rolling action applies repeated compressive loading in the contact area with stresses up to about 400,000 psi (2.8 GPa). In addition to causing elastic deformation of the rolling element and raceway, this stress induces shearing stress in the steel in a zone approximately 0.002–0.003 in. (0.05–0.075 mm) below the surface. This shearing stress induces fatigue cracks in the steel, which gradually grow and intersect. Small surface areas loosen and break away, forming pits. The actual time required for this to occur depends on many factors, including load, speed, and continuity of service, as well as the fatigue strength of the bearing steel.

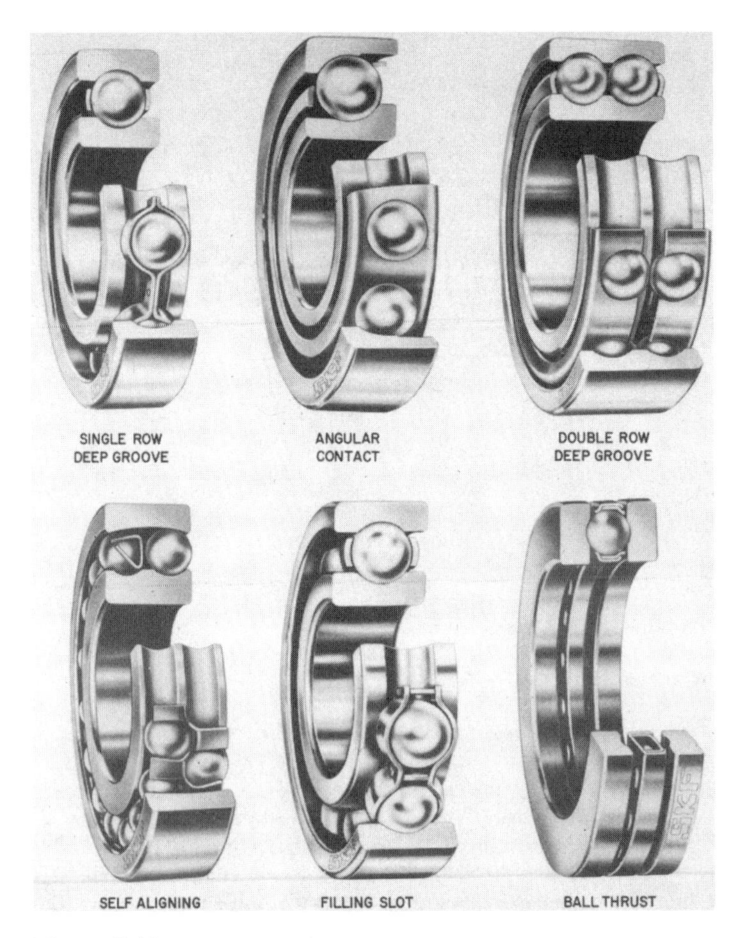

Figure 8.40 Popular types of ball bearing.

As discussed earlier in connection with elastohydrodynamic lubricating films (Section I.A.2), inadequate lubrication can greatly reduce the fatigue life of rolling element bearings. In addition, current research has shown that chemical effects have a considerable influence on fatigue life. The chemical composition of the lubricant additives, the base stock, and the contact surface materials are the major chemical variables. The water content of the atmosphere and the lubricant may also contribute significantly to chemical effects. At operating temperatures below 212°F (100°C), most industrial oils may contain dissolved water. Because the water molecules are small in comparison to oil molecules, they readily diffuse to the tips of the initial microcracks resulting from cyclic stressing. Although the precise mechanism by which this water accelerates fatigue cracking is not clear, there is evidence that the water in the microcracks breaks down and liberates atomic hydrogen, which attaches itself to the metal below the surface. This causes hydrogen embrittlement. Research data indicate that water-induced fatigue can reduce bearing life by 30–80%.

A. Need for Lubrication

Operation of rolling element bearings always results in the development of a certain amount of friction that comes from three main sources:

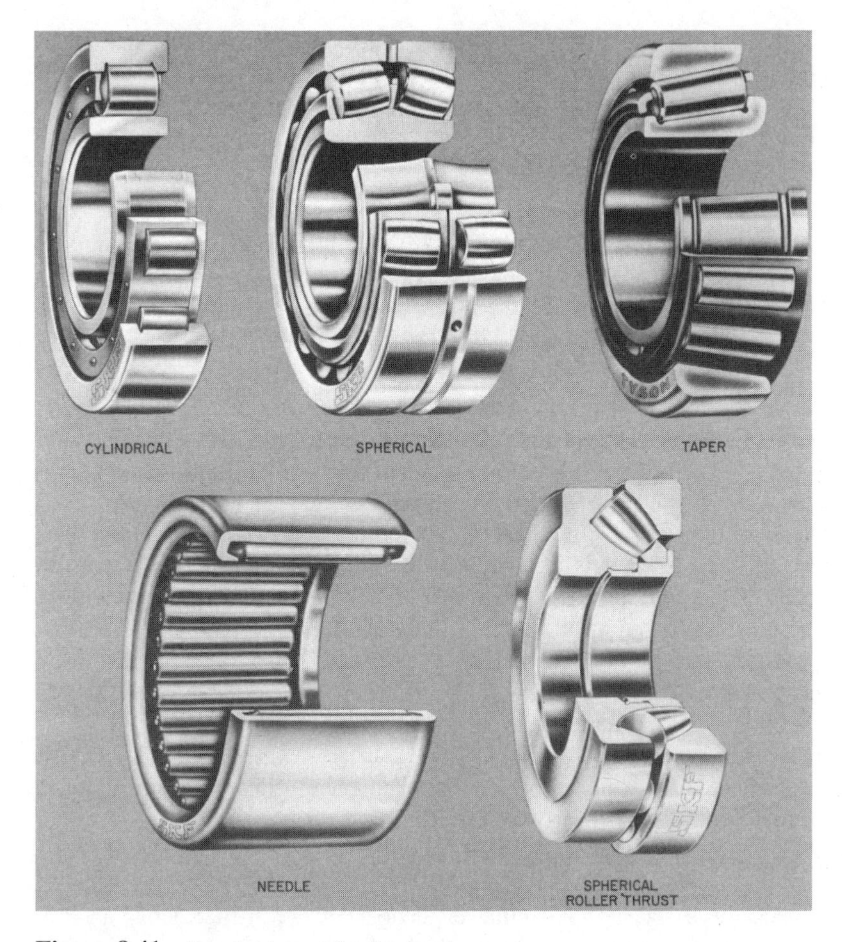

Figure 8.41 Popular types of roller bearing.

1. The fluid friction from shearing of the lubricating films
2. Displacement or churning of lubricant in the path of the rolling elements
3. Sliding or contact between various elements of the bearing

Sliding occurs in the following main areas:

1. Slipping between rolling elements and the raceways (sometimes referred to as skidding). Because of the elastic deformation of both the raceway and the rolling elements, true rolling does not occur and slight sliding results.
2. Sliding between the rolling elements and the separator (cage).
3. Sliding between the end of rollers and the raceway flanges of roller bearings. This type of sliding may be particularly severe in bearings of certain types designed for thrust loads.
4. Sliding between the shaft or spindle and contact-type housing seals.
5. Sliding between adjacent rolling elements in full-complement bearings.

Lubrication aims to maintain suitable films between all these sliding parts—EHL films for the contacts between the rolling elements and raceways, or thin films for other sliding parts. These films must be adequate to minimize friction and protect against wear. In addition, lubrication is expected to protect against rusting or other corrosive effects of contaminants and may provide part of the sealing against contaminants. In circulation systems, the lubricant also acts as a coolant to remove heat generated by friction, or heat conducted to the bearing from outside sources.

B. Factors Affecting Lubrication

Whether oil or grease is used for rolling element bearings depends on a number of factors. For bearings installed in machines that require oil for other elements, the same oil is often supplied to the bearings. In other cases, it may be more convenient to lubricate the bearings separately, and grease is often used. Groups of similar bearings may be lubricated with oil by a circulation system, an oil mist system, or with grease in a centralized lubrication system. Many bearings are now ''packed for life'' with grease by the bearing manufacturer and need no further lubrication in service. The characteristics of these various methods of application have some influence on the oil or grease selected.

The thickness of the elastohydrodynamic films formed in the contact areas between rolling elements and the raceways is a function of the speed at which the surfaces roll together, the load, the oil viscosity, and the operating temperature. The film thickness increases with increases in speed or oil viscosity and decreases with increases in load or operating temperature (since higher temperatures reduce the oil viscosity). For any given set of operating conditions, the oil viscosity (or the viscosity of the oil in a grease) should be selected to provide a safe minimum film thickness.

The lubrication requirements of the other sliding surfaces of a bearing must also be considered in the selection of a lubricant, but in general, the primary consideration in lubrication selection is the requirements of the EHL films.

1. Effect of Speed

The speed at which the surfaces of a rolling element bearing roll together is a fairly complex calculation, so an approximation called the bearing speed factor, nd_m, is usually used. The bearing speed factor is determined by multiplying the rotational speed in revolutions per minute n by the pitch diameter in millimeters d_m. For convenience, the pitch diameter is taken as half the sum of the bearing bore d and the bearing outside diameter D.* That is:

$$nd_m = \frac{n(d + D)}{2}$$

Manufacturers have established maximum bearing speed factors for both oil and grease lubricated bearings. These factors are shown in Table 8.1 for commercial grade bearings. Precision bearings are rated 5–50% higher. The higher speed factors allowed for oil lubrication generally reflect the better cooling available with oil, and the lower fluid friction of low viscosity oils.

---

* The older DN factor (bearing bore in millimeters, D, multiplied by the speed in rpm, N) is still found in some references, although it has not been officially used by the bearing manufacturers' associations for many years. It is not a satisfactory factor to use because for a given bearing bore there are bearings with considerable variations in outside diameter and size of the rolling elements. Thus, for any given bore diameter there can be considerable variation in the speeds with which the elements roll together.

Table 8.1 Bearing Speed Factors nd_m

Bearing type	Oil lubricated[a]	Grease lubricated
Radial ball bearings	500,000	340,000
Cylindrical roller bearings	500,000	300,000
Spherical roller bearings	290,000	145,000[b]
Thrust–ball and roller bearings	280,000	140,000

[a] Oil lubrication is preferred where heat dissipation is required.
[b] Grease lubrication is not recommended for spherical roller thrust bearings.

For bearings lubricated with grease, the considerations relative to bearing speed are mainly concerned with the physical properties of the grease. At low to moderate speeds, the grease must be soft enough to slump slowly toward the rolling elements, but it must not be so soft that excess material gets into the path of the rolling elements. Excess grease increases shearing friction and can cause high operating temperatures. For high speeds, a relatively stiff grease is sometimes used, but it must not be so stiff that the rolling elements, after cutting a channel through the grease, are unable to pick up and distribute sufficient amounts of grease to provide continuous replenishment of the lubricating films. Moreover, to keep shearing friction low and to prevent leakage from housing seals, the grease must have good resistance to softening due to mechanical shearing.

2. Effect of Load

Under EHL conditions, the effect of load on film thickness is not as great as the effect of speed or oil viscosity. For example, while doubling the speed or oil viscosity might result in a film thickness increase exceeding 50%, doubling the load might result in only about a 10% decrease in film thickness. As a result, under steady load conditions the viscosity of oil can generally be selected on the basis of the bearing speed factor and operating temperature without regard to the load.

When shock or vibratory loads are present, somewhat greater film thickness is necessary to prevent metal-to-metal contacts through the film; thus, higher viscosity oils are usually required. Under severe shock loading conditions, lubricants with enhanced antiwear properties may be desirable.

Pressures between the rolling elements and separators (cages) are not high enough to form EHL films. Additionally, these pressures act more or less continuously on the same spots in the separator pockets. As a result, thin film lubrication exists, and sometimes a higher viscosity oil not only improve the resistance to wiping of the oil films but can reduce wear of the separator, as well. When grease is the lubricant, wear can be reduced only if the lubricant can penetrate the clearances between the rolling elements and separators. This is especially important with machined separators where clearances are small.

In tapered and spherical roller thrust bearings, heavy loads sometimes force the ends of the rollers against the flanges of the raceways with considerable pressure. Severe wiping of the lubricant films and wear of the rollers and flanges may occur. In bearings of these types, a lubricant with enhanced antiwear properties may be required, even though the rollers and raceways can be lubricated satisfactorily with conventional products.

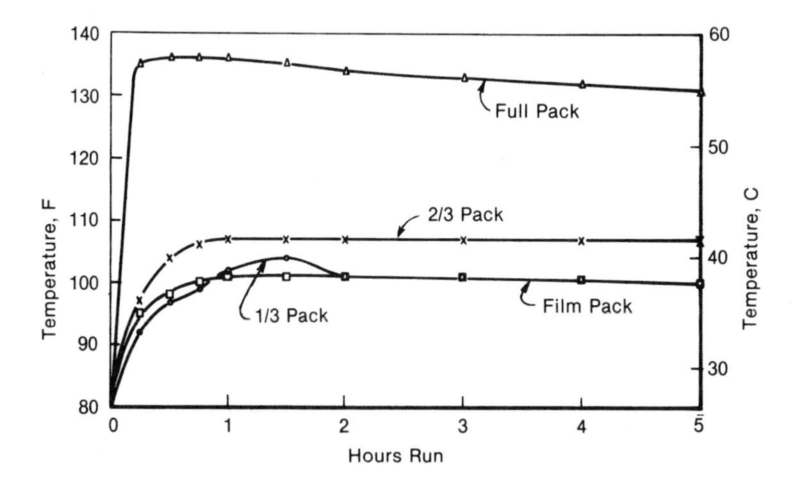

Figure 8.42 Temperature rise in grease-lubricated bearings. As shown by the curves, the final operating temperature of a bearing is a function of the amount of grease packed in it.

3. Effect of Temperature

Since both viscosity and grease consistency are functions of temperature, the operating temperature of bearings must always be considered during selection of lubricants. Bearing operating temperatures may be increased above normal by heat that is conducted to the bearing from a hot shaft or spindle, or by heat radiated to the housing from a hot surrounding atmosphere. Excessive churning of grease resulting from overfilling can also raise bearing temperatures (Figure 8.42). Higher than normal temperatures reduce oil viscosity, whereupon film thickness may decrease below a safe level, and grease may soften so much that excessive churning and further frictional heating occur.

High temperatures also increase the rate at which both oil and grease deteriorate due to oxidation. Oxidation may result in thickening of oil and eventually could lead to deposits that will interfere with oil flow or bearing operation. Oxidation may also lead to deposits with greases, and in severe cases, to hardening such that the grease cannot feed and lubricate.

Under low temperature conditions, the lubricant must be such that the bearing can be started with the power available, and it must distribute sufficiently to prevent excessive wear before frictional heating warms the bearing and lubricant.

4. Contamination

Solid particles of any kind that are trapped between the rolling elements and raceways are the most frequent cause of shortened bearing life. Consequently, dirt should be kept out of bearings as much as possible (including during periods of storage prior to installation), and lubricants should be changed before oxidation has progressed to a point at which deposits begin to form. The use of premium quality, oxidation-inhibited lubricants can greatly extend the period of time that lubricants may be left in service without excessive oxidation.

Water that gets into a bearing tends to reduce the fatigue life and cause rusting that can quickly ruin a bearing. Water, when mixed with some greases, may cause them to soften and perhaps leak from the bearings. Large quantities of water can wash the lubricant out. Sometimes fluids such as acids get into the bearings and cause corrosion. Any of these conditions usually involves special precautions and possibly specialty lubricants.

C. Lubricant Selection

Rolling element bearings generally require high quality lubricants, many of which are specially formulated for the purpose.

1. Oil Selection

Oils for the lubrication of rolling element bearings should have the following characteristics:

1. Excellent resistance to oxidation at operating temperatures, to provide long service without thickening or formation of deposits that would interfere with bearing operation.
2. Proper viscosity at operating speeds and temperatures to protect against friction and wear. Two charts (Figure 8.43) may be used to select the proper viscosity at the operating temperature from the type of bearing, the operating speed, and bearing outside diameter (D). The speed factor (SF) is determined for the speed and type of bearing from Figure 8.43a. The bearing size/speed factor ($BS/SF = SF \times D$) is used in Figure 8.43b to determine the correct oil viscosity for a given specific film thickness (λ). For steady loads, the minimum viscosity line, corresponding to a specific film thickness, $\lambda = 1.5$ (see discussion of EHL in Section I.A.2), will provide adequate film thickness. When shock or vibratory loads are present, higher viscosity and thicker films are desirable. Therefore, the optimum viscosity line, corresponding to a specific film thickness, $\lambda = 4.0$, should be used. ASTM D 341, Standard Viscosity–Temperature Charts for Liquid Petroleum Products, should be used to translate this viscosity to the viscosity at the standard reporting temperatures. EHL calculations should be used for more accurate determination of viscosity requirements.
3. Antirust properties to protect against rust when moisture is present.
4. Good antiwear properties where required because of heavy or shock loads, or where thrust loads cause heavy wiping between the ends of rollers and raceway flanges.
5. Good demulsibility to allow separation of water in circulation systems.

In addition to these characteristics, increasing emphasis is now being placed on the interrelationship between oil characteristics and fatigue. The specific lubricant properties that influence fatigue are not easily defined; but where average bearing life is lower than expected, special lubricants formulated to help minimize fatigue may be desirable. The combination of select base stocks and specific additives can enhance the antifatigue properties of lubricants.

2. Grease Selection

Grease for the lubrication of rolling element bearings should have the following characteristics:

Excellent resistance to oxidation to resist the formation of deposits or hardening that might shorten bearing life. Greases used in ''packed for life'' bearings are expected to provide trouble-free performance for the life of the bearing, assuming normal operating temperatures.

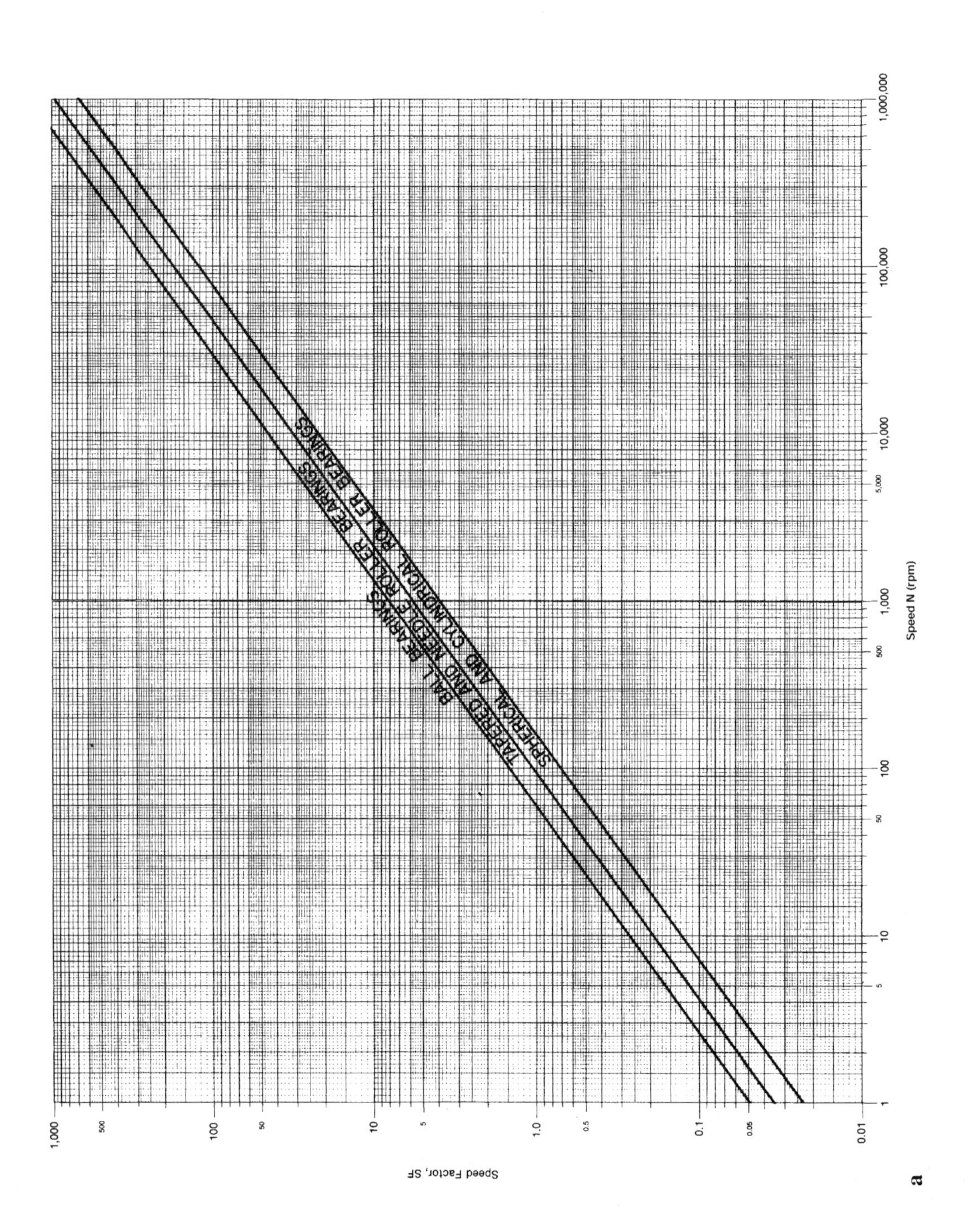
SPHERICAL AND CYLINDRICAL ROLLER BEARINGS
TAPERED AND NEEDLE ROLLER BEARINGS
BALL BEARINGS
Speed N (rpm)
Speed Factor, SF
1,000,000
100,000
10,000
5,000
1,000
500
100
50
10
5
1
1,000
500
100
50
10
5
1.0
0.5
0.1
0.05
0.01

a

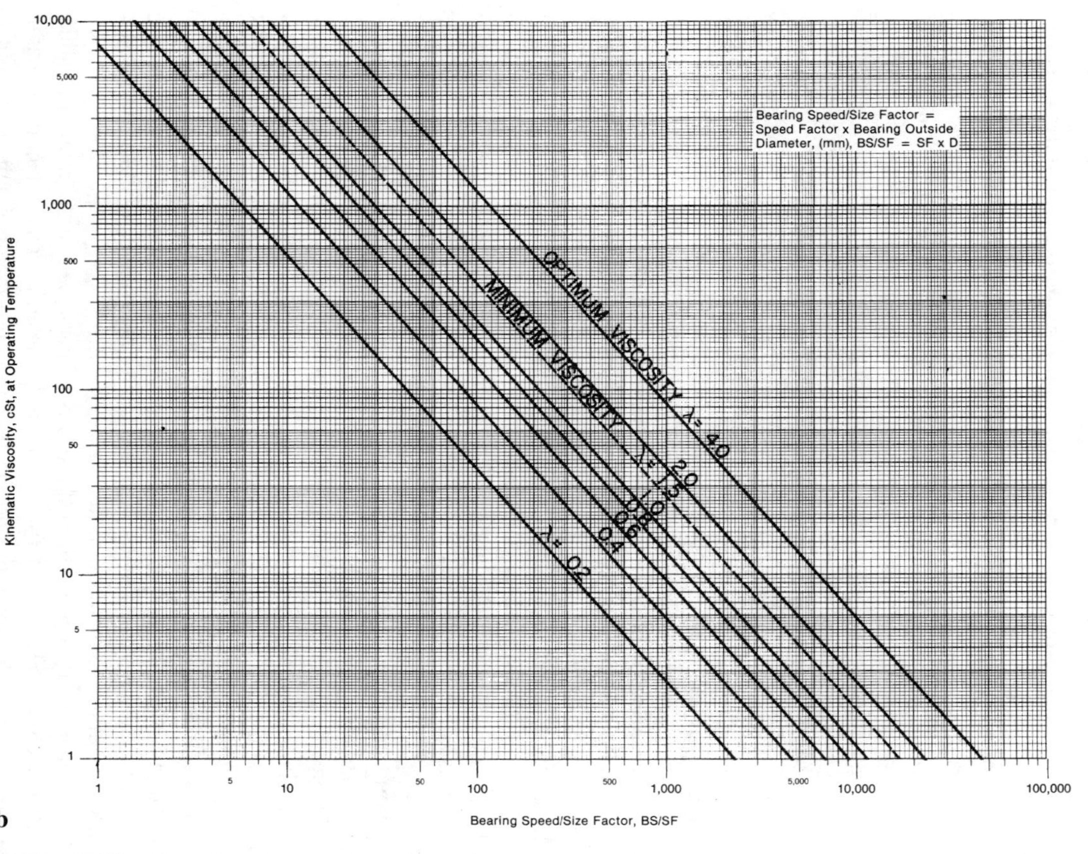

Figure 8.43 Viscosity selection charts for rolling element bearings. (a) For normal loads oil viscosity selected with the minimum viscosity line is generally satisfactory. (b) For shock or vibratory loads, somewhat thicker oil films are required and the optimum viscosity line should be used.

Mechanical stability to resist excessive softening or hardening as a result of shearing in service.

Proper consistency for the method of application and proper feedability at operating temperatures to maintain adequate lubricating films without excessive slumping, which would increase friction and bearing heating.

Controlled oil bleeding, particularly for high speed bearings, to supply the small amount of oil needed to form EHL films.

Enhanced antiwear properties to resist the wiping action between roller ends and raceway flanges in bearings carrying heavy radial or axial thrust loads.

Ability to protect surfaces against rusting. When small quantities of water contaminate a bearing, the grease should absorb the water without appreciable softening or hardening. When large quantities of water are present, the grease must resist being washed out.

Good compatibility with system and system components.

Depending on application, the ability to resist corrosion or deterioration where acids or caustic materials in small quantities can get into the grease.

IV. SLIDES, GUIDES, AND WAYS

The category of bearings sometimes called flat bearings includes all slides, guides, and ways used (1) on forging and stamping presses, (2) as crosshead guides of certain compressors, diesel and steam engines, and (3) on metalworking machines such as lathes, grinders, planers, shapers, and milling machines. The service conditions under which these bearings operate vary widely. Crosshead guides operate at relatively high speeds under conditions that permit the formation of fluid lubricating films most of the time. The requirements for lubrication usually are not severe, and the guides usually are designed to operate on the same oil used for the main and connecting rod bearings of the machine.

The lubrication of ways and slides of machine tools can present special problems. At low speeds and under heavy loads, the lubricant tends to be wiped off causing boundary lubrication to prevail. While this results in higher friction, boundary films have the advantage of being almost constant in thickness. With low loads and high traverse speeds, the oil viscosity must be high enough to allow the formation of fluid films that will lift and float the slide. With variations in speed or load, these fluid films can vary in thickness enough to produce wavy surfaces on the parts being machined, or cause them to run offsize. Thus, precision machining generally requires that the slides and ways operate under boundary conditions at all times. This frequently means that friction-reducing and antiwear additives must be included in the oil.

The phenomenon known as stick–slip can be encountered in the motion of slides and ways. If the static coefficient of friction of the lubricant is greater than the dynamic coefficient, more force is required to start the slide from rest than is required to maintain it in motion after it has started. There is always some amount of free play in the feed mechanism; so when force is applied to start the slide in motion, there will be initial resistance. When the force becomes high enough, the slide will begin to move. As soon as motion begins, the force required to maintain motion decreases so the slide will jump ahead until the free play in the feed mechanism is taken up. At low traverse speeds, this can be a continuous process, producing chatter marks on the workpiece. With cross-slides, stick–slip effects can make it extremely difficult to set feed depths accurately. Stick–slip effects can be overcome with additives that reduce the static coefficient of friction to a value equal to or less than the dynamic coefficient.

With vertical guides, the lubricant tends to drain from the surfaces. To resist this tendency and secure adequate films, special adhesive characteristics are needed.

A. Film Formation

When the two parallel surfaces move with respect to each other, there is some tendency to draw oil in and form a fluid film between the surfaces. However, such a film does not have high load-carrying capacity, and when the surfaces are long, the film is usually wiped or squeezed away rapidly. Traverse grooves in the slide divide it into a series of shorter surfaces, where lubricant supplied at the grooves can improve film formation and load-carrying ability. In some cases, with precision machine tools, hydrostatic bearings are being used. Pockets are machined in the slide and supplied from a constant pressure system with a flow restrictor for each pocket. This arrangement eliminates stick–slip effects and, if properly arranged, maintains the slide parallel to the way. More and more hydrostatic systems are being used in industry, particularly where stick–slip characteristics can contribute to reduced precision of machined parts.

1. Grease Lubrication

On some machine tools, grease is used for the lubrication of the slides and ways. Relative to oil, grease lubrication provides some advantages and disadvantages that should be recognized. The advantages are better sealing, improved stay-put properties, and less generally susceptibility to wash-off by water or metalworking fluids. The disadvantages are apparent when a high degree of accuracy in machined parts is necessary, and in applications in which the grease can act as a binder for debris and pose potential compatibility issues with metalworking fluids.

The potential disadvantage in accuracy of machined parts would be found in applications calling for tables or ways to operate with a specific film thickness at all times. Greases have a greater tendency to build thicker films when initially applied, and these films are reduced as the grease is squeezed out or temperatures rise. Greases can also act as a binder for chips, grinding compounds, and other materials present in machining operations. These could damage slides and ways if pulled into the contact areas. Excessive grease that could contaminate coolants or metalworking fluids may shorten batch life or reduce performance.

2. Lubricant Characteristics

On the basis of the foregoing requirements, the characteristics of suitable lubricants for slides, guides, and ways can be summarized as follows:

Proper viscosity at operating temperature for ready distribution to the sliding surfaces and for forming the necessary boundary films

High film strength to maintain the required boundary films under heavy loads and antiwear capability to control wear under these boundary conditions

Proper frictional characteristics to prevent stick–slip and chatter

Adequate adhesiveness to maintain films on intermittently lubricated surfaces, especially vertical surfaces, and to resist the washing by any metalworking coolants used

Slides and guides in some machines such as open-crankcase steam engines and forging machines may be exposed to considerable amounts of water. Lubricants for these applications must be specially formulated to resist being washed off.

V. GEARS

Gears are employed to transmit motion and power from one rotating shaft to another, or from a rotating shaft to a reciprocating element. With respect to lubrication and the formation and maintenance of lubricating films, gears can be classified as follows:

Spur (Figure 8.44), bevel (Figure 8.45), helical (Figure 8.46), herringbone (Figure 8.47), and spiral bevel (Figure 8.48) gears
Worm gears (Figure 8.49)
Hypoid gears (Figure 8.50)

The differences in the action between the teeth of these three classes of gears have a considerable influence on both the formation of lubricating films and the properties of the lubricants required for satisfactory lubrication.

A. Action Between Gear Teeth

As gear teeth mesh, they roll and slide together. The progression of contact as a pair of spur gear teeth of usual design engage as shown in Figure 8.51. The first contact is between a point near the root of the driving tooth (upper gear) and a point at the tip of the driven tooth. In view A, these points are identified as 0–0 lying on the line of action. At this

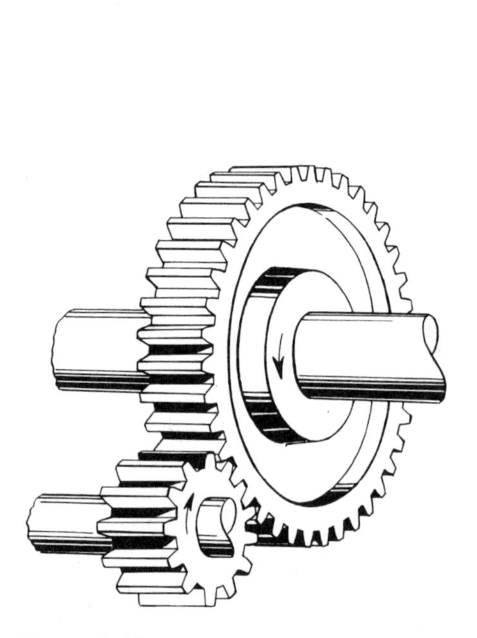

Figure 8.44 Spur gears.

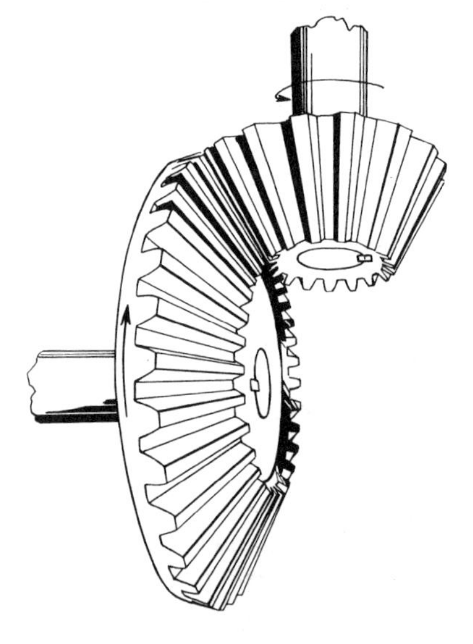

Figure 8.45 Bevel gears. These gears transmit motion between intersecting shafts. The shafts need not meet at a right angle.

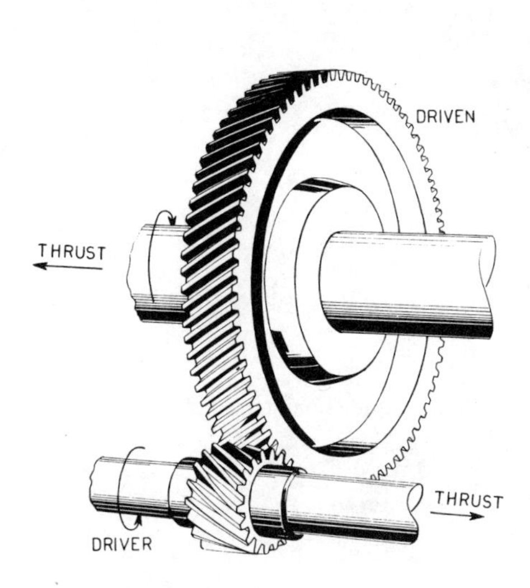

Figure 8.46 Helical gear and pinion.

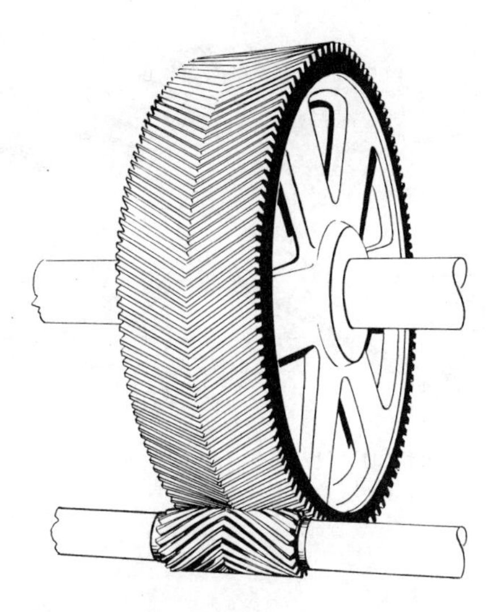

Figure 8.47 Herringbone gear and pinion.

Figure 8.48 Spiral bevel gears. These gears transmit motion between intersecting shafts.

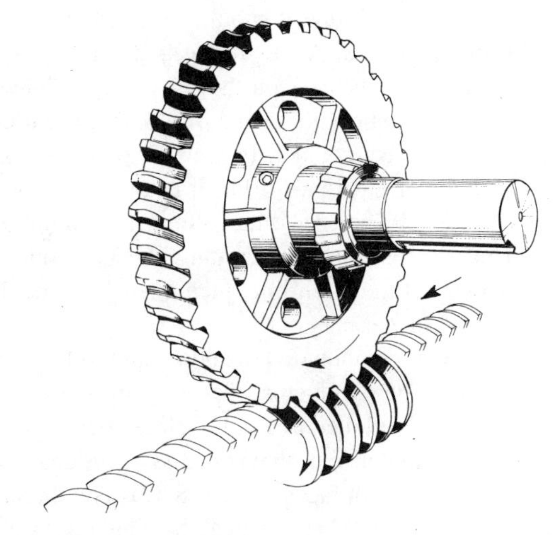

Figure 8.49 Worm gears; the worm here is represented by an endless rack.

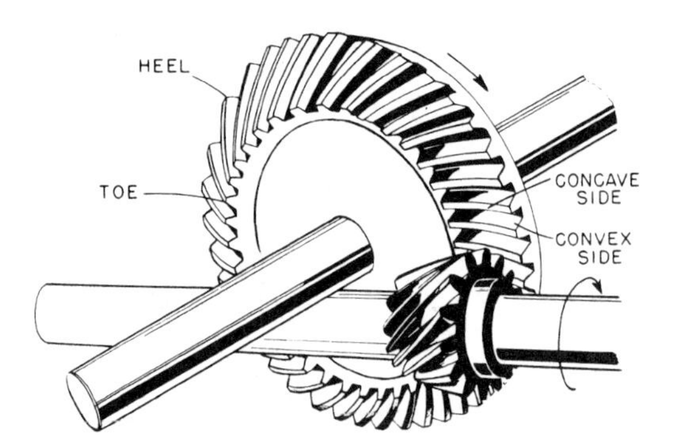

Figure 8.50 Hypoid gears. These gears transmit motion between nonintersecting shafts crossing at a right angle.

time, the preceding teeth are still in mesh and carrying most of the load. As contact progresses, the teeth roll and slide on each other. Rolling is from root to tip on the driver and from tip to root on the driven tooth. The direction of sliding at each stage of contact is as indicated by the small arrows.

In view B, contact has advanced to position 3–3, which is approximately the beginning of ''single tooth'' contact when one pair of teeth pick up the entire the load. It will be seen that to reach this point of engagement, since the distance 0–3 on the driven gear is greater than the distance 0–3 on the driver, there must have been sliding between the two surfaces. View C, position 4–4, shows contact at the pitch line, where there is pure rolling—no sliding. It should be noted, particularly, that the direction of sliding reverses at the pitch line. Also, sliding is always away from the pitch line on the driving teeth, and always toward it on the driven teeth. View D shows contact at position 5–5, which marks the approximate end of a single-tooth contact. As shown, another pair of teeth is about to make contact. In view E, two pairs of teeth are in mesh, but shown at position 8–8, the original pair of teeth is about to disengage.

It will be seen that rolling is continuous throughout mesh. Sliding, on the other hand, varies from a maximum velocity in one direction at the start of mesh, through zero velocity at the pitch line, then again to a maximum velocity in the opposite direction at the end of mesh.

This combination of sliding and rolling occurs with all meshing gear teeth regardless of type. The two factors that vary are the amount of sliding in proportion to the amount of rolling, and the direction of slide relative to the lines of contact between tooth surfaces.

With conventional spur and bevel gears, the theoretical lines of contact run straight across the tooth faces (Figure 8.52). The direction of sliding is then at right angles to the lines of contact. With helical, herringbone, and spiral bevel gears, because of the twisted shape of the teeth, the theoretical lines of contact slant across the tooth faces (Figure 8.53). Therefore, the direction of sliding is not at right angles to the lines of contact, and some side sliding along the lines of contact occurs.

With worm gears, as with spur gears, the same sliding and rolling action occurs as the teeth pass through mesh. Usually, this sliding and rolling action is relatively slow because of the low rotational speed of the worm wheel. In addition, rotation of the worm

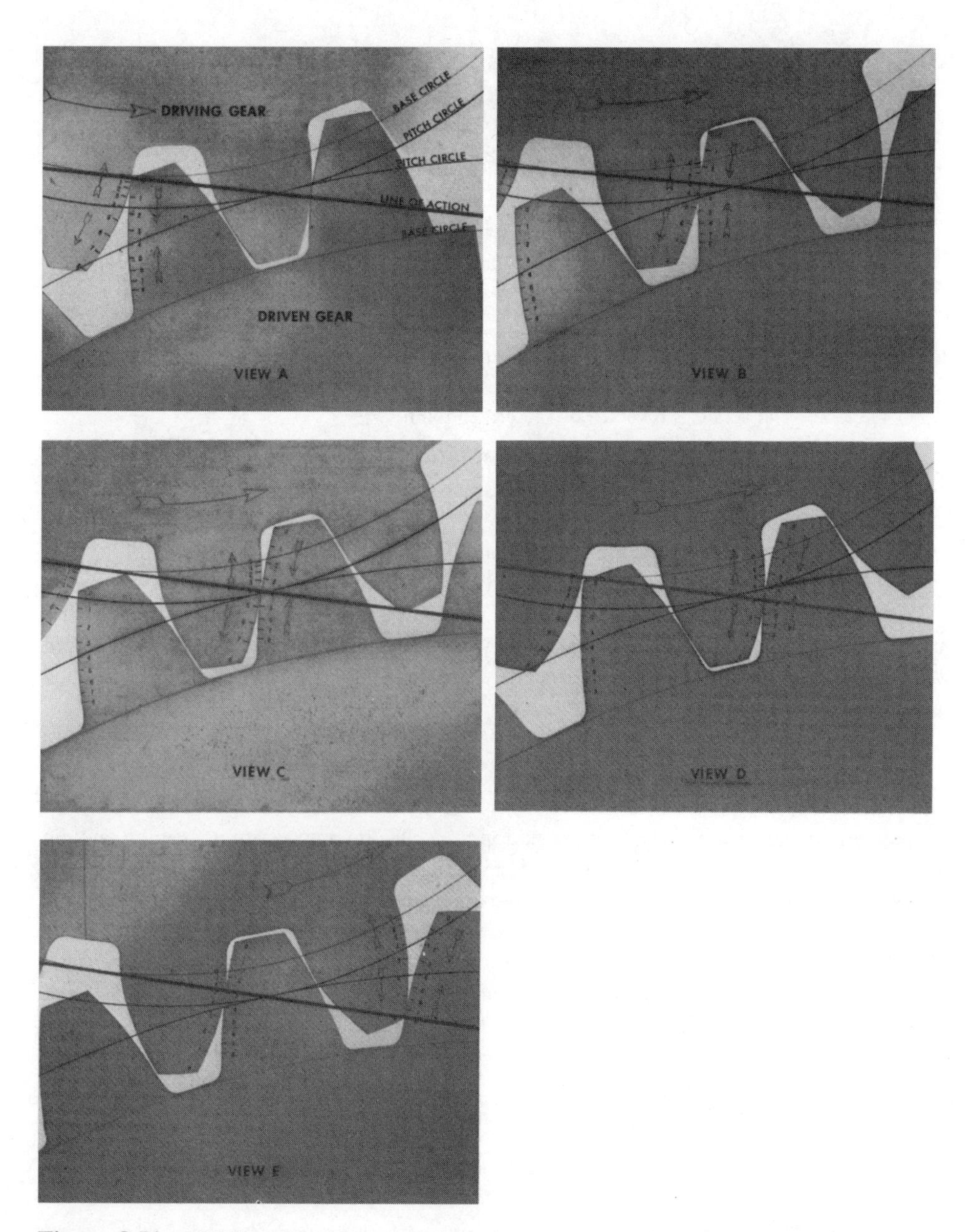

Figure 8.51 Meshing of involute gear teeth: the progression of rolling and sliding as a pair of involute gear teeth (a commonly used design) pass through mesh. The amount of sliding can be seen from the relative positions of the numbered marks on the teeth.

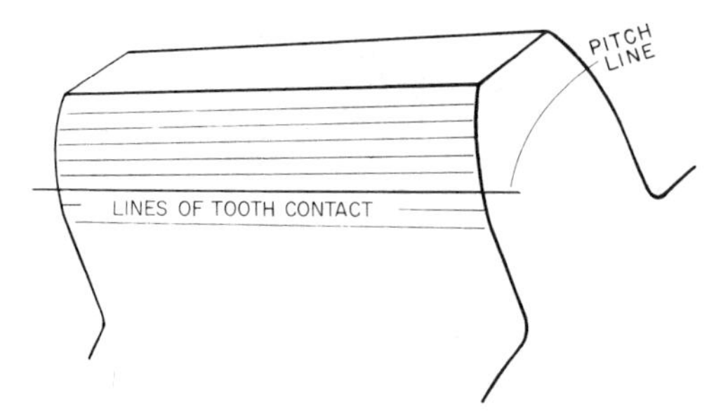

Figure 8.52 A spur gear tooth showing lines of tooth contact. On the driving tooth, this contact first occurs below the pitch line. As the gear turns, this contact progressively sweeps upward to the top of the tooth. The action is reversed on the driven tooth.

introduces a high rate of side sliding. The combination of two sliding actions produces a resultant slide, which in some areas is directly along the line of contact.

In addition to the usual rolling action, hypoid gears have a combination of radial and sideways sliding that is intermediate between the motions of worm gears and spiral bevel gears. The greater the shaft offset, the more nearly the sliding conditions approach those found in worm gears.

B. Film Formation

As pointed out earlier, for EHL film formation to occur, the pressure must be high enough to cause elastic deformation of the contacting surfaces. Thus, the theoretical lines or points of contact are expanded into areas, and a convergent zone exists ahead of the contact area. In the case of industrial gears, in all except very lightly loaded applications, tooth loading is high enough to produce elastic deformation along the line of contact. A convergent zone immediately ahead of the contact area also exists at all times (see Figure 8.54), and the conditions for the formation of thick EHL films are complex, and are beyond the scope

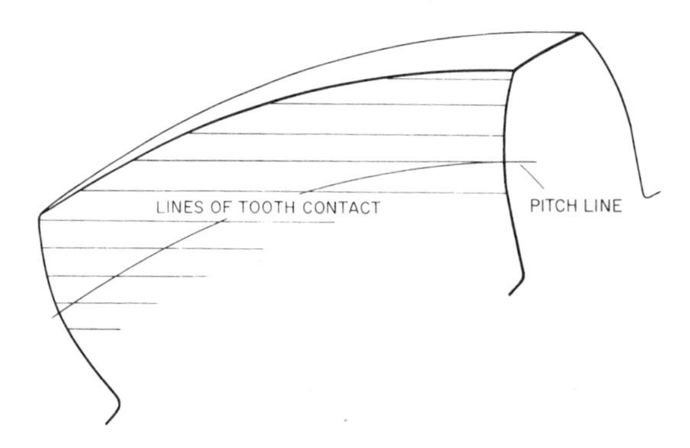

Figure 8.53 Tooth contact of low angle helical gear.

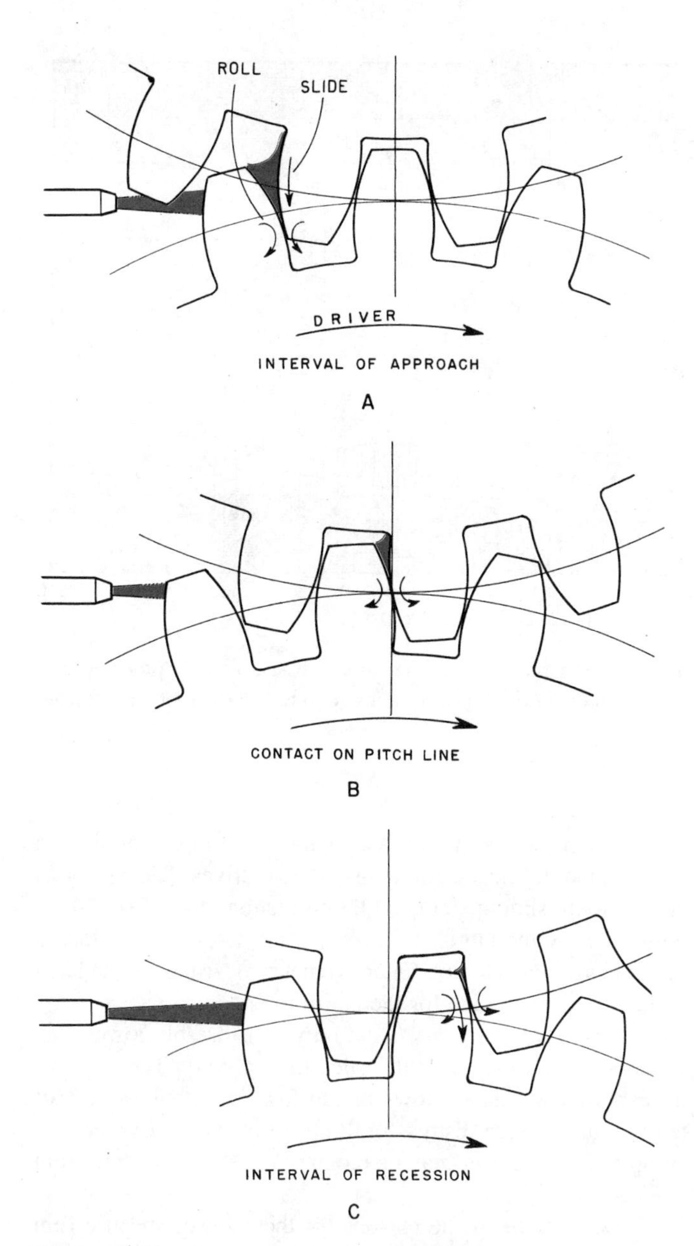

Figure 8.54 Convergent zone between meshing gear teeth. Clearly, if oil is present between meshing gear teeth, it will be drawn into the convergent zone between the teeth; the point of this wedge-shaped zone always points toward the roots of the driving teeth.

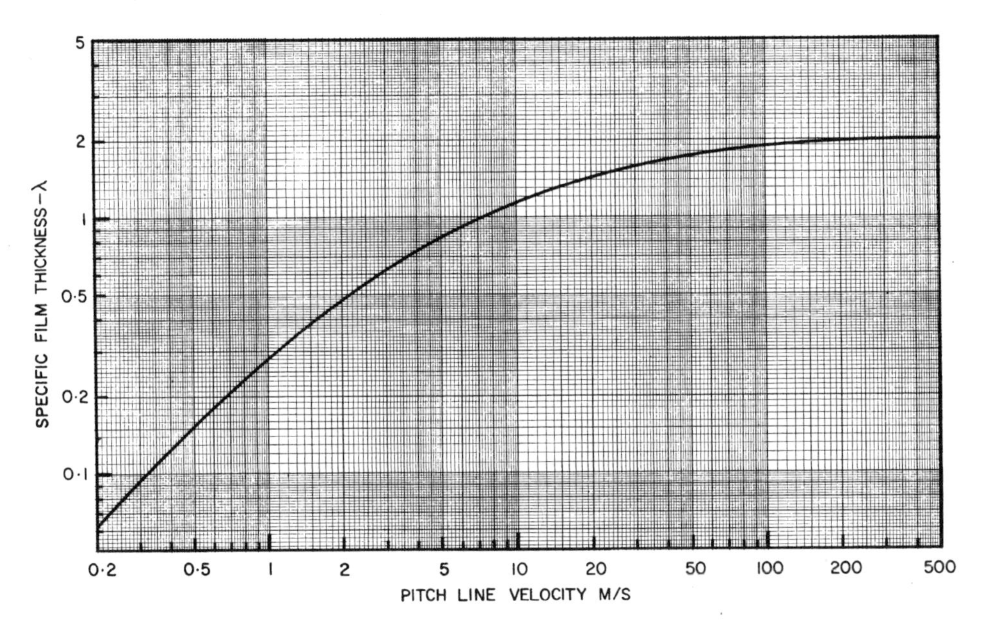

Figure 8.55 Critical specific film thickness for gears: the curve is based on a 5% probability of surface distress to define the target film thickness, which is adjusted to reflect the root mean square (rms) surface roughness, $\sigma = (\sigma_2^2 + \sigma_2^2)^{1/2}$.

of this publication.* However, certain factors in these calculations are of importance in the general consideration of selection of lubricants for industrial gear drives. The equations used do not consider the effect of tooth sliding action on the formation of the EHL films. The entraining velocity tending to carry the lubricant into the contact zone is considered to be the rolling velocity alone. The rolling velocity, for convenience, is usually calculated at the pitch line and is taken to be representative for the entire tooth.

The critical specific film thickness λ for gears is not only considerably lower than for rolling element bearings but is also a function of the pitch line velocity. The curve of Figure 8.55, developed from experimental data, shows that at low speed values of λ of 0.1 or lower can be tolerated without surface distress in the form of pitting or wear. At higher speeds, values of λ of up to 2.0 or higher may be required for equal freedom from tooth distress.

Currently, no analysis has been made of the reasons for these lower specific film thicknesses providing satisfactory results in gears. However, it is generally accepted that in the range where $\lambda < 1.0$, lubricants containing extreme pressure and antifatigue additives are required.

In the selection of lubricants for gears, tooth sliding is considered from two aspects:

1. It tends to increase the operating temperature because of frictional effects.
2. Sliding along the line of contact tends to wipe the lubricant away from the convergent zone; thus, it is more difficult to form lubricating films.

* An excellent reference on this subject is the Mobil EHL Guidebook.

C. Factors Affecting Lubrication of Enclosed Gears

The lubricant in an enclosed gear set, which represents the major portion of gear usage, is subjected to very severe service, being thrown from the gear teeth and shafts in the form of a mist or spray. In this atomized condition, it is exposed to the oxidizing effect of air. Fluid friction and, in some cases, metallic friction generate heat, which raises the lubricant temperature. The violent churning and agitation of the lubricant by the gears of splash-lubricated sets also raises the temperature. Raising the temperature increases the rate of oxidation. Sludge or deposits, formed as a result of oil oxidation, can restrict oil flow, or interfere with heat flow in oil coolers or heat dissipation from the sides of the gear case. Restrictions in the oil flow may cause lubrication failure, while heat-insulating deposits decrease cooling and cause further increases in the rate of oxidation. Eventually, lubrication failure and damage to the gears may result.

In selecting the lubricant for enclosed gear sets, in addition to the requirement for adequate oxidation resistance, the following factors of design and operation require consideration:

1. Gear type
2. Gear speed
3. Reduction ratio
4. Operating and start-up temperatures
5. Transmitted power
6. Surface finish
7. Load characteristics
8. Drive type
9. Application method
10. Contamination (water, metalworking fluids, dirt, etc.)
11. Lubricant leakage

1. Gear Type

With spur and bevel gears, the line of contact runs straight across the tooth face, and the direction of sliding is at right angles to the line of contact. Both these conditions contribute to the formation of effective lubricating films. However, only a single tooth carries the whole load during part of the meshing cycle, resulting in high tooth loads. Additionally, if one tooth wears, there is no transfer of load to other meshing teeth to relieve the load on the worn tooth, and wear of that tooth will continue.

Helical, herringbone, and spiral bevel gears always have more than one pair of teeth in mesh. This results in better distribution of the load under normal loading. Under higher loading, the individual tooth contact pressures may be as high as in comparable straight tooth gears under normal loading. The sliding component along the line of contact, because of time and high viscosity of the lubricant in the contact area, has little or no effect on the EHL film in the contact area. In the convergent zone ahead of the contact area, the sliding component tends to wipe the lubricant sideways. Therefore, not as much lubricant is available to be drawn into the contact area, and the resultant pressure increase in the convergent zone may not be as great. These effects may contribute to a need for slightly higher viscosity lubricants, although, in general, oils for gears of these types are selected on the same basis as for straight tooth gears.

An additional factor present with helical, herringbone, and spiral bevel gears is that if one tooth wears, the load is transferred simultaneously to other teeth in mesh. This relieves the load on the worn tooth and may make lubricant characteristics somewhat less critical for gears of these types. With all such gears, it is important that the lubricant have a viscosity high enough to provide effective oil films, but not so high that excess fluid friction will occur.

The high rate of side sliding in worm gears results in considerable frictional heating. Generally, the rolling velocity is quite low, so the velocity tending to carry the lubricant into the contact area is low. Combined with the sliding action tending to wipe the lubricant along the convergent zone, this makes it necessary to use high viscosity lubricants (typically ISO 460 or 680 viscosity grade). EP additive-type gear oils are not normally recommended for worm gears but, to help reduce the wiping effect and reduce friction, lubricants containing friction-reducing materials are usually used. Because of their friction-reducing and long life characteristics, synthetic lubricants (such as synthesized hydrocarbon or polyalkylene glycols) are the lubricant of choice for most worm gear applications.

Hypoid gears are of steel-to-steel construction and are heat-treated. They are designed to transmit high power in proportion to their size. Combined with the side sliding that occurs, these gears operate under boundary or mixed film conditions essentially all the time and require lubricants containing active extreme pressure additives.

2. Gear Speed

The higher the speed of meshing gears, the higher will be the sliding and rolling speeds of individual teeth. When an ample supply of lubricant is available, speed assists in forming and maintaining fluid films. At high speed, more oil is drawn into the convergent zone; in addition, the time available for the oil to be squeezed from the contact area is less. Therefore, comparatively low viscosity oils may be used (despite their fluidity there is insufficient time to squeeze out the oil film). At low gear speeds, however, more time is available for oil to be squeezed from the contact area and less oil is drawn into the convergent zone; thus, higher viscosity oils are required.

3. Reduction Ratio

Gear reduction ratio influences the selection of the lubricating oil because high ratios require more than one step of reduction. When the reduction is above about 3 : 1 or 4 : 1, multiple reduction gear sets are usually used and above about 8 : 1 or 10 : 1, they are nearly always used. In a multiple reduction set, the first reduction operates at the highest speed and so requires the lowest viscosity oil. Subsequent reductions operate at lower speeds so require higher viscosity oils. The low speed gear in a gear set is usually the most critical in the formation of an EHL film. In the case of a gear reducer, this would be the output gear. In very high speed gear reducers, both the lowest speed and highest speed gears should be checked to determine the more critical condition. In some cases, a dual viscosity system may be employed, using a lower viscosity oil for the high speed gears and a higher viscosity oil for the low speed gears. In some gear sets, this can be accomplished automatically by circulating the cool oil first to the low speed gears, and then, after it is heated and its viscosity decreased, to the high speed gears.

4. Operating and Start-Up Temperatures

The temperature at which gears operate is an important factor in the selection of the lubricating oil, since viscosity decreases with increasing temperature and oil oxidizes more rapidly at high temperatures. Both the ambient temperature where the gear set is located and the temperature rise in the oil during operation must be considered.

When gear sets are located in exposed locations, the oil must provide lubrication at the lowest expected starting temperature. In splash-lubricated units, this means that the oil must not channel at this temperature, while in pressure-fed gear sets, the oil must be fluid enough to flow to the pump suction. At the same time, the oil must have a high enough viscosity to provide proper lubrication when the gears are at their highest operating temperature. For gears exposed to low temperatures (<0°C) during start-up or in continuous operation, synthetic lubricants such as synthesized hydrocarbon oils (SHF) are most often recommended due to their very low pour point, high viscosity index, and excellent shear stability.

During operation, the heat generated by metallic friction, between the tooth surfaces and by fluid friction in the oil, will cause the temperature of the oil to rise. The final operating temperature is a function of both this temperature rise in the oil and the ambient temperature surrounding the gear case. Thus, a temperature rise of 90°F (32.2°C) and an ambient temperature of 60°F (15.6°C) will produce an operating temperature of 150°F (65.6°C), while the same temperature rise at an ambient temperature of 100°F (37.8°C) will produce an operating temperature of 190°F (87.8°C). In the latter case, an oil of higher viscosity and better oxidation stability would be required to provide satisfactory lubrication and oil life at the operating temperature. For gear sets equipped with heat exchangers in the oil system, both the ambient temperature and the temperature rise are less important, since the operating temperature of the oil can be adjusted by varying the amount of heating or cooling.

5. Transmitted Power

As noted in the discussion of EHL film formation, load does not have a major influence on the thickness of EHL films. However, it cannot be ignored. As load is increased, the viscosity of the lubricant may have to be increased to adjust for the small affect of load on film thickness, particularly where λ values were marginal prior to increasing load.

Load also has an influence on the amount of heat generated by both fluid and mechanical friction. Gears designed for higher power ratings will have wider teeth, teeth of larger cross section, or both. Regardless, a greater surface area is swept as the teeth pass through mesh, causing mechanical and fluid friction to be greater. At the same time, the relative area of radiating surface in proportion to the heat generated is usually less in a large gear set than it is in a small one. As a result, larger gear sets, transmitting more power, tend to run hotter unless they are equipped with oil coolers. However, if the operating temperature of a gear set is properly taken into account in the selection of lubricant viscosity, the heating effects based on the amount of power transmitted will be taken care of.

6. Surface Finish

As discussed, surface roughness has an important influence on the thickness of oil films required for proper lubrication. Rougher surfaces require thicker oil films to obtain complete separation, and higher viscosity oils. On the other hand, smoother surfaces can be lubricated successfully with lower viscosity oils. Since some smoothing of the surfaces results from running in, some authorities recommend using an estimated ''run in'' surface roughness rather than the ''as finished'' values in oil film thickness calculations and for selection of oil viscosity.

7. Load Characteristics

The nature of the load on any gear set has an important influence on the selection of a lubricating oil. If the load is uniform, the torque (turning effort) and the load carried by the teeth will also be uniform. However, excessive tooth loads due to shock loads may tend to momentarily rupture the lubricating films. Therefore, where the shock factor has not been considered in the design or selection of a gear set, a higher than normal oil viscosity may be required to prevent film rupture.

In some operations, the conditions may be more severe owing to overloads or to a combination of heavy loads and extreme shock loads, for instance, on rolling mill stands or in applications where gears are started under heavy load and/or have the capability of reversing direction. In such cases, it may be impossible to maintain an effective oil film. Hence, during a considerable part of mesh, boundary lubrication exists. This condition generally requires the use of extreme pressure (EP) oils.

Occasionally, owing to lack of space or other limiting and unavoidable factors, gears are loaded so heavily that it is difficult to maintain an effective lubricating film between the rubbing surfaces. Such a condition is quite usual for hypoid gears in the automotive field. When operating under this condition of extreme loading, the potential for metal-to-metal contact can be so severe that wear cannot be completely avoided. However, it can be controlled by the use of special extreme pressure lubricants containing additives designed to prevent welding and surface destruction under severe conditions. Only slow wear of a smooth and controlled character will then take place. Synthetic hydrocarbon lubricants formulated with extreme pressure additives have proven to be ideal lubricants for hypoid gears.

8. Drive Type

Electric motors, steam turbines, hydraulic turbines, and gas turbines are generally used in applications where the requirement is for uniform torque. Therefore, when the power transmitted by gears is developed by one of these prime movers, gear tooth loading is uniform. Reciprocating engines, however, produce variable torque, so some variation in gear tooth loading results. When gears are driven by prime movers that vary in torque, higher viscosity oils may be required to assure effective oil films. Higher viscosity oils may not be necessary when the type of drive has been considered and compensated for in the design or selection of the gear set.

9. Application Method

When lubricating oil is applied to gear teeth by means of a splash system, the formation of an oil film between the teeth is less effective than when the oil is circulated and sprayed directly on the meshing surfaces. This is particularly true of low speed, splash-lubricated units in which only a limited amount of oil may be carried to the meshing area. A higher viscosity oil is needed to offset this condition, since with higher viscosity, more oil clings to the teeth and is carried into the mesh.

When a gear set is lubricated by a pressure system rather than a splash lubrication system, there is better dissipation of heat. This is because the pressure tends to throw the oil against all internal surfaces of the gear case, and more heat is conducted away by these radiating surfaces. With a splash system, particularly a low speed unit, the oil may dribble over only a small part of the internal surface of the gear case, thus restricting heat dissipation. As a result, splash lubricated units usually run hotter and require higher viscosity oils.

10. Water Contamination

Water sometimes finds its way into the lubrication systems of enclosed gears. This water may come from cooling coils, condensed steam, washing of equipment, or condensation of moisture in the atmosphere. In the latter case, it is often an indication of inadequate venting of the gear case and oil reservoir. Water contamination is likely to occur in gear sets operated intermittently, with warm periods of operation alternating with cool periods of idleness causing moisture to condense. Applications in high humidity conditions where temperatures can drop to levels at or below the dew point may need to be equipped with desiccant breathers. Where moisture contamination may occur, it is necessary to use an oil with good demulsibility, that is, an oil that separates readily from water.

Water and rust also act to speed up deterioration of the oil. Water separates slowly, or not at all, from oil that has been oxidized or contaminated with dirt. In this respect, iron rust is a particularly objectionable form of contamination. Water in severely oxidized or dirty oil usually forms stable emulsions. Such emulsions may cause excessive wear of gears and bearings by reducing the lubricant's ability to provide proper lubrication and by restricting the amount of oil flowing through pipes and oil passages to the gears and bearings. Oxidized oil promotes the formation of stable emulsions, and this is another reason for using an oxidation-resistant oil in enclosed gears. Obviously then, to protect gear tooth surfaces and bearings, the oil must not only separate quickly from water when new but also must have the high chemical stability necessary to maintain a rapid rate of separation even after long service in a gear case.

11. Lubricant Leakage

Although most enclosed gear cases are oil tight, extended operation or more severe operating conditions may result in lubricant leakage at seals or joints in the casing. When the amount of leakage is high and cannot be controlled by other methods, it may be necessary to use special lubricants, such as semifluid greases, designed to resist leakage. Special considerations may be required when one is using antileak oils or semifluid greases, since these may not be consistent with the manufacturer's lubricant recommendations.

VI. LUBRICANT CHARACTERISTICS FOR ENCLOSED GEARS

The necessary characteristics of lubricants for enclosed gears may be summarized as follows.

1. Correct viscosity at operating temperature to assure distribution of oil to all rubbing surfaces and formation of protective oil films at prevailing speeds and pressures
2. Adequate low temperature fluidity to permit circulation at the lowest expected start temperature
3. Good chemical stability to minimize oxidation under conditions of high temperatures and agitation in the presence of air, and to provide long service life for the oil
4. Good demulsibility to permit rapid separation of water and protect against the formation of harmful emulsions
5. Antirust properties to protect gear and bearing surfaces from rusting in the presence of water, entrained moisture, or humid atmospheres
6. A noncorrosive nature to prevent gears and bearings from being subjected to chemical attack by the lubricant
7. Foam resistance to prevent the formation of excessive amounts of foam in reservoirs and gear cases

8. Good compatibility with system components such as seals and paints, and with gear metallurgy

In addition to these characteristics, many modern gear sets operating under severe service conditions or in applications where loads are heavy or shock loads are present require lubricants with extreme pressure (EP) properties to minimize scuffing and destruction of gear tooth surfaces. Worm gears usually require lubricants with mild wear- and friction-reducing properties. It is important to note that some highly additized EP gear

Table 8.2 Viscosity Ranges for AGMA Lubricants

Rust and oxidation-inhibited gear oils, AGMA lube no.	Viscosity range [mm^2/s (cSt) at 40°C][a]	Equivalent ISO grade[a]	Extreme pressure gear lubricants,[b] AGMA lube no.	Synthetic gear oils,[c] AGMA lube no.
0	28.8–35.2	32		0 S
1	41.4–50.6	46		1 S
2	61.2–74.8	68	2 EP	2 S
3	90–110	100	3 EP	3 S
4	135–165	150	4 EP	4 S
5	198–242	220	5 EP	5 S
6	288–352	320	6 EP	6 S
7, 7 Comp[d]	414–506	460	7 EP	7 S
8, 8 Comp[d]	612–748	680	8 EP	8 S
8A Comp[d]	900–1100	1000	8 A EP	—
9	1350–1650	1500	9 EP	9 S
10	2880–3520	—	10 EP	10 S
11	4140–5060	—	11 EP	11 S
12	6120–7480	—	12 EP	12 S
13	190–220 cSt at 100°C (212°F)[e]	—	13 EP	13 S
Residual compounds[f] AGMA lube no.	Viscosity ranges [cSt at 100°C (212°F)][e]			
14R	428.5–857.0			
15R	857.0–1714.0			

[a] Per ISO 3448, Industrial Liquid Lubricants—Viscosity Classification. Also ASTM D 2422 and British Standards Institution B.S. 4231.
[b] Extreme pressure lubricants should be used only when recommended by the gear manufacturer.
[c] Synthetic gear oils 9S–13S are available but not yet in wide use.
[d] Compounded with 3–10% fatty or synthetic fatty oils.
[e] Viscosities of AGMA lubricant 12 and above are specified at 100°C (210°F) because measurement of viscosities of these heavy lubricants at 40°C (100°F) would not be practical.
[f] Residual compounds—diluent types, commonly known as solvent cutbacks, are heavy oils containing a volatile, nonflammable diluent for ease of application. The diluent evaporates, leaving a thick film of lubricant on the gear teeth. Viscosities listed are for the base compound without diluent.
CAUTION: These lubricants may require special handling and storage procedures. Diluent can be toxic or irritating to the skin. Do not use these lubricants without proper ventilation. Consult lubricant supplier's instructions.
Source: From ANSI/AGMA 9005-D94 *Industrial Gear Lubrication,* with permission of the publisher, the American Gear Manufacturers Association, 1500 King Street, Suite 201, Alexandria, VA 22314.

oils may have negative effects on worm gears particularly where different metallurgy such as bronze on steel, are used.

Standard 9005-D94 (ANSI/AGMA 9005-D94) of the American Gear Manufacturers Association (AGMA) combines the specifications for enclosed and open gear lubricants. This specification supersedes AGMA standard 250.04 (Lubrication of Industrial Enclosed Gearing) and 251.02 (Lubrication of Industrial Open Gearing). This AGMA standard provides specifications for rust and oxidation (R&O), compounded (included in the R& O specification), extreme pressure (EP) gear lubricants and for synthetic gear lubricants for industrial gearing. The viscosity grade ranges correspond to those in ASTM D 2422 (Standard Recommended Industrial Liquid Lubricants—ISO Viscosity Classification) and B.S. 4231 from the British Standards Institution. The AGMA specification uses gear pitch line velocities as the primary parameter for determining lubricant selection in other than double-enveloping worm gears. Earlier specifications were based on gear center distances. The AGMA grades and the corresponding ISO viscosity grades are shown in Table 8.2. At the time of this publication, AGMA was in the process of revising and releasing a new AGMA standard.

Tables 8.3 and 8.4 provide AGMA lubricant number guidelines for enclosed gearing.

Table 8.3 AGMA Lubricant Number Guidelines for Enclosed Helical, Herringbone, Straight Bevel, Spiral Bevel, and Spur Gear Drives

	AGMA lubricant numbers,[c–e] ambient temperature °C/(°F)[f,g]			
Pitch line velocity of final reduction stage[a,b]	−40 to −10 (−40 to +14)	−10 to +10 (14 to 50)	10 to 35 (50 to 95)	35 to 55 (95 to 131)
Less than 5 m/s (1000 ft/min)[h]	3S	4	6	8
5–15 m/s (1000–3000 ft/min)	3S	3	5	7
15–25 m/s (3000–5000 ft/min)	2S	2	4	6
Above 25 m/s (5000 ft/min)[h]	0S	0	2	3

[a] Special considerations may be necessary at speeds above 40 m/s (8000 ft/min). Consult gear drive manufacturer for specific recommendations.
[b] Pitch line velocity replaces center distance as the gear drive parameter for lubricant selection. The corresponding table from the previous standard is included as Annex B[i] for reference.
[c] AGMA lubricant numbers listed refer to R&O and synthetic gear oil shown in Table 4.* Physical and performance specifications are shown in Tables 1 and 3[i] EP or synthetic gear lubricants in the corresponding viscosity grades may be substituted where deemed acceptable by the gear drive manufacturer.
[d] Variations in operating conditions (surface roughness, temperature rise, loading, speed, etc.) may necessitate use of a lubricant of one grade higher or lower. Contact gear drive manufacturer for specific recommendations.
[e] Drives incorporating wet clutches or overrunning clutches as backstopping devices should be referred to the gear manufacturer as certain types of lubricant may adversely affect clutch performance.
[f] For ambient temperatures outside the ranges shown, consult the gear manufacturer.
[g] Pour point of lubricant selected should be at least 5°C (9°F) lower than the expected minimum ambient starting temperature. If the ambient starting temperature approaches lubricant pour point, oil sump heaters may be required to facilitate starting and ensure proper lubrication (see 5.1.6).*
[h] At the extreme upper and lower pitch line velocity ranges, special consideration should be given to all drive components, including bearing and seals, to ensure their proper performance.
* AGMA Table 4 is shown in Table 8.2. AGMA Tables 1 and 3, Annex B, and paragraph (5.1.6) are not included in this book.
Source: From ANSI/AGMA 9005-D94, *Industrial Gear Lubrication,* with the permission of the publisher, the American Gear Manufacturers Association, 1500 King Street, Suite 201, Alexandria, VA 22314.

Table 8.4 AGMA Lubricant Number Guidelines for Enclosed Cylindrical Worm Gear Drives[a]

	AGMA lubricant numbers[a] ambient temperature [°C/(°F)][c,d]			
Pitch line velocity[b] of final reduction stage	−40 to −10 (−40 to +14)	−10 to +10 (14 to 50)	10 − 35 (50 to 95)	35 − 55 (95 to 131)
<2.5 m/s (450 ft/min)	5S	7 Comp	8 Comp	8S
>2.5 m/s (450 ft/min)	5S	7 Comp	7 Comp	7S

[a] AGMA lubricant numbers listed above refer to compounded R&O oils and synthetic oils shown in Table 4.* Physical and performance specifications are shown in Tables 1 and 3.† Worm gear drives may also operate satisfactorily using other types of oil. Such oils should be used, however, only with approval of the gear manufacturer.
[b] Pitch line velocity replaces center distance as the gear drive parameter for lubricant selection.
[c] Pour point of the oil used should be at least 5°C (9°F) lower than the minimum ambient temperature expected.
[d] Worm gear applications involving temperatures outside the limits shown, or speeds exceeding 2400 rpm or 10 m/s (2000 ft/min) sliding velocity, should be referred to the manufacturer. In general, for higher speeds a pressurized lubrication system is required along with adjustments in recommended viscosity grade.
* AGMA Table 4 is shown in Table 8.2.
† AGMA Tables 1 and 3 are not included in this book.
Source: From ANSI/AGMA 9005-D94, *Industrial Gear Lubrication,* with the permission of the publisher, the American Gear Manufacturers Association, 1500 King Street, Suite 201, Alexandria, VA 22314.

A. Factors Affecting Lubrication of Open Gears

In contrast to enclosed gears that are flood-lubricated by splash or circulation systems, there are many gears for which it is not practical or economical to provide oil-tight housings. These so-called open gears can only be sparingly lubricated and perhaps only at infrequent intervals. Gears of this type are lubricated by either a continuous or an intermittent method. Some of the more sophisticated open gearing systems may be lubricated with a full circulation system that captures and filters the oil for reuse.

The three most common continuous methods are splash, idler gear immersion, and pressure. In the first two the lubricant is lifted from a reservoir or sump (sometimes referred to as a slush pan) by the partially submerged gear or an idler. Pressure systems require a shaft or independently driven pump to draw oil from a sump and spray it over the gear teeth. With the continuous methods of application, lubrication of open gears is similar to that with enclosed gears. Since the gears are usually large and relatively slow moving, very high viscosity lubricants are required. Most gears lubricated in any of these ways are equipped with relatively oil tight enclosures. AGMA has published guideline for continuous methods of application (Table 8.5).

In addition, many different intermittent methods of application are used. Some are arranged for automatic timing, while others must be controlled manually. Methods used include automatic spray, semiautomatic spray, forced-feed lubricators, gravity or forced drip, and hand application by brush. Grease-type lubricants can be applied by means of hand or power grease guns or by a centralized lubrication system.

With the intermittent methods of application, fluid films may exist when lubricant is first applied to the gears. However, these films quickly become thinner as the lubricant is squeezed aside, whereupon only extremely thin films remain on the metal surfaces. During much of the time, therefore, these gears operate under conditions of boundary

Table 8.5 AGMA Lubrication Number Guidelines for Open Gearing (continuous method of application)[a,b]

		Pitch line velocity				
		Pressure lubrication		Splash lubrication		Idler immersion
Ambient temperature [°C (°F)][c]	Character of operation	<5 m/s (1000 ft/min)	>5 m/s (1000 ft/min)	<5 m/s (1000 ft/min)	5–10 m/s (1000–3000 ft/min)	≤1.5 m/s (300 ft/min)
− 10 to 15[d] (15–60)	Continuous	5 or 5 EP	4 or 4 EP	5 or 5 EP	4 or 4 EP	8–9 8 EP–9 EP
	Reversing or frequent "start–stop"	5 or 5 EP	4 or 4 EP	7 or 7 EP	6 or 6 EP	8–9 8 EP–9 EP
10 to 50[d] (50–125)	Continuous	7 or 7 EP	6 or 6 EP	7 or 7 EP	8–9[e] 8 EP–9 EP	11 or 11 EP
	Reversing or frequent "start–stop"	7 or 7 EP	6 or 6 EP	9–10[f] 8 EP–9 EP		

[a] AGMA lubricant numbers listed refer to gear lubricants shown in Table 4.* Physical performance specifications are shown in Tables 1 and 2.[g] Although both R&O and EP oils are listed, the EP is preferred. Synthetic oils in the corresponding viscosity grades may be substituted where deemed acceptable by the gear manufacturer.
[b] Does not apply to worm gearing.
[c] Temperature in vicinity of the operating gears.
[d] When ambient temperatures approach the lower end of the given range, lubrication systems must be equipped with suitable heating units for proper circulation of lubricant and prevention of channeling. Check with lubricant and pump suppliers.
[e] When ambient temperature remains between 30°C (90°F) and 50°C (125°F) at all times, use 9 or 9 EP.
[f] When ambient temperature remains between 30°C (90°F) and 50°C (125°F) at all times, use 10 or 10 EP.
* AGMA Table 4 is shown in Table 8.2. AGMA Tables 1 and 2 are not included in this book.
Source: From ANSI/AGMA 9005-D94, *Industrial Gear Lubrication* with the permission of the publisher, the American Gear Manufacturers Association, 1500 King Street, Suite 201, Alexandria, VA 22314.

lubrication. Under boundary conditions, the extremely thin film must resist being rubbed or squeezed off the surfaces of the teeth. The ability of the film to resist the rubbing action depends both on its viscosity and on the action that takes place between the lubricant and the metal surfaces. The lubricating film must bond to the tooth surfaces strongly enough to minimize metal-to-metal contact (and resultant wear). With a properly selected lubricant applied at sufficiently frequent intervals, wear may be kept to a negligible amount.

1. Temperature

Whether it is possible to maintain adequate boundary films under the pressures existing between meshing teeth depends, among other things, on the lubricant characteristics and the operating temperature to which the gears are exposed. Heat causes the lubricant to thin out and drop off or be thrown off the gear teeth, decreasing the amount of lubricant remaining on the rubbing surfaces. This results in thinner films. Increasing temperatures also decrease the resistance of the lubricant to stay on the contact surfaces. Thus, high temperatures require more viscous lubricants. Conversely, when gears operate at low temperatures, the lubricant must not become so viscous or hard that it will not distribute properly over the tooth surfaces. At the same time, the lubricant must be such that it does not harden and chip or peel from the teeth at the lowest temperatures encountered.

2. Dust and Dirt

Many open gears, whether operating outdoors or indoors, are exposed to dusty and dirty conditions. Abrasive dust, adhering to oil wetted surfaces, will form a lapping compound that causes excessive wear of the teeth. When viscous lubricants are used, the dirt may pack in the clearance space at the roots of the teeth, forming hard deposits. Packed deposits between gear teeth tend to spread the gears and overload the bearings. These deposits that build up in the tooth root area, if hard enough, can also lead to tooth wear and possible breakage.

3. Water

Open gears operating outdoors are often exposed to rain or snow, and outdoor and indoor gears contact the splash of process fluids. To protect gear tooth surfaces against wear or rust and corrosion, the lubricant must resist being washed off the gears by these fluids.

4. Method of Application

The method of application must be considered when one is selecting a lubricant for open gears. If the lubricant is to be applied by drip force-feed lubricator, or spray, it must be sufficiently fluid to flow through the application equipment. For brush application, the lubricant must be sufficient fluid to be brushed evenly on the teeth. In any case, during operation, the lubricant should be viscous and tacky to resist squeezing from the gear teeth. Very viscous lubricants of some types can be thinned for application by heating, or diluent-type products may be used. These latter products contain a nonflammable diluent that reduces the viscosity sufficiently for application. Shortly after application, the diluent evaporates, leaving the film of viscous base lubricant to protect the gear teeth. When open gears are lubricated by dipping into a slush pan, the lubricant must not be so heavy that it channels as the gear teeth dip into it. When open gears are lubricated by grease-type materials, the consistency and pumpability must permit easy application under the ambient conditions prevailing. Lubricant quantity guidelines for intermittent methods of application can be found in publication 9005-D94 from ANSI/AGMA (Table 8.6).

5. Load Characteristics

Open gears are often heavily loaded and shock loads may be present. These conditions usually require the use of lubricants with enhanced antiwear and EP properties.

VII. AGMA SPECIFICATIONS FOR LUBRICANTS FOR OPEN GEARING

AGMA Standard 9005-D94 also contains specifications for open gear applications. The specifications include R&O gear oils (compounded gear oils are included in this specification), EP gear oils, and synthetic gear oils. AGMA does publish specifications for *residual gear compounds* for open gearing included in Tables 8.2 and 8.6. Generally heavier grades of straight mineral oils and EP oils with viscosities ranging from 400 to 2000 cSt at 100°C (without diluent) are considered to be residual compounds. Other heavy-bodied lubricants designed for open gear applications would be included in this category. These are generally mixed with diluent for ease of application.

Table 8.6 AGMA Lubricant Number Guidelines for Open Gearing Intermittent Applications[a–c] [where gear pitch line velocity does not exceed 7.5 m/s (1500 ft/min)]

	Intermittent spray systems[e]			Gravity feed or forced drip method[f]	
Ambient temperature [°C (°F)][d]	R&O or EP lubricant	Synthetic lubricant	Residual compound[g]	R&O or EP lubricant	Synthetic lubricant
−10 to 15 (15–60)	11 or 11 EP	11 S	14 R	11 or 11 EP	11 S
5 to 40 (40–100)	12 or 12 EP	12 S	15 R	12 or 12 EP	12 S
20 to 50 (70–125)	13 or 13 EP	13 S	15 R	13 or 13 EP	13 S

[a] AGMA Viscosity number guidelines listed refers to gear oils shown in Table 4.*
[b] Does not apply to worm gearing.
[c] Feeder must be capable of handling lubricant selected.
[d] Ambient temperature is temperature in vicinity of the gears.
[e] Special compounds and certain greases are sometimes used in mechanical spray systems to lubricate open gearing. Consult gear manufacturer and spray system manufacturer before proceeding.
[f] EP oils are preferred, but may not be available in some grades.
[g] Diluents must be used to facilitate flow through applicators.
* AGMA Table 4 is shown in Table 8.2.
Source: From ANSI/AGMA 9005-D94, *Industrial Gear Lubrication* with permission of the publisher, the American Gear Manufacturers Association, 1500 King Street, Suite 201, Alexandria, VA 22314.

VIII. CYLINDERS

Cylinders are found in reciprocating compressors, reciprocating internal combustion engines, some older steam engines, hydraulic systems, some air tools, and linear actuators. For the purposes of this brief discussion of cylinder lubrication, reference is made only to compressors and internal combustion engines.

In understanding cylinder lubrication of reciprocating equipment such as compressors or internal combustion engines, pistons, and piston rings are an integral part. In engines and compressors, the pistons are the main component that is used to compress air or other gases. Resultant pressures and temperatures can be high, and therefore some method of sealing the compressed media from bypassing the piston and cylinder is needed. This sealing is provided by the rings and the lubricant. The rings also serve to maintain alignment of the piston in the cylinder bore, act as the bearing surface for controlling wear, and control the oil films on the cylinder walls.

The lubrication of cylinders involves conditions that range from boundary films to full fluid films. For example, at the top and bottom or piston travel in the cylinder, the piston and rings actually stop and reverse direction.* These areas are called *ring reversal areas.* In these areas, the full fluid films developed by the rings in motion are lost and boundary lubrication conditions exist until the motion of the pistons and rings reestablishes the films. In the ring reversal area at the bottom of travel, the reestablishing of the films

* Engines and compressors can have vertical, horizontal, or ''V'' configurations. Some reciprocating engines also have opposed pistons (i.e., an upper and lower piston in each cylinder). For discussion purposes, vertical travel of pistons has been selected, but it refers to the outward-most position in horizontal or other angles of the piston travel. Discussion also is limited to single-acting compressor cylinders.

is not as critical as at the top of travel (assuming vertical travel). At the top of travel, pressures and temperatures are much higher, and as a result, most of the wear in this type of equipment occurs at the top ring reversal area.

The lubricant's primary function is to reduce wear and provide sealing. In addition, the lubricant must minimize formation of deposits, provide protection against rust and corrosion, and handle moisture and other liquids and gases entering the cylinders as a result of compression or combustion. In the compression of some gases, the solubility of the gases in the lubricant films may result in decreasing the oil's viscosity. This may dictate use of higher viscosity lubricants. Depending on application and operating conditions, the selection of a correct lubricant can be complex. This selection process is discussed in greater detail in Chapter 10 (Internal Combustion Engines) and Chapter 17 (Compressors).

IX. FLEXIBLE COUPLINGS

When two rotating shafts are to be connected, some degree of misalignment is almost unavoidable. This is either because of static effects such as deflection of the shafts or thermal effects causing the shafts to change relative positions in their supporting bearings. Misalignment may be angular where the shafts meet at an angle that is not 180°, parallel where the shafts are parallel but displaced laterally, or axial, as results, for example, from endplay. In some cases, all three types of misalignment may be present. To accommodate misalignment, flexible couplings of various types are used to connect shafts together.

In addition to protecting the machines against stresses resulting from misalignment, flexible couplings transmit the torque from the driving shaft to the driven shaft and may help to absorb shock loads.

Universal joints and constant velocity joints may properly be considered as flexible couplings. Both types of joint will accommodate relatively large amounts of misalignment, and are used to some extent in industrial applications; see Chapter 16 for a detailed discussion. Here we consider only types of flexible coupling used in industrial applications, where the amount of misalignment is relatively small.

Several types of lubricated flexible coupling are in use. Gear-type couplings (Figure 8.56) have hubs with external gear teeth (or splines) keyed to the shafts. A shell or sleeve with internal gear teeth at each end meshes with the teeth on the hubs and transmits the

Figure 8.56 Typical Falk gear coupling components.

Figure 8.57 Chain coupling with cover.

Figure 8.58 Spring coupling, horizontally split cover.

torque from the driving shaft to the driven shaft. Misalignment is accommodated in the meshing gears. Where large amounts of misalignment must be accommodated, special tooth forms that permit greater angular motion are used.

Chain couplings (Figure 8.57) have sprockets connected by a chain wrapped around and joined at the ends. The chain may be of either the roller or silent type. Roller chain couplings depend on the relative motion between the rollers and sprockets for their flexibility. Barrel-shaped rollers and special tooth forms may be used where large amounts of misalignment must be accommodated. Silent chain couplings depend on the clearance between the chain links and sprocket teeth for their flexibility.

Spring-laced couplings (Figure 8.58), also called steel grid or spring-type couplings, are made of two serrated flanges laced together with metal strips. Flexibility is obtained from movement of the metal strips in the serrations.

There are two types of sliding couplings. Sliding block couplings have two C-shaped jaw members connected by a square center member. The jaws are free to slide on the surfaces of the block to accommodate misalignment. Sliding disk couplings have flanged hubs with slots machined in the flanges. These slots engage jaw projections on a center disk. The floating center disk allows relative sliding movement between the flange slots and the disk jaws.

Slipper-type couplings, used mainly on rolling mills, are actually a type of universal joint. A C-shaped jaw attached to one shaft is fitted with pads that can turn in their recesses. A tongue on the other shaft fits between the pads. The tongue can move sideways in the pads to accommodate lateral misalignment, and the pads can rotate in their recesses to accommodate angular misalignment.

Flexing member couplings compensate for misalignment both by flexing of the components and by sliding between them. The radial spoke coupling uses thin steel laminations to connect the coupling flanges. Clearance in the slots of the outer flange and flexing of the laminations allows for misalignment. Axial spoke couplings have flexible pins mounted in sintered bronze bushings connecting the flanges. The pins can flex or move in and out in their bushings to accommodate misalignment.

A. Lubrication of Flexible Couplings

All motion in flexible couplings that require lubrication is of the sliding type. Motion is more or less continuous when a coupling is revolving, but the amount of motion may be

Table 8.7 Grease Lubricated Coupling Operating Classifications[a]

Operating conditions	Operating groups		
	I	II	III
1. Rotational speed (rpm)			
d (in.)	≤ 3600	$\geq 2800^{b}/d^{1/2}$	$\leq 2800^{b}/d^{1/2}$
d (mm)	≤ 3600	$\geq 14{,}100/d^{1/2}$	$\leq 14{,}100/d^{1/2}$
2. Misalignment (degree)	≤ 0.75	≤ 0.5	≥ 0.75
3. Continuous torque			
T (lb-in.)	$\leq 1200d^3$	$\geq 1200d^{3\ c}$	$\geq 1200d^{3\ c}$
T (N·m)	$1200d^3/8.8\ (25.4)^3$	$\geq 8.3 \times 10^{-3}d^3$	$\geq 8.3 \times 10^{-3}d^3$
4. Peak torque	$\leq 2.5\ T$	$\leq 2.5\ T$	$\geq 2.5\ T$
5. Maximum coupling Surface temperature	150°F (65°C)	170°F (77°C)	212°F (100°C)
6. Normal relube Interval (months)[d]	6–12	12–36	1 or less

[a] Definitions: d, shaft diameter; T, torque; G, ratio of actual acceleration to gravitational acceleration [$G \approx 28.4d$ (rpm) $^2/10$ ^{6}when d is in inches, $1.12d$ (rpm) $^2/10$ ^{6}when d is in millimeters]; d is not an exact value because it relates approximately to its pitch radius of a coupling.
[b] Relates to centrifugal force on the lubricant of approximately 200 g.
[c] Relates to shaft torsional stress of approximately 6000 lb/in.^{2} (0.207 MPa).
[d] The actual relube interval is dependent on experience with the specific application.
Source: From AGMA *Standard for Lubrication of Flexible Couplings* (AGMA-9001-B97), with permission of the publisher, the American Gear Manufacturers Association, 1500 King Street, Suite 201, Alexandria, VA 22314.

small, varying with the amount of misalignment between the shafts. Contact pressure may be high, particularly when there is considerable misalignment, since misalignment tends to reduce the surface area over which the load is distributed. The thickness of the lubricating films formed and the type of lubricant required depend to a considerable extent on the type of coupling.

AGMA defines three major operating groups for grease-lubricated flexible couplings as a function of shaft diameter, rotational speed, misalignment, torque, and coupling surface temperature. For the three operating groups, three coupling grease specifications are defined: AGMA CG-1, CG-2, and CG-3. These specifications are defined in AGMA Standard 9001-A86 (Table 8.7), as are AGMA flexible coupling operating classifications and grease specifications (Table 8.8).

1. Gear Couplings

Gear couplings can transmit more torque than any other type of flexible coupling of equal size. Since they rotate continuously, a supply of lubricant is usually contained in the assembly. As the unit revolves, centrifugal force causes the lubricant to be thrown outward, forming an annulus of lubricant that should completely submerge the gear teeth. The relative motion between the gear teeth may be sufficient to create and maintain appreciably thick lubricating films.

Either oil or grease may be used for gear couplings. Greases are often used where speeds are normal to high and temperatures are moderate. For normal speeds, a soft grease

Table 8.8 AGMA Coupling Grease Specifications

Characteristic (test method)[a]	Type CG-1	Type CG-2	Type CG-3
Minimum base oil viscosity:			
In centistokes	198 at 40°C (104°F)	288 at 40°C (104°F)	30 at 100°C (212°F)
In SSU (approx.)	900 at 100°F (38°C)	1300 at 100°F (38°C)	150 at 210°F (99°C)
Separation characteristics[b]	K36 $\leq$ 60/24, or 8% maximum fluid insoluble material	K36 $\leq$ 24/24	No restriction
National Lubrication Grease Institute (NLGI) grade			
a) metallic grid	1–3	1–3	1–3
b) Gear or chain			
where rpm $\geq$ $200/\sqrt{\bar{d}}$ in $1008/\sqrt{\bar{d}}$ mm	0–3	0–1	1–2
where rpm $\leq$ $200/\sqrt{\bar{d}}$ in $1008/\sqrt{\bar{d}}$ mm[c]	0–1	Not applicable	1–2
Minimum dropping point	190°F (88°C)	195°F (91°C)	302°F (150°C)
Compatibility[d]	The coupling grease must be compatible with coupling seals and gaskets		
Oxidation resistance—max pressure drop at 100 h	20 lb/in.^{2} (13,790 Pa)	20 lb/in.^{2} (13,790 Pa)	20 lb/in.^{2} (13,790 Pa)
Antirust properties	Not required	ASTM rating pass	ASTM rating pass
Antiwear additives[d]	Not required	Not required	Required[e]
Extreme pressure (EP) additives[d]	Not required[f]	Not required[f]	Required
Timken OK load	Not required	Not required	40 lb minimum
Four-ball EP test	Not required	Not required	Weld point 250 kg minimum

[a] Accepted test methods;
Viscosity, ASTM D 445
Grease composition, ASTM D 128
Centrifuge test, ASTM D 4425
NLGI grade, ASTM D 217
Dropping point, ASTM D 566 or D 2265
Antirust properties, ASTM D 1743
Oxidation resistance, ASTM D 942
Four-ball EP test, ASTM D 2596
Timken OK load, ASTM D 2509.

[b] ASTM centrifuge test.

[c] Relates to a centrifugal force on the lubricant of approximately 10 *g*.

[d] No test method.

[e] Experience has shown that a minimum of 5% (by weight) MoS_2 (molybdenum disulfide) is beneficial for couplings with hardened teeth.

[f] EP additives recommended by some coupling manufacturers.

Source: From AGMA *Standard for Lubrication of Flexible Couplings* (AGMA-9001-B97), with permission of the publisher, the American Gear Manufacturers Association, 1500 King Street, Suite 201, Alexandria, VA 22314.

made with a high viscosity oil, often with extreme pressure properties, is usually used. For higher speeds, a stiffer grease may be required (see Tables 8.7 and 8.8).

2. Chain Couplings

Chain couplings (see Figure 8.57) are usually grease-lubricated. Couplings without dust covers are generally lubricated by brushing with grease periodically. The grease must resist throwoff and must penetrate the chain joints to provide lubrication. Generally, soft greases with good adhesive properties are required. Couplings with covers are packed with grease, usually of a type similar to those used in geared couplings.

3. Spring-Laced Couplings

Couplings like those shown in Figure 8.58 are grease-packed. Again, greases similar to those used in geared couplings are usually required.

4. Sliding Couplings

Sliding block couplings may have the block formed from an oil-impregnated, sintered metal. In other cases the block contains a reservoir. In the latter case, high viscosity oils with good adhesiveness are required to minimize throwoff. Sliding disk couplings require similar oils.

5. Slipper Joint Couplings

Slipper joint couplings are usually large and carry both heavy loads and shock loads, so the lubrication requirements are severe. All-loss systems are usually used to apply the lubricant. High viscosity oils with enhanced film strength are required.

6. Flexible Member Couplings

Axial spoke couplings have prelubricated bushings for the pins. When such couplings are enclosed in a dust cover, a soft, adhesive grease is required for packing. Radial spoke couplings require high viscosity oil or semifluid grease.

7. Lubrication Techniques

The preferred method of lubrication is to manually pack flexible couplings before closing. This procedure should include a coating of grease on all working surfaces, including seals.

To lubricate a coupling that is assembled, remove the highest lube plug. Slowly pump the quantity of grease specified by the coupling manufacturer via the lowest lube plug. It is preferred that the coupling be rotated so that the higher lube plugs are located at three and nine o'clock to assure adequate filling of the bottom half of the coupling. All working surfaces including seals should be coated with grease.

Key fits and keyways should be sealed with an oil-resistant sealing compound.

The proper quantity of grease is specified by the manufacturer, since couplings of varying series and shaft diameter require different amounts. If in doubt, the coupling should not be filled more than 75% of capacity to allow for thermal expansion.

Seals that become damaged or distorted should be replaced to avoid premature coupling failure from lack of lubrication.

Oil is generally preferred for gear couplings operating at normal to high speeds and high temperatures when the coupling is large enough to hold a reasonable supply of oil. Couplings may be oil-filled, in which case they require only enough oil to cover the gear teeth during operation. Continuously supplied couplings may be designed to recirculate the oil, or to collect it to permit drainage and disposal after use. Continuous supply systems generally are required where operating temperatures are high. Regardless of the method of applying the oil, quite high viscosity oils with good antiwear properties are required.

X. DRIVE CHAINS

Drive chains are used to transmit power in a wide variety of applications. Chains fall into two general categories:

1. Machined surface chains suitable for high speed, precision drives
2. Cast or forged link chains, usually made without machined surfaces and suitable for lower speed and power or where environmental conditions dictate the use of low cost drives.

A. Silent and Roller Chains

Silent and roller chains are precision mechanisms that, when properly lubricated, will transmit high power at high speeds and will give long service life. Silent chains, also called inverted tooth chains, operate in such a manner that the links of the chain almost completely fill the clearance space in the sprocket throughout the arc of contact, greatly reducing backlash and minimizing noise. Roller chains are available in various designs, and with 20 strands or more for use where high powers must be transmitted.

With silent chains, the lubricant must penetrate between the links and distribute along the seat pins to provide lubricating films on the rubbing surfaces. With roller chains, the lubricant must coat the outsides of the rollers to lubricate the contact surfaces of the rollers and sprockets and must also penetrate and distribute along the rubbing surfaces between the rollers and bushings. As a result, in both types of chain, the viscosity of the lubricating oil used is extremely important. Oils with enhanced film strength are also generally desirable, both to improve protection against wear and to permit the use of lighter oils with better penetrating properties.

Silent and roller chains usually are installed in the oil-tight housings, where the lubricant is applied by force feed, oil mist, slinger, or oil ring, or by the dipping of the chain or sprocket into a bath of oil. Occasionally these chains are used without housings for low speed, low power drives. For applications of this type, the lubricant is applied by one of the all-loss methods such as drip, wick feed oiler, brush, oil mist, aerosol, or pour can. The characteristics of the method of application must be taken into account in lubricant selection, as well as the requirement for controlling drip and throwoff when the chain is not enclosed.

Normally drive chains of these types are not grease-lubricated. Under certain environmental conditions, such as corrosive atmospheres, a properly selected grease may provide better protection than oil. If maintaining proper lubrication is difficult because of chain location or if excess lubricant is frequently thrown off, a lube-free chain should be considered when replacement becomes necessary.

B. Cast or Forged Link Chains

Detachable link, pintle, fabricated, and cast roller types are among the chain forms available. Detachable link chains may be of cast malleable iron, or medium carbon steel hardened to improve strength and wear characteristics. Pintle links may be either cast or forged.

Table 8.9 Viscosity of Oil for Silent and Roller Chains (guideline)

Ambient temperature			
°F	°C	Recommended viscosity	ISO viscosity grade
−20 to +20	−30 to −7	SAE 10	46
20 to 40	−7 to −4	SAE 20	68
40 to 100	4 to 38	SAE 30	100
100 to 120	38 to 49	SAE 40	150
120 to 140	49 to 60	SAE 50	220

Source: American Chain Association.

These chains usually operate at low speeds, transmitting relatively low power; although fabricated chains suitable for speeds up to 1000 ft (300 m) per minute are available. The chains may be open, or fitted with dust covers. Under most conditions, they require viscous, tacky oils, but under dusty or dirty conditions lower viscosity oils may reduce dirt pickup. Occasionally, diluent-type open gear lubricants that have a dry, rubbery surface after application may provide better performance.

C. Viscosity Selection

The American Chain Association guidelines for lubricant viscosities for silent and roller chains are shown in Table 8.9. Where drip or throwoff may be a problem, higher viscosity oils or oils with special adhesive properties may be desirable. Where water washing occurs, oils compounded to resist water washing and protect against rusting are required.

XI. CAMS AND CAM FOLLOWERS

In many machines, rotary motion must be converted to linear motion. For linear motion that is comparatively short, this conversion is commonly accomplished with a cam and cam follower combination (cam follower is sometimes referred to as a tappet). Probably the best-known applications of these mechanisms is for operating intake and exhaust valves in a reciprocating internal combustion engine, but many other applications also exist. Since the cam will lift only the cam follower, some arrangement must be made to keep the cam follower in contact with the cam. This may be accomplished by transmitting load from the mechanism being actuated, or directly with a loading mechanism such as a return spring.

There are three general types of cam follower, to be selected according to such factors as loads, speeds, and the complexity of the cam shape.

1. Flat surface cam followers are used in valve systems of most older and quite a few newer internal combustion engines. To conserve fuel, many newer automotive internal combustion engines are using roller tappet camshafts designs.
2. Roller cam followers may be used when loads are extremely high when reduced frictional characteristics are desired.
3. Spherical cam followers may be used where precise conformity to a complex cam is required.

Of these, the roller generally has the least severe lubrication requirements, since the rolling contact tends to develop and maintain films.

In many cases with flat cam followers, the cam lobe is tapered and the cylindrical cam follower, which is free to rotate, is located with one axis toward one side of the cam. With this arrangement, the cam follower will be rotated as it is forced upward by the cam. This spreads the wear more uniformly over the surface of the cam follower and may promote the formation of oil films.

In heavily loaded applications, both flat and spherical cam followers operate on EHL films at least part of the time. Even then, lubricants with antiwear additives are often necessary if rapid wear and surface distress are to be avoided. In the case of four-cycle gasoline engines, for example, one of the functions of the oil additive zinc dithiophosphate, now used almost universally, is to provide antiwear activity for the camshaft and valve lifters (cam followers). With the increased use of roller follower cams in automotive-type engines, the requirements for antiwear additives may be changed, mainly to prolong life of emission control devices such as catalytic converters by reducing the phosphorus levels in oils.

Most cam and cam follower machine elements are included in lubrication systems supplying other machine elements as well. For this reason, lubricants are rarely selected to meet the requirements of cams and cam followers alone. Generally, the lubricant is selected to meet the basic requirements of the system, and if special requirements exist for the lubrication of cams and followers, as in the case of a gasoline automobile engine, these must be superimposed on the basic requirements.

XII. WIRE ROPES

Wire rope or cable is used for a great variety of purposes ranging from standing (stationary) service, such as guys or stays and suspension cables, to work involving drawing or hoisting heavy loads. In these various services, all degrees of exposure to environmental conditions are encountered. These range from clean, dry conditions in applications such as elevator cables in office buildings, to full exposure to the elements on outdoor equipment, immersion in water (e.g., on dredging equipment), and exposure to corrosive environments such as acid water found in many mining applications. These and other operating factors require that wire ropes be properly lubricated to provide long rope life and maximum protection against rope failure where the safety of people is at stake.

A wire rope consists of several strands laid (helically bent, not twisted) around a core. The core is usually a rope made of hemp or other fiber, but an independent wire rope or strand may be used. Each strand consists of several wires laid around the core, which usually consists of one or more wires but may be a small fiber rope. The number of wires per strand typically ranges from 7 to 37 or more.

A. Need for Lubrication

A number of factors contribute to the need for proper lubrication of wire ropes.

1. Wear

Each wire of a wire rope can be in contact with three or more wires over its entire length. Each contact is theoretically along a line, but this line actually widens to become a narrow band because of a deformation under load. As load is applied and as a rope bends or flexes over rollers, sheaves, or drums, stresses are set up that cause the strands and individual wires to move with respect to each other under high contact pressures. Unless lubricating films are maintained in the contact areas, considerable friction and wear result from these movements.

2. Fatigue

One of the principal causes of wire rope failures is metal fatigue. Bending and tension stresses, repeated many times, causes fatigue. Eventually, individual wires break and the rope is progressively weakened to the extent that it must be removed from service. If lubrication is inadequate, the stresses are increased by high frictional resistance to the movement of the wires over one another, fatigue failures occur more rapidly, and rope life is shortened.

3. Corrosion

Another principal cause of rope failure is corrosion. This covers both direct attack by corrosive materials, such as acid water encountered in mines, and various forms of rusting. To protect against corrosion, lubricant films that resist displacement by water must be maintained on all wire surfaces.

4. Core Protection

Wear, deterioration, or drying out of the core result in reduction of the core diameter and loss of support for the strands. The strands then tend to overlap, and severe cutting or nicking of the wires may occur. The lubricant applied in service must be of a type that will penetrate through the strands to the core to minimize friction and wear at the core surface, seal the core against water, and keep it soft and flexible.

B. Lubrication in Manufacture

During manufacture, wire rope cores are saturated with lubricant. A second lubricant, designed to provide a very tenacious film, is usually applied to the wires and strands to lubricate and protect the wires and to help keep (seal) the lubricant in the core as they are laid up. These lubricants protect the rope during shipment, storage, and installation.

C. Lubrication in Service

Much of the core lubricant applied during manufacture is squeezed out when the strands are laid, and additional lubricant is lost from both the core and strands as soon as load is applied to a rope. As a result, in-service lubrication must be started almost immediately after a rope is placed in service.

Proper lubrication of wire ropes in service is not easy to accomplish. Some of the types of lubricant required for wire ropes may not be easy to apply, and often wire ropes are somewhat inaccessible. Various methods of applying lubricants are used, including brushing, spraying, pouring on a running section of the rope, reliance on drip or force-feed applicators, and running the rope through a trough or bath of lubricant. Generally, the method of application is a function of the type of lubricant required to protect a rope under the conditions to which it is exposed.

D. Lubricant Characteristics

Wire rope lubricants should be able to do the following:

1. Form a durable, adhesive coating that will not be thrown or wiped from the rope as it operates over pulleys or drums
2. Penetrate between adjacent wires, to lubricate and protect them against wear and to keep the rope core from drying out and deteriorating
3. Provide lubrication between pulleys and sheaves and the rope
4. Resist being washed off by water
5. Protect against rusting or corrosion by acid, alkaline, or salt water
6. Form nonsticky films so that dust and dirt will not build up on ropes
7. Remain pliable and resist stripping at the lowest temperatures to which the rope will be exposed
8. Resist softening or thinning out to the extent that throwoff or drippage occurs at the highest temperatures at which the rope will operate
9. Be suitable for application to the rope under the service conditions encountered

These requirements necessitate some compromises. Wire rope lubricants may be formulated with asphaltic or petrolatum bases and usually contain rust preventives and materials to promote metal wetting and penetration. Diluent products are used in some cases for ease of application. Grease products containing solid lubricants such as graphite or molybdenum disulfide are also used.

BIBLIOGRAPHY

Mobil Technical Books

Plain Bearings, Fluid Film Lubrication
Gears and Their Lubrication
Mobil EHL Guidebook

Mobil Technical and Service Bulletins

Rolling Element Bearings: Care and Lubrication
Flexible Coupling Lubrication

9
Lubricant Application

After the proper lubricant for an application has been selected, it must be delivered, in the correct quantities and under the specific application conditions, to the elements that require lubrication. Two categories of lubricant application are prevalent: *all-loss* methods, where a relatively small amount of lubricant is applied periodically and after use leaks or drains away to waste; and *reuse* methods, where the lubricant leaving the elements is collected and recirculated to lubricate again. Application methods such as some pedestal bearings and enclosed splash-lubricated gear cases are considered to be forms of reuse. Various centralized application systems, such as centralized grease systems and mist systems (not to be confused with circulation systems), are actually specialized all-loss systems and are discussed separately. Reuse application systems are preferred because they conserve lubricant, minimize waste, and help control environmental pollution. They also provide other advantages, as discussed in this chapter.

I. ALL-LOSS METHODS

Most open gears and wire ropes, many drive chains and rolling element bearings, and some cylinders, bearings, and enclosed gears are lubricated by all-loss methods. Nearly all grease lubrication (except sealed, packed rolling element bearings) represents the all-loss approach. Only relatively small amounts of grease are applied, mainly to replenish the lubricating films, but in some cases to flush away some or all of the old lubricant and contaminants.

A. Oiling Devices

The oldest known method of applying lubricants is by using an oil can. With high viscosity lubricants such as are used on open gears and some wire ropes, a paddle, swab, brush, or caulking gun may be required instead, all these are variations of hand oiling.

While still widely used, hand oiling has several disadvantages. Immediately after application there is usually an oversupply of oil, and excessive leakage or throwoff occurs. Then follows a period when more or less of the proper quantity of oil is present; and finally, depending on the frequency of application, there is usually a period when too little oil is present (starvation). During this latter period, wear and friction may be high. Also, with hand application, lubrication points may be neglected, either because they get overlooked or because they are difficult or hazardous to reach. Oil leakage onto machine parts, floors, or goods being processed can be hazardous and costly in terms of safety and/or materials wasted. Hand oiling is costly both in terms of labor and because of lost production if machines must be shut down for the purpose. Many devices have been developed and are in wide use to overcome some of the disadvantages of hand oiling. The objective of these devices is to feed oil continuously or at regular, frequent intervals in small amounts with as little attention as possible.

1. Drop Feed and Wick Feed Cups

Drop feed and wick feed cups, shown in Figures 9.1 and 9.2, are often used to supply the small amount of oil required by high speed rolling element bearings, thin film plain bearings and slides, and some open gears. The rate of oil feed from the drop feed cup can be adjusted with a needle valve, while the wick feed can be adjusted by changing the number of strands in the wick. Both devices have the disadvantage of requiring to be started and stopped by hand when the machine is started or stopped.

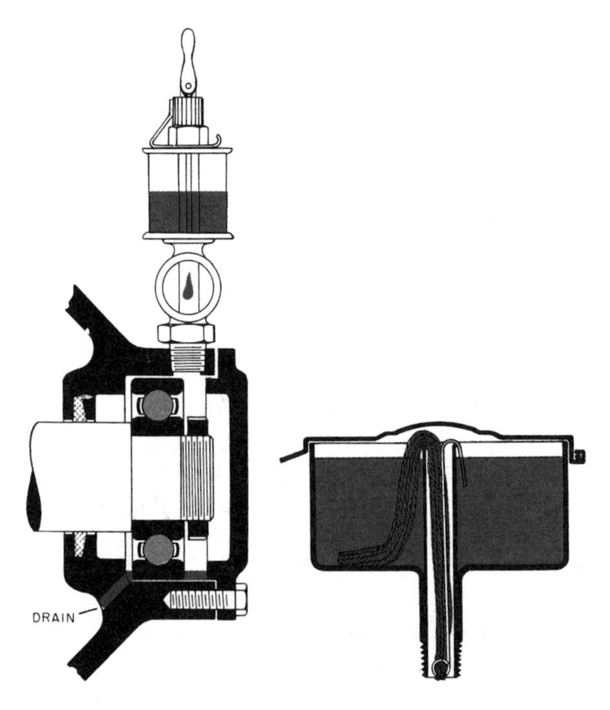

Figure 9.1 Drop feed (left) and wick feed (right) cups. In the drop feed application, the oil drops fall on the lock nut, which throws a spray of oil into the bearing.

Figure 9.2 Drip oiling system.

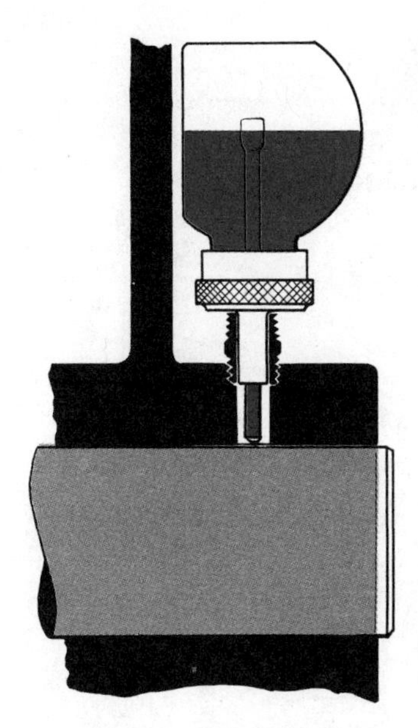

Figure 9.3 Bottle oiler.

Some drop feed cups are controlled by solenoid-operated valves, which eliminate the problems of manual actuation.

2. Bottle Oilers

In a typical bottle oiler (Figure 9.3), the spindle of the oiler rests on the journal and is vibrated slightly as the journal rotates. This motion results in a pumping action, which forces air into the bottle, causing minute amounts of oil to feed downward along the spindle to the bearing. The oil feed is more or less continuous but stops and starts when the machine is stopped or started.

3. Wick and Pad Oilers

In one type of wick oiler (Figure 9.4), a felt wick is held against the journal by a spring. The wick draws oil up from the reservoir by capillary action, and the turning journal wipes oil from the wick. No or little oil is fed when the journal is not turning. In another variation (Figure 9.5), the wick carries oil up to the slinger, which throws it into the bearing in the form of a fine spray. In both cases, oil leaking out along the shaft drains back to the reservoir, so the devices have some elements of a reuse system.

Pad oilers are sometimes used for long, open bearings. One end of the pad rests in an oil reservoir, while the other end rests along the journal. Capillary action draws oil up the pad, where it is wiped off by the turning journal.

4. Mechanical Force-Feed Lubricators

Force-feed lubricators are used in applications requiring positive feed of lubricants under pressure. A variety of force-feed lubricators are in use. In the type shown in Figure 9.6, oil is drawn from the reservoir in the base on the downstroke of the single plunger pump

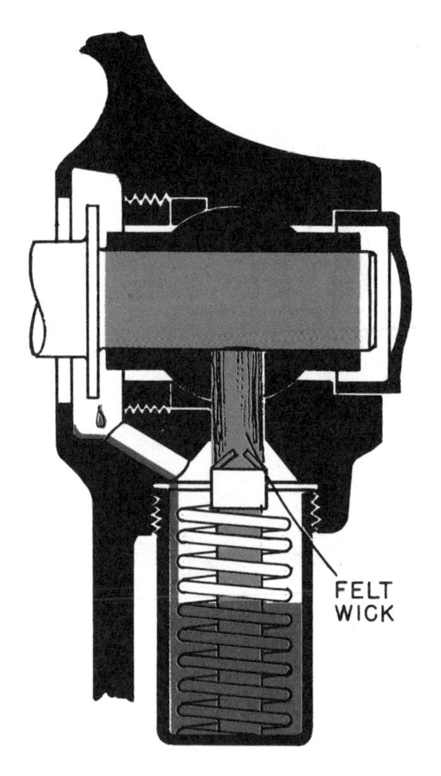

Figure 9.4 One type of wick oiler.

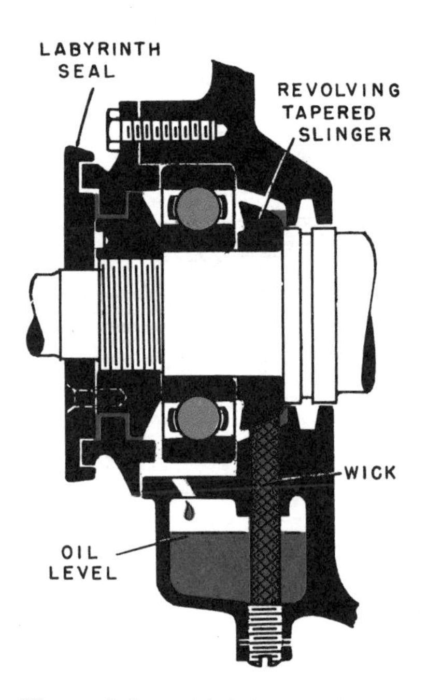

Figure 9.5 Wick-fed spray oiler.

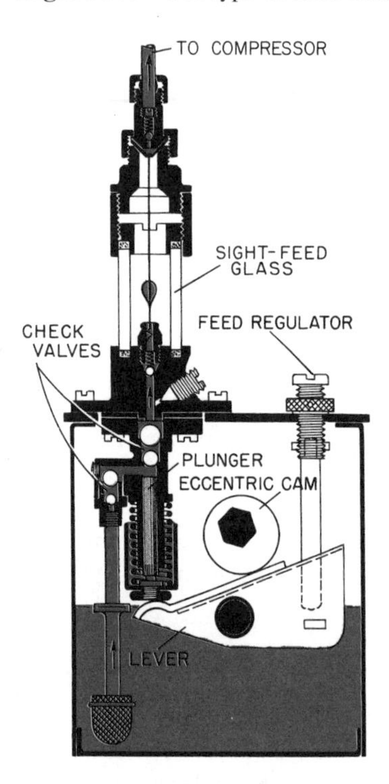

Figure 9.6 Force-feed lubricator with liquid-filled sight glass.

and forced under pressure on the upstroke, through the liquid-filled sight glass to the delivery line. The pump is operated by an eccentric cam and lever, which can be driven from a shaft on the machine or operated by a hand crank. The stroke of the pump can be regulated to adjust the oil feed rate, which can be estimated by counting the drops as they pass through the liquid-filled sight glass.

Some modern additive oils can cloud the liquid in the sight glasses of this type of lubricator, so lubricators with dry sight glasses are sometimes used. One of these is shown in Figure 9.7. The pump shaft is cam shaped in cross section and causes the plunger to move up and down. Specially shaped surfaces (*A*) at right angles to the pump shaft axis also act on the eccentric head of the plunger, causing it to oscillate about its axis. On the upstroke of the plunger, the head is turned so that the suction slot registers with the suction channel port, while the delivery slot, which is 90° away, is blanked off. Oil is drawn up the suction tube and over to the sight glass, where the falling drops may be observed. From the sight glass, the oil flows to the space below the plunger. On the downstroke of the plunger, the head rotates to align the delivery slot with the delivery channel port; the suction slot is blanked off. As the plunger is pushed downward, it forces oil out through the delivery tube. The plunger stroke, and amount of oil pumped, can be adjusted by means of the lift-adjusting screw. The lubricator can be driven by the machine or turned by a hand crank.

Mechanical force-feed lubricators are used on large stationary and marine diesel engines, cylinders of reciprocating steam engines, gas engines, and reciprocating compressors to supply lubricant to the cylinders. One pumping unit is used for each cylinder feed, and all the pumping units are usually mounted on a single reservoir.

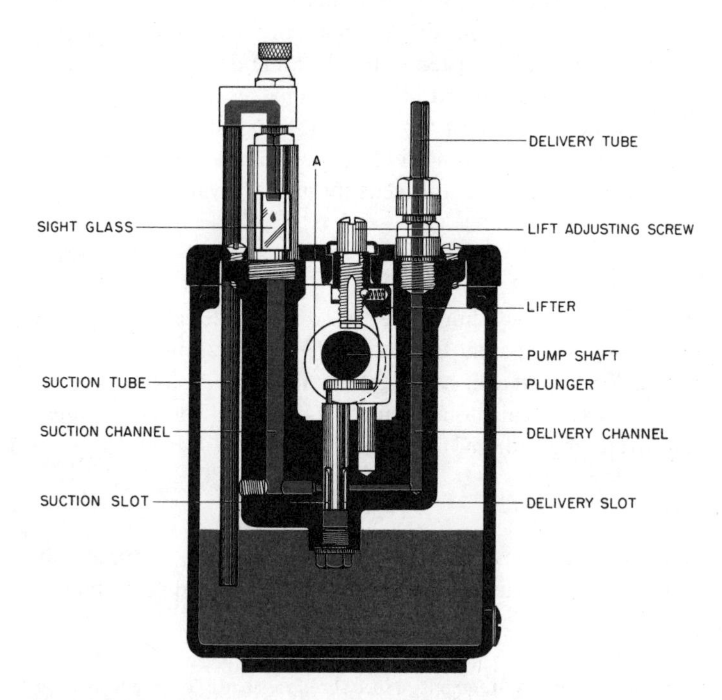

Figure 9.7 Force-feed lubricator with air drop sight glass.

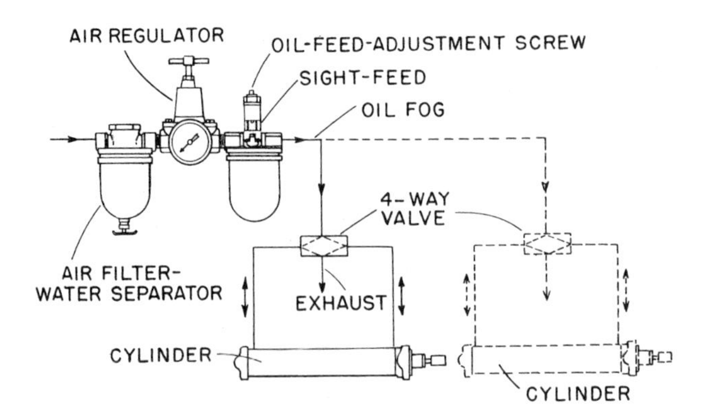

Figure 9.8 Fog lubricator for pneumatic cylinders.

5. Air Line Oilers*

Air-powered cylinders and tools are often lubricated by ''lubricating'' the compressed air supply. Pneumatic cylinders used for actuating parts of machines may be lubricated by means of an oil fog lubricator such as that in Figure 9.8. The flow of air through the lubricator creates an air–oil fog that carries sufficient oil to lubricate the cylinders.

Air tools such as rock drills and paving breakers are often lubricated by means of air line oilers. The air line oiler consists of an oil reservoir that contains a device for feeding a metered amount of atomized oil into the airstream. It is coupled into the air hose a short distance from the drill, and a fine spray of oil is carried from it to lubricate the drill wearing surfaces that are reached by the air. A method of varying the rate of oil feed is provided so that the oil feed can be adjusted to the drill and drilling conditions.

In the air line oiler (Figure 9.9), air for rock drill actuation passes through the center tube, and line pressure is applied to the oil reservoir via port *A* and the vertical drilled passages shown on the left. As the result of a venturi effect, a reduced pressure prevails at port *B*. Because of the difference in pressure, oil feeds through the valve (right view) and port *B* into the air stream.

6. Air Spray Application

Some open gear and wire rope lubricants, including some grease-type materials, are applied by means of hand-operated air spray equipment that uses either external mixing nozzles or airless atomizing equipment. This approach, however, has the same disadvantages as hand oiling. A number of automatic or semiautomatic units such as that shown in Figure 9.10 have been developed to overcome some of these problems.

B. Grease Application

Greases are used mainly for rolling element bearings, flexible couplings, and thin film plain bearings and slides. Greases are also used in some open and enclosed gear applica-

* Air line oilers and air spray lubrication (Section 1.A.6) comprise a specialized variation of oil mist lubrication, which is discussed at the end of this chapter.

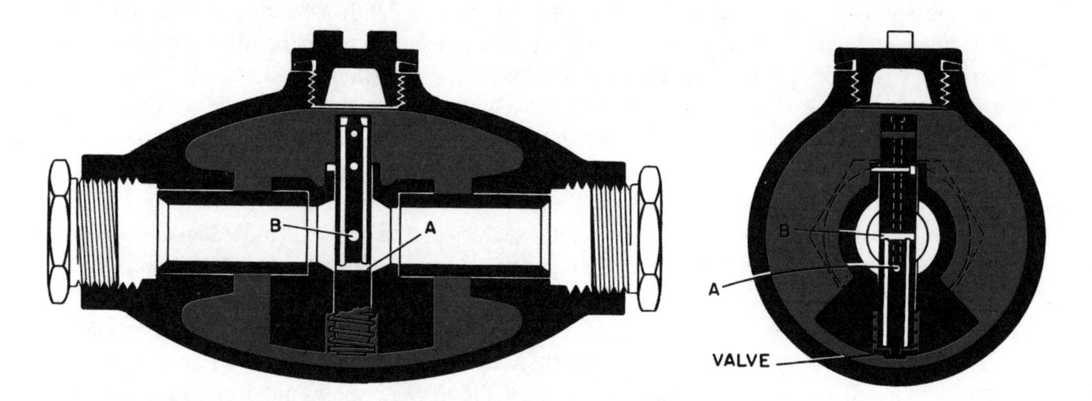

Figure 9.9 Air line oiler. Oil is drawn into the airstream through the port *B* by venturi effect. The enlarged section around the valve acts as a pendulum to swing the valve and keep the oil intake submerged in oil regardless of the position of the outer housing.

tions. Because of the lower leakage tendencies of greases, periodic application with a hand or air-powered grease gun overcomes many of the disadvantages of hand oiling; but, as with any manual lubrication method, fittings will be overlooked and the cost of application is usually high. Therefore, central grease lubrication systems are being used more frequently, and a number of bearings are being "packed for life."

Many rolling element bearings are manually packed at the time they are installed, and periodically thereafter, as is the case with automotive wheel bearings. Grease is forced

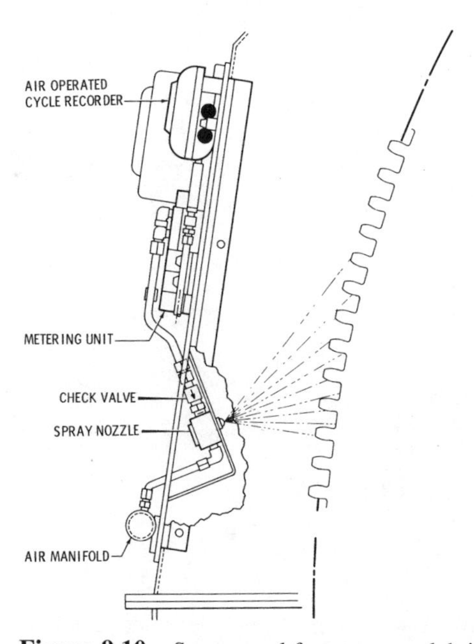

Figure 9.10 Spray panel for open gear lubrication.

into the spaces between the rolling elements by hand or with a bearing packer, and a moderate amount of grease is placed in the hub. The housing must not be filled completely, for there would be no place for the grease between the raceways to go when the bearing was operated. Excessive churning of the grease and overheating would then result.

Bearings intended for power grease gun lubrication must have provision for pressure relief to prevent pressure buildup and permit purging of old grease. This is sometimes accomplished by allowing sufficient clearance in the seals, but the same effect can also be achieved with a relief valve or a drain plug such as that in Figure 9.11. When a bearing of the latter type is relubricated, the drain plug should be removed and left out until the bearing has been operated long enough to expel all excess grease.

A wide range of self-contained grease lubrication devices ranging from spring-loaded reservoirs that feed small amounts continuously to battery- or gas-operated units that can be programmed to activate at specified intervals. Figure 9.12 shows an electromechanical lubricator that contains a motor, a piston pump, a gearbox, and a microprocessor capable of delivering precise lubricant quantities at specified intervals. These units are available in a range of sizes capable of single- or multiple-point distribution.

''Packed for life'' bearings are used for many light- to medium-duty services. These bearings have seals and are packed at the factory with a special long life grease intended to last for the life of correctly applied bearings. In some cases, the service range of bearings of this type is extended by making provision for relubrication by one of the following methods: removing one seal and applying grease through the seal by means of a gun with a hollow needle, or forcing grease through holes in the outer ring. Care is necessary with the latter method because only a small amount of pressure will dislodge some seals.

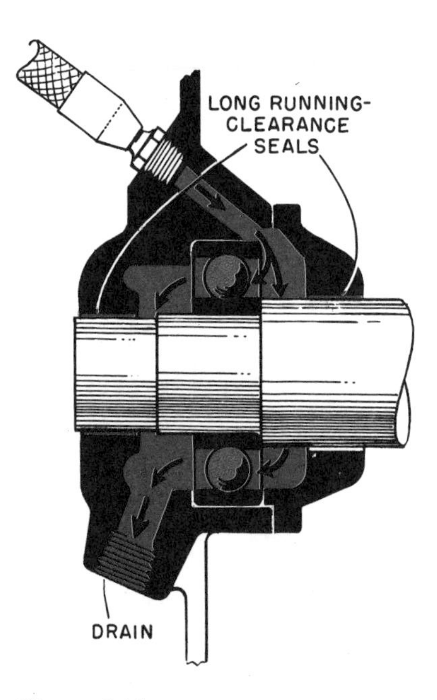

Figure 9.11 Free purging housing design, which permits good purging of old grease from both sides of the bearing when new grease is applied.

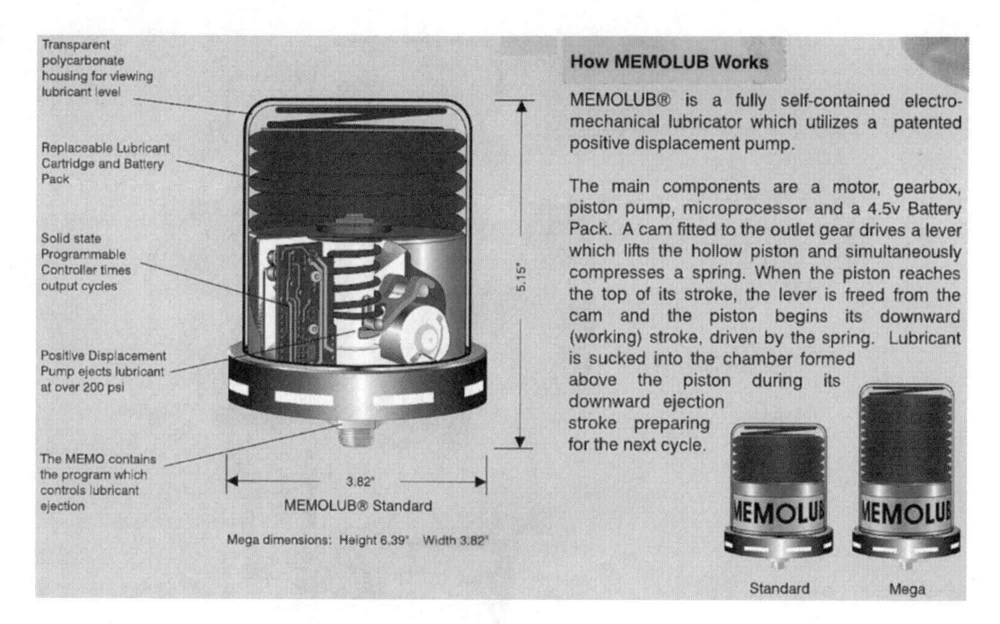

Figure 9.12 Electromechanical lubricator.

II. REUSE METHODS

Reuse methods of oil application include circulation systems supplying lubricant for one or more machines and self-contained systems such as bath, splash, flood, and ring oiling.

A. Circulation Systems

The term ''circulation system'' generally refers to a system in which oil is delivered from a central reservoir to all bearings, gears, and other elements requiring lubrication. All the oil, disregarding minor leakage, drains back to a central sump and is reused. Two principal variations of this type of system are used—namely, pressure feed and gravity feed. In pressure feed systems, a separate sump and reservoir may be used, or the two may be combined (Figure 9.13). The oil is pumped directly to the parts requiring lubrication. Where a separate sump is used, it may be either ''wet'' (drain is located so that a certain amount of oil remains in the sump at all times) or ''dry,'' (drain is located and sized so that the sump remains essentially empty at all times). In gravity feed systems, oil is pumped to an overhead tank and then flows under gravity head to the elements requiring lubrication.

In either type of system, very often the rate of flow is determined primarily by what is needed for cooling. This amount of oil usually will be more than is needed for lubrication. Means of cooling and purifying the returned oil are often included in circulation systems.

Although the many types of application for circulation systems require considerable variations in size, arrangement, and complexity, in a general way, circulation systems can be considered in one of three groups:

1. Systems comprising a compact arrangement of pump, reservoir, and oil passages built into the housing of the lubricated parts such as that in Figure 9.13

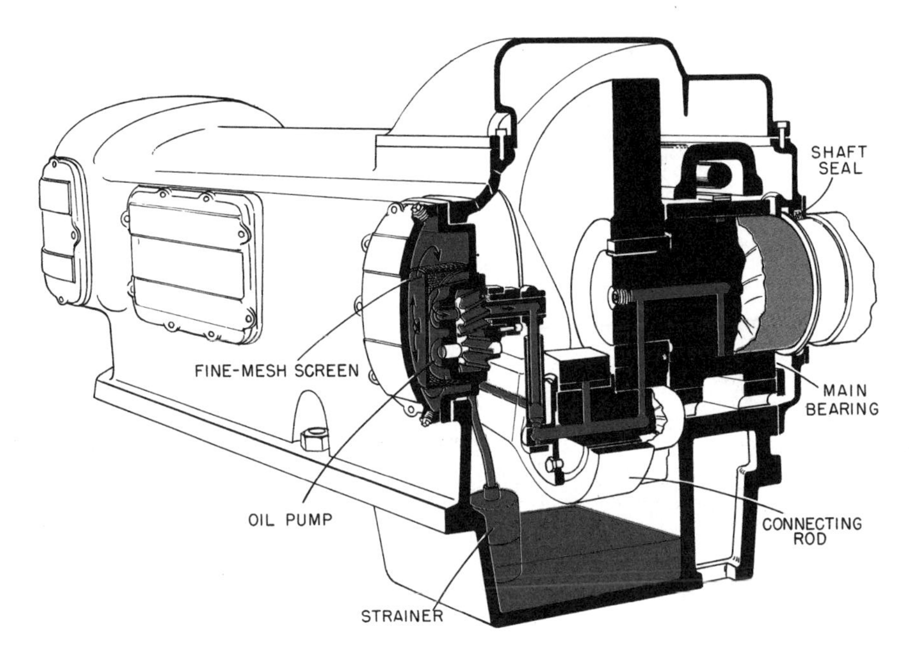

Figure 9.13 Pressure-feed circulation system for horizontal, duplex two-stage compressor: cutaway view through the main and crankpin bearings of the right-hand frame. The spiral gear oil pump draws oil from the reservoir through a strainer and forces it through a fine mesh screen and the hollow pump arm to the crankpin bearing. Excess oil is bypassed to the reservoir through a relief valve (not shown). From the crankpin bearing, oil flows under pressure through internal passages to the main bearing, to the crosshead pin bearing and crosshead guides.

2. Systems employing a multicompartment tank combining reservoir and purification facilities (Figure 9.14)
3. Systems comprising an assembly of individual units (reservoir, oil cooler, oil heater, oil pumps, purification equipment, etc.)

The system illustrated in Figure 9.15 is fairly typical of the third type. Returning oil drains to a settling compartment, entering the reservoir at or just above the oil level. Water and heavy contaminants settle, and the sloping bottom of the reservoir helps to concentrate these impurities at a low point from which they can be drained. Partially purified oil overflows a baffle to the clean oil compartment. In some systems, especially where large reservoirs are used, baffles may be omitted. The clean oil pump takes oil, usually through a suction strainer, and pumps it to a cooler, optional oil filter, and then to bearings, gears, and other lubricated parts. The pressure desired in the oil supply piping is maintained by means of a relief valve, which discharges to the reservoir at a point below the oil level. A continuous bypass purification system is shown. The pump takes 5–15% of the oil in circulation from a point above the maximum level of separated water in the reservoir and pumps it through a suitable filter back to the clean oil compartment. The following discussion of good practices in circulation system design refers primarily to systems of this type, but the ideas presented are fundamental to most systems.

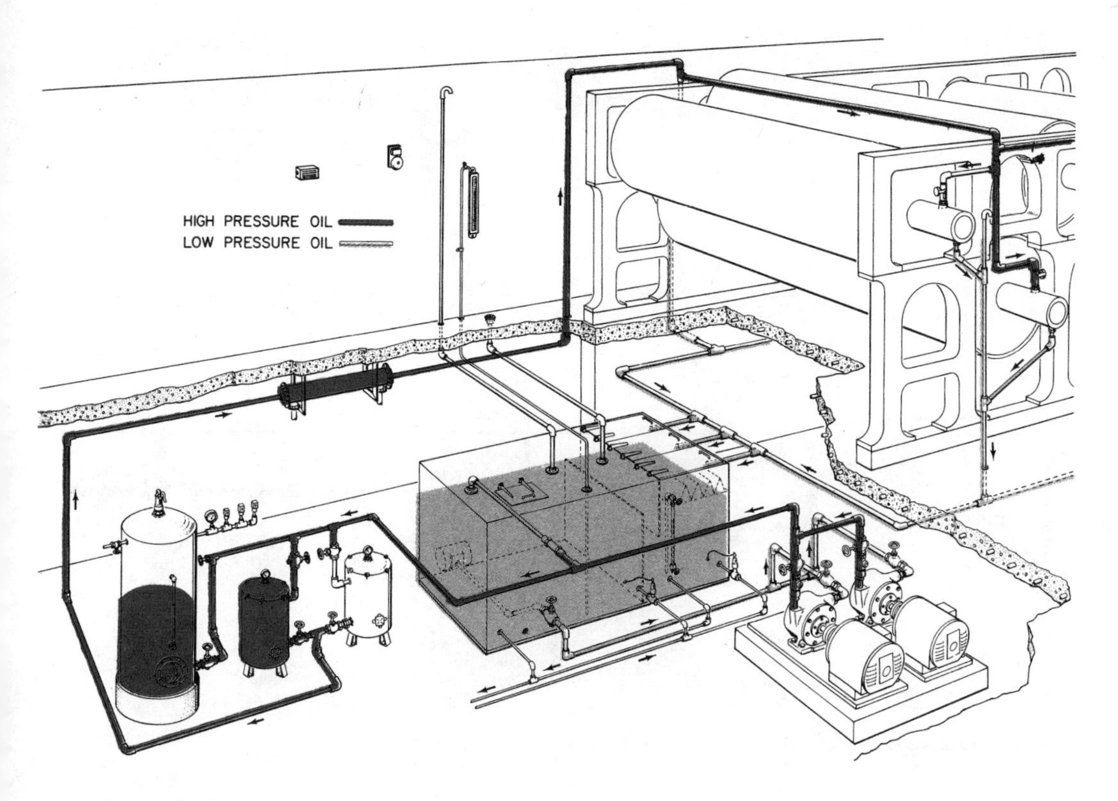

Figure 9.14 Paper mill dryer circulation system.

1. Oil Reservoirs

The bottom of an oil reservoir should slope at a ratio of more than 1 to 25 (4%) toward a drain connection, which should be located at the lowest point in the reservoir. This construction promotes the concentration of water and settled impurities and permits their removal without excessive loss of oil. In addition, it permits more complete removal of flushing oil or solvents used in cleaning operations prior to start-up or during an oil change. Where oil is taken directly from the reservoir for purification, it should be removed close to the low point but above any water or settled impurities. An opening or openings above the oil level, adequate for inspection and cleaning, should be provided. Large reservoirs should have an opening large enough for a person to enter. These and any other openings should have well-secured, dust-tight covers. Before personnel enter enclosed reservoirs, all safety precautions should be taken.

A connection should be provided at the highest point in the reservoir for ventilation. Proper ventilation results in the removal of moisture-laden air and thereby reduces condensation on cooler surfaces above the oil level and subsequent rusting of these surfaces. Ventilation fixtures should be designed with air filters or desiccant breathers to prevent entrance of airborne contaminants and excessive moisture. Instead of natural ventilation, where a source of water contamination is common—for example, large steam turbine systems or paper mill dryer systems—medium-sized and large reservoirs should be pro-

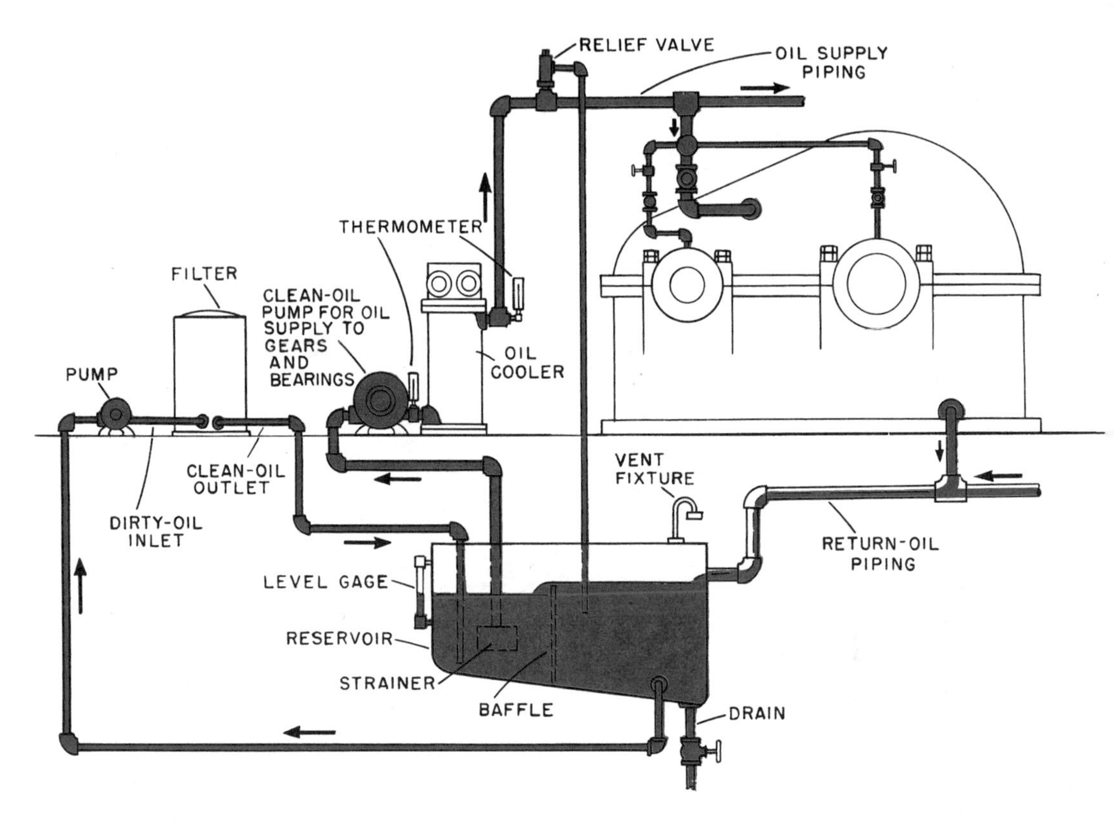

Figure 9.15 Circulation system diagram: elements illustrated include one type of pressure circulation system and several commonly accepted practices.

vided with a ''vapor extractor,'' capable of maintaining a slight vacuum in the air space above the oil level. Too high a vacuum should be avoided, however, since it might pull plant atmospheric contaminants into the lubrication system.

The main oil return connection should be located at or slightly above the oil level and remote from the oil pump suction. Returning oil should not be permitted to drop from a considerable height directly into the oil body, since this action tends to whip air into the oil and cause foaming or to hold water and contaminants in suspension. Instead, the fall of oil should be broken and dispersed by means of a baffle, sheet metal apron, or fine screen. A line that may carry any air should never discharge below the oil level. A connection for the return to the reservoir of a ''solid'' stream of oil, as from a pressure relief valve, should be placed about 6 in. (150 mm) below the oil level.

It is convenient to consider reservoir size in terms of the oil volume flowing in the system. The reservoir should be large enough to ensure that oil velocity in it will be low and that the oil will have sufficient rest time to allow adequate separation of water and entrained solids, separation of entrained air, and collapse of any foam that may exist. In practice, reservoir sizes range from a suggested minimum of 2 times system flow per minute to over 40 times. Many representative systems have reservoirs of 5–10 times system flow per minute. Newer design large steam turbine systems and large paper mill dryer systems will typically be in the range of 40 plus times total pumping capacity.

2. Pump Suction

The clean-oil pump suction opening should be above the bottom of the reservoir to avoid picking up and recirculating settled impurities. However, it must be below the lowest oil level that may occur during operation. Where there is considerable variation of the oil level in the reservoir and it is desired to take oil at or near the surface, a floating suction may be used. Floating suction is frequently used in reservoirs of systems exhibiting constant, extreme water contamination. The oil supplied to the circulation systems from the top of the reservoir will have the least water contamination.

3. Bearing Housings

The floors of bearing housings should have a slope of about 1 in 50 (2%) toward the drain connection. The design should be such that there are no pockets to trap oil and prevent complete drainage. Shaft seals should be adequate to prevent loss of oil or the entrance of liquid or solid contaminants. Any type of breather or vent fixture on a bearing housing should be provided with an air filter to keep out dust and dirt.

4. Return Oil Piping

Gravity return oil piping should be sized to operated about half-full under normal conditions, and it should have a slope of at least 1 in 60 (1.7%) toward the reservoir. Any unavoidable low spots should have provision for periodic water removal. Severe piping bends, such as 90° or more out of the bearing housings, should be avoided to minimize the potential for oil backing up into the bearings causing overheating and increased oil leakage. Smooth flow passage of return oil is also important in avoiding buildup of deposits in return piping.

5. Circulation System Metals

Exclusive of bearings, the parts of a circulation system should preferably be made of cast iron or carbon steel. Fittings of bronze are acceptable. Stainless steel tubing is very good. Aluminum alloy tubing is acceptable as far as chemical inertness with oil is concerned, but it may not have sufficient structural strength for high pressure lines. No parts should be galvanized. As a general rule, no parts, exclusive of bearings, should be made of zinc, copper, lead, or other materials that may promote oil oxidation and deterioration. Copper tubing may be used for oil lines in some installations, but should not be used in systems such as those for steam turbines, where extremely long oil life is desired. The use of copper should be confined to those applications for which the oil is formulated specifically to inhibit the catalytic effects of copper.

6. Oil Filtration

Much of the older equipment was equipped with coarse filtration (40 μm or larger) or in some cases, no filtration at all. As machinery became more complex, the importance of oil filtration in helping provide long equipment life was well recognized. This is particularly true in close-tolerance equipment such as servovalves in the machine tool industry or other precise control mechanisms such as governor controls in large turbines. Figure 9.15 showed filtration in a bypass loop, but more commonly this function is found in the pressure side of the supply line to equipment components. In addition to bypass (or kidney loop) filtration and high pressure side filtration, filtration can be in low pressure return

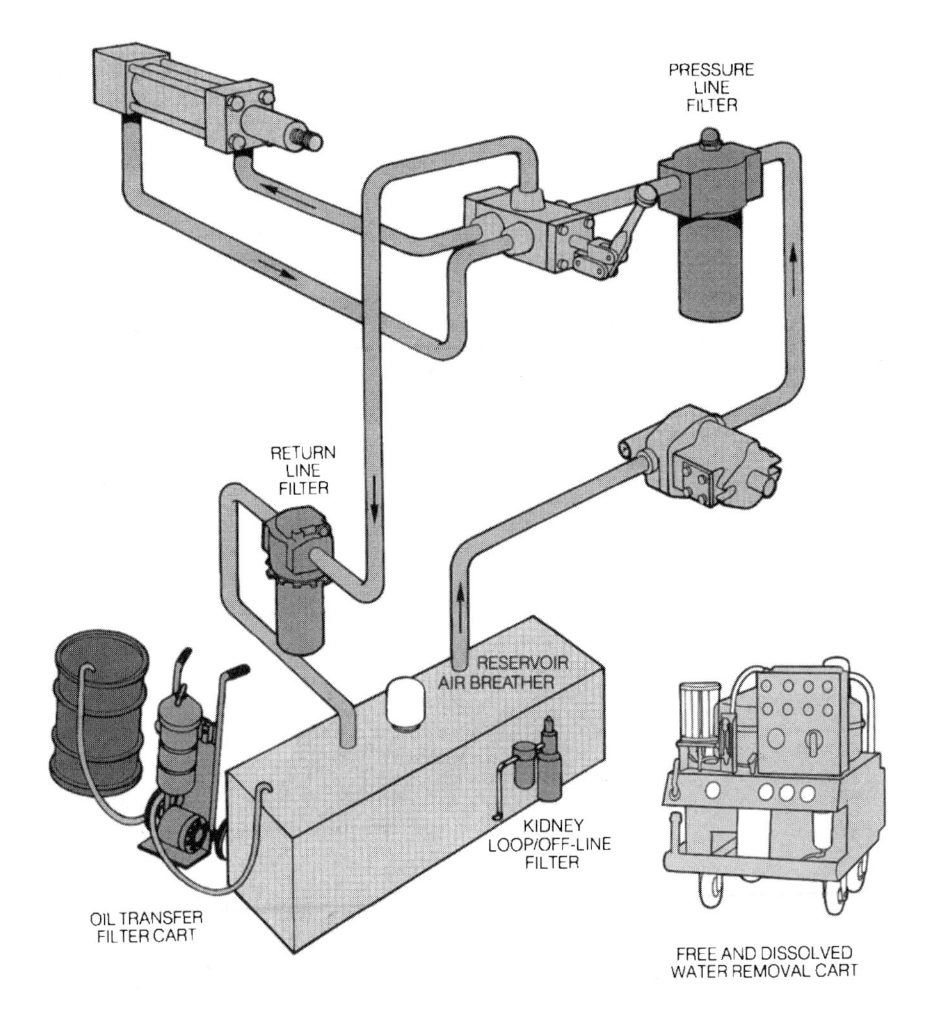

Figure 9.16 Oil filtration and purification.

lines (except not generally used in gravity return systems). Figure 9.16 shows these three main types of filtration and two alternative methods for keeping system oils clean: oil transfer equipment and a freestanding reclamation unit.

7. Oil Coolers

Oil coolers should be located so that all connections and flanged covers are accessible and the tube bundles can be removed conveniently for cleaning. Cooler capacity should be adequate to prevent oil temperature from rising above a safe maximum during the hottest conditions. Means should be provided to control oil temperature at all times by regulating the flow of cooling medium. Where practical, the oil pressure should always be higher than the water pressure to prevent water entering the lubricant in the event of a cooler leak.

8. Oil Heating

Heating of the oil is desirable or required in circulation system-perhaps the oil is too cold, at start-up condition, to provide adequate flow or lubrication to critical components, or perhaps certain thermal conditions need to be maintained within the system.

The two most common methods of heating oils in industrial applications are steam and electricity. Steam is readily available in many large plants such as fossil fuel fired power plants or paper mills and, most often, steam is used for oil heating requirements. When steam is used, caution should be taken to prevent the exposure of stagnant oil to the full temperatures of the steam. Even saturated condensate steam at 15 psi has a temperature of 335°F. Superheating of these steams can raise temperatures considerably. It is a good practice to maintain heating element surface temperatures below 200°F unless higher quality oils are used and/or oil flow across the heating elements can be maintained.

Oil degradation caused by contact with high heater skin temperatures can take various forms, including the following:

Additive depletion
Additive decomposition
Oxidation
Hydrocarbon cracking

If electrical immersion heaters are used, maximum safe heater watt densities should be determined. This information is available through oil suppliers and manufacturers of immersion heaters. As a general rule, a safe watt density to keep surface element temperatures in the 200°F range is about 5 W/in.^{2}. In many applications it may be desirable to heat oil quickly, and this will necessitate either multiple heating elements or fewer elements with much higher watt densities (higher heating element surface temperatures). In these instances, it will be necessary to maintain sufficient oil velocity across the heating elements to minimize the time that the thin films of oil are in contact with the high temperature surfaces.

9. Monitoring Parameters

The two most common parameters to measure on circulating oil systems are oil temperature and oil level. Oil temperatures should be measured in the oil reservoir, in the supply to components, and on the discharge side of main system operating components, as well as at the inlet and outlet of heat exchangers. Changes in ''normal'' operating temperatures, which may signal a malfunction in the system, could be used to predict a pending component failure.

Maintaining oil at the proper level in the reservoir is important in several respects. It allow adequate retention time in the reservoir to drop out contaminants such as water and abrasive materials, and to dissipate air, while providing some radiant cooling of the system return oil. Maintaining proper levels and temperatures will go a long way to improving oil life, reducing filter costs, and protecting equipment components.

The more sophisticated systems will monitor many more parameters such as pressures, follows, and differential pressures across filters and heat exchangers. Alarms may be used to indicate low levels, high temperatures, low flows, or low pressure. These alarms can be audible (bells) or visual (warning lights) and can be tied into computer systems to monitor operations or to alert personnel remote from the equipment location.

III. OTHER REUSE METHODS

In addition to circulation systems, a number of other methods of oil application involve more or less continuous reuse of the oil. These are differentiated from integral circulation systems primarily in that pumps are not used to lift the oil.

A. Splash Oiling

Splash oiling is encountered mainly in gear sets or in compressor or steam engine crankcases. Gear teeth, or projections on connecting rods, dip into the reservoir and splash oil to the parts to be lubricated or to the casing walls, where pockets and channels are provided to catch the oil and lead it to the bearings (Figure 9.17). In some systems, oil is raised from the reservoir by means of a disk attached to a shaft, removed by a scraper, and led to a pocket from which it is distributed (Figure 9.18). This variation may be called a flood lubrication system. In either case, the oil returns to the reservoir for reuse after it has flowed through the bearings or over the gears. Accurate control of the oil level is necessary to prevent either inadequate lubrication or excessive churning and splashing of oil.

B. Bath Oiling

The bath system is used for the lubrication of vertical shaft hydrodynamic thrust bearings and for some vertical shaft journal bearings. The lubricated surfaces are submerged in a bath of oil, which is maintained at a constant level. When necessary, cooling coils are placed directly in the bath. The bath system for a thrust bearing may be a separate system or may be connect into a circulation system.

C. Ring, Chain, and Collar Oiling

In a ring-oiled bearing, oil is raised from a reservoir by means of a ring that rides on and turns with the journal (Figure 9.19). Some of the oil is removed from the ring at the point

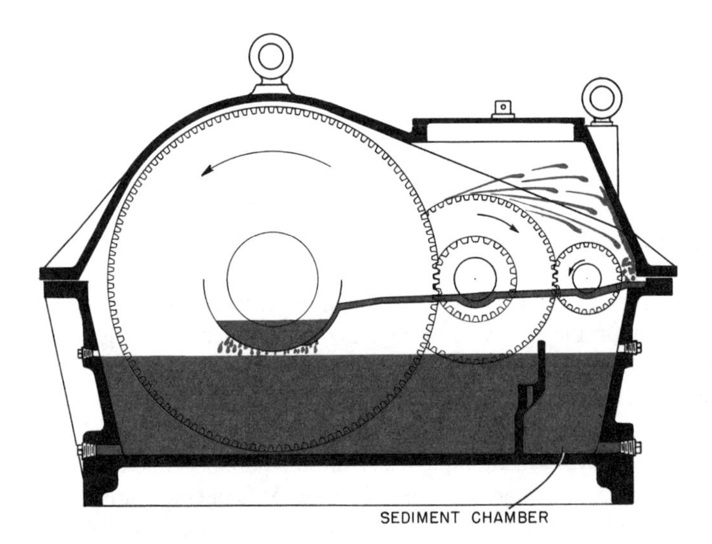

Figure 9.17 Splash oiling system. The gear teeth carry oil directly to some gears and splash it to others and to collecting troughs that lead it to bearings not reached by splash.

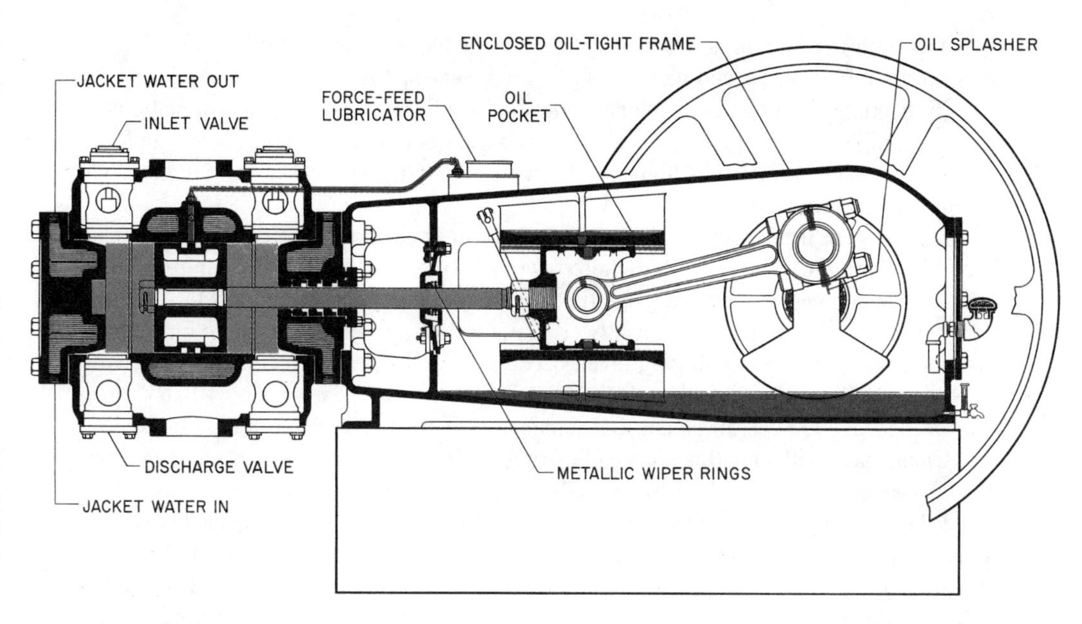

Figure 9.18 Flood lubrication system. In this single-cylinder, water-cooled compressor, running parts are lubricated by the splash and flood method. Crankshaft counterweights and an oil splasher dip into the oil reservoir and throw oil to all bearings. Oil pockets over the crosshead and over each double-row, tapered roller main bearing assure an ample supply for these parts.

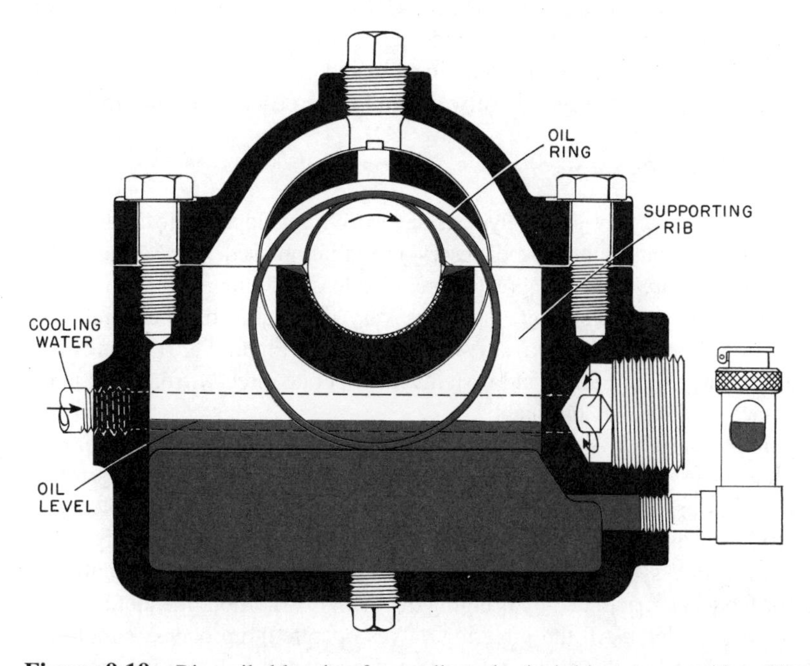

Figure 9.19 Ring-oiled bearing for small mechanical drive steam turbine. The plug in the cap can be removed to observe the turning of the oil ring. The running oil level (indicated) is raised during shutdown as a result of oil draining from the bearing and walls of the housing. The bearing is cooled by means of water passages through supporting ribs on either side of the oil ring slot. The spring cap on the fill and level gage fitting (right) helps keep out dirt.

of contact with the journal and is distributed by suitable grooves in the bearing. The oil flows through the bearing and drains back to the reservoir for reuse.

Ring oiling is applied to a wide variety of medium speed bearings in stationary service. At high surface speeds, too much slip occurs between ring and journal and not enough oil is delivered. Also, at high speeds, in large, heavily loaded bearings, not enough cooling may be provided.

Oil rings are usually made about 1.5–2 times journal diameter. Bearings more than about 8 in. (200 mm) long, usually required two or more rings. The oil level in reservoirs is usually maintained so that the rings dip less than one-quarter their diameter. The oil level, within a given range, is not usually critical; too low a level may result in inadequate oil supply, however, and too high a level, because of excessive viscous drag, may cause ring slip or stalling. As a result, too little oil may reach the bearing, and ''flats'' may wear on the rings to such an extent that satisfactory performance is no longer possible.

Chains are used sometimes instead of rings in low speed bearings, since they have greater capacity for lifting oil at low speeds.

Where oils of very high viscosity are required for low speed, heavily loaded bearings, a collar that is rigidly attached to the shaft may be used instead of a ring or chain. A scraper is required at the top of the collar to remove the oil and direct it to the distribution grooves in the bearing.

IV. CENTRALIZED APPLICATION SYSTEMS

A number of factors have contributed to the growing use of centralized lubricant application systems. Among these are improved reliability, reduced cost of labor for lubricant application, reduced machine downtime required for lubrication, and, generally, a reduction in the amount of lubricant used through reduction of waste and more efficient use of lubricants.

A. Central Lubrication Systems

A number of types of central lubrication system have been developed. Most can apply either oil or grease, depending on the type of reservoir and pump used. Greases generally require higher pump pressures because greater pressure losses occur in lines, metering valves, and fittings. Pump and reservoir capacities vary depending on the number of application points to be served, ranging from small capacities (Figure 9.20) to units that install on standard drums (Figure 9.21) and systems that operate directly from bulk tanks or bins requiring large volumes of lubricant.

In some systems, called *direct* systems, the pump serves to pressurize the lubricant and also to meter it to the application points. In *indirect* systems, the pump pressurizes the lubricant but valves in the distribution lines meter it to the application points.

Two basic types of indirect systems are in common use, and in turn each type has two variations. In *parallel* systems, also called header or nonprogressive systems, the metering valves or feeders are actuated by bringing the main distribution line up to operating pressure (Figure 9.22, left). All the metering valves operate more or less simultaneously. This type of system has the disadvantage that if one valve fails, no indication of failure is given at the pumping station. However, all the other application points will continue to receive lubricant. In series, or progressive, systems the valves are ''in'' the main distribution line (Figure 9.22, right). When the main distribution line is brought up to pressure, the first valve operates. After it has cycled, flow passes through it to the

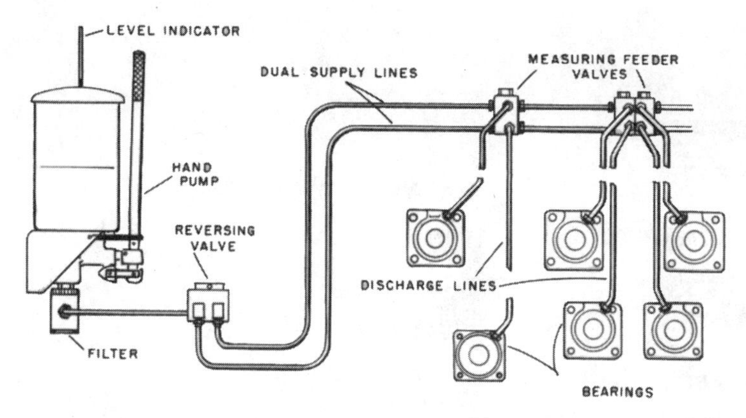

Figure 9.20 Typical hand-operated centralized lubrication system for lubrication of a limited number of points. It can be either a dual-line system (as shown) or a single-line system.

Figure 9.21 Drum-mounted pump and control unit. A standard 400 lb drum serves as the reservoir. The pump can be of one air, electric, or hydraulic type; it develops 3500 psi (24 MPa) pressure. The unit is designed for a single-line, spring return lubrication system.

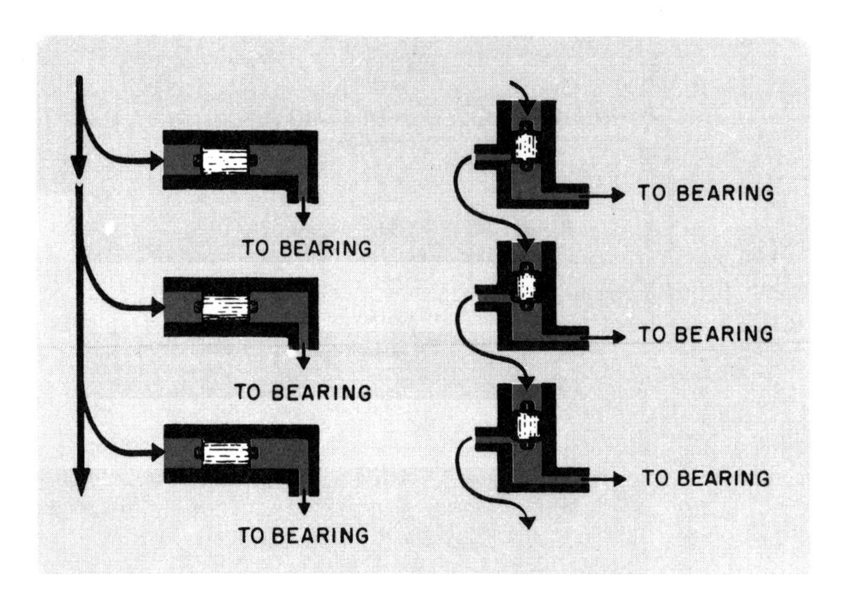

Figure 9.22 Parallel and series lubrication system. In the parallel system (left) the valves are ''off'' the supply line, while in the series system (right) the valves are ''in'' the supply line.

second valve, and then to each succeeding valve in turn. Thus, if one valve fails, all fail, and the pressure rise at the pump or the distribution block can be used to signal that a failure had occurred. The two variations of each of these basic types are as follows.

1. Two-Line System

One variation of the parallel system uses two supply lines (Figure 9.23). The four-way valve directs pressure alternately to the two lines, at the same time relieving pressure in the line that is not receiving flow from the pump. The valve can be operated manually from the machine, cycled by a timer, or controlled by a counter that measures the volume delivered by the pump. The valves are designed to deliver a charge of lubricant to the application point each time the flow in the lines is reversed.

2. Single-Line Spring Return

In the second variation of the parallel system, only a single distribution line is used, the layout is shown in Figure 9.24. As with the four-way valve of the preceding system, the three-way valve may be operated manually from the machine, by a timer, or by a counter measuring pump output. The valves deliver a charge of lubricant when system pressure is applied to them and reset themselves when the system pressure is relieved.

3. Series Manifolded System

In the series manifolded type of system, a single supply line is used. No relief valve is required, since the manifold-type valves automatically reset themselves and continue cycling as long as pressure is applied to them through the supply line. The system can be cycled by starting and stopping the pump.

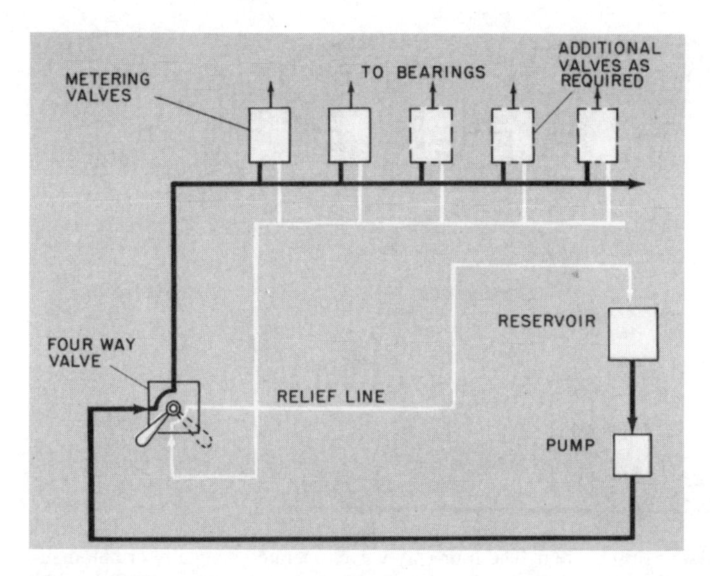

Figure 9.23 Two-line parallel system. The four-way valve, operated manually or automatically, alternately directs pump pressure to one line and then the other. When one line is pressurized, the other line is relieved.

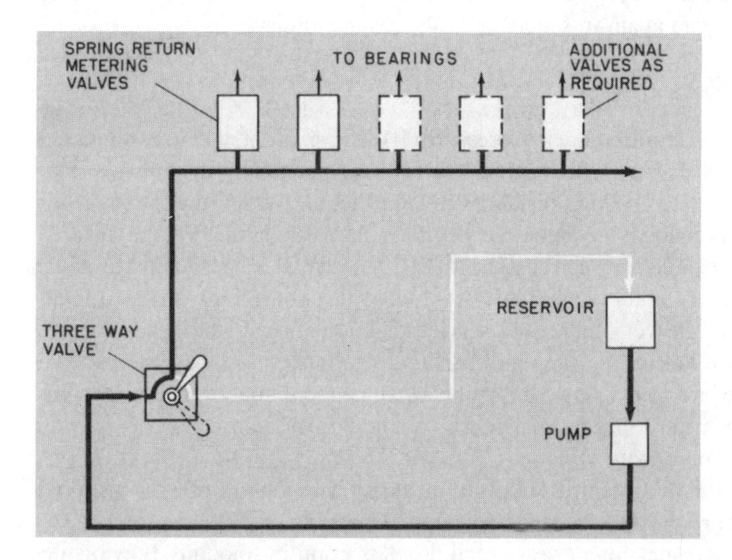

Figure 9.24 Single-line spring return system. The three-way valve, operated manually or automatically, either directs pump pressure into the supply line or relieves the pressure in the line to permit the spring return valves to reset.

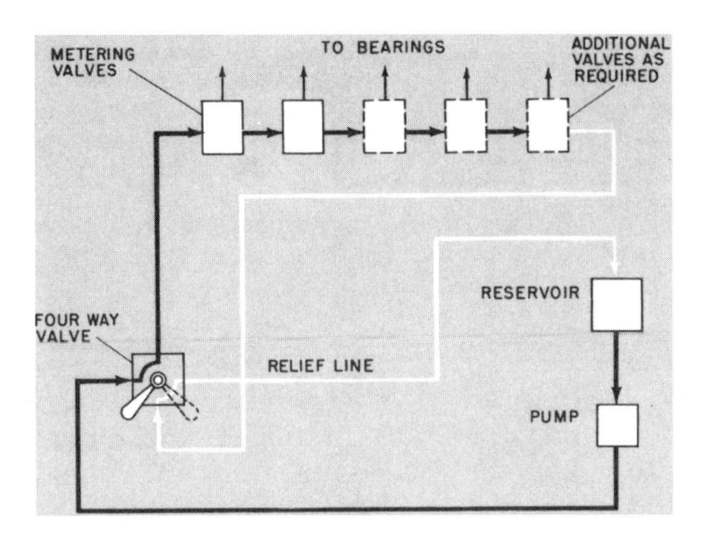

Figure 9.25 Series reversing flow system. The four-way valve, operated manually or automatically, directs pump pressure to one end of the closed-loop supply line while relieving the pressure at the other end.

4. Series System, Reversing Flow

The second series system uses a single supply line with a four-way valve to reverse the flow in it (Figure 9.25). The valves are designed to deliver a charge of lubricant, then permit lubricant flow to pass through to the next valve. When the flow in the supply line is reversed, the valves cycle again in sequence in the reverse order.

B. Mist Oiling Systems

In oil mist lubricators, oil is atomized by low pressure (10–50 psi, 70–350 kPa) compressed air into droplets so small that they float in the air, forming practically dry mist, or fog, that can be transported relatively long distances in small tubing. When the mist reaches the application point, it is condensed, or coalesced, into larger particles that wet the surfaces and provide lubrication. Condensing can be accomplished in several ways. Oil mist systems have proven their reliability in an increasing variety of applications. They are used in all types of industry—from the very light duty service of lubricating dental handpieces to the heavy-duty service of lubricating steel mill backup rolls. In the past, the systems were usually built onto or adapted to existing equipment. Machine tool builders are now designing them into their newer machines, primarily for spindle bearing lubrication, to provide greater reliability and productivity.

An oil mist lubrication system is simply a means of distributing oil of a required viscosity from a central reservoir to various machine elements.

A true oil mist is a dispersion of very small droplets of oil in smoothly flowing air. The size of these droplets averages from 1–3 μm (1 μm = 0.000039 in.) in diameter. In comparison, an ordinary air line lubricator produces an atomized mixture of droplets up to 100 μm in diameter, which are suspended (temporarily in turbulent air flowing at high velocity and pressure). In an air line lubricator system, the air is a working fluid that is transmitting power, whereas in an oil mist system, air is used only as a carrier to transport the oil to points where it is required.

The droplet size is a very important consideration in the proper design of an oil mist system. The larger the droplets, the more likely they are to wet out and form an oily film at low impingement velocities. At practical, low flow rates, the size limit is taken to be 3 μm. Droplets over this size will wet or spread out on surfaces quite readily, while particles less than this diameter will not.

A dispersion of droplets less than 3 μm in diameter will form a stable mist and can be distributed for long distances through piping. At the points requiring lubrication, these drops can be made to wet metal surfaces by inducing a state of turbulence, causing small droplets to collide and form into large diameter drops. These larger drops wet metal surfaces to provide the necessary lubricant film.

This formation of larger drop sizes that will wet metal surfaces is referred to as condensation, although other terms (reclassification, condensing, coalescing, etc.) are also used.

Different degrees of condensation may be achieved by using different adapters at the points requiring lubrication. These adapters are usually classified as mist nozzles, spray or partially condensing nozzles, and completely condensing, or reclassifying, nozzles. When high speed rolling element bearings create sufficient turbulence in the bearing housing to cause the droplets to join and wet out, a mist-type nozzle may be used. When gears are being lubricated, it is usually necessary to partially condense the oil mist to ensure that the limited amount of agitation within the gear housing will cause the droplets to coalesce and wet out. When slow moving slides or ways are being lubricated, it is usually necessary to completely condense the oil mist into a liquid which is then applied to the bearing surface.

In a typical oil mist system (Figure 9.26) compressed air enters through a water separator, a fine filter, and an air regulator to the mist generator (a). From the generator the mist is carried to a manifold (b) and then to the various application points (c).

To produce an oil mist, liquid oil is blasted with air to mechanically break it up into tiny particles. Droplets over 3 μm are screened or baffled out of the flow and returned to the sump or reservoir. The resultant dispersion (containing oil droplets averaging 1–3 μm in diameter) is the oil mist to be fed into the distribution system.

The sizes of the venturi throat, oil feed line, and pressure differentials impose physical limits on the viscosity of oil that can be misted. By the judicious use of oil heaters in the reservoir, and in some designs air line heaters to heat incoming air, the viscosity of normally heavy-bodied oils can be lowered to make misting possible. Systems without heaters usually can handle oils up to approximately 800–1000 SUS at 100°F (173–216 cSt at 38°C). If ambient temperatures are much below 70°F (21°C), heat is likely to be needed to reduce the oil's effective viscosity. Also, oils of over 1000 SUS at 100°F (216 cSt at 38°C) usually require heating to lower their effective viscosity and make possible the formation of a stable oil mist.

If the immersion elements in the oil reservoir are not properly adjusted, additional heating of the oil by the heated air will raise the bulk oil temperature until it is able to oxidize quite readily, and varnish or sludge may form in the generator. When heated air is being used, it should be no hotter than necessary to allow easy misting (usually below a maximum temperature of 175°F or 80°C). Also, the oil reservoir immersion elements used in conjunction with heated air should be used primarily at start-up and later, only if necessary to maintain oil temperature during operation. Naturally, the immersion elements

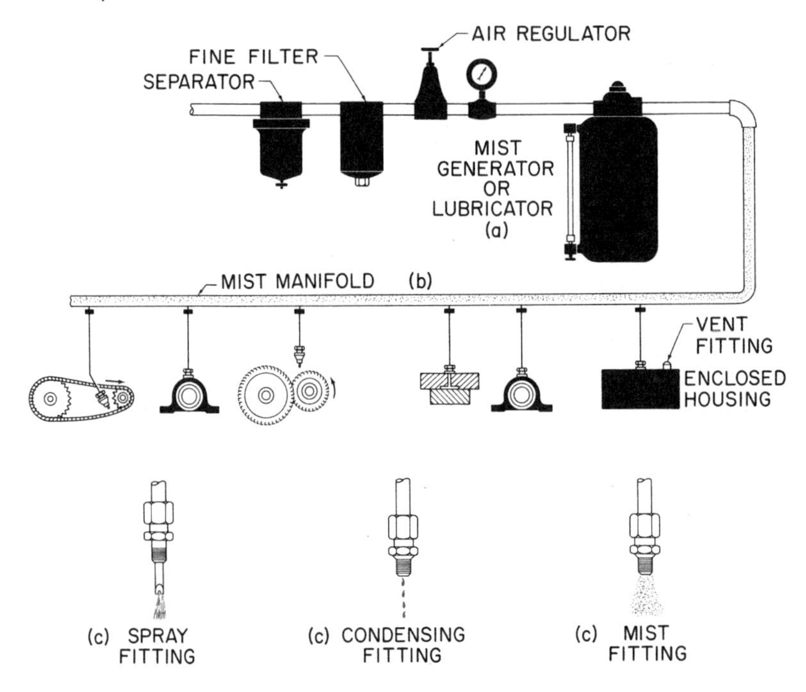

Figure 9.26 Oil mist lubrication system.

should be designed for oil heating and should have a low watt density (e.g., 5 W/in.^{2}, or 0.8 W/cm^2).

In addition to lubrication, mist systems provide other benefits: (1) the flow of air through bearing housings provides some cooling effect, and (2) the outward flow of air under positive pressure helps prevent the entrance of contaminants. Proper venting must be provided to permit this air to flow out of housings.

BIBLIOGRAPHY

Mobil Technical Books

Compressors and Their Lubrication
Gears and Their Lubrication
Plain Bearings, Fluid Film Lubrication

Mobil Technical Bulletins

Handling Storing and Dispensing Industrial Lubricants
Oil Mist Lubrication
Rock Drill Lubrication
Rolling Element Bearings, Care and Maintenance
Mobil Lubrication Service Guides
Centralized Oil or Grease Lubrication Systems
Oil Mist Systems

10

Internal Combustion Engines

The term ''internal combustion'' describes engines that develop power directly from the gases of combustion. This class of engines includes the reciprocating piston engines, used in a wide variety of applications, and most gas turbines. However, since closed-system gas turbines are not truly internal combustion engines, gas turbines are discussed separately. This chapter is concerned primarily with reciprocating piston engines.

Piston engines range in size from the fractional horsepower units used to power toys and prototype equipment, such as model airplanes, to engines for marine propulsion and industrial use that develop power in the order of 50,000 hp or more. While this wide range of engine sizes and the types of application of the engines present a variety of lubrication challenges, certain factors affecting lubrication are more or less common to all reciprocating engines.

The primary objectives of lubrication of reciprocating engines are the prevention of wear and the maintenance of power-producing ability and efficiency. These objectives require that the lubricant function effectively to lubricate, cool, seal, and maintain internal cleanliness. How well these factors can be achieved depends on the engine design, fuel, combustion, operating conditions, the quality of maintenance, and the engine oil itself.

I. DESIGN AND CONSTRUCTION CONSIDERATIONS

Among the design and construction features that affect lubrication are the following:

1. **Combustion cycle:** whether two stroke or four stroke
2. **Mechanical construction:** whether trunk, piston, or crosshead type
3. **Supercharging:** whether the engine is supercharged (via supercharger, turbocharger, or blower) or naturally aspirated
4. **General characteristics:** describing the lubricant application system as a whole

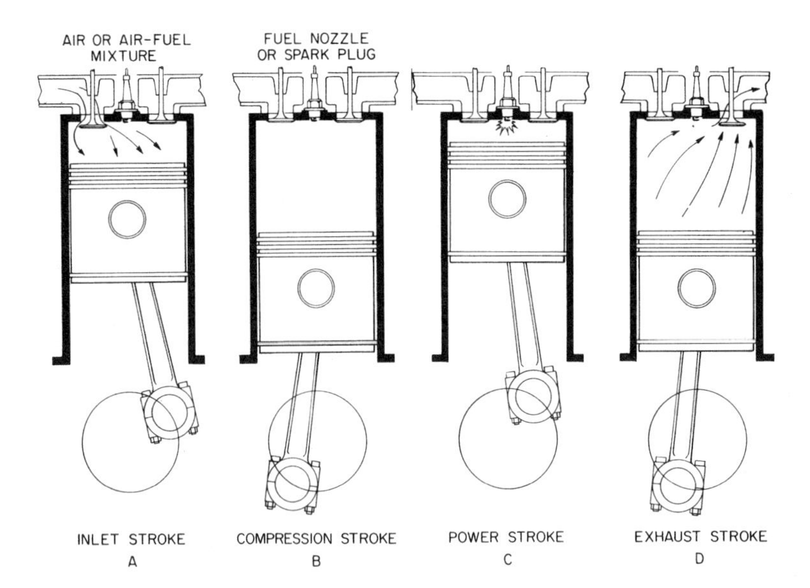

Figure 10.1 Four-stroke cycle. (A) The piston is moving downward, drawing in a charge or air or air–fuel, mixture. (B) The valves are closed and the piston compresses the charge as it moves upward. (C) Fuel has been injected and ignited by the high temperature compressed air, or a spark is passed across the spark plug igniting the air–fuel charge, and the piston is pushed downward on the power stroke by the expanding hot gases. (D) The burned gages are forced out of the cylinder through the open exhaust valve.

A. Combustion Cycle

In a reciprocating engine, the combustion cycle in each cylinder can be completed in one revolution of the crankshaft (i.e., one upstroke* and one downstroke of the piston) or in two revolutions of the crankshaft (i.e., two upstrokes and two downstrokes). The first engine is referred to as a two-stroke-cycle, or more simply, a two-cycle engine, while the second is referred to as a four-stroke-cycle, or four-cycle engine. Either cycle can be used for engines operating with spark ignition (gasoline or gas) or compression ignition (diesel). The four-stroke-cycle engine, which is more widely used, is described first.

1. Four-Stroke Cycle

The sequence of events in the four-stroke cycle is illustrated in Figure 10.1. On the inlet or intake stroke (Figure 10.1A), the intake valve is open and the piston is moving downward. Air or an air–fuel mixture is drawn in through the cylinder that fills the intake valve. In diesel engines only air is drawn or forced into the cylinder intake stroke; fuel is introduced through a high pressure injector at the top of the compression stroke. In most gas and gasoline engines, an air–fuel mixture is introduced through the intake valve. Newer four-cycle engine designs use injection of the fuel directly into the cylinder similar

* Since not all reciprocating engines have vertical cylinders, this term is not strictly accurate; however, it does describe the stroke in which the piston approaches the head end of the cylinder.

to diesel engines. As the piston starts moving up (Figure 10.1B), the intake valve closes and the air (or charge) is compressed in the cylinder. Near the top of this compression stroke, fuel is injected and/or a spark is passed across the spark plug. The fuel ignites and burns, and as it expands, it forces the piston down on the power stroke. Near the bottom of this stroke, the exhaust valve opens so that on the next upward stoke of the piston the burned gases are forced out of the cylinder. The assembly is then ready to repeat the cycle.

Most four-cycle engines are equipped with poppet valves in the cylinder head for both intake and exhaust. Various arrangements are used to operate these valves. In the conventional arrangement, a camshaft is located along the side or center of the cylinder block depending on engine configuration. It is driven from the crankshaft by gears, by a silent chain or, in some passenger cars engines, by a toothed belt. Cam followers (often called valve lifters) of either the roller, solid, or hydraulic type ride on the cams and operate push rods, which in turn operate the rocker arms to open the valves. Valve closing is accomplished by springs surrounding the valve stems. This type of arrangement results in some mechanical lag in valve operation at high speeds; thus, some high speed automotive engines have the camshafts located above the cylinder head so that the cams bear directly on the valve stems or on short rocker arms. This arrangement is called overhead camshaft construction. The cam drive for many of these units utilizes a toothed rubber belt, but a few designs incorporate gear drives for the overhead cams. Some large, medium, and low speed diesel engines now are equipped for direct operation of the valves in a somewhat similar manner, and fully hydraulic valve actuation is also used on a few engines. The complete valve operating mechanism is often referred to as the valve train.

Loading on the rubbing surfaces in the valve train may be high, particularly in high speed engines, where stiff valve springs must be used to ensure that the valves close rapidly and positively. This high loading can result in lubrication failure unless special care is taken in the formulation of the lubricant.

2. Two-Stroke Cycle

The sequence of events in the two-stroke cycle is illustrated in Figure 10.2. Near the bottom of the stroke, the exhaust valves open and the piston uncovers the intake ports, allowing the scavenge air to force the exhaust gases from the cylinder. As the piston starts on the upstroke (Figure 10.2B), the exhaust valves* close and the piston covers the intake ports so that air (or charge) is trapped in the cylinder and compressed. Near the end of this compression stroke, the fuel is injected and begins to burn (or the charge is ignited). The expanding gases then force the piston down on the power stroke.

Two-cycle engines are usually built either with ports for intake and valves for exhaust, or with ports for both intake and exhaust. With the combination of ports and valves (see Figure 10.2), the scavenge air and exhaust gases flow more or less straight through the cylinder, so this arrangement is referred to as uniflow scavenged. With ports for both intake and exhaust, if the intake ports are on one side of the cylinder and the exhaust ports are on the other side, the engine is referred to as cross-scavenged. The scavenge air flows more or less directly across the cylinder. Where the exhaust ports are located on the same side of the cylinder as the intake ports, the engine is referred to as loop-scavenged. The scavenge air must flow in a loop into the cylinder and then back to the exhaust ports.

---

* Many larger two-cycle engines do not use exhaust valves in the cylinder heads but have exhaust ports on the opposite side of intake ports. Example are some large two-cycle gas engines, discussed later in the chapter.

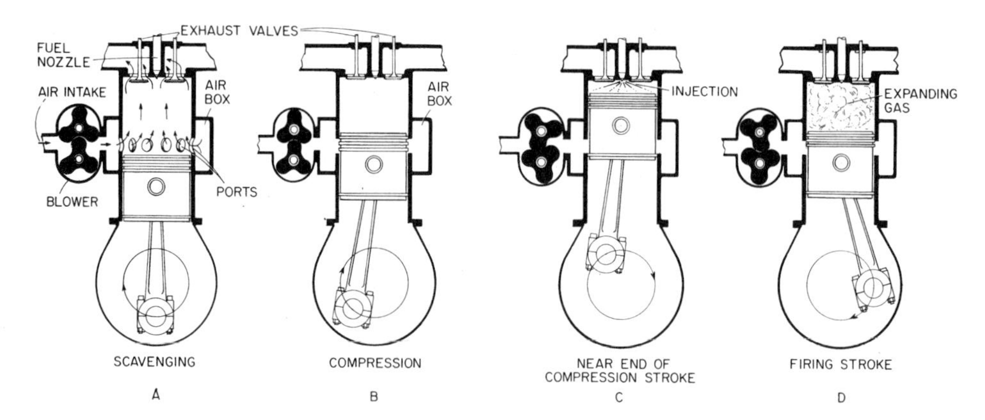

Figure 10.2 Two-stroke cycle. (A) Air from the blower is driving exhaust gas from the cylinder. (B) The charge of fresh air is trapped in the cylinder and compressed as the piston moves upward. (C) Fuel is injected; the fuel is ignited when the piston is close to the top of its stroke. (D) The hot expanding gases of combustion force the piston down on the power stroke.

Most small, two-cycle gasoline engines are of the cross-scavenged type, while all three types of scavenging are used for two-cycle diesel engines. In large engines, the type of scavenging has some influence on piston temperatures and the type of piston cooling (oil or water) that can be used; therefore, indirectly, it has an influence on lubricant selection.

Since the two-stroke cycle does not have a full positive exhaust stroke to rid the cylinder of combustion gases, the scavenging process must be assisted by pressure developed outside the cylinder. This can be accomplished by means of a separate blower or compressor, driven either by the engine or an outside source, or by what is known as crankcase scavenging. With crankcase scavenging, the section of the crankcase under each cylinder is sealed, except for a check valve to admit air (or charge) on the upstroke of the piston, and a transfer passage to permit the air (or charge) to be delivered to the intake ports as the piston approaches and goes through its bottom dead center position. A variation of this principle is used on some large crosshead diesel engines. The lower end of the cylinder is closed and a packing gland installed where the piston rod passes through. With the addition of a check valve and transfer piping to carry the air to the intake ports, the piston then will pump air for scavenging or supercharging purposes. This may be referred to as pulse charging.

In theory—for the same bore, stroke, and rational speed—a two-stroke-cycle engine will develop twice the power of a four-stroke-cycle engine. As a result, more fuel is consumed per unit of time, the cylinder and average temperatures tend to be higher, and more contaminants may find their way into the oil. On many two-stroke-cycle engines, the oil is mixed with the fuel.

B. Mechanical Construction

Most piston engines, including all automotive engines, are of the trunk piston type. Large, low speed diesel engines for marine propulsion and industrial applications are usually of the crosshead type.

In a trunk piston engine, the piston is connected directly to the connecting rod by a piston pin or wrist pin. Thus, the side thrust from the crankpin and connecting rod must be carried through the piston and rings to the cylinder wall. This may result in high rubbing pressures on one side of the piston (thrust side) and on the mating section of cylinder wall.

In a crosshead type of engine, the piston is connected rigidly to a piston rod, which is connected to the crosshead containing the wrist pin. The crosshead also has a sliding guide bearing to absorb the side thrust from the crankshaft and connecting rod. This bearing is usually generously proportioned so that these thrust loads are readily carried by the lubricating films. No side thrust is carried by the piston. The cylinder assembly is usually completely separated from the crankcase, whether by means of a diaphragm containing a packing gland or by having the lower end of the cylinder closed so that it can be used for pulse charging.

Crosshead-type engines are also built with double-acting pistons, or with two pistons acting in opposite directions in the same cylinder—engines having this configuration are said to be opposed-piston engines. One of the chief disadvantages of crosshead construction is that it results in an engine that has a greater overall height than a trunk piston engine of the same horsepower. However, most engines with bores above 600 mm (24 in.) are of the crosshead type.

C. Supercharging

One of the factors that limits the power developed by an internal combustion engine is the amount of air it can ''breathe.'' It would be easy to supply more liquid fuel, but it would be undesirable to exceed the quantity the available air can burn more or less completely. Supercharging is a method of increasing the available combustion air by supplying air at higher pressure, thus making it possible to burn more fuel and produce more power.

The air for supercharging is provided by a supercharger, turbocharger, or blower, which may be driven by an engine, a motor, or an exhaust gas turbine. The latter is probably most frequently used now. With some large engines, supercharging by the blower may be supplemented by pulse charging.

Since supercharging increases the amount of air in the cylinders, and thus the amount of fuel that can be burned, it tends to raise combustion temperatures and pressures, and to increase the deteriorating influences on the lubricating oil. To increase supercharging efficiency, the compressed air is sometimes cooled between compression and engine intake.

D. Methods of Lubricant Application

Small and medium-sized trunk piston engines usually are lubricated by a combination of pressure and splash. Oil under pressure is fed to the main and crankpin bearings and, from the crankpin bearings, through drilled passages to the wrist pin bearings. Oil under pressure is also fed to an oil gallery, from which it is distributed to the camshafts, valve lifters, and rocker arms. Oil is splashed to the lower cylinder walls to lubricate the rings and cylinders and to the undersides of the pistons for cooling. In some larger engines, oil is sprayed under pressure to cool piston undersides. Eventually, all this oil, neglecting the small amount that finds its way into the combustion chambers and is burned, drains back to the crankcase and is reused.

In some cases, the crankcase is arranged for dry sump operation with a reservoir, and possibly an oil cooler for the main supply. Where this arrangement is used, cylinder lubricant is supplied from the wrist pin supply.

Trunk piston engines with bores approximating 250 mm (~10 in.) usually have separate cylinder lubricators of the mechanical force-feed type. Any excess cylinder lubricant then drains back into the crankcase supply, carrying contaminants with it. Crosshead engines are equipped with at least two separate lubrication systems. One system supplies the lubricant for the cylinders, while a second system supplies the main and crankpin bearings and crossheads, and usually the piston rod packing glands. A third system may be provided to lubricate the supercharger, or the supercharger may be lubricated from the crankcase system. Piston cooling oil is also supplied from the crankcase system in engines with oil cooled pistons. There is little or no contamination of the crankcase oil of trunk piston engines by excess cylinder oil, other than the minor amounts that may leak past the packing glands. However, any used cylinder oil that is collected above the diaphragms may present a disposal problem because it is generally unsuitable for further use and is extremely difficult to recondition satisfactorily.

A complicating factor in high output, late model, crosshead engines with oil-cooled pistons is that piston temperatures may be high enough to cause thermal cracking of the oil while it is in the piston cooling spaces. This can lead to buildup of deposits in the piston cooling spaces, overheating of the pistons, and eventual cracking or other mechanical failure.

Most small two-cycle gasoline engines, such as are used for marine outboard and utility purposes, are arranged for crankcase scavenging. To control the amount of oil carried into the cylinders with the charge, these engines are designed for dry sump operation. A small proportion of oil is premixed with the fuel, or injected, so the charge consists of a mixture of air, fuel, and oil. Some of the oil condenses on the crankcase surfaces, where it provides lubrication of the main, crankpin, and wrist pin bearings. Also, some of it is carried into the combustion chambers, where it is burned with the fuel. Incomplete combustion of this oil, or the use of the wrong type of oil, can cause spark plug fouling and combustion chamber and port deposit buildup.

An important consideration with any internal combustion engine is the volume of oil in the system in proportion to the amount of contaminants that must be carried by the oil. Generally, large engines in marine and stationary applications can be arranged so that the volume of oil is large, and since operation is at essentially constant speed, combustion efficiency can be kept high. The rate of contaminant buildup is then quite low and usually can be controlled to some extent by means of sophisticated purification equipment. With automotive engines, the amount of oil that can be carried into the system is proportionally much lower, for a variety of reasons. With automotive gasoline engines, particularly, it is desirable to restrict the volume of oil to promote more rapid warm-up. In addition, the lower oil volume is more economical when frequent periodic changes must be performed. Having a lower oil volume helps to evaporate both the water of condensation and that diluting fuel, which result from start–stop operations and cold starting. However, the variable speed conditions under which automotive engines operate, along with frequent starts and stops, tend to reduce combustion efficiency, with the result that the rate of contaminant buildup in the oil is relatively high. The newer, more fuel-efficient engine designs incorporating fuel injection systems are helping reduce contaminant levels in the oils by providing cleaner, more efficient combustion even in start–stop operation. Use of smaller oil quantities also leads to higher operating temperatures for the oil under extended operation, which in turn means that the exposure of the oil to oxidizing conditions may be more severe.

II. FUEL AND COMBUSTION CONSIDERATIONS

When a hydrocarbon fuel is burned, certain products and residues are formed. One of the most important of these is water. In the case of liquid fuels, the volume formed is slightly greater than the volume of fuel burned. If the engine is fully warmed up, virtually all this water passes out of the exhaust as vapor. If the engine is relatively cold, or under certain pressure and temperatures conditions in the cylinders, some of the water may condense and eventually find its way into the crankcase, where it will mix with the engine oil. The amount of condensation that occurs and the other residues that are formed are largely dependent on the type of fuel and the operating conditions of the engines it is burned in. High temperature thermostats in the cooling system both help reduce the levels of moisture and improve combustion conditions.

A. Gasoline Engines

Most of the four-cycle gasoline engines have automotive or similar applications. In these applications, the engine is generally started cold and may operate much of the time at below-normal temperatures, either in stop-and-go conditions or when carrying a light load. Under these conditions, a significant amount of the water formed from combustion of the gasoline may condense on the cylinder walls and eventually reach the crankcase by way of blowby past the pistons and rings. In doing so, it tends to wash the lubricant films off the cylinder walls and promote rusting or corrosion. The rust formed is scuffed off almost immediately, but in the process metal is removed, and what appears to be mechanical wear occurs. In the crankcase and in other areas where temperatures are low, the water can combine with the oil and other contaminants to form sludge that is usually of a soft, sticky consistency.

When a gasoline engine is cold, the air–fuel mixture must be enriched by choking to provide enough vaporized fuel for starting. Some of the excess liquid fuel is blown out the exhaust, but a proportion of it drains past the pistons and into the crankcase, usually carrying with it some partially burned components and fuel soot. As the engine warms up, much of the gasoline is evaporated off, but some of the heavier ends, some of the partially decomposed materials, and any solid residues remain in the oil. Even when the engine is fully warmed up, some of these fuel decomposition products blow by into the oil.

Fuel soot and partially burned fuel and water, when present in the oil, can lead to the formation of varnish, sludge, and deposits. Rusting of ferrous surfaces and corrosion of bearings can be promoted, particularly by the residues. Upon combination with small amounts of oil residue burned or partially burned in the combustion chambers, the solid residues, such as fuel soot, can form deposits that adhere to piston tops and combustion chamber surfaces. Deposits of this type can significantly increase the fuel octane number requirement of an engine by reducing the combustion chamber volume (increased compression ratio).

The use of unleaded gasolines required for engines equipped with catalytic converters may have the effect of reducing the amount of rust and corrosion that occurs. In some older engines, the absence of the lead antiknock additives in the gasoline may result in accelerated valve wear, called valve recession. Generally, this relates more to metallurgy than to lubrication. Somewhat higher operating temperatures result from the emission controls, but these have not presented a major problem from the lubrication point of view. The use of exhaust gas recirculation (EGR) to control emissions of nitrogen oxides (NO_x), along with other emission control changes, may result in a greater tendency to form valve stem and engine deposits.

Two-stroke-cycle gasoline engines have such applications as outboard motors and motors for snowmobiles, chain saws, and utility purposes such as pumps and lighting plants. These engines are lubricated by oil mixed with the fuel, so some oil is always present in the charge. Under low speed, low load conditions, such as trolling in a boat, poor combustion can result in heavy buildup of port and piston crown deposits. Ashless and low ash oils tend to help control the level of deposit formation that is attributable to the ash-containing additives in the formulation.

B. Diesel Engines

Diesel engines in trucks generally operate at more nearly constant speed and higher load factors than gasoline engines. Thus, combustion conditions are more nearly optimum, and water condensation and fuel dilution are not as serious problems. Where excessive fuel dilution occurs, it is usually the result of a mechanical problem, such as a faulty injector or defective turbocharger. Diesel fuel is not as readily evaporated from the engine oil as is gasoline; thus, if a problem exists, the concentration of diesel fuel will tend to increase steadily. This can lead to deposits and a reduction of the oil viscosity sufficient to promote mechanical wear.

One of the major problems with diesel engines used to be the level of sulfur in the fuel. Sulfur can be present in diesel fuels in a considerably higher concentration than in gasolines, and although some current regulations require sulfur levels in diesel fuels to be below 500 ppm, in many areas sulfur levels are higher. When sulfur burns, it forms sulfur dioxide, part of which may be further oxidized to sulfur trioxide. In combination with water, these sulfur oxides form strong acids that not only are corrosive in themselves but also have a strong catalytic effect on oil degradation. Since piston temperatures are also high, this may result in heavy deposits of carbon and varnish on the pistons and in the ring grooves. Under severe conditions, deposits may build up in the piston ring grooves until the rings cannot function properly and may even stick, causing higher oil consumption, high wear, blowby, and loss of power. Manufacturers of diesel engine passenger cars are particularly concerned about soot deposits and recommend more frequent oil drains—more than twice as many as an required for gasoline-fueled car engines.

Many large diesel engines in marine and stationary industrial service are operated on residual-type fuels with sulfur content in the 2–4% range. The strong acids formed from the combustion of these high sulfur content fuels can be extremely corrosive to rings and cylinder liners, with the result that metal removal may be rapid and wear rates excessive. Oil selection becomes more critical as fuel sulfur levels rise.

C. Gaseous Fueled Engines

Engines burning clean gaseous fuels, such as liquefied petroleum gas (LPG: propane, butane, etc.) or natural gas, are comparatively free of the contaminating influences encountered in liquid-fueled engines. Although water is formed from combustion, most of it passes out the exhaust as vapor. Products of partial combustion that blow by to the crankcase tend to polymerize with the engine oil and cause an increase in viscosity; eventually, these may result in varnish and lacquer formation, as discussed in greater detail toward the end of this chapter.

III. OPERATING CONSIDERATIONS

The operating conditions in an engine have a major influence on the severity of the service the lubricating oil is exposed to in the process of performing its functions of minimizing wear, reducing corrosion, reducing friction, assisting in cooling and sealing, and controlling deposits.

A. Wear

Normally, most engine wear is that experienced by cylinders, piston rings, and pistons. Less frequently, wear may be experienced in bearings and on camshafts. The three principal causes are abrasion, metal-to-metal contact, and corrosion.

Much of the dust and dirt carried into the cylinders with the intake air is hard and abrasive. Some diesel fuels (particularly the residual fuels used in many large engines) may also contain abrasive materials. Abrasive particles carried by the oil onto load-supporting surfaces of the cylinder walls and other areas can cause wear to the rings and cylinder, no matter how persistent the lubricating oil films. Fortunately, wear due to abrasives can be held to almost negligible values by the use of effective air, fuel, and oil filters. Proper maintenance of these filters is important, since a ruptured or damaged filter can allow free passage of abrasive materials.

As a result of the conditions of boundary lubrication that exist on the upper cylinder walls, metal-to-metal contact cannot be avoided entirely. The rate of wear from this cause depends to a large degree on the suitability of the lubricating oil. When the oil is of the correct viscosity and has adequate antiwear characteristics, wear due to metal-to-metal contact is kept at a low rate. However, reduced rates of oil flow and poor oil distribution, caused by deposits, can contribute to increased rates of metallic wear.

In large, highly supercharged, trunk piston engines, the side thrust on the cylinder walls is high. It has been found that some of these engines require oils with high load-carrying ability to keep metallic wear rates to an acceptable level.

Corrosion and corrosive wear result from water or a combination of water, severe oil degradation products, and corrosive end products of combustion. When operating temperatures are low, either during the warm-up period or as a result of low load or stop-and-go operation, condensation is increased and corrosive wear may be rapid. Where the use of high sulfur diesel fuels is unavoidable, the compounds formed from their combustion promotes corrosive wear. Generally, it has been found that the use of alkaline additives in the lubricating oil acts to retard this type of wear. These materials neutralize the acidic substances, reducing their corrosive and catalytic effects.

Some bearings of the hard alloy type are susceptible to corrosive attack by certain oxidation products resulting from oil degradation. Oils for engines where corrosion-susceptible bearings are used are formulated with additives that provide protection against these oil oxyproducts.

Metallic wear may be encountered on highly loaded valve train parts and fuel pump cams of some engines. This problem is most often encountered in passenger car engines. Antiwear agents are incorporated in oils for these engines to minimize wear. With this exception, metallic wear is not generally a problem with engine bearings as long as sufficient oil of the proper viscosity is available, and there is no interruption of oil flow. Current engine technology using rollerized valve train components and electric fuel pumps will help reduce the wear.

B. Cooling

Engine cooling is necessary to avoid engine damage and failure through overheating and thermal distortion. This is primarily a function of the cooling system, but the engine oil also has an important role to play in cooling, especially in large diesel engines, where forced oil cooling of the pistons is used. Heat picked up by the oil is dissipated by natural radiation from the walls of the crankcase, or by means of an oil cooler if the oil is relied on for substantial engine cooling.

The specific heat of all-petroleum oils is essentially the same and, of itself, has no relevance in the choice of a lubricant. Of importance is chemical stability and ability to resist the formation of deposits that might interfere with heat transfer, either from the engine parts to the oil or from the oil to its cooling medium. Under the severe conditions encountered in large diesel engines with oil-cooled pistons, thermal stability—that is, the ability to resist cracking of oil and deposit formation at high temperatures—is also important.

C. Sealing

Effective cylinder sealing is necessary to minimize blowby and thereby maintain power and economy. Blowby, which cannot be entirely prevented, is a function of engine design (speed, ring, and cylinder conformity, and provision for cylinder lubrication) and oil. The greater responsibility lies with the rings and their ability to adjust themselves to the varying cylinder contours throughout the length of ring travel, but the oil has an important complementary role. Generally, as far as the oil is concerned, there will be maximum contribution if the ring grooves are clean and unobstructed so that the rings are free to move as required, and if there is no excessive removal of oil from the cylinder walls by the oil control rings. Too little oil on the cylinder walls not only will result in poor sealing but may result in rapid wear. This, in turn, leads to even poorer sealing. Too much oil on the cylinder walls, on the other hand, results in the exposure of more of the oil to combustion conditions, and in turn, oil consumption and rate of contaminant buildup in the oil may be high.

D. Deposits

Control of deposits in an engine is a fundamental need if long life and efficient engine performance are to be realized. Deposit formation is affected by engine design, operating conditions, maintenance, fuel combustion, and the performance of the oil. In turns, deposits affect engine power output, noise, smoothness, economy, life, and maintenance cost.

Two important sources of engine deposits have been discussed briefly, namely, dirt entering with the fuel and combustion air, and the fuel combustion process. Dirt in air or fuel causes abrasive wear, as already noted. In addition, it contributes to deposits on piston crowns, in ring grooves, and on valves. Deposits in these areas are often referred to as ''carbon,'' but this is a loose term. Usually, analysis will show that the deposits consist of dirt, solid combustion residues such as fuel soot, lubricating oil in various stages of decomposition, and residues from lubricating oil additives. It is important to note that any analysis of an engine deposit will show additive metals. This does not necessarily mean that an ''oil'' problem exists.

Temperatures in the combustion zone are high; thus, continued exposure of oil reaching this area causes it to oxidize, crack, and polymerize to heavier hydrocarbons. Large amounts of deposits in the combustion zone may be caused by excess amounts of oil reaching the upper cylinder area. Other possible causes are maladjustments in the fuel system, improper fuels that cause smoky combustion, poor ignition, faulty or worn valve guides, or worn cylinders and rings.

Diesel engines are not particularly sensitive to deposits in the combustion zone, but gasoline engines are, especially those with high compression ratios. Depending on the character of the deposits, continued exposure to the combustion process may cause them to glow, which may provide an unwanted source of ignition. Undesirable combustion phenomena such as knock, preignition, rumble, and run-on can result. Use of a higher octane fuel is usually beneficial in alleviating these problems, as is use of an oil that has less tendency to form adhering deposits in the combustion chambers.

Deposits in ring grooves are similar in origin to those in combustion chambers, but since they are not exposed to the direct flame of combustion, they may be more carbonaceous. Toward the bottoms of the pistons (cooler areas), they also tend to have a higher oil content. In severe cases, groove deposits can be packed so tightly in the clearance spaces behind the compression rings that the rings cannot operate freely. Compression pressure cannot then be maintained, and blow by becomes excessive. Also, oil control ring slots may become plugged; thus, oil control is lost.

Piston varnish is also formed from fuel and oil decomposition products. It may vary from a smooth, shiny, almost transparent coating often called lacquer to a dark, opaque coating that becomes progressively more carbonaceous with continued operation.

Valve deposits (Figure 10.3) are sometimes observed. These are generally the decomposition products of both fuel and oil, formed somewhat in the same manner as combustion chamber deposits. Intake valves may show more deposits than exhaust valves, particularly when they run cool because of low load operation. Under this condition, there is a greater opportunity for droplets of gasoline to form gums on the valve stems. Dirt and solid residues will then adhere to these gum deposits. In some designs of overhead valve engines, oil flow down the valve stems is excessive, and this speeds up deposit formation, as well as being wasteful of oil.

The other major type of deposit is the emulsion or sludge formed by water, fuel decomposition residues, and solid residues. Sludge generally deposits on cooler engine surfaces, such as the bottom of the crankcase pan, the valve chambers, and the top decks (Figure 10.4). The main problem with this type of deposit is that it can be picked up by the oil and carried to areas such as the oil pump inlet screen or oil passages, where it can obstruct oil flow and cause lubrication failure.

Although the physical appearance of deposits varies greatly throughout an engine, basically these deposits result from the combustion process and lubrication oil deterioration. The exact chemical and physical nature of deposits depends on the area where they are formed, the duration of exposure, and any relative motion present. In the case of sludge, water is an essential factor, and any condition that encourages the entrance and retention of water in an engine oil promotes sludge formation.

The contribution of the engine oil to the control of deposits is discussed more fully in Section IV.

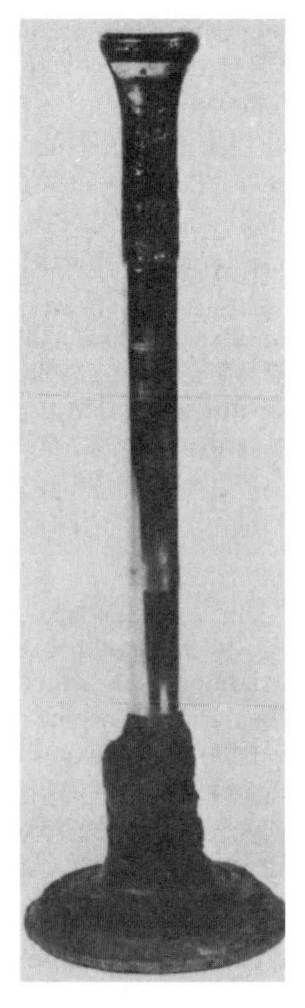

Figure 10.3 Engine valve deposits.

Figure 10.4 Top deck (cylinder head) sludging: heavy sludge buildup resulting from cold engine operation combined with an oil of inadequate quality.

IV. MAINTENANCE CONSIDERATIONS

The quality of maintenance of engine components that affect the lubrication process is of considerable importance. It is of course necessary to have clean combustion and crankcase ventilation air, which means that air cleaners and the positive crankcase ventilation (PCV) system must be serviced regularly. A clogged air cleaner, while it may be effective in cleaning the air, can restrict the volume of air reaching the engine enough to reduce power output significantly. To maintain power, the natural tendency is to open the throttle, which only adds to the difficulty (since the extra fuel is not burned). Diesel engines are quite sensitive to such a condition and react by smoking and rapid buildup of engine deposits and lubricating oil contamination (soot). In addition to keeping the filters serviced, it is important that the piping connecting the filter to the engine be unobstructed and leak free. This is particularly important in installations having the air filter located a considerable distance from the engine.

The carburetor and ignition systems of spark ignition engines and the fuel injection systems of both gasoline and diesel engines should be in good working order and properly adjusted to assure the cleanest, most complete combustion possible. Malfunction or incorrect adjustment of these systems can result in increased amounts of unburned fuel in the cylinders, and dilution or more rapid buildup of contaminants in the oil, as well as increased emissions.

Proper engine oil drain schedules should be followed, to ensure that oil is drained before the contaminant load becomes so great that the oil's lubricating function is impaired, or heavy deposition of suspended contaminants occurs. In principle, the correct drain interval for a given engine in a given service is best established by means of a series of used-oil analyses and inspections of engine condition. As a practical matter, this is an economical approach only for very large engines or fleets of similar engines in similar service. For most smaller engines, such as those used in passenger cars, drain intervals are usually established by the manufacturer on the basis of experience accumulated over the years with similar engines operating under typical conditions.

Drain intervals can vary considerably depending on the engine and service. For example, a large diesel engine in central station use, with a relatively large crankcase oil supply, may operate for thousands of hours between oil changes. Such engines are usually in good adjustment, temperatures are moderate, and the contamination rate is low in comparison to the volume of oil in the system. On the other hand, a passenger car engine may require an oil change every few thousand miles. Such engines are physically small with relatively small crankcase capacity, and operate under conditions conducive to rapid oil contamination. Not only are load factors often low, but these engines may be engaged mostly in short runs and start–stop service, as well operating over wide ranges of ambient temperature, all of which favor the accumulation of oil contaminants and the risk of deposits.

The presence of an oil filter does not necessarily permit an extension of the oil drain interval. Filters do not remove oil-soluble contaminants and water, which are important factors in deposit formation. Regular filter changes are, however, important in keeping the filter operable so that it can perform its function of removing insoluble contaminants from the oil.

V. ENGINE OIL CHARACTERISTICS

Some of the more important characteristics of engine oils are discussed in the following sections. The reader is also referred to the discussions of physical and chemical characteristics, additives, and evaluation and performance tests in Chapter 3.

A. Viscosity, Viscosity Index

In reciprocating engines, viscosity of the engine oil is extremely important. It has a bearing on wear prevention, sealing, oil economy, frictional power losses (fuel economy), and deposit formation. For some engines, particularly vehicle engines, it also is a factor in cranking speed and starting ease. Too high a viscosity may cause excessive viscous drag, reduction in cranking speed, and increased fuel consumption after the engine is started.

The viscosities of engine oils are usually reported according to the SAE J300 (Engine Oil Viscosity Classification), as discussed in Chapter 3. While this system was originally intended for automotive engine oils only, its use has now been extended to include most oils for internal combustion engines. Engine manufacturers normally specify the viscosities of oils for their engines, according to ambient temperature and operating conditions, by SAE grade.

Viscosity index (VI) is important in engines that must be started and operated over a wide temperature range. In these cases, all other factors being equal, oils with higher viscosity indexes give less viscous drag during starting and provide thicker oil films for better sealing and wear prevention; moreover, oil consumption, at operating temperatures is lower.

In past years, the viscosity index of an oil was of limited importance for an oil that was used for engines not subject to frequent cold starts. This remains partially true today except that engine manufacturers are recommending the high VI oils for year-round service, primarily because the viscous drag of an oil is proportional to its viscosity, and higher viscosities at start-up or during operation will reduce efficiency (fuel economy). Other materials such as friction modifiers can be added to the oil to help reduce friction. At present, however, the single oil characteristic that has the biggest effect on fuel economy is viscosity.

B. Low Temperature Fluidity

When an oil is to be used in engines operating at low ambient temperatures, the oil must have low temperature fluidity adequate to permit immediate flow to the oil pump suction when the engine is started. The pour point of an oil is an adequate indication of whether the oil will flow to the pump suction. Most conventional oils will flow to the pump suction at temperatures below their pour points because the pump suction creates a considerably greater pressure head than is present in the pour point test. However, many multigrade oils will not circulate adequately at temperatures considerably above their pour points. The correlation between pour point and flow in instrumented engines is poor, and at best, pour point is only a rough guide to the minimum temperature at which an oil may be used safely.

ASTM has introduced two tests that measure the low temperature performance of an oil. The test that simulates cold cranking of an engine is the cold cranking simulator (CCS), and the test for measuring low temperature pumping is the mini rotary viscosimeter (MRV). Both the CCS and MRV show good correlation to low temperature performance. The maximum values for these tests are listed for the ''W'' grades in the SAE engine oil viscosity classification. The CCS is designed to reproduce the elements of viscous drag that effects cranking speed. The MRV is designed to predict the low temperature pumpability and therefore the ability of the oil to reach critical components under low temperature starting conditions.

C. Oxidation Stability (Chemical Stability)

High resistance to oxidation is an important requirement of a good engine oil in view of the high temperatures the oil is exposed to and, in the crankcase, the agitation of the oil in the presence of air. Deterioration of an engine oil by oxidation tends to increase viscosity, create deposit-forming materials, and promote corrosive attack on some hard alloy bearings. Where the engine oil capacity is relatively small, the rate of deterioration tends to be higher, other factors being equal.

An oil's natural oxidation stability is determined in part by the crude oil from which it is made and the refining processes to which it is subjected. Where engine design or operating conditions require a high degree of oxidation stability, oxidation inhibitors are used. As a general rule, the need for greater oxidation stability increases as oil service temperatures and drain intervals increase. Among the principal factors that make enhanced oxidation stability necessary are high engine specific power output (high horsepower per unit of displacement), small crankcase charge volume, long oil drain intervals, and modifications and devices to control emissions that result in high operating temperatures. For example, a heavily loaded truck engine requires an oil with excellent oxidation stability because of the high operating temperatures involved, while a large, low speed diesel engine in central station (stationary) service, with a large crankcase oil supply at a moderate temperature, requires an oil with good stability because the oil is expected to remain in service for thousands of hours. Regardless of the necessity for it, good stability is always desirable in view of its helpful influence on engine cleanliness.

Oxidation stability plays only a minor role in the process by which combustion chamber deposits are formed. Under the conditions just described, however, combustion of the lubricating oil to such deposits can be controlled partly by base stock refining and selection. See Section II (Physical and Chemical Characteristics), Chapter 3.

D. Thermal Stability

Thermal stability, or resistance to cracking and decomposition under high temperature conditions, is a fundamental characteristic of the lubricating base oil that cannot be substantially improved by means of additives. However, careful selection of additives is important in formulating thermally stable oils, since decomposition of the additives can contribute to the formation of deposits under operating conditions that promote thermal cracking of the oil.

As noted earlier, thermal stability of the engine oil is of special concern in certain large, highly supercharged, two-cycle diesel engines used for marine propulsion. Thermal cracking of the system oil, experienced in the piston cooling spaces, has resulted in deposits that interfere with heat transfer. Oils for these applications must be manufactured from base stocks made from selected crudes by carefully chosen refining processes.

E. Detergency and Dispersancy

The natural detergent and dispersant properties of most oils is slight, and where these characteristics are important, they are obtained through the use of additives.

In nearly all current internal combustion engine applications, oils with enhanced detergency and dispersancy are necessary to control engine deposits and maintain engine performance. The levels of detergency and dispersancy required depend on a number of factors such as engine design, operating temperatures, type of fuel, continuity of operation, and exposure to low ambient temperatures. In general, conditions that tend to promote oil oxidation, such as supercharging or the use of high sulfur fuels, dictate the use of oils with higher levels of detergency. Conditions that promote condensation of water and unburned or partially burned fuel in the engine require the use of oils with higher dispersancy.

As noted in the preceding discussion, detergency and dispersancy are not clearly differentiated properties. There is a trend in additive development to improve the dispersancy of the so-called detergents and the detergency of the "dispersants." Thus, some oils formulated with dispersants but not detergents may give entirely adequate control of high temperature deposits in some services, while other oils formulated with detergents only may give good control of low temperature emulsions and sludges in some services. In general, however, most engine oils contain a mixture of the two types of material, with the concentrations and relative proportions depending on the type of engine service an oil is designed for.

One of the main functions of both detergents and dispersants is to suspend potential deposit-forming materials in the oil. In suspended form, these materials are relatively harmless and may be removed from the system by draining the oil. Regular oil drains for this purpose are important, particularly for engines with a relatively small oil capacity. If the oil drain intervals are too long, the ability of the additives to suspend the deposit-forming materials may be exceeded, whereupon deposits will begin to form on engine surfaces and engine performance will deteriorate. This is why manufacturers specify more frequent oil drains for "severe service" such as short trips, sustained high speeds, and trailer towing.

There are no absolute measures of detergency and dispersancy. It is now customary to describe engine oils on the basis of performance testing, as was discussed in Chapter 3 in Section III, Evaluation and Performance Tests. While this description includes an evaluation of detergency and dispersancy, the intent is to provide a more comprehensive description that includes all the factors that make a particular oil suitable for a particular type of engine service.

F. Alkalinity

Most detergents, and to a lesser extent many dispersants, have some ability to neutralize the acidic end products of fuel combustion and oil oxidation. When a considerable ability to neutralize acids is required, as in oils for diesel engines burning high sulfur fuels, however, highly alkaline (overbased) detergent-type materials are used. The concentration of these materials in an oil, and an indication of the oil's ability to neutralize acids, is given by the total base number (TBN), also called the alkalinity value. There is only a general relationship between TBN and the ability of an oil to control wear and corrosion caused by strong acids, since it has been found that some newer additive systems are more effective in this respect than would be predicted by consideration of the TBN level alone.

Many of the highly alkaline oils serve as cylinder oils in large engines, which is a once-through use. In these cases, the changes in the oil in service are of little or no concern. In engines in which the same oil serves as both the cylinder and crankcase oil, it is desired to monitor the alkalinity of the oil as a method of determining whether it is still capable of performing its neutralization function, particularly if high sulfur or other acid-producing constituents are present in the fuel.

G. Antiwear

In addition to the corrosive wear caused by acidic products of combustion, metallic wear may occur in areas where loads or operating conditions prevent the maintenance of effective lubricating films. The main areas of concern is this respect are cylinder walls and rings, particularly of large, high output trunk piston engines, and the valve train mechanisms of small, high speed engines. In these cases, it is usually necessary to use oils that are formulated with additives that provide enhanced protection against wear and scuffing under boundary lubrication conditions.

H. Rust and Corrosion Protection

All petroleum oils have some ability to prevent rusting and corrosion of engine metals. However, in most cases, this natural ability is not sufficient to do the following:

1. Protect hard alloy bearings from corrosion caused by oil oxyacids
2. Prevent rusting and corrosion due to condensation of water and combustion products in low temperature or stop-and-go service
3. Control the corrosive wear caused by acidic end products of combustion

Since one or more of these conditions, which can cause troublesome corrosion, are encountered to some extent in nearly all internal combustion engine service, most oils for internal combustion engines are formulated to provide additional protection against corrosion.

Automotive engine oils usually are formulated to provide protection against corrosion and particularly corrosion of hard alloy bearings. In the case of oils intended for gasoline engine service, protection against corrosion and rusting due to condensation of water and unburned or partially burned fuel components is emphasized. Diesel engine oils are usually formulated to provide protection against corrosion due to acidic end products of combustion. As stated earlier, in these latter oils, the function of corrosion protection is closely related to detergency and alkalinity.

Special preservative oils are available for the protection of engines that are to be laid up seasonally or stored for extended periods. These oils are formulated to provide protection against rust and corrosion due to atmospheric conditions and, usually, are also formulated to be suitable for short-term use in the engines under moderate operating conditions. As a result, an engine can be run safely on the preservative oil prior to being laid up to distribute the oil over the internal engine surfaces. Also, an engine can be run for a reasonable period after being taken out of storage before the preservative oil is drained and replaced by the normally recommended type of oil. This assumes that excessive contamination has not entered the engine or oil during storage.

I. Foam Resistance

All oils will foam to some extent when agitated. If excessive foaming occurs in an internal combustion engine, several problems may result. Overflow and spillage of oil is, of course, one of the most obvious, but foaming can also result in starvation at the oil pump inlet, or slugs of foam being drawn into the pump with the oil. Foam entrained in the oil can cause failure of lubricating films, and noisy, erratic operation of hydraulic valve lifters. To reduce foaming, particularly of oils intended for small, high speed engines where agitation is severe, a defoamant is often included in the formulation. Overfilling the crankcase can cause foaming even with defoamants.

J. Effect on Gasoline Engine Octane Number Requirement (ONR)

Although ONR is not classed as a lubricating function directly, the characteristics of the lubricating oil in fact have a considerable influence on the octane number requirement for gasoline engines. Some oil always finds its way past the rings into the combustion chambers, where it is partially burned to form deposits that adhere to the piston tops and combustion chamber surfaces. As these deposits build up, the ONR of the engine gradually increases. The increase in ONR is a function of both the quantity and nature of the deposits. Heavy, ragged deposits (Figure 10.5, left) cause the octane number requirement to increase more than smooth, even deposits (Figure 10.5, right), primarily because the sharp projections on the ragged deposits can be more easily heated to incandescence. In this condition, the deposits may ignite the fuel charge, either before the spark plug fires or before the

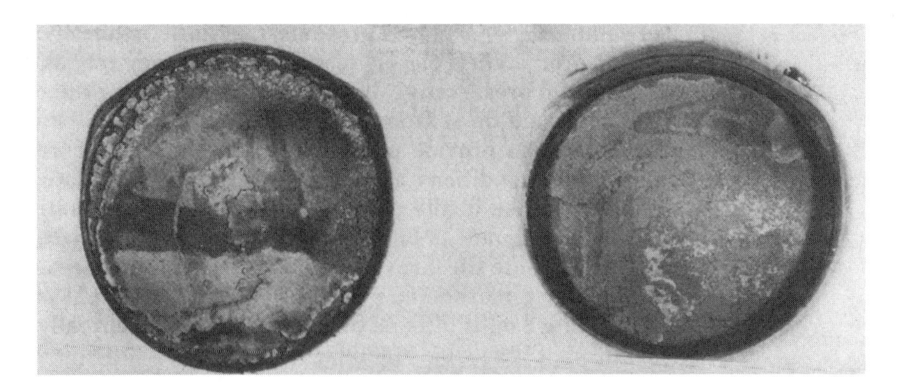

Figure 10.5 Combustion chamber deposits: the heavy, ragged deposits on the left, formed when a conventional oil was used, generally contribute significantly more to the engine octane number requirement than the smooth deposits on the right.

normal flame front reaches portions of the charge. In either case, knock and power loss result.

It has been found that the contribution of the lubricating oil to the increase in engine ONR may be reduced considerably by careful attention to such factors as crude oil source, refining processes, and the use of base stocks having a minimum of heavy hydrocarbons. One of the major factors in accomplishing the latter is the use of all-distillate base stocks, rather than the older practice of incorporating some bright stock in the blends. Many multiviscosity automotive engine oils are manufactured to provide maximum benefits in the control of octane number requirement increase.

Although oil formulation can help control combustion chamber deposits, the type of service, operating conditions, engine conditions, and maintenance practices generally are the major contributors to combustion chamber deposits. These conditions also affect engine emissions and fuel economy.

K. Identification

The use of the SAE viscosity classification system was discussed in Chapter 3. It is important to remember that this system is concerned only with viscosity. Thus, it may be used to describe the viscosity characteristics of oils intended for widely different services. Several other systems in general use for describing the performance qualities of automotive engine oils are the API's Engine Service Classifications for Engine Oils, the European Engine Oil Specification of ACEA, and the ILSAC (International Lubricant Standardization and Approval Committee) GF series.

1. API Engine Service Classifications for Engine Oils

The original system set up by the American Petroleum Institute (API) included three gasoline engine service classifications (ML, MM, and MS) and two diesel engine classifications (DG and DS). A third diesel classification, DM, was added later, and in 1969 the ML classification for gasoline engines was dropped. However, certain inherent difficulties with the system resulted in a joint effort by API, ASTM (American Society for Testing and Materials), and SAE (Society of Automotive Engineers) to develop a new system that would be more effective in communicating engine oil performance and engine service classification information between the petroleum and automotive industries. This system, which was finalized during 1970 and remains effect, consists of two categories: S (service station) and C (commercial), is described in API Bulletin 1509 and SAE Recommended Practice SAE J183. Table 10.1 shows the API service categories.

2. The API Classification System

The API system is open ended, so that additional classifications can be added when needed. When the letter designations are used by oil marketers or engine manufacturers to indicate the service for which oils are suitable or required, it is the intent that they be preceded by the words ''API Service.'' To illustrate, a service station oil suitable for servicing new 2001 model year cars under warranty would be referred to as ''for API Service SJ.'' If oils are suitable for more than one service, it is appropriate that these oils be so designated: for example, ''for API Services CG-4/SH.''

There is a relationship between API engine service classifications and some of the oil quality standards outlined earlier, reflecting existence, for each classification, of minimum performance requirements defined in terms of performance in prescribed engine tests (Tables 10.2 and 10.3). The prescribed engine tests for any service classification include all those required for the corresponding related designations, or acceptable alternates. Also, the minimum performance requirements in each test are at least equal to those required by any of the corresponding related designations. Thus, for example, the engine test requirements for an oil for API service SJ include all the engine test requirements for Ford Spec M2C153-G, and the minimum performance levels in those tests are at least equal to those required to meet the Ford specification.

This approach of providing a complementary definition of minimum performance requirements for the service classifications offers the user better assurance that oils marked as being suitable for a particular service classification will give an acceptable minimum performance level in service. However, performance differences can still exist among oils designated as belonging to the same service classification, since many oils are, and will be, formulated to exceed the minimum performance levels. These oils will, of course, provide additional benefits in service.

3. ILSAC Performance Specifications

In addition to API service categories, the International Lubricant Standardization and Approval Committee, in conjunction with the automobile manufacturers, created the GF (gasoline-fueled) series for passenger car engine oils. Designed principally to measure fuel economy benefits of engine oils, the GF categories also address other critical areas of performance. The initial introduction of the GF series was in 1992. The first category, the GF-1, used the Sequence VI test engine (1986 Buick V-6), and to claim ''Energy Conserving'' in the API donut (Figure 10.6), a fuel economy improvement of 1.5% or greater was required. If fuel economy improvements were 2.7% or better, an oil could claim ''Energy Conserving II'' and get a GF-1 Starburst license. Fuel economy improvements are measured against a reference oil in a standard test. The ILSAC GF-2 was introduced in 1996 and uses the Sequence VIA test engine (1993 Ford 4.6-liter V-8). With the introduction of the GF-2, the Energy Conserving II category was eliminated and replaced by a single Energy Conserving rating. The Starburst license for a GF-2 oil requires a 0.5% fuel economy improvement for 10W-30 oils or a 1.1% improvement for 5W-30 oils. The GF-3 is

Table 10.1 API Service Category Chart

Gasoline engines[a]	Diesel engines[a]
SA (obsolete) For older engines, no performance requirement. Use only when specifically recommended by manufacturer.	CA (obsolete) For light-duty engines (1940s and 1950s).
SB (obsolete) For older engines. Use only when specifically recommended by manufacturer.	CB (obsolete) For moderate-duty engines for 1949–1960.
SC (obsolete) For 1967 and older engines.	CC (obsolete) For engines introduced in 1961.
SD (obsolete) For 1971 and older engines.	CD (obsolete) Introduced in 1955. For certain naturally aspirated and turbocharged engines.
SE (obsolete) For 1979 and older engines	CD-II (obsolete) Introduced in 1987; for two-stroke cycle engines.
SF (obsolete) For 1988 and older engines	CE (obsolete) Introduced in 1987; for high-speed, four-stroke, naturally aspirated, and turbocharged engines. Can be used in place of CC and CD oils.
SG (obsolete) For 1993 and older engines.	CF-4 (current). Introduced in 1990; for high-speed, four-stroke, naturally aspirated and turbocharged engines. Can be used in place of CE oils.
SH Introduced in 1993; discontinued as an API service symbol except when used in combination with certain C categories.	CF (current). Introduced in 1994; for off-road, indirect-injected, and other diesel engines including those using fuel with over 0.5% weight sulfur. Can be used in place of CD oils.
SJ (current). Introduced as an API service symbol in October 1996. For all engines presently in use.	CF-2 (current). Introduced in 1994; for severe-duty, two-stroke cycle engines. Can be used in place of CD-II oils.
	CG-4 (current) Introduced in 1995; for severe-duty, four-stroke cycle engines using fuel with less than 0.5% weight sulfur. Can be used in place of CD, CE, and CF-4 oils.
	CH-4 (slated for introduction in 1998); severe-duty four-stroke diesel engines. Will be used in place of GG-4. Emphasis is on emission and extended drains.

[a] Each gasoline engine category exceeds the performance properties of all the previous categories and can be used in place of the lower one. For example, an SJ oil can be used for any previous "S" category.

Table 10.2 Engine Oil Classification System for Automotive Gasoline Engine Service-Service Oils (S)

API Automotive gasoline engine service categories	Previous API engine service categories	Industry definitions	Engine test requirements
SA	ML	Straight mineral oil	None
SB	MM	Inhibited oil only	CRC L-4[a] or L-38; Sequence IV[a]
SC	MS (1964)	1964 Models	CRC L-38 Sequence IIA[a] Sequence IIIA[a] Sequence IV[a] Sequence V[a] Caterpillar L-1 (1% sulfer fuel)[a]
SD	MS (1968)	1968 Models	CRC L-381 Sequence IIB[a] Sequence IIIC[a] Sequence IV[a] Sequence VB[a] Falcon Rust[a] Caterpillar L-1[a] or 1H[a]
SE	None	1972 Models	CRC L-38 Sequence IIB[a] Sequence IIIB[a] or IIID[a] Sequence VC[a] or VD[a]
SF	None	1980 Models	CRC L-38 Sequence IID Sequence IIID[a] Sequence VD[a]
SG	None	1980 Models	CRC L-381 Sequence IID Sequence IIIE Sequence VE Caterpillar 1H2
SH	None	1994 Models	CRC L-38 Sequence IID Sequence IIIE Sequence VE
SJ	None	1997 Models	CRC L-38 Sequence IID Sequence IIIE Sequence VE

[a] Obsolete test.

Table 10.3 Engine Oil Classification System for Commercial Diesel Engine Service-Commercial Oils (C)

API commercial engine service categories	Previous engine service categories	Related military or industry designations	Engine test requirements
CA	DG	MIL-L-2104A	CRC L-38 Caterpillar L-1 (0.4% sulfur)[a]
CB	DM	MIL-L-2104A, Supplement 1	CRC L-38 Caterpillar L-1 (0.4% sulfur)
CC	DM	MIL-L-2104B MIL-L-45152B	CRC L-38 Sequence IID Caterpillar 1H2[a]
CD	DS	MIL-L-45199B, Series 3 MIL-L-2104C/D/E	CRC L-38 Caterpillar IG2
CD-II	None	MIL-L-2104D/E	CRC L-38 Caterpillar 1G2 Detroit Diesel 6V53T
CE	None	None	CRC L-38 Caterpillar 1G2 Cummins NTC-400 Mack T-6; Mack T-7
CF-4	None	None	CRC L-38 Cummins NTC-400 Mack T-6; Mack T-7 Caterpillar 1K
CF-2	None	None	CRC L-38 Caterpillar 1M-PC Detroit diesel 6V92TA
CF	None	None	CRC L-38 Caterpillar 1M-PC
CG-4	None	None	CRC L-38 Sequence IIIE GM 6.2L Mack T-8 Caterpillar 1N
CH-4	None	None	CRC L-38 CAT 1P GM 6.5L Mack T-9 Cummins M11

[a] Obsolete tests.

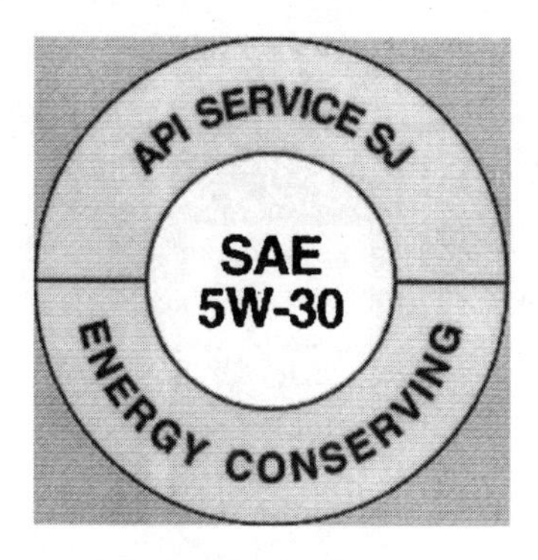

Figure 10.6 The API donut.

slated for introduction around mid-year 2001. Several new engine and bench tests will be used to evaluate GF-3 candidates by measuring fuel economy, emissions, and improvements in oil performance.

4. ACEA European Engine Oil Specifications

The Association des Constructeurs Européens D'Automobiles (ACEA) introduced new sequences for engine oils as of September 1999. These new ACEA sequences replaced the CCMC (Comité des Constructeurs du Marché Commun) specifications previously used by the European engine manufacturers. The ACEA sequences currently cover three ranges of engines and applications: ''A'' sequence for service fill oils for gasoline engines; ''B'' sequence for service fill oils for light-duty diesel engines, and ''E'' sequence for service fill oils for heavy-duty diesel engines.

Table 10.4 shows the engine tests used for each of the current sequences. For these sequences, five new engine tests were introduced to help further identify oil performance characteristics above the older CCMC specifications. In addition to the 1998 sequences, ACEA created two new categories for 1999. These are ''E4 and ES'' for superhigh performance diesel engines. New ACEA sequence releases are anticipated for 2001.

5. U.S. Military Specifications

In the past, the Department of the Army, U.S. Department of Defense, had responsibility only for issuing specifications for oils for use in military vehicles. However, to provide standardization of lubricating oils used in the many commercial vehicles operated by other branches of the government, the army was also given the responsibility for preparing an oil specification for those vehicles.

For more than 50 years, the U.S. Army developed and maintained specifications for lubricants to be used in military equipment. The lubricant specifications are commonly referred to as mil specs. Currently, some of the mil specs are being converted to commercial item description (CID) format and to performance specifications. These are still entirely military specifications. There are plans to turn over the more than 5 million lubricant specifications to a civilian technical group for developing and maintaining needed lubricants to satisfy all branches of the military. An SAE–military–industry lubricants task force is being formed to review the possibility of a joint military–civilian committee to handle existing and future military specifications and potential correlation to API, ACEA,

Table 10.4 ACEA 1999 European Engine Oil Sequences[a]

Sequence	Partial listing of engine tests	Performance parameter
	European gasoline engines	
A1-98	Sequence IIIE	High temperature oxidation
A2-96, Issue 2	CEC-L-55-T-95	High temperature deposits, ring sticking, oil thickening
A3-98		
	Sequence VE	Low temperature sludge
	CEC-L-38-A-94 (TU3MS)	Valve train scuffing wear
	CEC-L-53-T-95 (M111SL)	Black Sludge
	CEC-L-55-T-95 (M111SL)	Fuel economy
	European light-duty diesel engines	
B1-98	VWIC TD (except B4-98)	Ring sticking, piston cleanliness
B2-98	CEC-L-56-T-95 (XUD11ATE)	Medium temperature viscosity
B3-98	CEC-L-51-T-95 (OM 602A)	Wear
B4-98	CEC-L-78-T-97 (VW TDI) (only B4-98)	Piston cleanliness
	European heavy-duty diesel engines	
E2-96, Issue 3	CEC-L-42-A-92 (OM 364A) (only E2-96 Issue 3 and E3-96, Issue 3)	Bore polish, piston cleanliness
E3-96, Issue 3	CEC-L-51-T-95 (OM 602A)	Wear
E4-99	Mack T-8	Soot
E5-99	CEC-L-52-T-97 (OM 441LA) (only E4-99, E5-99)	Bore polishing, turbocharger deposits

[a] All the engine tests listed for each sequence apply to all categories under each engine classification unless noted.

ILSAC, and other specifications. These could possibly be published as an ''SAE J'' number.

Sections (a)–(k) give brief descriptions of the current engine oil specifications, along with obsolete specifications that may be still used to some extent as performance references.

(a) U.S. Military Specification MIL-PRF-2104G. This specification describes lubricating oils for internal combustion engines used in combat tactical service. There is no comparable API or ILSAC specification that matches this specification. The API equivalent to this specification is represented approximately by CG-4, CF, and CD II.

(b) CID A-A-52306A. This specification describes lubricating oils for heavy-duty diesel engines used in nontracked vehicles. The API equivalent to this specification is represented approximately by CG-4, CF, and CF-2.

(c) CID A-A-52039 B. This specification describes lubricating oil for use in military automotive engines. It is equivalent to API service category SH and is soon to be upgraded to SJ.

(d) U.S. Military Specification MIL-PRF-46167. This specification describes lubricating oil for use in engines exposed to arctic conditions.

(e) U.S. Military Specification MIL-PRF-21260. This specification describes preservative oils to be used in equipment that will be subject to long-term storage.

(f) U.S. Military Specifications MIL-L-2104 D–F (obsolete.) This specification for lubricating oils for internal combustion engines used in combat tactical service has been superseded by MIL-PRF-2104G.

(g) U.S. Military Specification MIL-L-46152 (obsolete.) This specification described oils for both gasoline and diesel engines in commercial vehicles used in U.S. federal and military fleets. In contrast to earlier U.S. military specifications, it places strong emphasis on gasoline performance in addition to diesel performance. The gasoline engine performance requirements are the same as API Service SE, and the diesel engine performance requirements are the same as the former MIL-L-2104B (API service CC): It covers oils in the SAE 10W, 30, and 20W-40 viscosities.

(h) U.S. Military Specification MIL-L-2014C (obsolete.) This specification described oils for gasoline and diesel engines in tactical vehicles in U.S. military fleets. The diesel engine performance requirements are the same as the former Military Specification MIL-L-45199B (API service CD). Gasoline engine performance in Sequences IID and VD is required at a level approximately intermediate between API Services SC and SD. The U.S. military operates very few gasoline engines in tactical service, so the gasoline engine performance requirements were set at a moderate level that would provide adequate performance without risk of compromising the severe diesel requirements desired.

(i) MIL-L-2104A (obsolete December 1, 1964). The performance requirements of this specification were the same as those now required for API Service CA.

(j) Supplement 1. This terminology dates back to the time of Military Specification 2-104B, which preceded MIL-L-2104A. The performance requirements for Supplement 1 were the same as those now required for API Service CB.

(k) MIL-L-45199B (obsolete November 20, 1970). The performance requirements of this specification were the same as those now required for API Service CD.

6. Manufacturer Specifications

Many engine manufacturers issue specifications for oils for their engines. Some of these specifications apply only during the break-in or warranty period, while most others represent the type of oil the manufacturer believes should be used for the life of the engine. These specifications are compatible with standard oil qualities available in the marketplace. For example, Chrysler passenger car engine oils are of API Service SJ, SJ/CD, ILSAC GF-2 quality, while Ford and General Motors specify API ''Starburst'', ILSAC GF-2. Still others specify special formulations designed to meet conditions that are specific to certain engines in certain applications. In the latter category is the oil required for certain railroad diesel engines. Where specifications of this type exist, special oils are usually developed and marketed to meet the specification requirements, assuming the volume required in the marketplace is sufficient to justify the costs of the development.

VI. OIL RECOMMENDATIONS BY FIELD OF ENGINE USE

The remainder of this chapter constitutes a guide to the types and viscosities of oils usually recommended for internal combustion engines used in the various major fields of application. It must be remembered that a particular engine may have different requirements for use in different fields of application.

A. Passenger Car

Most U.S. passenger cars have four-stroke-cycle gasoline engines. A small number have diesel engines. Two-stroke-cycle gasoline engines were phased out many years ago but may see some resurgence around the world. The usual recommendations for model year 2000 production, four-stroke gasoline engines are oils for API service SJ and ILSAC GF-2. These oils provide good protection against low temperature deposits and corrosion, protect against wear, and provide excellent protection against oxidation, thickening, and high temperature deposits under the most severe conditions of high speed operation or trailer towing. They also provide improved fuel economy and help assure the long-term performance of emission control systems. More emphasis is also being placed on oil quality aspects to provide longer drain intervals.

Many older model four-cycle gasoline engines can be satisfactorily operated on oils for API service SG or SH (now obsolete specifications), although oils for API service SJ will provide somewhat better overall performance and are more readily available.

The viscosities usually recommended by the automotive manufacturers for passenger car engines are SAE 10W-30 or 5W-30 for year-round service. The main reasons for choosing these lower viscosity products is to help achieve CAFE requirements and to help assure quick oil supply to critical areas such as rocker arms at cold starts. For extreme low temperature operation, SAE OW-XX oils are available.

The recommendations for passenger car diesel engines are generally similar to those for four-cycle gasoline engines, that is, oils for API Service CD or SJ/CD. Some manufacturers permit the use of multiviscosity oils, while others prefer single-viscosity types.

Two-cycle gasoline engines are used to some extent in various areas of the world. These engines are lubricated either by premixing the oil with the fuel or by injecting oil into the fuel at the carburetor. The oil used is usually of either SAE 30 or 40 viscosity and is formulated specifically for this service. Generally, four-cycle engine oils are not satisfactory because the additives in them will form undesirable deposits in the combustion chambers of two-cycle engines.

B. Truck and Bus

A significantly larger proportion of diesel engines are used in trucks and buses than in passenger cars. Both two-cycle and four-cycle diesel engines are used; the gasoline engines are all four cycle.

The recommendations for gasoline engines in trucks and buses are similar to those of passenger cars, that is, oils for API Service SJ. Multiviscosity SAE 10W-30, 15W-40, 20W-40, and 20W-50 oils are being used to take advantage of the improved starting and fuel economy they provide.

There is considerable variation in the oils recommended for diesel truck and bus engines. As shown in Table 10.3, there are several current API service classifications for commercial diesel engines. These service classifications are differentiated by a suffix (-2 or -4) to signify two- or four-cycle diesel engine requirements. The "C" categories without a suffix can be used in either engine type. The higher performance and quality levels for diesel engines are API CF, CF-2, CG-4, and CH-4. The two-cycle engines are somewhat sensitive to ash content in the oil; thus, the usual recommendation for them—whether supercharged or naturally aspirated—is an oil for API Service CF-2 with certain restrictions on the ash content. Some of the four-cycle engines also have shown sensitivity to the additive system; therefore, various manufacturers may have special requirements over and above the basic API service classifications.

Where liquefied petroleum gas engines are used in trucks or buses, oils for API Service SJ are often used for convenience, although oils of somewhat lower quality may be satisfactory. In some cases, special oils containing ashless additive systems (no metallo-organic detergents) are recommended.

C. Farm Machinery

The engines used in farm machinery include gasoline, diesel, and LPG. Some two-cycle gasoline engines are used for utility purposes such as water pumps and lighting plants. In general, the oil recommendations parallel those of passenger cars and trucks. There is a trend toward the recommendation of oils for API Service CF or CG-4 for naturally aspirated diesel engines as well as supercharged engines. Some manufacturers express a preference for special oils without metallo-organic detergents (low ash formulations) for LPG engines. Viscosities recommended closely parallel those for truck and bus engines.

The small two-cycle gasoline engines used for utility purposes are lubricated by oil mixed into the fuel. For convenience, these engines are sometimes lubricated with SAE 30 or 40 oils of API Service SH or SJ quality, but, where available, special oils designed for use in two-cycle gasoline engines should be used. These latter oils are formulated for the conditions encountered in the two-cycle engines and generally give lower port, combustion chamber, and spark plug deposits. Viscosities are usually in the SAE 30 or 40 viscosity range, and there is a trend to prediluting the oils with approximately 10% of a petroleum solvent so they will mix more readily with the fuel. Usually, fuel oil mixtures on the order of 16:1 to 40:1 are used, but some of the newer engines are designed for the use of higher ratios.

D. Contractor Machinery

Contractor machinery covers a wide variety of engines, ranging from small, two-cycle gasoline engines through automotive-type gasoline and diesel engines to larger diesel engines used to power giant cranes and earthmoving machines. Generally, oil recommendations closely parallel those for similar engines in other automotive equipment.

E. Aviation

Primarily, reciprocating piston engines for aviation use are of the four-stroke-cycle gasoline type. By far the majority of these engines are relatively small and are used in personal and civil aircraft.

For many years aircraft engines were operated on high quality, straight mineral oils. In recent years, however, dispersant-type oils have been developed which offer benefits in engine cleanliness, and most aircraft engine manufacturers now accept or recommend the use of these oils. Straight mineral oils may still be recommended for the break-in period.

Aircraft engine oils are usually formulated from special base oils made from selected crudes by carefully controlled refining processes. Quality is usually controlled to meet U.S. and U.K. government specifications, although some of the engine manufacturers have their own closely related specifications. Dispersant-type oils are usually manufactured by combining an approved additive system with proven straight mineral aircraft engine oils.

Table 10.5 Aircraft Engine Oil Specifications

Common grade designation	U.S. Specifications		Approximate SAE grade
	Straight mineral (formerly MIL-L-6082E)	Dispersant (formerly MIL-L-22851D)	
80	SAE J1966	SAE J1899	40
100	SAE J1966	SAE J1899	50
120	SAE J1966	SAE J1899	60

The viscosities of aircraft engine oils may be designated by SAE viscosity grade, or by a grade number which is the approximate viscosity in SUS at 99°C (210°F). The specification numbers and viscosity grades covered are shown in Table 10.5.

F. Industrial

Nearly every type of internal combustion engine has some industrial application. The smaller engines are usually of the automotive type, so the oils used are similar to those recommended for automotive applications. Our discussion pertains to the larger engines used as prime movers in power plants, mills, pipeline pumping stations, refineries, and so forth. Both two- and four-stroke-cycle engines are used in these applications. Diesel engines are used with fuels ranging from clean distillates to heavy residuals. Spark ignition engines are usually run on natural gas, producer gas, or LPG. Dual-fuel engines, which are operated on a combination of diesel fuel (about 5%) as an igniter and natural gas, are also used.

1. Diesel Engines

Diesel engines range in size from just a few horsepower up to engines rated 50,000 hp or more. Generally, they can be divided into three classes: high speed engines, medium speed engines, and low speed engines.

The lubrication requirements differ considerably for these classes, to some extent because of the different fuels that the various classes of engines can burn satisfactorily.

(a) High Speed Engines. Generally, engines are considered to be "high speed" if they are designed to operate at speeds of 1000 rpm or higher. Engine sizes range up to about 300 mm (11.8 in.) bore, with power outputs up to about 400 hp per cylinder. Multicylinder engines with outputs up to about 7000 hp are available. All these engines are of the trunk piston type; they may be either supercharged or naturally aspirated, and either two-stroke or four-stroke cycle. The rings and cylinder walls are lubricated by oil splashed from the crankcase sprayed on piston undercrowns or from oil supplied to wrist pins through passages in connecting rods.

High speed engines require high quality fuels, and, therefore, are usually operated on distillate fuels similar to those used in automotive diesel engines. As a result, the sulfur content of the fuel rarely exceeds 1% and often is considerably lower. With fuels of this quality, corrosive wear of cylinders and rings can be controlled satisfactorily by oils of the types developed for automotive diesel service, such as oils for API Service CG, or

CH (followed by the appropriate suffix). For engines that belong to one of the types developed for railroad service, one of the special railroad engine oils may be used. Viscosities recommended are usually either SAE 30, SAE 40, or SAE 20W-40.

(b) Medium Speed Engines. These engines are designed to operate at speeds ranging between 375 and 1000 rpm. Engine sizes and outputs in this classification range from about 225 mm (8.85 in.) bore with an output of 125–135 hp per cylinder, to 600 mm (23.6 in.) or larger bore developing 1500 hp or more per cylinder. Large medium speed engines with power outputs to 30,000 hp or more from V-type configurations of up to 20 cylinders are available. All engines in this class are of the trunk piston type, generally four cycle. Newer engines are usually supercharged.

Smaller engines in this class have the rings and cylinder walls lubricated by oil splashed or thrown from the crankcase to the lower parts of the cylinder walls. Larger engines have separate cylinder lubricators to supply supplemental oil to the cylinder walls.

Medium speed engines are operated on a wide range of fuels. Many of the smaller engines are operated on high quality distillate fuels. Somewhat heavier fuels, such as distillates, may be used in some engines. The disadvantages of this fuel include a higher boiling range and a higher sulfur content. Also, some residual components may be included in the blend. Many of the large engines are designed to operate on residual fuels, although the residual fuels used are somewhat lower in viscosity than the heavy, bunker-type fuels that are often used in low speed engines. Sulfur contents of the residual fuels may range upward to about 2.5% and in some cases even higher.

While some medium speed engines may be operated on automotive diesel engine oils of about API Service CG or CH quality, in general, the oils used in these engines are developed specifically for them and for similar engines used in marine propulsion service.

For convenience, these oils can be described in terms of their total base number (TBN). Smaller engines, and larger engines burning high quality fuels, are usually lubricated with oils of 10–20 TBN. Where residual fuels are used or operation is severe with better quality fuels, oils of 30–40 TBN are usually used. The 30–40 TBN oils are frequently used for the cylinders of engines with separate cylinder lubricators. In some cases with poor quality, high sulfur fuels, oils in the 50–70 TBN range may be used. Some care should be taken with this latter approach, since after use the cylinder oil drains into the crankcase. Any incompatibility between cylinder oil and crankcase oil could, therefore, cause difficulties.

Viscosity grades used are usually SAE 30 or 40 for the crankcases, and SAE 40 or 50 for cylinders with separate lubricators. For engines in intermittent service in exposed locations, subject to low temperatures, multiviscosity oils or synthetic oils may be used in the crankcase.

(c) Low Speed Engines. The low speed engine class consists of the large crosshead-type engines, most of which operate at speeds below 400 rpm. Engines in this class usually range from about 700 to 1060 mm (27.5–41.5 in.) bore, with outputs for the largest engines as high as 4500 hp per cylinder, or 54,000 hp from a 12-cylinder engine, the largest built. Almost all these engines operate on the two-stroke cycle. Separate cylinder lubricators are used.

Low speed engines are usually operated on residual fuels, some with sulfur content of 4.0% or more. Since the combustion of fuels of this quality results in the formation of large amounts of strong acids, highly alkaline oils are needed to control corrosive wear and corrosion of rings and cylinder liners. Where the fuel sulfur content is relatively low, oils in the 20–40 TBN range may be used; with higher sulfur fuels, oils in the 60–80 TBN range are usually used. The viscosity grade is usually either SAE 40 or 50.

The crankcase, or system, oil in these engines lubricates the bearings and crossheads and may lubricate the supercharger bearings. Since a large volume of oil is involved, extremely stable oils designed for long service life are desirable. Little or no contamination of the system oil by combustion products occurs; therefore, high levels of detergency and alkalinity are not required. Many engines are still operated on straight mineral oils, or inhibited straight mineral oils. Other engines are operated on the lower alkalinity oils used for trunk piston engines, usually oils of 10 TBN or less. A number of newer, large engines have oil-cooled pistons. The high temperatures to which the system oil is exposed in the piston cooling spaces of these engines may cause thermal cracking of the oil with a build up of deposits. In turn, these deposits may interfere with heat transfer, potentially resulting in excessive piston temperatures. This had led to the development of special system oils designed to provide better thermal stability and better ability to control the buildup of deposits due to thermal cracking. Usually, these oils are manufactured from selected crudes by carefully controlled refining processes. The additives used in them are also carefully selected for their ability to resist thermal decomposition at high temperatures.

Usually, system oils are of SAE 30 grade, although SAE 40 oils may occasionally be used.

2. Natural Gas Fired Engines

Reciprocating, spark-ignited internal combustion engines burning natural gas (methane) as the fuel present a wide variety of configurations, designs, and applications. These engines can be two-stroke cycle, four-stroke cycle, stoichiometric, lean-burn, naturally aspirated, or turbocharged, and they operate over a wide range of loads and speeds. Speeds range from as little as 200 rpm to over 3000 rpm and range in size from 100 hp to well over 20,000 hp. The lubrication needs of these engines vary significantly based on the designs, applications, operating conditions, and fuel quality used to fire them. The applications range includes mainline gas compression, field gathering gas compression, power generation, and driving pumps.

Low speed gas engines are designed to operate at speeds below 500 rpm and high speed gas engines are designed to operate above 900 rpm. All engines designed to operate at speeds between 500 and 900 rpm will be defined as medium speed engines. For example, most high speed gas engines are designed to operate at speeds of 1000 rpm or higher but can satisfactorily operate at speeds below 900 rpm. These engines are still defined as high speed engines. The definitions of speed used here vary slightly depending on industry sources but such discrepancies do not affect the validity of the representations made in this section.

With respect to selecting lubricants for gas engines, there are no widely accepted industry specifications to define performance requirements. Although there are specifications for internal combustion engines using gasoline and diesel fuel, these specifications do not apply to gas-fueled engines. Engines operating on gaseous fuels, other than converted automotive engines using LPG or propane, require lubricating oils designed and formulated to meet the unique requirements of the gas engine. Both base stocks and additive combinations are critical in balancing the performance needs of these engines.

Most lubricating oils used in gas engines today were developed specifically for this type of service. Most contain dispersants to control varnish-type deposits resulting from oxidation and nitration. Other additives include detergents, antiwear agents, oxidation inhibitors, corrosion inhibitors, and metal deactivators. Some engine manufacturers recommend only oils containing ashless additives (no metallo-organic detergents), while others recommend the use of oils containing metallo-organic oxidation and corrosion inhibitors in combination with ashless dispersants. Still others prefer that metallo-organic detergents be included in the formulations. The amount of metallo-organic detergent required, as measured by sulfated ash content (TBN is also an indication of detergent level), varies considerably depending on engine manufacturer, engine design, fuel quality, and operating conditions. Where clean dry natural gas is burned as the fuel, the main purpose of including ash-containing additives in the formulations is to control valve and valve seat wear in four-cycle engines. The residue from burning the ash-containing additives (mainly detergents) during combustion produces a solid lubricant to help protect the valve and seat surfaces. Depending on such factors as metallurgy and operating conditions, varying the amount of ash in the oil and the residue it produces during combustion has been found to be effective at controlling wear in different engines. However, using oils with too high an ash level can have negative consequences on engine performance. Selecting the optimum oil for a given application requires balancing many factors, such as engine makes and models, with operating conditions and the fuel qualities. Reflecting the various needs of the different engines, premium gas engine oils can be classified as follows:

Ashless oils: oils with sulfated ash levels of 0.00% and containing ashless inhibitors (oils below 0.11% sulfated ash are considered to be in the same category as ashless oils).

Low ash oils: oils with sulfated ash levels between 0.11 and 0.30%. They contain ashless dispersants and can contain small amounts of metallo-organic oxidation inhibitors along with antiwear additives.

Medium ash oils: oils with sulfated ash levels between 0.30 and 0.90%. These oils contain metallo-organic detergents in combination with other inhibitors.

High ash oils: oils with sulfated ash levels above 0.90%. These oils contain higher levels of detergents.

Landfill gas oils: a class of gas engine oils specially formulated to handle the unique requirements and often severe engine conditions resulting from burning landfill gas in the engines. Ash levels range from 0.50 to >1.00%. For very severe fuel conditions, higher ash level products may be required.

The oil viscosity typically recommended and used is usually SAE 40 grade. However, some engine manufacturers recommend SAE 30 grades for their engines. An SAE 30 grade or multiviscosity oil (e.g., 15W-40) can be used in low temperatures. Where extremes of temperatures exist, synthetic gas engine oils will provide the best protection and the most reliable service. Applications such as remotely started and stopped engines subjected to low oil temperatures at start-up benefit from the use of synthetic gas engine oils. Because of their higher level of oxidation and thermal stability, in many applications oil and filter change intervals can be extended with synthetic gas engine oils.

3. Two-Stroke-Cycle Engines

Large, slow speed, two-stroke-cycle gas engines, such as those used for gas compression in mainline transmission stations, generally operate at around 300 rpm. These engines depend on the use of high quality, low ash oils for maximum performance. The selection of proper base stocks for two-stroke-cycle gas engines is at least as critically important to performance as is the selection of additives. The oil from the crankcase is generally used to lubricate the power cylinders. Hence, lower quality lubricants, oils with too high an ash level, or excessive power cylinder feed rates will result in engine operating problems. These problems consist of spark plug fouling, combustion chamber deposits, ring sticking, and the plugging of exhaust and intake ports, which in turn, cause losses in power and efficiency. In some instances, improper selection of the lubricant can cause cylinder and piston scuffing.

The oils recommended for large two-stroke-cycle engines are SAE 40, ashless to low ash (0.00–0.1% ash) oils formulated with high quality base stocks that provide very low carbon formation characteristics. Although not generally recommended for the large two-stroke-cycle engines, multiviscosity oils have been used in some instances. Some smaller two-stroke-cycle engines used in field gas gathering operations subjected to low temperatures may benefit from the use of multiviscosity gas engine oils.

4. Four-Stroke-Cycle Low to Medium Speed Gas Engines

The low speed four-stroke-cycle gas engines are used in mainline transmission stations for gas compression. The medium speed engines are used for mainline transmission and to drive electric generators for station power. Occasionally, these medium speed engines are also used in packaged compressor units in field gas gathering. The fuel is clean natural gas with the exception of gas gathering, where wellhead gas is used. The wellhead gas is generally clean dry fuel, but fuel quality needs to be known for these applications before a lubricant is selected. The oils recommended for the engines burning clean natural gas are low to medium ash SAE 40 or SAE 30 grade oils depending on the particular engine manufacturer. Engines with higher brake mean effective pressure (BMEP) ratings require higher levels of antiscuff protection. If valve life is a concern, the medium ash oils should be used.

5. Four-Stroke-Cycle High Speed Gas Engines

The four-stroke-cycle high speed gas engines are generally used to drive compressors in field gas gathering operations. These engines are also used for power generation in applications such as cogeneration. Sometimes the fuel is clean natural gas, while at other times the fuel can contain liquids and sulfur compounds. Landfill gas can contain high levels of chloride compounds where conventional low ash oils will not provide adequate protection against the corrosive materials formed during the combustion process when chlorides are present. The quality and contents of the fuel are one key to making proper lubricant choices. Once the fuel issues have been addressed, the next concern, since the newer low emissions engines have significantly lower levels of oil consumption, is to protect valve life in the high speed engines. Since in turn, the valves depend on a certain amount of oil ash residue to provide a solid film between the valve faces and seats, higher ash level oils may be required to offset the reduced combustion ash levels caused by the lower oil consumption. The oils recommended for the four-stroke-cycle high speed gas engines are generally SAE 40 grade oils with ash levels that range from low ash to high ash formulations. The level of ash required varies by engine manufacturer and even by model from the same manufacturer. SAE 30 grade oils can be used for lower temperature applications, and although monograde oils are preferred, multiviscosity oils can be used in some engines.

6. Oil Selection

Some of the items that need to be considered in the selection of an oil for natural gas engine applications are as follows:

Engine manufacturers' recommendations
Oil suppliers' recommendations
Makes and models of engines
Fuel used
Operating conditions
  Engine loads
  Ambient conditions
  Oil and water temperatures
  Engine speeds
Engine condition
Past experience
Type of service (continuous, or frequent start–stops)
Add-on equipment
  Catalytic converters
  Ebullient cooling
  Lean-burn conversions
  Stratified charge systems
Oil consumption rates
Desired oil drain interval

7. Dual-Fuel Engines

To some extent, the lubricating oil in a dual-fuel engine is subjected to the deteriorating influences found in both diesel and gas engines. The diesel fuel burned as a pilot fuel tends to produce soot and varnish deposits, while the gas fuel, combined with high operating temperatures, tends to cause oxidation and nitration of the oil. While these conditions are not necessarily conflicting, there is some indication that oils specially developed for the service may provide superior engine cleanliness with both diesel and gas engine oils. In some applications where only 5% pilot high quality diesel fuel and clean natural gas can use gas engine oils with good results. SAE 30 or 40 oils are used as crankcase oils, with SAE 50 or even higher viscosity oils preferred for cylinder lubrication of engines with separate cylinder lubricators.

8. Marine Engines

(a) Marine Inboard (Small Craft). The engines used in small marine craft are often modified versions of automotive engines, both gasoline and diesel. As a result, the qualities and viscosities of oils recommended closely parallel those used in comparable automotive engines.

(b) Marine Outboard. The majority of outboard motors used for boats are two-stroke-cycle gasoline engines because of the high power-to-weight ratio of this design. Oil is mixed with the fuel for lubrication. In spite of a trend to very lean oil-to-fuel ratios, some oil is always carried into the combustion chambers where deposits can be formed. Excessive deposit buildup can affect combustion and obstruct the ports, causing loss of power. Metallo-organic additives in the oil may aggravate this deposit problem; thus, most oils for outboard engines are formulated with ashless dispersants. A corrosion inhibitor to provide additional protection against rusting during shutdown periods is usually included. The base oils are selected to minimize carbonaceous deposits in the combustion chambers. Data obtained in a standardized outboard motor test may be submitted to the National

Marine Manufacturers Association (NMMA) for certification that an oil is suitable for Service TC-W3 (TC, two cycle). Oils certified in this manner are generally acceptable to the outboard motor manufacturers.

The current NMMA TC-W3 specification is the only specification licensed by NMMA, but the older obsolete specifications such as TC-WC, TC-W, and TC-WII are still marketed and may appear on oil containers. The NMMA replaced the Boating Industry Association (BIA); it is primarily a U.S. spec but is somewhat recognized worldwide. In addition, ISO (International Standards Organization) and JASO (Japanese Automotive Standards Organization) are in the process of establishing global oil specifications for two-cycle (2T) and four-cycle (4T) engines for motorcycles. The ''2T'' specification will also cover two-cycle engines used in other applications such as outboard engines, chain saws, and snowmobiles. Fuel-to-oil ratios of 50 : 1 or higher are now recommended, particularly for high output engines. Many oils now are prediluted with a small proportion (about 10%) of a special petroleum solvent to facilitate mixing with gasoline. At some locations, where there is a high volume of fuel sales for outboard motors, premixed oil and fuel may be offered.

(c) Marine Propulsion. The propulsion engines used in vessels range from small automotive type engines used in fishing boats, harbor craft, and other small craft, to the largest diesel engines now manufactured. In general, the discussion of industrial diesel engines applies to marine propulsion engines.

(d) Marine Auxiliary. Auxiliary engines used on ships are medium or high speed diesel engines of moderate size. The engines are operated on distillate fuels of fairly good quality. In some cases, convenience dictates the type of oil used in these engines. Where the main propulsion engines are operated on a 10–20 TBN oil, for example, the same oil can be used satisfactorily in the auxiliary engines. In other cases, the oils used in the propulsion engines are not suitable for use in the auxiliary engines, and a special oil must be carried for the auxiliaries. This oil will be either one of the 10–20 TBN marine diesel engine oils or an automotive diesel engine oil, depending on the preference of the engine manufacturer.

9. Railroad Engines

Because of space limitations, diesel engines used in railroad locomotives often have specific ratings high enough to give high power output from a relatively small engine. Combined with restrictions on cooling water capacity, this results in both crankcase and cylinder temperatures being high. Long idling periods and rapid speed and load changes also contribute to severe operating conditions for the lubricating oil.

Both two- and four-stroke-cycle engines are used in this service, and cylinder lubrication is supplied by oil throw from the crankcase.

U.S. railroad engine manufacturers recommend oils developed specifically for their engines. These oils are usually SAE 40 grades, although SAE 20W-40 multigrade oil is receiving greater acceptance. Some builders have special requirements because of specific design issues. EMD, for example, recommends zinc-free oils because of the adverse effects of the phosphorus portion of the zinc compounds on silver bearings.

10. Specialty Applications

Large numbers of small gasoline engines are used in such applications as snowmobiles, chain saws, lawn mowers, garden tractors, home lighting plants, and trail bikes. Most of these engines are air-cooled, two-cycle units. Four-cycle engines are used where weight is not a major concern or, in larger engines, where the higher fuel consumption of the two-cycle engines might be a concern. The two-cycle engines are lubricated by oil mixed with the fuel. Both oils developed for outboard motors and oils developed specially for air-cooled, two-cycle engines are used. Special oils are used most frequently in snowmobiles and chain saw engines. Mixing is critical with snowmobile oils because of the low temperatures at which the engines are operated, so prediluted oils are frequently used. Operating temperatures in snowmobiles also may be high because the engine is shrouded to provide heat for the operator, and this may cause piston varnish problems. Mixing is also a problem with chain saws, as is high temperature operation with piston varnish and port deposits.

BIBLIOGRAPHY

Mobil Technical Books

Diesel Engine Lubrication in Stationary Service
Mobil Technical Bulletins
Engine Oil Specifications and Tests—Significance and Limitations
Diesel Engine Operation

11

Stationary Gas Turbines

Four commercially important types of prime movers provide mechanical power for industry: the steam turbine, the diesel engine, the spark-ignited gas or gasoline piston engine, and the gas turbine. All types achieve their result by converting heat into mechanical energy; any machine that does this (i.e., uses chemical reactions to supply thermal energy) is properly called a heat engine.

Of the basic prime movers, the gas turbine is the most direct in converting heat into usable mechanical energy. The steam turbine introduces an intermediate fluid (steam), so the products of combustion (from the chemical reaction) do not act directly on the mechanism creating motion. Piston engines initially convert heat into linear motion, which must be transmitted through a crankshaft to produce usable shaft power. But the gas turbine, in its simplest form, converts thermal energy into shaft power with no intermediate heat or mechanical redirection.

The theory of gas turbines was well understood by about 1900, and although development work started shortly after that, it was not until 1935 that engineers overcame low compressor efficiency and temperature limitations imposed by available materials and succeeded in building a practical gas turbine.

Just before World War II, the Swiss built and operated successful gas turbine plants for industrial use. In the United States, the first large-scale gas turbine application was in a process for cracking petroleum products.

Near the end of World War II, both Germany and England succeeded in developing turbojet-propelled aircraft. Turbojet and related designs for aircraft propulsion are an extension of the industrial gas turbine, and ironically enough, aircraft turbines have now been adapted for stationary power applications.

Gas turbines have been used in power generation since the 1940s. The early units had the capacity to produce 3500 kW·h of power output with thermal efficiencies in the 20–25% range. Today's gas turbines can produce power outputs of more than 200 MW and attain thermal efficiencies as high as 40% for a simple cycle open system, making them competitive with other forms of prime mover. They have gained a place in stationary applications because of several advantages: light weight, small size, ease of installation, quickness in starting, multifuel capabilities, no cooling water requirements (for some designs), and availability of large amounts of exhaust heat. Recovery of this exhaust heat in a steam turbine or other in-plant heating requirements is raising the efficiency of the combined cycle application to over 60%.

Gas turbines in power generation will operate at speeds that range from 3000 to 12000 rpm and drive generators that rotate at 1800 or 3600 rpm for 60-cycle power (1500 or 3000 rpm for 50-cycle power). Some are direct-connected to the generators, while others go through gear reduction units to operate the generators at the appropriate speeds.

I. PRINCIPLES OF GAS TURBINES

Gas turbines basically consist of an axial compressor to compress the intake air, a combustor section, and a power turbine. The compressed air is mixed with fuel (liquid or gas) and burned in combustion chambers (combustors). The hot gases expand through a turbine or turbines to drive the load. There may be a single shaft with a single turbine to drive both the compressor and the load, or two shafts with a high pressure turbine to drive the compressor and a low pressure turbine to drive the load.

A. The Simple Cycle, Open System

The simple cycle, open system gas turbine consists of compressor, combustor, and turbine (Figure 11.1). The compressor draws in atmospheric air, raises its pressure and temperature, and forces it into the combustor. In the combustor, fuel is added, which burns on contact with the hot compressed air, boosting its temperature and heat energy level. The hot, compressed mixture travels to the turbine, where it expands and develops mechanical energy (i.e., torque applied to a shaft). A portion of this energy is needed to drive the

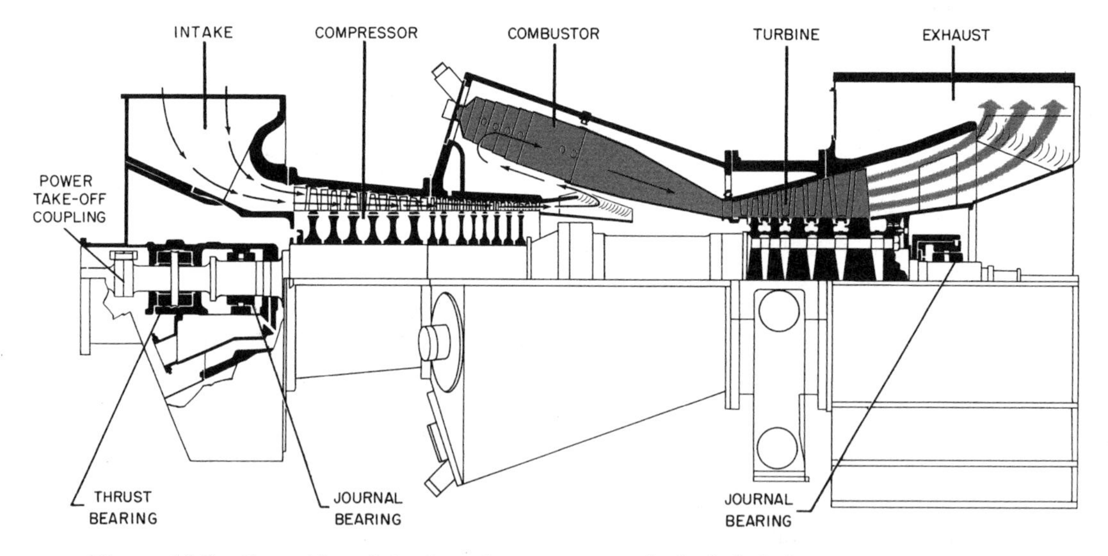

Figure 11.1 Gas turbine of simple cycle, open system single-shaft design.

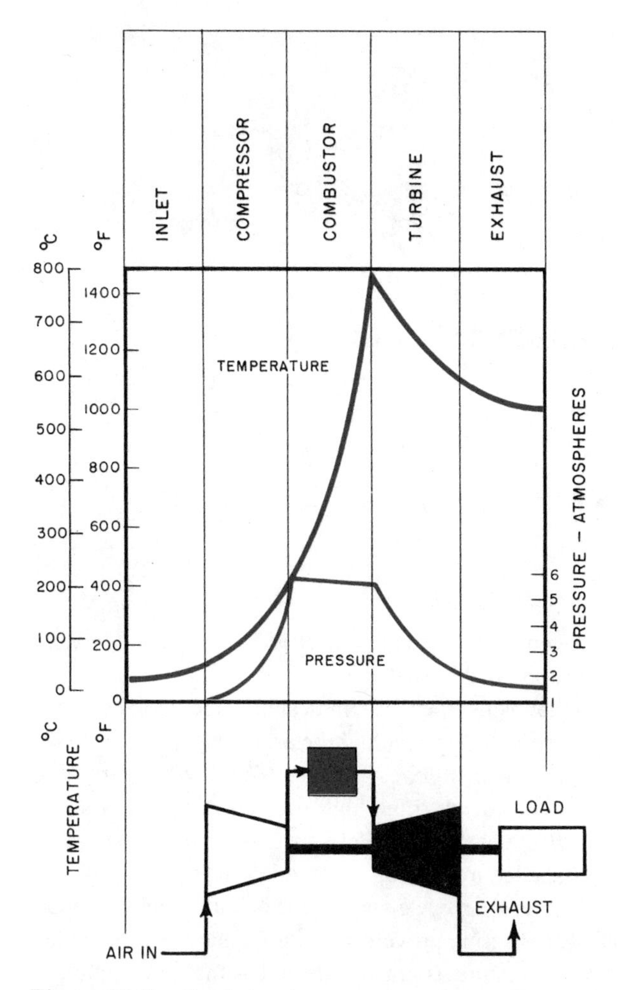

Figure 11.2 Graph showing approximately how gas temperatures and pressures change in flowing through a gas turbine of type shown in Figure 11.1.

compressor; the rest is available to drive a useful load such as a generator or compressor. Figure 11.2 illustrates how temperature and pressure rise and fall within this type of gas turbine system. It is the gaseous working fluid that gives the gas turbine its name—not the fuel, which may be liquid or gaseous (or even solid).

The flow of energy through the cycle is shown in Figure 11.3. The entering air contributes zero energy. (Actually, shaft work into the compressor is transferred to the compressed air to add some energy to it as it enters the combustor.) Fuel introduced into the combustor represents net energy input, balanced by work output (useful shaft work) plus exhaust energy vented to the atmosphere. The turbine work required to drive the compressor simply circulates within the cycle—from shaft to compressed air to combustor energy, and back to turbine work. Surprisingly, of the total shaft work developed by the turbine, almost two-thirds must go to drive the compressor, leaving only a little more than

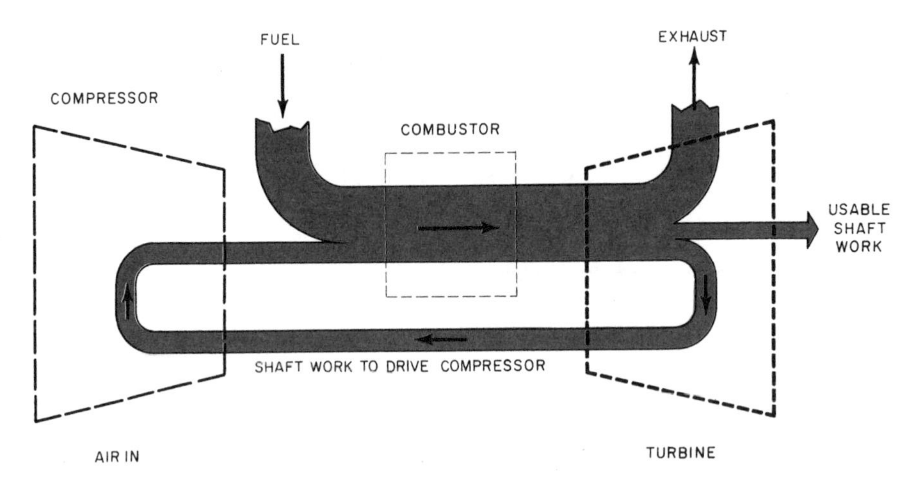

Figure 11.3 Diagram showing flow of energy in simple cycle, open system gas turbine.

one-third for useful shaft work. Improvements in compressor efficiencies and component metallurgy are increasing the amount of available shaft power.

A common variation of the simple cycle, open system design is the two-shaft type (Figure 11.4). The turbine is divided into two stages: a high pressure stage, which drives the compressor, and a low pressure stage, which drives the load. The low pressure (free or power) turbine is not mechanically connected to the high pressure turbine. Its speed can be controlled independently to suit the speed of the driven unit, while the speed of the high pressure turbine can vary as needed to develop the required power.

The thermal efficiency of the simple, open cycle gas turbine is improving, currently reaching a maximum of about 40%. Efforts to improve thermal efficiency have included designing for as high a turbine inlet temperature as practicable and as high a compressor pressure ratio as practicable up to an optimum value. Improvement can also be made through recovery of exhaust heat, which is discussed later.

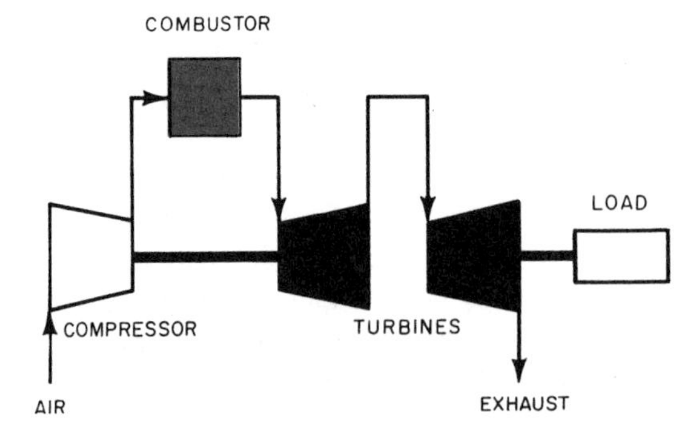

Figure 11.4 Simple cycle, open system two-shaft gas turbine.

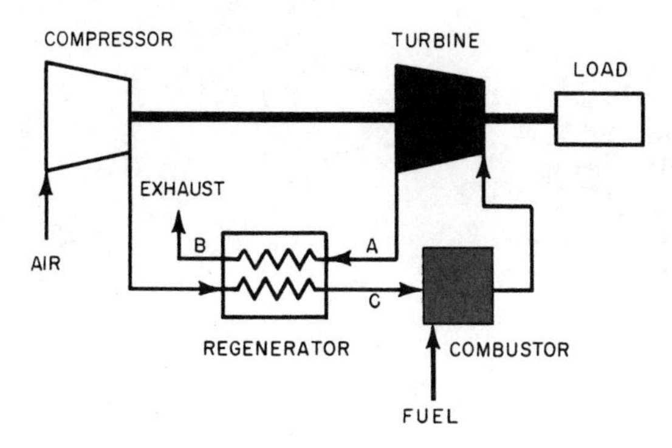

Figure 11.5 Regenerative cycle open system showing use of exhaust to heat entering air and improve thermal efficiency.

B. Regenerative Cycle, Open System

The regenerative cycle was designed to make use of exhaust heat (Figure 11.5). The regenerator receives hot, turbine exhaust gas at *A* and rejects it a *B*. Compressed air passes through the unit in a counterflow direction and picks up heat from the turbine exhaust; then the heated air leaves the regenerator at *C* and passes into the combustor. Hotter air in the combustor needs less fuel to reach maximum temperature before flowing into the turbine; thus, thermal efficiency is improved. This conforms to a basic rule for heat engines, which states that the effect of raising the mean effective temperature at which heat is added while lowering the mean effective temperature at which heat is rejected increases a cycle's thermal efficiency.

C. Intercooling, Reheating

The compressor tends to be an inefficient machine; the ratio of energy at its outlet to that at its inlet is low. A detailed discussion of compressors appears later. Elementary theory shows, however, that compression in which the temperature of the gas does not rise requires less work than compression with increasing temperature. To partially achieve the former, the compression system is sometimes divided into two stages, and the working fluid (air) is withdrawn and cooled between stages, as shown in Figure 11.6.

Intercooling gives more net work per pound of working fluid, allowing a smaller turbine plant for the same output or a greater output from the same size turbine. Also, where combustor temperature is maintained, the portion of usable shaft work developed increases because less shaft work is required at the compressor.

Just as the compression process may be divided, so may the expansion process, with reheating between stages instead of intercooling (Figure 11.7). Reheating, ideally to the same top temperature in each stage, offers a straightforward way of boosting output with little increase in plant size, although a combustor is needed for each stage, and the turbine is correspondingly more complex. With higher temperatures at the exhaust, the regenerative cycle is almost always used. Without a regenerator, there is little gain in thermal efficiency.

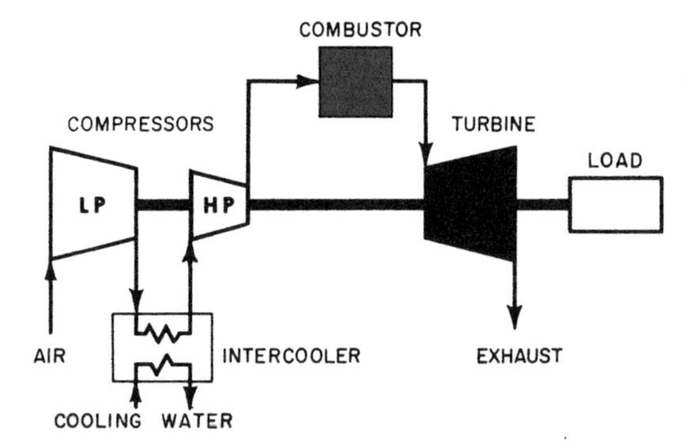

Figure 11.6 Intercooling between compressor stages to reduce work of compression and size of machine.

D. Essential Gas Turbine Components

1. Compressor

The key to successful gas turbine operation is an efficient compressor. Early turbine plants were troubled with units that could not handle large-volume airflows efficiently. Today, almost all compressors in industrial gas turbine applications are axial flow or centrifugal, or a combination of the two.

In the axial flow compressor, blades on the rotor have airfoil shapes to provide optimum airflow transmission. Moving blades draw in entering air, speed it up, and force it into following stationary vanes, shaped to form diffusers that convert the kinetic energy of moving air to static pressure (Figure 11.8). A row of moving blades and the following row of fixed blades are considered a stage; air flows axially through these stages. The number of axial flow compressor stages in a modern industrial gas turbine varies from 12 to 22. Pressure ratios range from 5.5:1 to 30:1. With intake air at 14.7 psia (1.03 kP/cm^2) and 60°F (15.6°C), and a pressure ratio of 6.45:1, air will be compressed, in an uncooled compressor, to about 94.7 psia (6.66 kP/cm^2) and 426°F (219°C). Blade heights

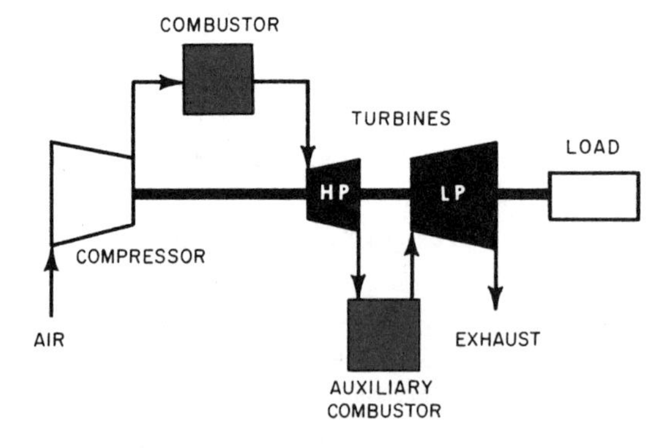

Figure 11.7 Reheating between turbine stages to increase power output.

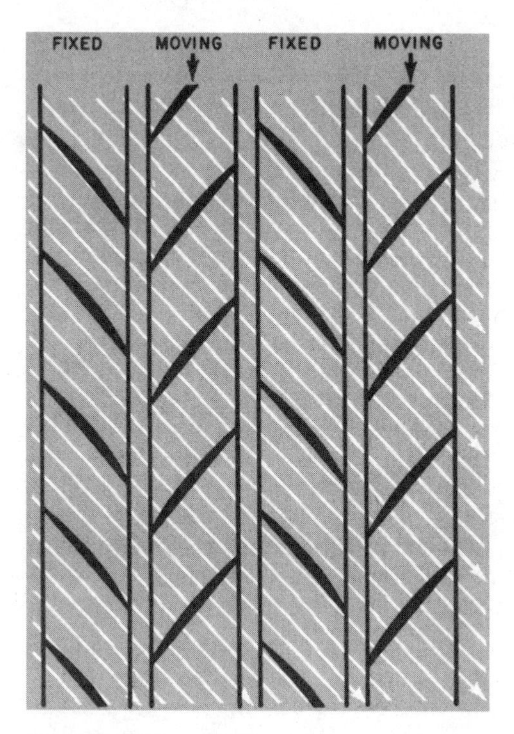

Figure 11.8 Axial flow compressor blading.

decrease as air travels through the axial flow compressor because the air's specific volume diminishes at the increasingly higher pressures.

Air displacement of axial flow compressors varies from 8 to 20 ft^3/min (13.6–34 m^3/h) of working fluid per horsepower. Large-volume, multistage machines may displace more than 300,000 $ft/^3$/min (510,000 m^3/h), with temperature rises of 400°F (204°C) and more. Compressor efficiencies range from 85 to 92%, although performance hinges on minimum fouling of blades and diffusers.

Though the centrifugal compressor can be multistaged, its best performance is as a single-stage machine operating at a pressure ratio not exceeding 4 or 5 : 1. In comparison to the axial flow compressor, the centrifugal machine operates more efficiently over a wider range of mass flow rates, is more rugged in construction, and is less susceptible to contaminants in the air that may form deposits in the airflow passages. However, the machine's application in industry is limited to small gas turbines because it cannot handle large volumes of air efficiently at higher pressure ratios.

2. Combustor

For efficient combustion, the combustor, or combustion chamber, must assure low pressure losses, low heat losses, minimum carbon formation, positive ignition under all atmospheric conditions, flame stability with uniform outlet temperatures, and high combustion efficiencies.

The combustion section may comprise one or two large cylindrical combustors, several smaller tubular combustors, an annular combustor with several fuel nozzles compactly surrounding the turbine, or an annular combustor in which several tubular liners, or cans, are arranged in an annular space (Figure 11.1). For all these, the design is such that less than a third of the total volume of air entering the combustor is permitted to mix with the fuel; the remainder is used for cooling. The ratio of total air to fuel varies among engine types from 40 to 80 parts of air, by weight, to one of fuel. For a 60:1 ratio, only about 15 parts of air is used for burning; the rest bypasses the fuel nozzles and is used downstream to cool combustor surfaces and to mix with and cool the hot gases before they enter the turbine.

3. Turbine

The turbine is an energy converter, transforming high temperature, high pressure gases into shaft output. Gas turbines follow impulse and reaction designs—terms that describe the blading arrangement. The distinction depends on how the pressure drop at each stage is divided. If the entire drop takes place across the fixed blades and none across the moving blades, it is an impulse stage (Figure 11.9A). If the drop takes place in the moving blades as well as the fixed blades, it is a reaction stage (Figure 11.9B).

A row of fixed blades ahead of a row of moving blades constitutes a stage. Many turbines are multistaged, since the amount of power that can be developed by a single stage of reasonable size, shape, and speed is limited.

For best performance, the gas turbine must work with high inlet temperatures. Manufacturers continually strive to build turbines that can accommodate higher temperatures, either through improved materials or through new designs. For the most part, design improvements focus on cooling the parts subjected to highest temperatures and stresses, particularly the first stage or two of the turbine. One cooling technique is to take air bled from the compressor outlet and channel it over parts holding the blades. Another method is to extend the shank linking the blade bucket to the rotor, thus removing rotor parts from the hottest concentration of gases.

Over 30 years ago, inlet temperatures of 1200°F (649°C) were considered to be extreme; today, as a result of continuing research, the ability of turbine materials to with-

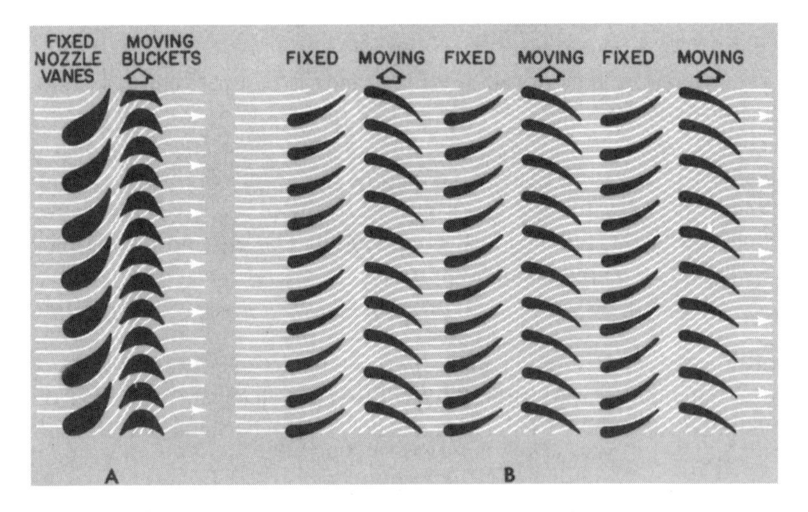

Figure 11.9 Impulse (A) and reaction (B) turbine blading.

stand extremely high temperatures permits more efficient operation. Temperatures in the range 1600–1700°F (871–927°C) are seen in fully loaded units, with some designs operating with firing temperatures at 2300°F (1260°C). By the time these gases reach the turbine exhaust, their temperature has dropped at least 600°F (333°C), and possibly by as much as 1200°F (649°C)—depending on inlet temperature and the number of stages. Today's high output gas turbines can have exhaust temperatures of about 1100°F (594°C).

Such increased turbine efficiencies are a result of substantial improvements in compressor efficiencies over the recent years. Compressors can have 22 stages and compression ratios that exceed 30:1. This increase in compressor efficiency has helped raise firing temperatures, which in turn has increased overall turbine efficiencies. Firing temperatures range from 1700°F for sequential combustion units (ABB GT 24, where there are two burn stages) to more than 2300°F for single combustion units (GE MS 7001 F) that are fully loaded.

II. JET ENGINES FOR INDUSTRIAL USE

In several applications, aircraft jet engines are being used for industrial service. Initially, instead of providing propulsion power directly, the hot compressed gases from the engine were fed to a power or free turbine, which converted the heat energy into rotative power (Figure 11.10). Used in this manner, the jet engine acts as a gas generator, providing the working fluid (gases) to power a turbine. As many as 10 jet engines have been used as gas generators for one large turbine. This provided a compact, lightweight design and the ability to replace components rapidly.

The compactness and light weight of the jet engine are partially explained by its relatively high speed—usually in the range 8000–20,000 rpm (Figure 11.11). Large industrial gas turbines, on the other hand, usually run at speeds of 3000–12,000 rpm.

In more recent years, the use of direct-connected aircraft jet engine designs has been adapted for industrial service. Since jet engines are mass-produced, they offer several advantages. Replacement parts are readily available, and even entire engines can easily be replaced. This makes possible interchangeability and factory overhaul instead of maintenance in the field. The light weight reduces foundation needs and makes installation easier. Another advantage: in most applications cold starts to full load operation can be achieved in less than 2 min. Also, simple cycle thermal efficiencies of 40% or more are possible with some of the newer design jet engine derivatives.

However, bearings of jet engines (practically always of rolling element type) run hotter than those of typical industrial gas turbines. Most of the bearings are "buried" within the engines and surrounded by hot gases. Bearing housings are sealed and pressurized with hot air bled from appropriate compressor stages. The resulting problem is twofold: (1) removing the heat rejected at the bearing and (2) finding lubricants that can stand up to bearing temperatures of 400°F (204°C) and more. The answers are high lubricant flow rates and the use of synthetic, high temperature lubricants, discussed in Chapter 5.

Aircraft-type gas turbines will never, however, completely replace the rugged heavy-duty industrial gas turbines. The latter offer several advantages: they can use plain bearings with long life capability not limited by eventual fatigue failure, as in the case of rolling element bearings; they can operate on a wide range of low cost fuels; and their bearings are usually located in areas relatively low in temperature. This last factor not only means that the lubricant requirement is less strict, it also means that special high temperature construction materials need not be used.

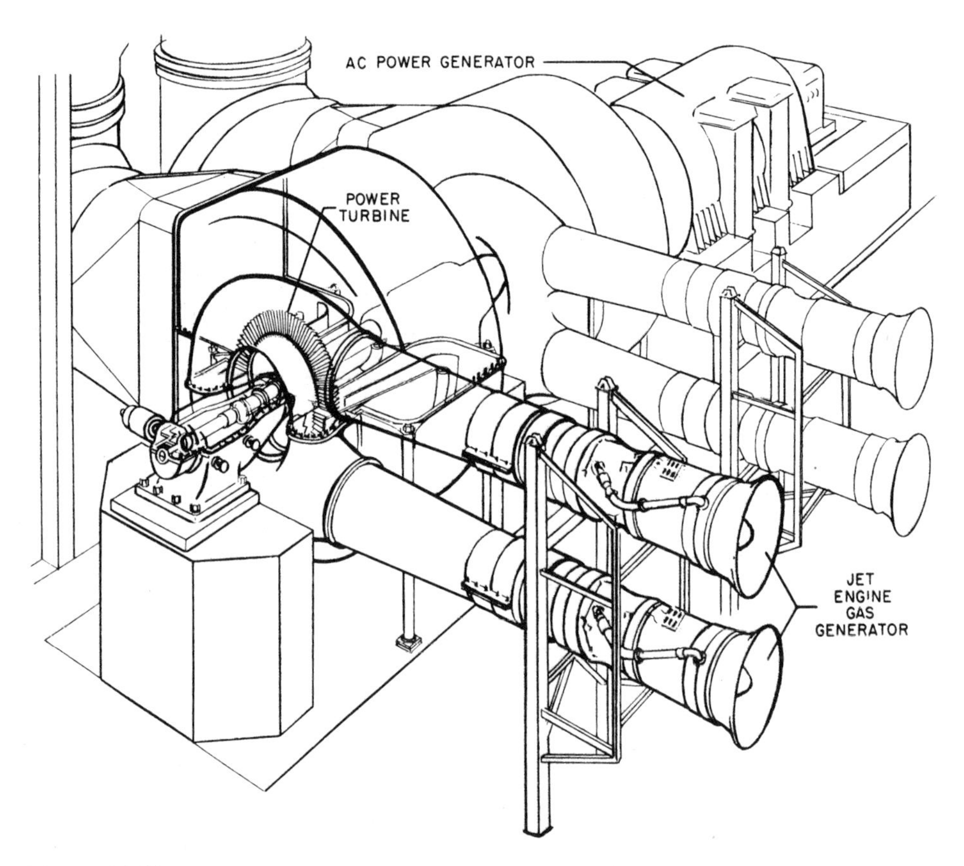

Figure 11.10 Two jet engines provide hot gas to each of the two power turbines coupled to a single generator.

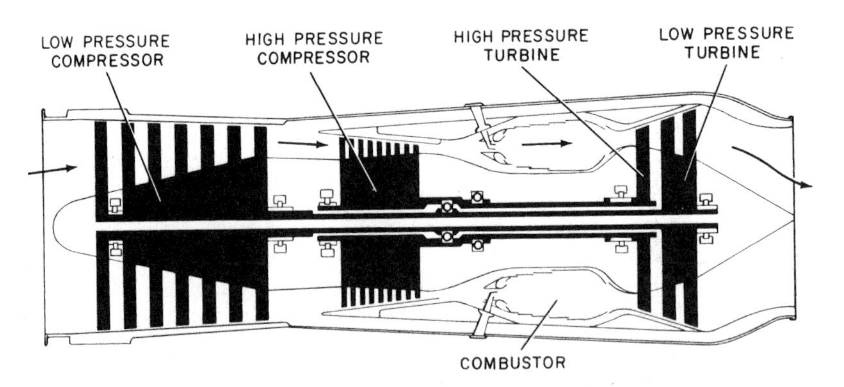

Figure 11.11 Aircraft-type generator of dual-axial or twin-spool design.

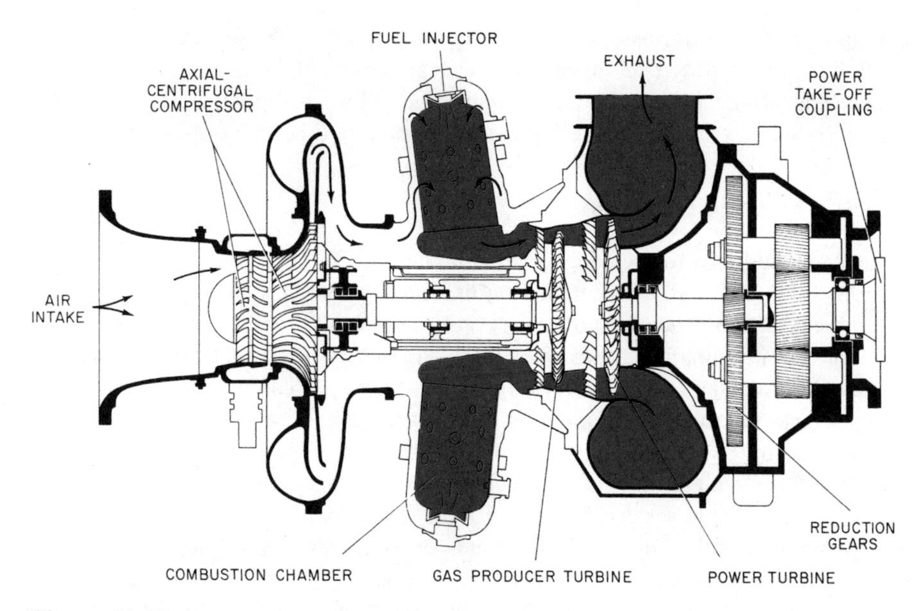

Figure 11.12 Small (360 hp max) two-shaft gas turbine for mechanical or generator drive. Shaft bearings of the gas producer and the power section are of the plain, or sleeve, type. Both plain and rolling element bearings are used in the reduction gearbox.

A. Small Gas Turbine Features

Not all industrial turbines are large machines delivering thousands of horsepower. A great many are machines of less than 1000 hp rating, and a surprisingly large number of units are under 100 hp in size.

Many different makes of small gas turbines are available. Most follow design practices of larger industrial turbines, using plain (journal) bearings located in relatively cool areas. Others, however, including those designed for frequent start–stops, especially at low temperatures, use rolling element bearings, which are also located in relatively cool areas. Some makes are available with either plain or rolling element bearings, depending on the type of service anticipated.

Some relatively small gas turbines use axial flow compressors and turbines. Others, to achieve compactness, use centrifugal compressors (usually one but sometimes two stages) or, as in Figure 11.12, an axial stage followed by a centrifugal stage. Some use a radial in-flow turbine rather than an axial one. Small turbines usually run at speeds ranging from about 18,500 to 60,000 rpm.

Several small gas turbines being developed for vehicle use rely on regenerators, high turbine inlet temperatures, and other means to increase thermal efficiency and reduce fuel consumption. Most of these machines are of the two-shaft type, with gas generator speeds ranging from 22,000 to 65,000 rpm and power turbine speeds ranging from 12,000 to 46,000 rpm. When mass-produced versions of these turbines are available, it is likely that they will find many applications for stationary purposes as well as vehicle drives.

III. GAS TURBINE APPLICATIONS

Gas turbines, in sizes from 50 hp to more than 200,000 hp, have wide application in industry today. Four broad areas of usage stand out: electric power generation, pipeline transmission, process operations, and total energy applications. These four areas indicate the gas turbine's versatility and suggest its wide-ranging potential as an efficient power source. The gas turbine's success in these applications, along with increased thermal efficiencies, will doubtless make this design more desirable in other areas as well.

A. Electric Power Generation

Utilities use gas turbines for peaking generation and for emergency service. Most units exceed 10,000 hp, and both conventional turbines and jet engines adapted for industrial service are used. System loads in a utility vary daily and seasonally. For efficient operation, base loads are usually supplied by steam turbines, with peak loads handled by gas turbines. In emergency shutdowns of base load equipment, gas turbines can take over. In peaking applications, a gas turbine may operate only a few hundred hours a year; this arrangement is the most efficient for year-round generation. As gas turbine operation has become more economical and convenient, the units are being used for longer periods because they are capable of giving reliable service. More and more gas turbines are being used as base-loaded units because of the improved efficiencies and reliability, as well as considerations of fuel costs and environmental emissions.

Outside the utility, gas turbines are used to drive generators in industrial plants and buildings on an emergency standby basis. Sometimes they are used as peaking units to supplement on-site steam plant power. Along with diesel and gas engines, gas turbines are finding application in a new concept called continuous duty/standby, in which they are normally coupled to a nonessential load like the compressors in an air conditioning system. In emergency blackout situations, the nonessential load is dropped and a generator supplying essential electrical loads such as elevators, exit lighting, and critical equipment is picked up. Hospitals, universities, and retirement facilities are finding this form of emergency power extremely reliable.

B. Pipeline Transmission

In the transmission of natural gas, centrifugal compressors are often used, with gas turbines, fueled by natural gas, driving them. This is an ideal arrangement because this preferred turbine fuel is readily available and the large loads are well suited for gas turbine drives. Units range in size from 1000 to over 20,000 hp, with capacities in the 5000–10,000 hp range being the most prevalent.

C. Process Operations

In the area of process operations more than in any other, the full potential of the gas turbine has yet to be realized. In fact, if properly designed, the gas turbine can become an integral part of the process, as essential as any other key element in the system. In process applications, the turbine is designed not so much to satisfy unit efficiency as to satisfy system efficiency. For example, in addition to generating shaft power, a turbine with an oversized compressor can produce pressurized air for use in steel mill applications. Instead of attempting to utilize exhaust gas to raise turbine efficiency, the gas can be channeled to a process unit requiring heat, for example, a dryer. Such uses may be accomplished with supplementary firing using the hot exhaust gas (which contains up to 17% of oxygen) as preheated combustion air for kilns or boilers. Hot, pressurized gases from certain processes can be ducted to the turbine's combustor, reducing the amount of fuel needed; the turbine itself may drive process equipment.

D. Combined-Cycle Operation

Increasing use is being made of a recent concept of combining a gas turbine and a steam turbine as shown in Figure 11.13. Combined cycles to generate power achieve thermal efficiency as much as 65%. The high temperature energy in the gas turbine's exhaust gas is extracted in a waste heat boiler. The steam is fed to a steam turbine, and the exhaust steam either is used for process heating or is returned to a condenser as boiler feed water.

Another method of combined-cycle use passes the exhaust gases through two exhaust heat boilers in series. The first boiler produces superheated steam at 400 psia and the second, saturated steam at 104 psia. The high pressure steam is fed to the inlet of a mixed pressure steam turbine, and the low pressure steam is fed at an intermediate stage of the same steam turbine. The combined steam flow exits the turbine to the main condenser. The gas turbine, the steam turbine, and the compressor are mounted on a single shaft.

Combined-cycle plants are offered ranging from a net plant base load output of 12–594 MW and a heat rate of approximately 5300 Btu/kW·h.

E. Total Energy

In installations like that of Figure 11.14, all building energy is provided at the site. Typically, natural gas supplies a gas turbine, which drives a generator to carry the electrical load. Waste heat from the turbine exhaust supplies a waste heat boiler, which generates steam for absorption air conditioning, space heating, and domestic hot water. Some installations use diesels or reciprocating gas engines instead of gas turbines. Plant capacity depends on the size and type of building—shopping center, school, offices, apartment and so on. Power requirements today range from 300 hp to about 5000 hp. Compact turbines, either single shaft for constant speed applications or two shaft for variable speed applications, are selected. The units are small and light (frequently under 1500 lb) and can accept full load from start-up in a minute or less. Well-designed total energy systems are capable of thermal efficiencies exceeding 65%.

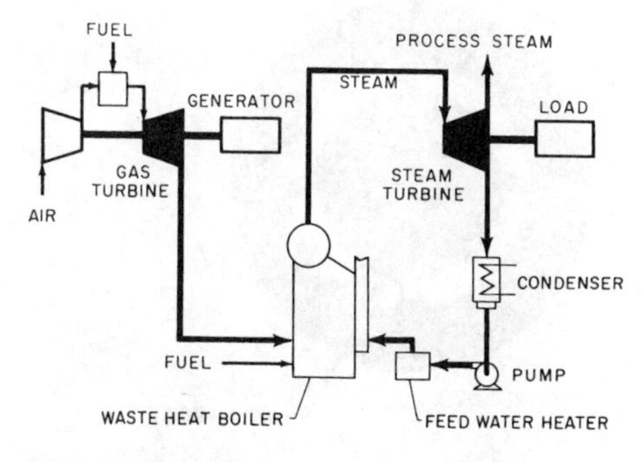

Figure 11.13 Gas turbine, steam turbine combined-cycle plant.

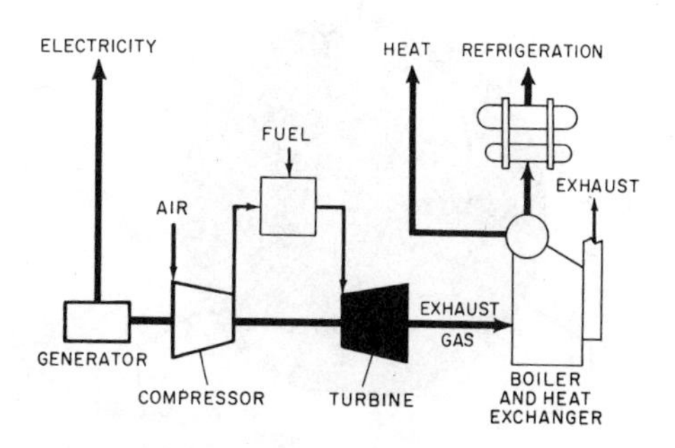

Figure 11.14 Use of exhaust heat from gas turbine to provide for absorption refrigeration in "total energy" plant.

IV. LUBRICATION OF GAS TURBINES

The principal purposes of lubrication are to reduce wear, reduce friction, keep systems clean, remove heat, and prevent rust. The elements of the gas turbines that need lubrication include plain and rolling element bearings, thrust bearings, gears, couplings, and contact type seals.

As noted earlier, aircraft-type gas turbines are equipped with rolling use plain (sliding) bearings. Plain bearings for radial loads may be of the following type:

1. Conventional heavy steel-backed, babbitt-lined cylindrical type
2. Precision-insert type (as in reciprocating engine practice)
3. ''Antiwhip'' type, to suppress self-induced vibration, which is sometimes a problem in high speed machinery

Two antiwhip gas turbines are shown schematically in Figure 11.15.

Gears for speed reduction and for accessory drive are usually of the spur, helical, herringbone, or bevel types. Some accessory drives are of the worm type.

A. Large Industrial Gas Turbines

Lubrication systems for the large, heavy-duty industrial gas turbines are generally similar to those for steam turbines or other high speed rotating machines.

A typical lubrication system may be described as a pressure circulation system, complete with reservoir, pumps, cooler, filters, and protective devices, and connecting tubing, fittings, and valves. Its purpose is to provide an ample supply of cooled, cleaned lubricating oil to the gas turbine, accessory and reduction gearing, driven equipment, and a hydraulic control system.

An example of the arrangement of these parts is shown in Figure 11.16. A gear-type, positive displacement main pump, driven by the power turbine, supplies pressurized oil to each of the main compressor and turbine bearings and to accessory gearboxes. Return oil from the bearings and gearboxes drains by means of gravity to the reservoir. Foot valves in the tank prevent oil in the supply system from draining back to the sump during shutdown periods.

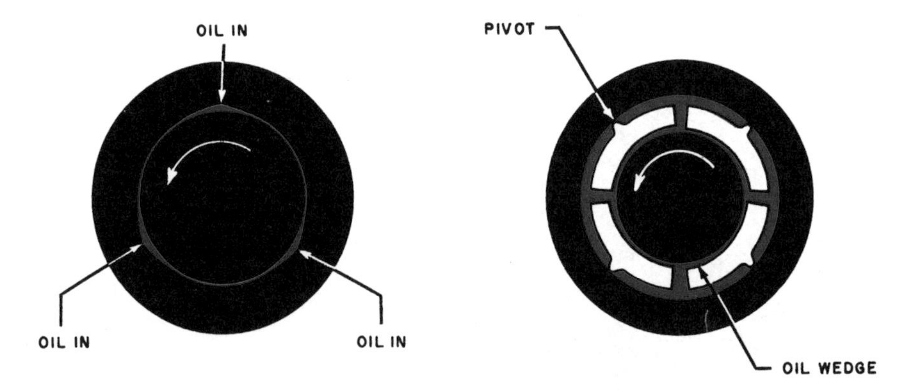

Figure 11.15 Antiwhip bearings: three-lobe type (left) and tilting pad type (right). Multiplicity of fluid films suppresses tendency for self-induced vibration of lightly loaded high speed shafts.

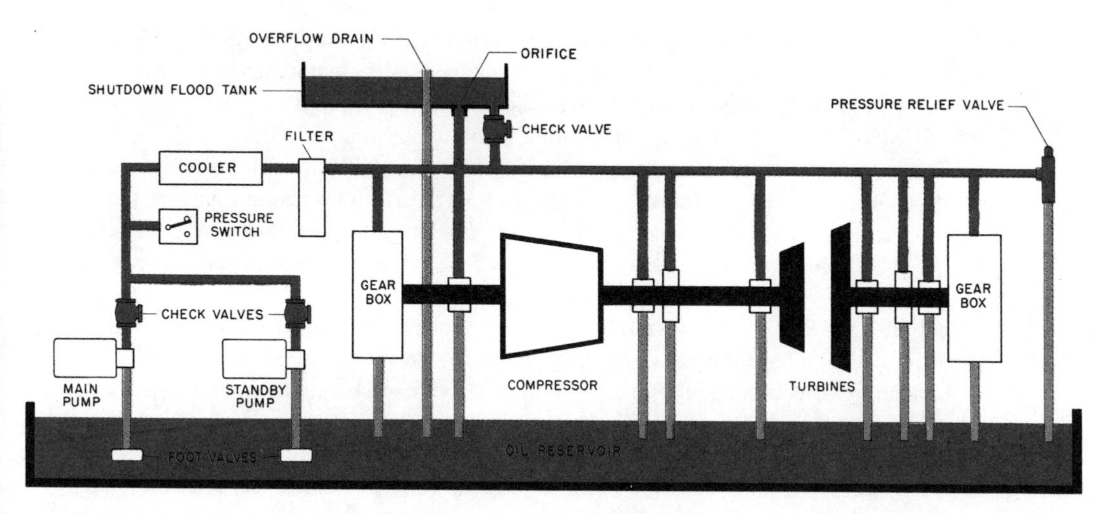

Figure 11.16 Lubrication system for large industrial gas turbine.

The pressurized oil passes through an oil cooler and filter (specified 5 μm by some manufacturers) before reaching the bearings. A pressure relief valve installed in the supply line maintains pressure at a constant level.

A motor-driven standby pump automatically supplies pressurized oil to the bearings before starting and for a few minutes after shutdown to prevent ''heat soak'' damage to the bearings. The standby pump also takes over automatically if the main pump fails during operation. Some lubrication systems, as shown in Figure 11.16, also include a flood tank to build up a reserve supply of oil during operation. This supply can flood the bearings by means of gravity for 15–20 min after system failure due to emergency power loss.

Protective devices are incorporated into the lubrication system to guard equipment against low oil supply, low oil pressure, and high oil temperature. The devices sound a warning or shut down the unit if any of these conditions occurs. In some installations, in the event of emergency power loss a pump driven by a battery-powered motor is available for supplying pressurized oil on a short-term basis. Also, some installations include an additional cooler and filter so that oil flow can be directed to one filter and cooler while the other filter and cooler are cleaned during operation of the gas turbine, avoiding shutdown. In some installations, oil heaters are also added.

Bearing oil header pressures are usually controlled at 15–40 psi (1.1–2.9 kP/cm^2). Hydraulic control pressures are often in the 50–150 psi (3.5–10.5 kP/cm^2) range, but much higher pressures are used in some systems.

Bearing and gear loads are generally moderate and shaft speeds are high, permitting use of circulating-type oils of ''light'' or ''medium'' viscosity.* Oil temperatures entering bearings and gears are usually in the 120–160°F (49–71°C) range, and temperature rise in these elements is 25–60°F (14–33°C). Bulk oil temperatures in the reservoir will usually be 155–200°F (68–93°C). These temperatures are well within the thermal limitations of

* ISO viscosity grade 32, 46, or 68 is commonly recommended.

high quality petroleum lubricating oil of suitable type. Starting temperatures may be low, so that oils of high viscosity index (like some synthetics), which do not thicken excessively at low temperatures, are recommended. One manufacturer specifies a maximum viscosity of 800 SUS (172.3 cSt) for reliable circulation of oil at start-up conditions.

The lubricating oil is recirculated in contact with air. Portions of it are broken into small drops or mist to ensure intimate mixing with air. Under these conditions, oil tends to oxidize, and the tendency is increased by high operating temperatures, by excessive agitation or splashing of oil, and by the presence of certain contaminants that act as catalysts. The rate of oil oxidation also depends on the ability of the oil to resist this chemical change. The use of thermally stable basestocks, such as hydroprocessed API group II or synthetic base oils, has been proven to provide excellent oxidation resistance.

Oxidation is accompanied by the formation of both soluble and insoluble products and by a gradual thickening of the oil. The insoluble products may be deposited as gum or sludge in oil passages, and their accumulation will restrict the supply of oil to bearings, gears, or hydraulic control devices. In some critical control system servovalves, even minute levels of oil degradation materials can contribute to malfunctioning of valves and loss of speed or load control. For long time performance in reuse systems, the lubricating oil must have high chemical stability to resist oxidation and the formation of sludge.

Water may find its way into bearing housings as a result of condensation of moisture from the atmosphere during idle periods or from other sources. The churning of oil with moisture in circulation systems tends to create emulsions. To resist emulsification, the oil should have the ability to separate quickly from water. With such an oil, any water that enters the bearing housings or circulation systems will collect at low points, from which it may be drained. It is also important that the oil selected have good antirust and anticorrosion characteristics.

B. Adapted Aircraft Gas Turbines

The lubrication of aircraft gas turbines adapted for stationary service differs substantially from that of large, heavy-duty industrial gas turbines. The lubrication systems are quite different, being of necessity much more compactly designed. The bearings, which are rolling element types, and gears, in some instances, are "buried" within the turbines and require more intricate passageways and other arrangements for supplying and removing oil.

The bearings and contact seals run much hotter than the plain bearings and clearance seals of the industrial turbines, usually requiring the use of synthetic lubricating oil and the placing of greater emphasis on the cooling function of the oil. Because of the need for compactness, bearing and gear loading is much higher in aircraft than in stationary practice, requiring extra load-carrying ability in the lubricant.

The lubrication system for a dual axial compressor gas generator is shown schematically in Figure 11.17. Oil from the tank is gravity-fed to boost pump *D*, which supplies oil at constant pressure to oil pressure pump *A* regardless of changes in pressure drop through the oil cooler and piping. During operation, pressure at the suction of *A* acts through pressure-sensing line *L* (1) on boost pump regulating valve *G*, regulating the pressure rise of oil boost pump *D*, and (2) on pressure-operated valve *N*, keeping it closed. During shutdown, pressure in line *L* falls, permitting valve *N* to open and vent antisiphon loop *M*. Since loop *M* extends above oil level, this prevents leakage of oil from the tank through pumps *D* and *A* into the auxiliary gear case. Oil pressure pump *A* forces oil—at

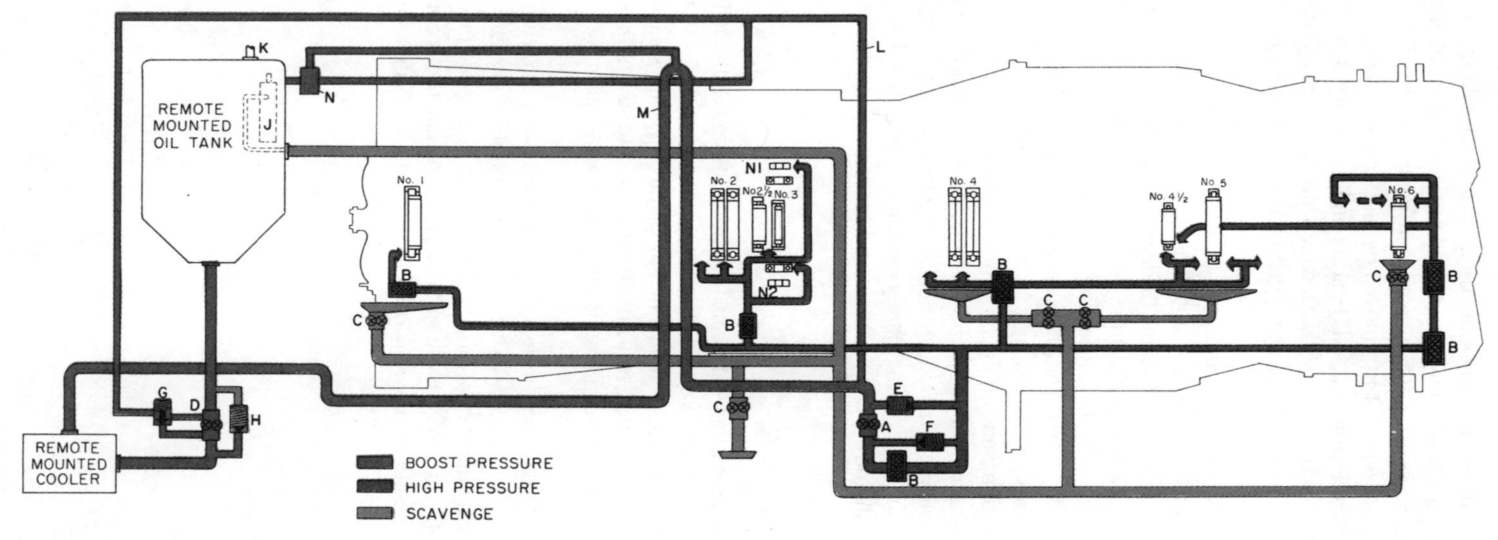

A · OIL PRESSURE PUMP
B · OIL STRAINER
C · SCAVENGE PUMP
D · OIL BOOST PUMP
E · MAIN PRESSURE REGULATING VALVE
F · MAIN SCREEN BY-PASS
G · BOOST PUMP REGULATING VALVE
H · BOOST PUMP RELIEF VALVE
J · OIL DEAERATOR
K · BREATHER CONNECTION
L · PRESSURE-SENSE LINE
M · ANTI-SIPHON LOOP
N · PRESSURE-OPERATED VALVE
NI · ACCESSORY-DRIVE BEARINGS
N2 · ACCESSORY-DRIVE BEARINGS

Figure 11.17 Lubrication system for aircraft gas turbine in stationary service.

about 45 psi (3.2 kP/cm^2)—through strainers or filters *B* to main bearing and accessory drive locations. Here the oil is jetted through calibrated orifices to parts requiring lubrication and cooling, including bearings, contact seals, and gears. The oil flows by gravity from these parts to sumps from which it is removed by individual scavenge pumps *C* and returned to the tank through a deaerator *J*. All bearing and accessory drive housings are vented by means of a breather system (not shown), which includes a rotary breather pump.

Power turbines used with aircraft-type gas generators have their own separate lubrication systems.

Thermal conditions in land-based, aircraft gas turbines are not only more severe than in heavy-duty industrial gas turbines, as noted earlier, but also are more severe than in their airborne counterparts. In aircraft use, the turbines operate at full power only during takeoff and climb—a small fraction of the time—while in stationary service they may operate for long periods at full load. The lubricating oil comes in contact with hot surfaces in the 400–600°F (204–316°C) range and mixes with hot gases that leak into bearing housings. Bulk oil temperatures in the tank may be in the 160–250°F (71–121°C) range. This is a very severe condition and requires a lubricant having excellent thermal and oxidative stability to resist degradation and deposit formation. Measures that may be taken to moderate this severe thermal condition include derating engines, increasing oil reservoir sizes, and improving oil cooling equipment. In any case, it is common to idle gas turbines for several minutes before shutdown to cool them and prevent exposing the oil to high ''soakback'' temperatures.

C. Small Gas Turbines

Small gas turbines, as indicated earlier, vary considerably in design. Lubrication systems, however, are usually relatively simple versions of the type of circulation system described for large industrial turbines. Essential elements include sump, strainer, shaft-driven main pump, pressure-regulating valve, cooler, full-flow filter, manifold to bearings, gears, and so on, and gravity return to sump. Some systems employ scavenge pumps to return oil to the main sump or tank.

Most small turbines are designed to use either synthetic or petroleum lubricating oils. Turbine-type oils similar to those described for large industrial gas turbines may be used, but oils of lower viscosity may be specified for some very high speed machines. In all cases, the gas turbine manufacturer specifies a suitable lubricant, and this recommendation should be followed unless specific approval is obtained for the use of some other product.

12

Steam Turbines

In a steam turbine, hot steam at a pressure above atmospheric is expanded in the nozzles, where part of its heat energy is converted to kinetic energy. This kinetic energy is then converted to mechanical energy in the turbine runner, either by the impulse principle or by the reaction principle. If the nozzles are fixed and the jets directed toward movable blades, the jet's *impulse* force pushes the blades forward. If the nozzles are free to move, the *reaction* of the jets pushes against the nozzles, causing them to move in the opposite direction.

In small, purely impulse turbines, steam is expanded to exhaust pressure in a single set of stationary nozzles. As a result of this single expansion, the steam issues from the nozzles in jets of extremely high velocity. To obtain maximum power from the force of the jets' impact on a single row of moving blades, the blades must move at about half the velocity of the jets. Thus, single-stage impulse turbines operate at very high rotative speeds. To reduce rotative speed while maintaining efficiency, the high velocity can be absorbed in more than one step, which is called *velocity compounding*.

Such a turbine (Figure 12.1) has one velocity-compounded stage followed by four pressure-compounded stages. In the velocity-compounded stage, steam is first expanded in the stationary nozzles to high velocity. This velocity is reduced in two steps through the first two rows of moving blades. The steam is then expanded again in a set of stationary nozzles and delivered to the first pressure stages. In each case, velocity is increased and pressure is decreased in the stationary nozzles. In the moving blades, velocity decreases but the pressure remains constant. The graph at the bottom of Figure 12.1 shows the changes in pressure and velocity as the steam flows through the various rows of nozzles and moving blades.

Another method of reducing rotor speed while maintaining efficiency is to decrease the velocity of the jets by dividing the drop in steam pressure into a number of stages. This is called *pressure compounding*. Refer again to Figure 12.1: the steam pressure is reduced by somewhat less than half in the velocity-compounded stage. The steam then passes to the pressure-compounded stages, where it is reduced in four steps to the final

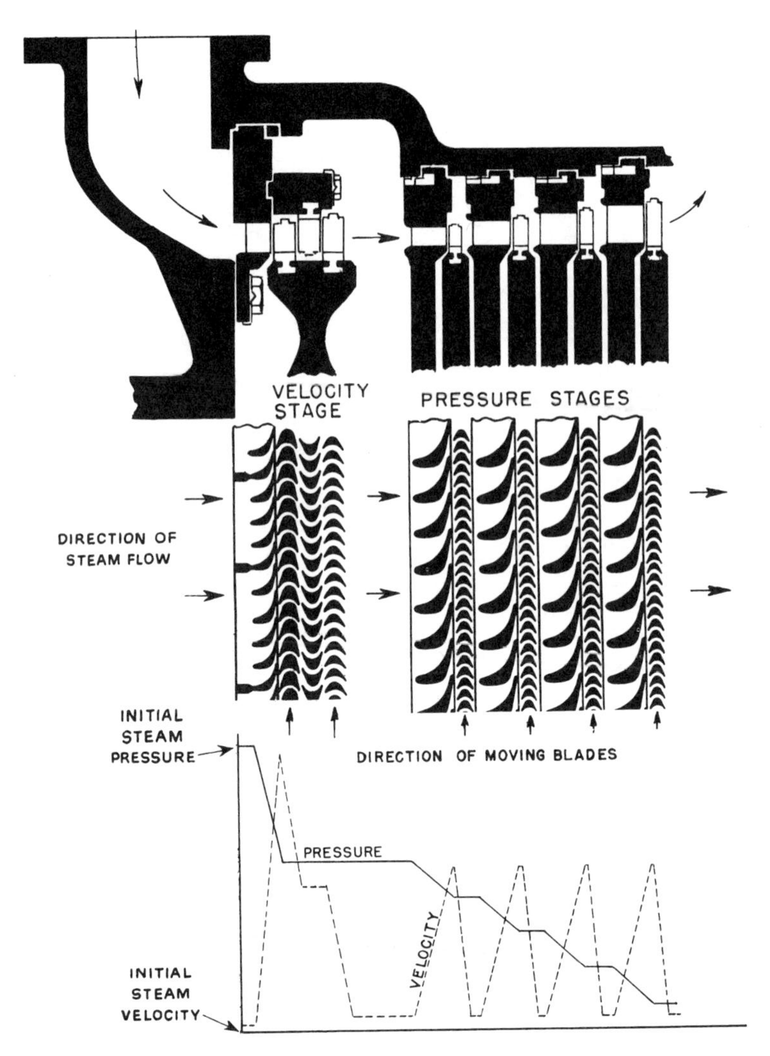

Figure 12.1 Turbine with impulse blading. Velocity compounding is accomplished in the first two stages by two rows of moving blades, between which is placed a row of stationary blades that reverses the direction of steam flow as it passes from the first to the second row of moving blades. Velocity compounding also can be accomplished by redirecting the steam jets so that they strike the same row of blades several times with progressively decreasing velocity.

exhaust pressure. Each pressure stage consists of a row of stationary nozzles and a row of movable blades, and the whole assembly is equivalent to mounting four single-stage impulse turbines on a common shaft. The arrangement of Figure 12.1, using velocity compounding in the first stage followed by pressure compounding in the remaining stages, is an approach used in many large turbines.

In reaction turbines, steam is directed into the blades on the rotor by stationary nozzles formed by blades that are designed to expand the steam enough to give it a velocity somewhat greater than that of the moving blades. (Because of this expansion to provide

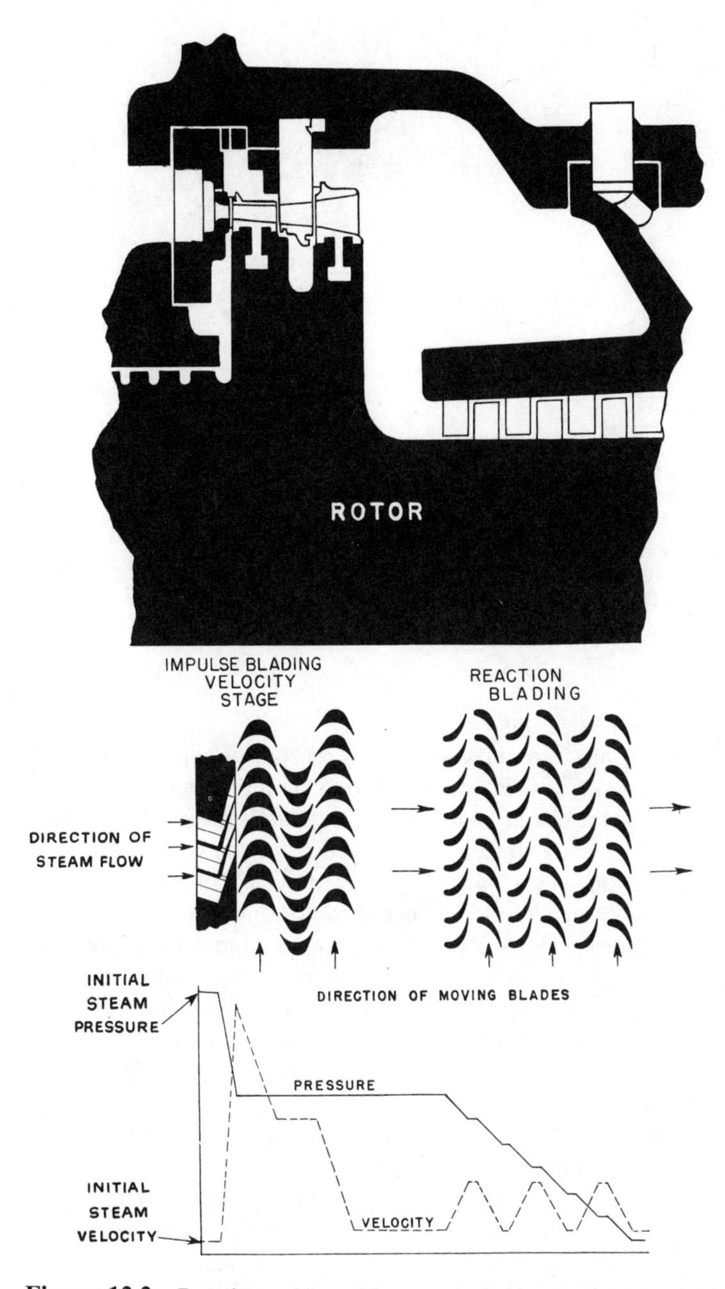

Figure 12.2 Reaction turbine with one velocity-compounded impulse stage. The first stage of this turbine is similar to the first, velocity-compounded stage of Figure 12.1. However, in the reaction blading of this turbine, both pressure and velocity decrease as the steam flows through the blades. The graph at the bottom shows the changes in pressure and velocity through the various stages.

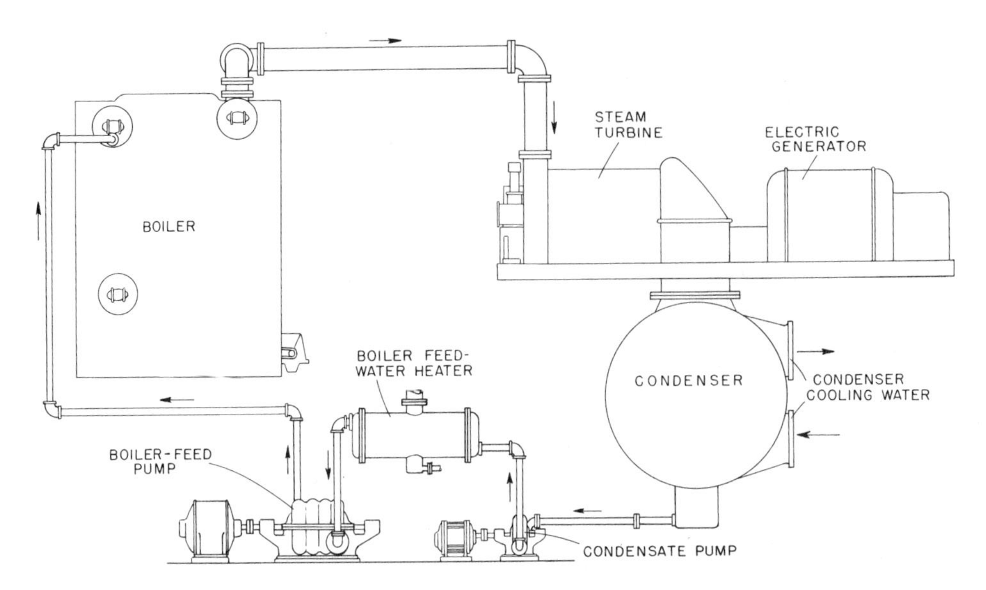

Figure 12.3 Simple power plant cycle: the working fluid, here steam and water, travels a closed loop in the typical power plant cycle.

steam velocity, this arrangement may be referred to as a 50% reaction turbine.) The moving blades form the walls of moving nozzles that are designed to permit further expansion of the steam and to partially reverse the direction of steam flow, which produces the reaction on the blades. The distinguishing characteristic of the reaction turbine is that a pressure drop occurs across both the moving and stationary nozzles, or blades (Figure 12.2). Normally reaction turbines employ a considerable number of rows of moving and stationary nozzles through which steam flows as its initial pressure is reduced to exhaust pressure. The pressure drop across each row of nozzles is, therefore, relatively small, and steam velocities are correspondingly moderate, permitting medium rotating speeds.

Reaction stages are usually preceded by an initial velocity-compounded impulse stage, as in Figure 12.2, in which a relatively large pressure drop takes place. This results in a shorter, less costly turbine.

In the radial flow reaction, or Ljungstrom, turbine, the steam does not flow axially through alternating rows of fixed and moving blades; rather, it flows radially through several rows of reaction blades. Alternate rows of blades move in opposite direction. They are fastened to two independent shafts that operate in opposite directions, each shaft driving a load.

After expansion in the turbine, the steam usually exhausts to a condenser, where it is condensed to provide a source of clean water for boiler feed. This simple cycle (Figure 12.3) forms the basis on which most steam power plants operate.

I. STEAM TURBINE OPERATION

Steam turbines are made in a number of different arrangements to suit the needs of various power plant or industrial installations. Turbines up to 40–60 MW capacity are generally single-cylinder machines. (The term *cylinders, chests, casings,* and *shells* are used interchangeably in this industry.) Larger units ranging in size up to 1250 MW are usually of compound type; that is, the steam is partially expanded in one cylinder then passed to one or more additional cylinders where expansion is completed. The simple cycle shown in Figure 12.3 is water to steam to power generation, and steam to water. This forms the basis on which most steam power plants operate.

A. Single-Cylinder Turbines

Single-cylinder turbines are of either the *condensing* or *backpressure* (noncondensing) type. These basic types and some of their subclassifications are shown in Figure 12.4.

When the steam from a turbine exhausts to a condenser, the condenser serves two purposes:

1. By maintaining a vacuum at the turbine exhaust, it increases the pressure range through which the steam expands. In this way, it materially increases the efficiency of power generation.
2. It causes the steam to condense, thus providing clean water for the boilers to reconvert into steam.

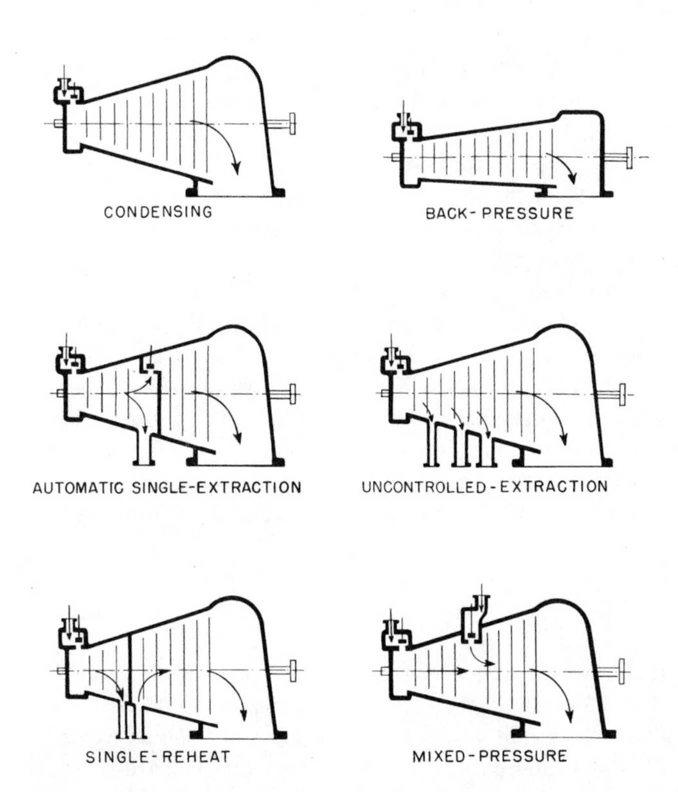

Figure 12.4 Typical single-cylinder turbine types: in comparison to backpressure turbines, condensing turbines must increase more in size toward the exhaust end to handle the larger volume of low pressure steam.

Industrial plants frequently require steam at low to moderate pressures for process use. One of the more economical ways of generating this steam is with a *combined-cycle* plant, where high pressure steam is used to power equipment such as generators and the exhaust steam from this equipment is used for heating or other services. The steam is generated at high pressure and, after expansion through the turbine to the pressure desired for process use, it is delivered to the process application. This permits power to be generated by the turbine without appreciably affecting the value of the steam for process use. It may be done with a backpressure turbine designed to exhaust all the steam against the pressure required for process use, or it may be done with an *automatic extraction* turbine in which part of the steam is withdrawn for process use at an intermediate stage (or stages) of the turbine and the remainder of the steam exhausted to a condenser. Such a turbine requires special governors and valves to maintain constant pressure of the exhausted steam and constant turbine speed under varying turbine load and extraction demands.

Steam can be also extracted without control from various stages of a turbine to heat boiler feedwater (*regenerative heating*). Such turbines are called *uncontrolled extraction* turbines, since the pressure at the extraction points varies with the load on the turbine.

To obtain higher efficiency, large turbines (called reheat turbines) are arranged so that after expanding partway, the steam is withdrawn, returned to the boiler, and reheated to approximately its initial temperature. It is then returned to the turbine for expansion through the final turbine stages to exhaust pressure.

High pressure noncondensing turbines have been added to many moderate pressure installations to increase capacity and improve efficiency. In such installations, high pressure boilers are installed to supply steam to the noncondensing turbines, which are designed to exhaust at the pressure of the original boilers and supply steam to the original turbines. The high pressure turbines are called *superposed* or *topping* units.

Where low pressure steam is available from process work, it can be used to generate power by admitting it to an intermediate stage of a turbine designed for the purpose and expanding it to condenser pressure. Such a machine is a *mixed-pressure* turbine, and is another form of *combined-cycle* operation.

Compound turbines have at least two cylinders or casings, a high pressure one and a low pressure one. To handle large volumes of low pressure steam, the low pressure cylinder is frequently of the double-flow type. Very large turbines may have an intermediate pressure cylinder, and two, three, or even four double-flow, low pressure cylinders. The cylinders may be in line using a single shaft, which is called *tandem compound,* or in parallel groups with two or more shafts, which is called *cross compound.* Reheat between the high and intermediate pressure stages may be employed in large turbines. Steam may be returned to the boiler twice for reheating. Some of these arrangements shown diagrammatically in Figure 12.5.

II. TURBINE CONTROL SYSTEMS

Although the trend is toward electronic speed sensing and control, all steam turbines are provided with at least two independent governors that operate to control the flow of steam. One of these operates to shut off the steam supply if the turbine speed should exceed a predetermined maximum. It is often referred to as an emergency trip. On overspeed, it closes the main steam valve, cutting off steam from the boiler to the turbine. The other, or main, governor may operate to maintain practically constant speed, or it may be designed for variable speed operation under the control of some outside influence. Extraction, mixed

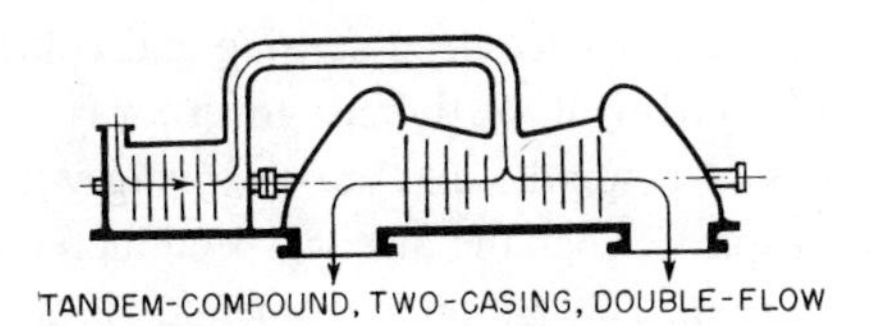

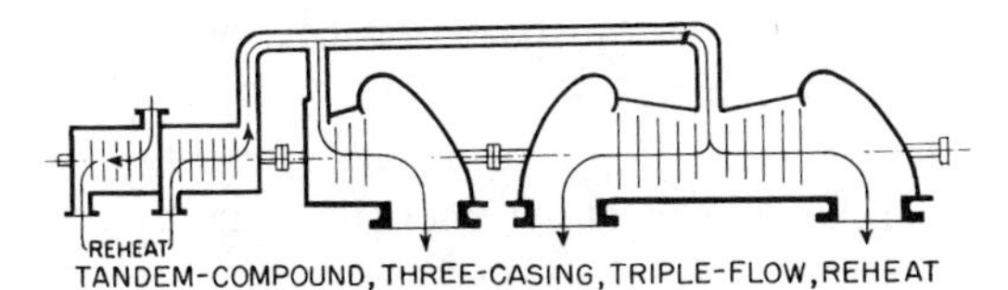

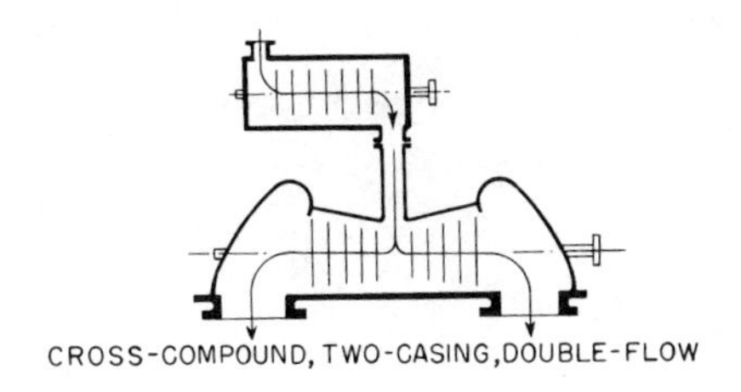

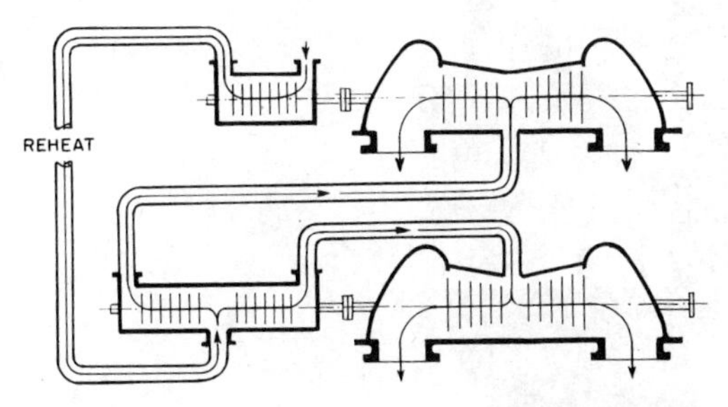

Figure 12.5 Some arrangements of compound turbines. While many arrangements are used, these diagrams illustrate some of the more common ones.

pressure, and backpressure turbines are provided with governors that control steam flow in response to a combination of speed and one or more pressures. The governors of such units are extremely complex, whereas the direct-acting speed governors on small, mechanical drive turbines are relatively simple.

A. Speed Governors

Speed governors may be divided into three principal parts:

1. Governor or *speed-sensitive element*
2. *Linkage,* or force-amplifying mechanism that transmits motion from the governor to the steam control valves
3. *Steam control valves*

Although most new turbine installations use electronic speed sensing and control, a commonly used speed-sensitive element is the centrifugal, or flyball, governor (Figure 12.6). Weights that are pivoted on opposite sides of a spindle and are revolving with it are moved outward by centrifugal force against a spring when the turbine speed increases, and inward by spring action as the turbine slows down. This action may operate the steam admission valve directly through a mechanical linkage, as shown, or it may operate the pilot valve of a hydraulic system, which admits and releases oil to opposite sides of a power piston, or on one side of a spring-loaded piston. Movement of the power piston opens or closes steam valves to control turbine speed.

Moderate-sized and large high speed turbines are provided with a double relay hydraulic system to further boost the force of the centrifugal governor and to increase the speed with which the system responds to speed changes.

A second type of speed-sensitive element is the oil impeller. Oil from a shaft-driven pump flows through a control valve to the space surrounding the governor impeller. The

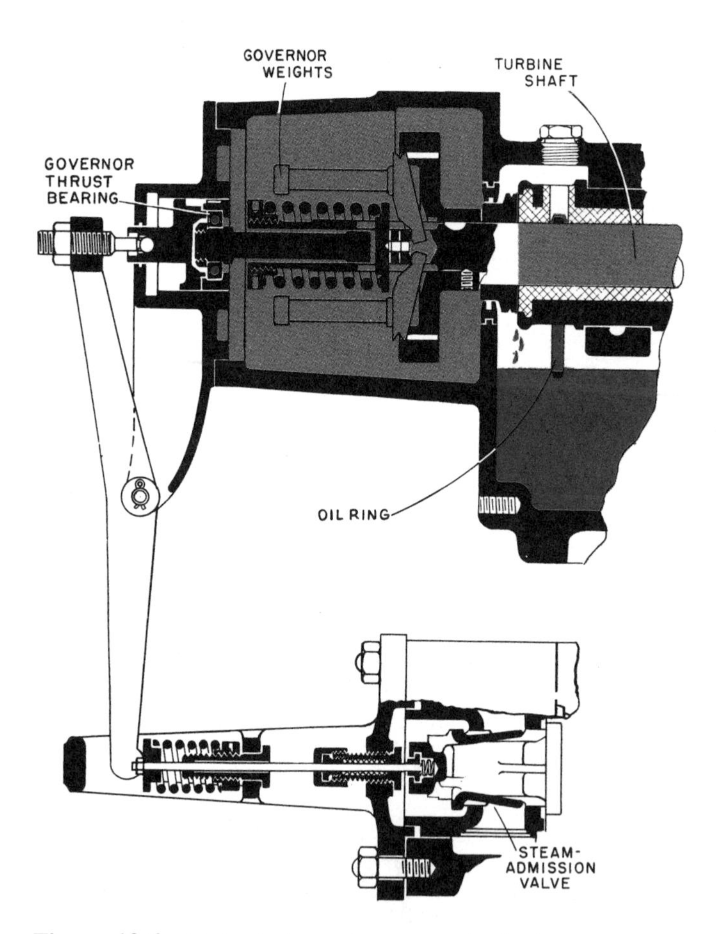

Figure 12.6 Mechanical speed governor. A simple arrangement such as this using a flyball governor is suitable for many small turbines.

impeller, mounted on the turbine shaft, consists of a hollow cylindrical body with a series of tubes extended radially inward. As the oil flows inward through the tubes it is opposed by centrifugal force and a pressure is built up that varies with the square of the turbine speed. This pressure is applied to spring-loaded bellows, which positions a pilot valve. The pilot valve, in turn, controls the flow to a hydraulic circuit that operates the steam control valves.

Newer turbines are equipped with electrical or electronic speed-sensitive devices. Signals from these devices, along with signals derived from load, initial steam pressure, and other variables, are fed to computer, which compares them and sends the appropriate signals to hydraulic servovalves to adjust the steam control valves.

As indicated, the linkage between the speed-sensitive element and the steam control valves may be anything from a simple lever to an extensive hydraulic system controlled by a computer.

In small turbines, the flow of steam is controlled by a simple valve, usually of the balanced type, to reduce the operating force required. In large units, a valve for each of several groups of nozzles controls steam flow. The opening or closing of a valve cuts in or out a group of nozzles. The number of open valves, and thus the number of nozzle groups in use, is varied according to the load. The valves may be operated by a barlift arrangement, by cams, or by individual hydraulic cylinders.

Additional control valves, called intercept valves, are required on reheat turbines. These are placed close to the intermediate, or reheat, cylinder and are closed by a governor system if the turbine starts to speed up as a result of a sudden large load reduction. This design is intended to prevent large volumes of high energy steam found in the piping of the high pressure turbine exhaust, the reheat boiler, and the intermediate pressure turbine from continuing to flow and possibly cause overspeeding and emergency tripping of the turbine. In older turbines, the intercept valves were controlled by a separate governor system, but the newer machines have the intercept valves operated by the main hydraulic control system. As an additional safety measure, intercept valves are preceded by stop valves, which are actuated by the main emergency overspeed governor (or other speed-sensing control system).

Automatic extraction and backpressure turbines are provided with governors arranged to maintain constant extraction or exhaust pressure irrespective of load (within the capacity of the turbine). The pressure-sensitive element consists of a pressure transducer, and its response to pressure changes is communicated through the control system to the valves that control steam extraction, and to the speed governor that controls admission of steam to the turbine. On automatic extraction turbines, the action of the pressure- and speed-responsive elements is coordinated so that turbine speed is maintained. This may not be the case for backpressure turbines.

III. LUBRICATED COMPONENTS

The lubrication requirements of steam turbines can be considered in terms of the parts that must be lubricated, the type of application system, the factors affecting lubrication, and the lubricant characteristics required to satisfy these requirements.

A. Lubricated Parts

The main lubricated parts of steam turbines are the bearings, both journal and thrust. Depending on the type of installation, a hydraulic control system, oil shaft seals, gears, flexible couplings, and turning gear may also require lubrication.

1. Journal Bearings

The rotor of a single-cylinder steam turbine, or of each casing of a compound turbine, is supported by two hydrodynamic journal bearings. These journal bearings are located at the ends of the rotor, outside the cylinder. In some designs, there may be one large journal bearing between the casings that supports both turbine rotors or a turbine and generator rotor (rigid coupling) instead of a separate bearing at the ends of each casing. Clearances between the shaft and shaft seals and between the blading and the cylinder, are extremely small, so to maintain the shaft in its original position and avoid damage to shaft seals or blading, the bearings must be accurately aligned and must run without any appreciable wear.

Primarily, the loads imposed on the bearings are due to the weight of the rotor assembly. The bearings are conservatively proportioned and thus pressures on them are moderate. Horizontally split shells lined with tin-based babbitt metal are usually used. The bearings are enclosed in housings and supported on spherical seats or flexible plates to reduce any angular misalignment.

The passages and grooves in turbine bearings are sized to permit the flow of considerably more oil than is required for lubrication alone. The additional oil flow is required to remove frictional heat and the heat conducted to the bearing along the shaft from the hot parts of the turbine. The flow of oil must be sufficient to cool the bearing enough to maintain it at a proper operating temperature. In most turbines, the temperature of the oil leaving the bearings is on the order of 140–160°F (60–71°C), but in special cases it may exceed 180°F (82°C).

When a turbine is used to drive a generator, the generator bearings are similar in design to the turbine bearings and are normally supplied from the same system.

Large turbines are now frequently provided with "oil lifts" (jacking oil) in the journal bearings to reduce the possibility of damage to the bearings during starting and stopping, and to reduce metal-to-metal contact during turning gear operation. Oil under high pressure from a positive displacement pump is delivered to recesses in the bottoms of the bearings. The high pressure oil lifts the shaft and floats it on a film of oil until the shaft speed is high enough to create a normal hydrodynamic film. For a shaft that is rotated for several hours or days to prevent rotor-sag, the oil lift is also required after turbine shutdown.

A phenomenon that occurs in relatively lightly loaded, high speed journal bearings, such as turbine bearings, is known as "oil whip" or "oil film whirl." The center of the journal of a hydrodynamic bearing ordinarily assumes a stable, eccentric position in the bearing that is determined by load, speed, and oil viscosity. Under light load and high speed, the stable position closely approaches the center of the bearing. There is a tendency, however, for the journal center to move in a more or less circular path about the stable position in a self-excited vibratory motion having a frequency of something less than half the shaft speed. In certain cases, such as some of the relatively lightweight, high pressure rotors of compound turbines that require large-diameter journals to transmit the torque, this whirling has been troublesome and has required the use of bearings designed especially to suppress oil whip.

Bearings designed to suppress oil whip are available in several types. Among the common types are the pressure or pressure pad bearings (Figure 12.7), the three-lobed bearing (Figure 12.8), and the tilting pad antiwhip bearing (Figure 12.9). The pressure pad bearing suppresses oil whip because oil carried into the wide groove increases in

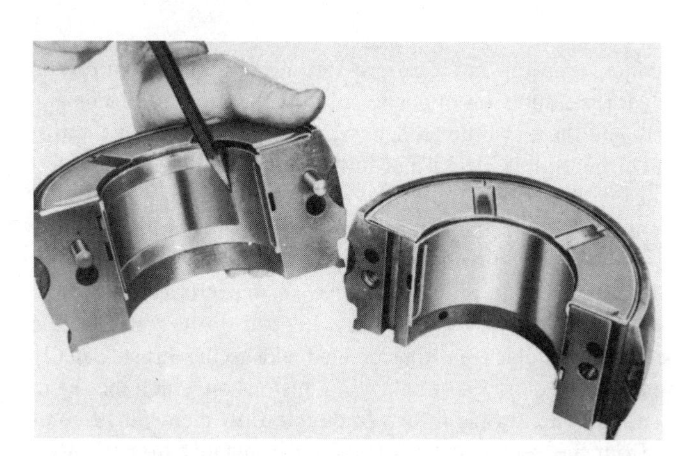

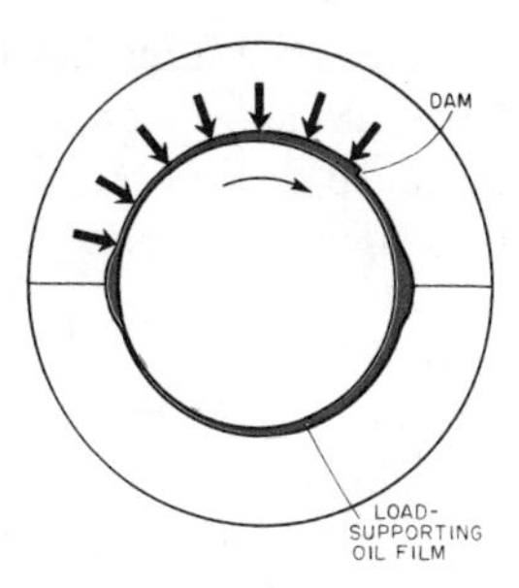

Figure 12.7 Pressure bearing. The wide grooves in the upper half end in a sharp dam at the point indicated; this causes a downward pressure (inset), which forces the journal into a more eccentric position that is more resistant to oil whip.

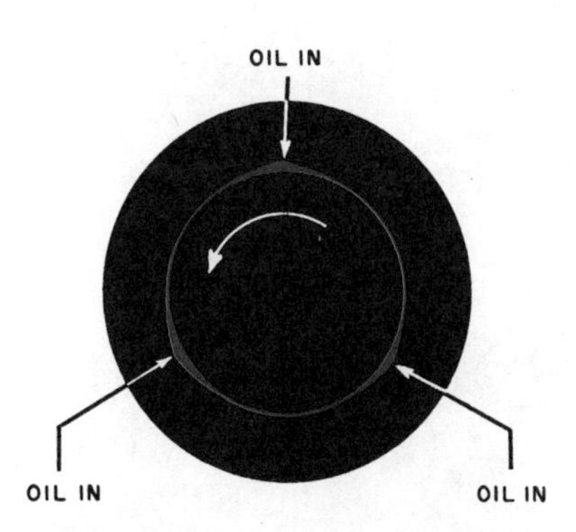

Figure 12.8 Three-lobed bearing. The shape of the bearing is formed by three arcs having a radius somewhat greater than the radius of the journal. This has the effect of creating a separate hydrodynamic film at each lobe, and the pressure in these films tend, to keep the journal in a stable position.

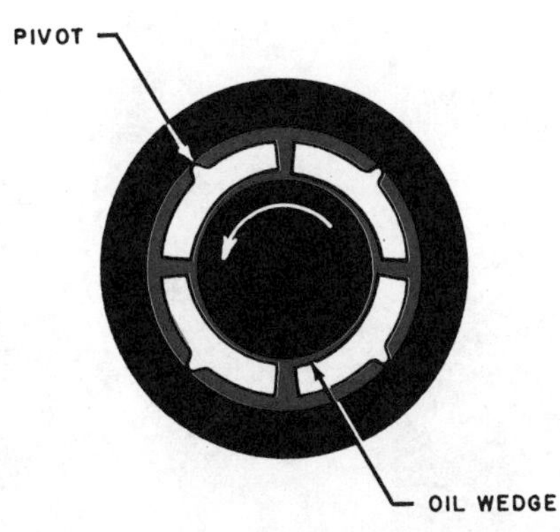

Figure 12.9 Tilting pad antiwhip bearing. As in the three-lobe bearing, the multiple oil films formed tend to keep the journal in a stable position.

pressure when it reaches the dam at the end. This increase in pressure forces the journal downward into a more eccentric position that is more resistant to oil whip. The other types illustrated depend on the multiple oil films formed to preload the journal and minimize the tendency to whip.

2. Thrust Bearings

Theoretically, in impulse turbines, the drop in steam pressure occurs almost entirely in the stationary nozzles. The steam pressures on opposite sides of the moving blades are, therefore, approximately equal, and there is little tendency for the steam to exert a thrust in the axial direction. In actual turbines, this ideal is not fully realized, and there is always a thrust tending to displace the rotor.

In reaction turbines, a considerable drop in steam pressure occurs across each row of moving blades. Since the pressure at the entering side of each of the many rows of moving blades is higher than the pressure at the leaving side, the steam exerts a considerable axial thrust toward the exhaust end. Also, when rotors are stepped up in diameter, the unbalanced steam pressure acting on annular areas thus created adds to the thrust. Usually the total thrust is balanced by means of dummy, or balancing, pistons on which the steam exerts a pressure in the opposite direction to the thrust. In double-flow elements of compound turbines, steam flows from the center to both ends, ensuring that thrust is well balanced.

Regardless of the type of turbine, thrust bearings are always provided on each shaft to take axial thrust and, thus, hold the rotor in correct axial position with respect to the stationary parts. Although thrust caused by the flow of steam is usually toward the low pressure end, means are always provided to prevent axial movement of the rotor in either direction.

The thrust bearings of small turbines may be babbitt-faced ends on the journal bearings, or rolling element bearings of a type designed to carry thrust loads. Medium-

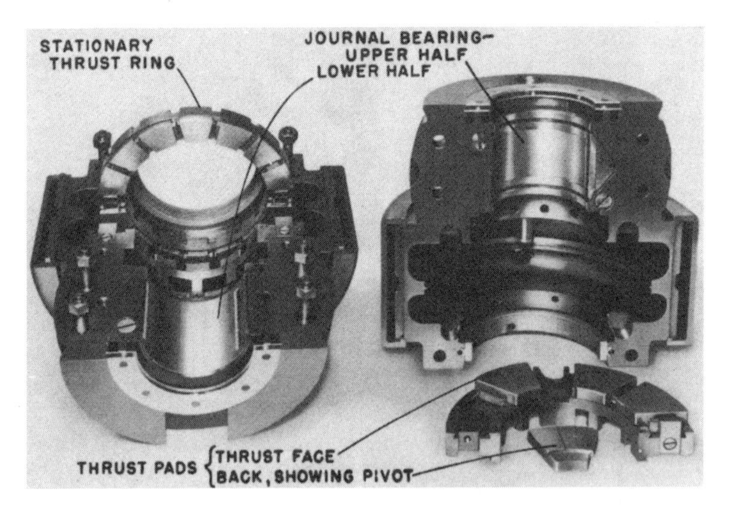

Figure 12.10 Combined journal and tilting pad thrust bearing. A rigid collar on the shaft is held centered between the stationary thrust ring and a second stationary thrust ring (not shown) by two rows of titling pads.

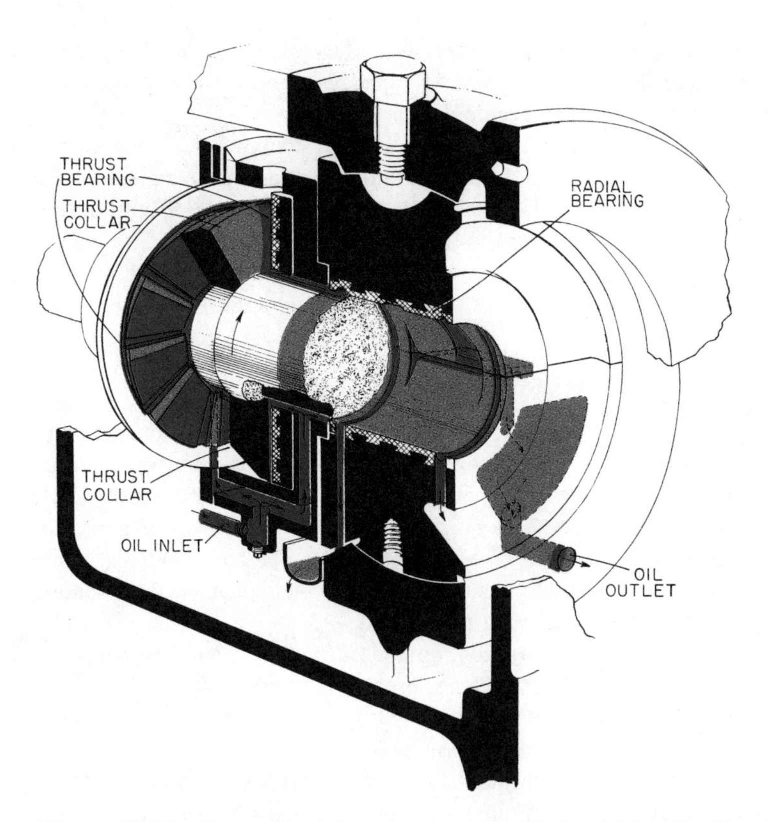

Figure 12.11 Tapered land thrust bearing and plain journal bearing. The thrust bearing consists of a collar on the shaft and two stationary bearing rings, one on each side of the collar. The babbitted thrust faces of the bearing rings are cut into sectors by radial grooves. About 80% of each sector is beveled to the leading radial groove, to permit the formation of wedge oil films. The unbeveled portions of the sectors absorb the thrust load when speed is too low to form hydrodynamic films.

sized and large turbines are always equipped with thrust bearings of the tilting pad (Figure 12.10), or tapered land (Figure 12.11) type.

3. Hydraulic Control Systems

As discussed earlier, medium-sized and large turbines have hydraulic control systems to transmit the motion of the speed or pressure-sensitive elements to the steam control valves. Two general approaches are used for these systems.

In mechanical hydraulic control systems, the operating pressure is comparatively low (<150 psi), and oil from the bearing lubrication system may be used safely as the hydraulic fluid. Separate pumps are provided to supply the hydraulic requirement. An emergency tripping device is provided to shut down the turbine if there is any failure in the hydraulic system.

Larger turbines now being installed are equipped with electrohydraulic control systems. To provide the rapid response needed for control of these units, the hydraulic systems operate at relatively high pressures, typically in the range of 1500–2000 psi.

The systems consist of an independent reservoir and two separate and independent pumping systems. The large fluid flow required for rapid response to sudden changes in load is usually provided by gas-charged accumulators.

The critical nature of the servovalves used in these systems requires that careful attention be paid to the filtration of the fluid, and strict limits on particulate contamination usually are observed. The need for precise control also calls for one use of both heaters and coolers to maintain the temperature of the fluid and, thus, its viscosity, in a narrow range. Since a leak or a break in a hydraulic line could result in a fire if the high pressure fluid sprayed onto hot steam piping or valves, fire-resistant hydraulic fluids are widely used in these systems.

4. Oil Shaft Seals for Hydrogen-Cooled Generators

Because it is a more effective coolant than air, hydrogen is commonly used to cool medium-sized and large generators. Shaft-mounted blowers circulate the gas through rotor and stator passages, then through liquid-cooled hydrogen coolers. Gas pressures up to 60 psi (413 kPa) are used. A further development, which has permitted increases of generator ratings over hydrogen cooling alone, is the direct liquid cooling of stator windings. Some liquid systems use transformer oil while others use water. Even with water-cooled stators, the interior of the generator is still filled with hydrogen.

The main connection between type of cooling and turbine lubrication is that when hydrogen is used for cooling, some of the oil is exposed to the hydrogen. Oil shaft seals (Figure 12.12) are used to prevent the escape of the hydrogen. Turbine oil for these seals may be supplied from an essentially separate system having its own reservoir, pumps, and so forth, or may be supplied directly from the main turbine lubricating system. In either case, before entering the reservoir, oil returning from the seals must be passed through a special tank to remove any traces of hydrogen. Otherwise hydrogen could accumulate in the reservoir and form an explosive mixture with air. In addition, the main turbine oil reservoir of all units driving hydrogen-cooled generators must be equipped with a vapor extractor to remove any traces of hydrogen that may be carried back by the sealing oil or the oil from the generator bearings.

5. Gear Drives

Efficient turbine speed is often higher than the operating speed of the machine being driven. This may be the case, for example, when a turbine drives a direct current generator, paper machine drives, centrifugal pumps, or other industrial machines. It is also the case when a turbine is used for ship propulsion. In these applications, reduction gears are used to connect the turbine to the driven unit.

Reduction gears used with moderate-sized and large turbines are usually of the precision-cut, double-helical type. Double reduction gear sets are required with marine propulsion turbines, and epicyclic reduction gears are sometimes used instead of conventional gear sets. Usually, the gear sets are enclosed in a separate oil-tight casing and are connected to the turbine and the driven machine through flexible couplings. Small machines may have the gear housing integral with the turbine housing and the pinion on the turbine shaft.

Reduction gears may have a circulation system that is entirely separate from the turbine system, or circulation may be supplied from the turbine system. In the latter case, a separate pump (or pumps), is provided for the gears. Some older small-geared turbines have ring-oiled turbine bearings and splash-lubricated gears.

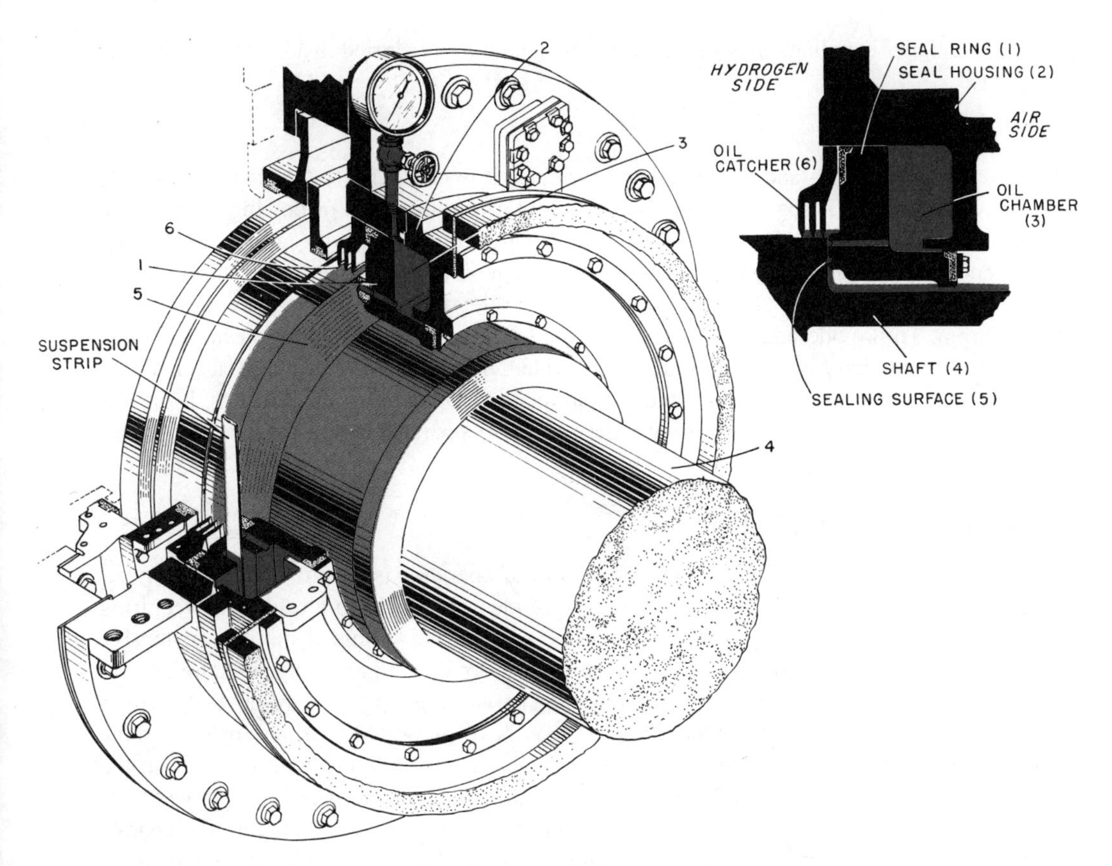

Figure 12.12 Shaft seal for hydrogen-cooled generator. Oil under pressure in the annular chamber formed between the seal ring and housing (see inset) forces the seal ring against the shaft shoulder. From the annular chamber, the oil flows through the passage shown to the sealing face formed by a shoulder on the generator shaft.

In marine propulsion applications, the bearing through which the propeller thrust is transmitted to the ship's hull is frequently an integral part of the gear set. It may be placed just behind the low speed bull gear, or at the forward end of the bull gear shaft. It is usually of the tilting pad type. With a fixed-pitch propeller, the thrust bearing must be designed to operate with either direction of rotation, since turbine rotation (and propeller rotation) is reversed for astern movement of the ship.

6. Flexible Couplings

Most modern turbogenerator sets are rigidly coupled. Many other turbines, especially geared units, are connected to their loads by flexible couplings. Gear-type couplings are used most commonly.

If gear-type couplings are enclosed in a suitable housing, they may be lubricated from the turbine circulation system. A coupling of this type acts as a centrifugal separator, and means are provided for a flow-through of oil so that any separated solids will be flushed out.

Smaller couplings may be lubricated by a bath of oil carried inside the case, or with grease.

7. Turning Gear

When one is starting and stopping large turbines, it is necessary to turn the rotor slowly to avoid uneven heating or cooling, which could cause distortion or bowing of the shaft. This is done with a barring mechanism or turning gear. The turning gear usually consists of either an electric or hydraulic motor that is temporarily coupled to the turbine shaft through reduction gears. Rotor speed, while the turning gear is operating, is usually below 100 rpm. To provide adequate oil flow to the bearings during this low speed operation, a separate auxiliary oil pump usually is provided. The oil coolers are used at maximum capacity to increase oil viscosity and to help maintain oil films in the bearings. If oil lifts are provided in the turbine bearings, they are also operated while the turning gear is operating.

B. Lubricant Application

One of the essential factors in reliable steam turbine operation is the provision of a lubricating system that will assure an ample supply of clean lubricant to all the parts requiring it. The size of the turbine generally determines whether the system is simple or extensive and complex. Small turbines, such as those used to drive auxiliary equipment, are usually equipped with ring-oiled bearings, other moving parts being lubricated by hand. Moderate-sized units driving through reduction gears may have ring-oiled bearings for the turbine and a circulation system (see Chapter 9) to supply oil to the gears and bearings in the reduction gear set. Most moderate-sized units and all large turbines are equipped with circulation systems that supply oil to all parts of the unit requiring lubrication. Separate circulation systems may be provided for the seal oil for hydrogen-cooled turbogenerators and for the hydraulic control systems.

C. Factors Affecting Lubrication

Steam turbines in themselves do not represent particularly severe service for petroleum-based lubricating oils. Because of the costs and time involved in shutting down most turbines to change the oil and clean the system, however, and because of the relatively large volume of oil contained in large turbine systems, turbine operators expect extremely long service life from the turbine oil. To achieve long service life, oils must be carefully formulated for the specific conditions encountered in steam turbine lubrication systems.

1. Circulation and Heating in the Presence of Air

The temperature of the oil in steam turbine systems is raised both by the frictional heat generated in the lubricated parts and by heat conducted along the shaft from the rotor. As the oil flows through the system, it is broken into droplets or mist, which permits greater exposure to air. System designs are usually conservative, to ensure that maximum oil temperatures are not excessive, but in long-term service some oxidation of the oil does occur. This oxidation may be further catalyzed by finely divided metal particles resulting from wear or contamination and water.

Slight oxidation of the oil is harmless. The small amounts of oxidation products formed initially can be carried in solution in the oil without noticeable effect. As oxidation continues, some of the soluble products may become insoluble, or insoluble materials may be formed. As an oil oxidizes, it should also show an increase in viscosity; but in most turbine systems the rate of oxidation is very slow, and viscosity increases are rarely noted.

Insoluble oxidation products may be carried with the oil as it circulates. Some may then settle out as gum, varnish, or sludge on governor parts, in bearing passages, on coolers, on strainers, and in oil reservoirs. Their accumulation may interfere with the supply of oil to bearings and with governors or control of the unit. Under moderate to severe oil oxidation, there is also the potential for the oil degradation materials to plate out on bearing surfaces, reducing clearances, and increasing bearing temperatures.

Some oxidation products that are soluble in warm oil become insoluble when the oil is cooled. This can result in insulating deposits forming on cooling coils or other cool surfaces. The resultant reduction in effectiveness of cooling may cause higher oil temperatures and contribute to more rapid oxidation.

An increase in viscosity within a range that has proved satisfactory in service is not necessarily harmful. However, excessive viscosity increase may reduce oil flow to bearings, increase pumping losses, and increase fluid friction and heating in bearings. All these effects tend to increase the operating temperature of the oil and contribute to more rapid oxidation of it.

2. Contamination

Water is the contaminant that is most prevalent in steam turbine lubrication systems. Common sources of water contamination are as follows:

1. Steam from leaking shaft seals or the shaft seals of turbine-driven pumps
2. Condensation from humid air in the oil reservoir and bearing pedestals
3. Water leaks in oil coolers
4. Steam leaks in oil heating elements (where used)

The lubrication systems of turbines that operate intermittently are more likely to become contaminated with water than are the systems of turbines that operate continuously.

When a turbine oil is agitated with water, some emulsion will form. If the oil is new and clean, the emulsion will separate readily. The water will then settle in the reservoir, where it can be drawn off or removed by the purification equipment. Oxidation of the oil, or contamination of it with certain types of solid material such as rust and fly ash, may increase the tendencies of the oil to emulsify and to stabilize emulsions after they are formed. Persistent emulsions can join with insoluble oxidation products, dirt, and so forth to form sludges. The character of these sludges may vary; but, if they accumulate in oil pipes, passages, and oil coolers, they can interfere with oil circulation and cause high oil and bearing temperatures.

Water in lubricating oil, combined with air that is always present, can cause the formation of common red rust and also black rust, similar in appearance to pipe scale. Rusting may occur both on parts covered by oil and on parts above the oil level. In either case, in addition to damage to the metal surfaces, rusting is harmful for a number of reasons. Particles of rust in the oil tend to stabilize emulsions and foam and to act as catalysts that increase the rate of oil oxidation. Rust is abrasive and when carried by the oil to the bearings may scratch the journals and cause excessive wear. If carried into the small clearances of governor or control mechanisms, it can cause sluggish operation or, in extreme cases, sticking, overspeed, or tripping the unit off-line.

Solid materials of many types can contaminate turbine oil systems. Pulverized coal, fly ash, dirt, rust, pipe scale, and metal particles are typical examples. These solid materials may enter the system in the following ways:

1. During erection (possibly remaining even after the initial cleaning and flushing)
2. Through openings in the bearing pedestals or reservoir covers
3. Carried in with water
4. Generated within the system
5. While adding makeup oil
6. During performance of system maintenance or repairs

Solid contaminants contribute to deposit or sludge formation, and some tend to reduce the ability of the oil to separate from water. Some are abrasive and may be the direct cause of scoring or excessive wear. Fine metal particles may act as catalysts to promote oil oxidation.

Air is also a contaminant. Air entrained in the oil will cause sponginess in hydraulic controls and may reduce the load-carrying ability of oil films. Entrained air also increases the exposure of the oil to air, therefore increasing the rate of oxidation. Excessive amounts of air in the system may lead to foaming in the reservoir or bearing housings. This could result in overflow, with oil loss and unsightly and hazardous spills. In some cases, foam escaping from bearings adjacent to a generator may be drawn into the windings or onto the collector rings, where it may cause breakdown of the insulation, short circuits, or arcing.

D. Lubricating Oil Characteristics

The first requirement of a steam turbine oil is that it have the proper viscosity at operating temperature to provide effective lubricating films, and adequate load-carrying ability to protect against wear in heavily loaded mechanisms and under boundary lubrication conditions. Other characteristics are concerned with providing long service life, protecting system metals, and maintaining the oil in a condition to perform its lubricating function. In formulating turbine oils, it is very important to select specific base stocks and additives and balance the formulation to provide all the performance characteristics required to assure good turbine lubrication as well as long life.

1. Viscosity

In direct-connected steam turbines, the main lubrication requirement is for the journal and thrust bearings. Higher viscosity oils provide a greater margin of safety in these bearings, but at the same time, increase pumping losses and friction losses due to shearing of the lubricant films. In high speed machines, particularly, the latter can become an important cause of power loss and heating of the lubricant. General practice is to use oils low in viscosity as possible within a range that has proved suitable service. Most larger units are designed to operate on oils of ISO viscosity grade 32 (28.8–35.2 cSt at 40°C). Some applications require somewhat higher viscosity oils, ISO viscosity grade 46 (41.4–50.6 cSt at 40°C) or ISO viscosity grade 68 (61.2–74.8 cSt at 40°C). These grades of oil also provide excellent performance in hydraulic control systems.

For geared turbines with a common circulation system supplying both the bearings and gears, higher viscosity oils or oils formulated with antiwear additives may be required to provide satisfactory lubrication of the gears. Oils of ISO viscosity grade 68 (61.2–74.8 cSt at 40°C) are used in many of these systems. Oils of viscosity grade 100 (90–110 cSt at 40°C) are also used in some machines, particularly marine propulsion turbines. With some geared turbines, the oil is passed through a cooler immediately before entering the gears. The resultant cooler oil has a higher viscosity, thus it provides better protection of the gears.

Reservoir temperatures of small ring-oiled turbine bearings vary widely from one design to another, depending principally on size and whether water cooling is used. The oils used for these range in viscosity from ISO viscosity grade 32 for the cooler units, up to as high as ISO viscosity grade 320 (288–352 cSt at 40°C) for the latter units.

2. Load-Carrying Ability

As noted earlier, steam turbine lubrication systems are conservatively designed. Bearing loads are moderate. Under these conditions, mineral oil lubricants of the correct viscosity normally provide adequate load-carrying ability. However, in turbines not equipped with oil lifts, boundary lubrication conditions occur in the bearings during starting and stopping. Under boundary lubrication conditions, some wear will occur unless lubricants with enhanced film strength are used.

To some extent, the increased load-carrying ability of the oil films needed during start-up is provided by the higher viscosity of the cool oil. However, the additive systems in turbine oils are frequently selected to provide some improvement in film strength to ensure an additional margin of safety.

In some cases, particularly in marine propulsion installations, the need for size and weight reduction has led to the use of heavily loaded gear reducers. Performance of these gears on conventional turbine oils is not satisfactory, and lubricants with extreme pressure properties are used. These lubricants are formulated to have the properties of good turbine oils with the addition of the extreme pressure properties; thus they may be used throughout the turbine lubrication system.

E. Oxidation Stability

The most important characteristic of turbine oils from the standpoint of long service life is their ability to resist oxidation under the conditions encountered in the turbine lubrication system. Resistance to oxidation is important from the standpoint of retention of viscosity, resistance to the formation of sludges, deposits, and corrosive oil oxyacids, and retention of water-separating ability, foam resistance, and ability to release entrained air.

Turbine oils usually are manufactured from base oils refined from selected crudes and/or special refining processes such as hydroprocessing. These oils are selected both for their natural oxidation stability and their response to oxidation inhibitors. Processing is carefully controlled and is often the most extensive applied to any base oils. Inhibitors are selected to be effective under the temperature conditions encountered in steam turbines and to provide the optimum improvement in the particular base oils used.

Oxidation life of steam turbine oils is frequently specified in terms of the turbine oil stability test (TOST: ASTM D 943; see Chapter 3). Manufacturer's specifications typically call for a test life to reach a neutralization number of 2.0 of 1000 h minimum. Commercial turbine oils of ISO VG 32 typically run from 4000 h to over 10,000 h. Higher viscosity products may run somewhat less. In Europe, the CIGRE test has been used with the limits for total oxidation products (TOP) of 1.0 mass percent and sludge of 40% of the TOP are typical. Commercial inhibited oils available today will generally give many years of service in well-maintained systems.

In addition to the ASTM D 943 for the evaluation of oxidation stability of turbine oils, the rotary bomb oxidation test (RBOT, ASTM D 2722) may also be used; The RBOT, as discussed in Chapter 3, has historically been used to evaluate gas turbine oils but is seeing some use for steam turbine oils. This is because the test is much shorter (less costly) than the TOST and because of the increased use of combined-cycle oils for both gas and steam turbines.

1. Protection Against Rusting

Straight mineral oils have some ability to protect against rusting of ferrous surfaces, but this ability is inadequate for the conditions encountered in steam turbine lubrication systems. Effective rust inhibitors are required in steam turbine oils.

In action, rust inhibitors ''plate out'' on metal surfaces, forming a film that resists displacement and penetration of water. This results in gradual depletion of the rust inhibitor. Normal makeup of new oil usually maintains an adequate level of rust inhibitor in the system. In certain cases, where excessive water contamination has been experienced, special treatment to replenish the rust inhibitor may be necessary.

Rust inhibitors must be carefully selected to provide adequate protection without affecting other properties of the oil, especially water separating ability.

2. Water-Separating Ability

New, clean, highly refined mineral oils will generally resist emulsification when water is mixed with them, and any emulsion that is formed will break quickly. Certain additives, such as some rust inhibitors, some contaminants, and oxidation products, can both increase the tendency of an oil to emulsify and make any emulsion that is formed more stable. As a result, careful selection of additives is necessary if a turbine oil is to have good initial water-separating ability, and excellent oxidation stability is necessary if this water-separating ability is to be maintained in service.

3. Foam Resistance

Turbine circulation systems are, in general, designed and constructed to minimize or eliminate conditions that have been found to cause foaming. However, turbine oils may contain defoamants to reduce the tendency to foam, and the stability of any foam that does form. Since oxidation can increase the foaming tendency and also the stability of the foam, good oxidation stability is an important factor in maintaining foam resistance in service.

F. Entrained Air Release

Rapid release of entrained air is particularly important in systems supplying turbines with hydraulic governors. Excessive amounts of entrained air in these systems can cause sponginess, producing delayed or erratic response.

The rate at which entrained air is released from a mineral oil is an intrinsic characteristic of the base stock itself. It is dependent on factors such as the source of crude, and type and degree of refining. Currently there are no known additives that will significantly improve the ability of an oil to release entrained air, but there are many additives that will degrade this ability. Thus, formulating a steam turbine oil with good air release properties is a process of selecting base oils with good air release properties, and then selecting additives that will perform the functions desired of them without seriously degrading the air release properties of the base oil.

G. Fire Resistance

Fire resistance fluids are usually used in electrohydraulic governor control systems of large turbines. The fluids used most commonly are based on either phosphate esters or blends of phosphate esters. As noted also, these systems are extremely critical of solid contaminants in the fluid, so considerable attention must be paid to filtration of the fluid.

The wide use of steam at temperatures as high as 1200°F (649°C) has increased the possibility of fires from oil leaks in turbine lubrication systems. This has led to some attempts to use fire-resistant fluids in turbine lubricating systems. Fluids used are the same as, or similar to, those used in hydraulic governor systems. However, these fluids cannot be used in older machines because they have a deteriorating effect on conventional generator insulating materials.

BIBLIOGRAPHY

Mobil Technical Book

Steam Turbines and Their Lubrication

13
Hydraulic Turbines

The primary application of hydraulic turbines is to drive electric generators in central power stations. Since the turbine shaft is usually rigidly coupled to the generator shaft with one set of bearings supporting both, for lubrication purposes the two units can be considered to be one machine. There is a wide range of sizes and operational characteristics for hydraulic turbines depending on the volume of water available and the pressure head of that water. Hydraulic turbines in commercial applications range in capacity from less than 1 MW to more than 750 MW. These units operate from as low as 40 rpm to as high as 2200 rpm, but the typical operating range is 100–200 rpm.

Many of the locations suitable for large hydroelectric installations are in remote often mountainous areas, with difficult access. Under these conditions, capital costs for plant construction and transmission lines are high. Therefore, for the power generated to be competitive in cost, plants must be designed for minimum maintenance and long service life. Generally, this has been accomplished, and the majority of hydroelectric units installed during the last 90 years are still in service.

Although many of the large hydroelectric installations are in remote locations, many are comparatively accessible. In recent years, the development of some of these more accessible locations has been aided by the introduction of the pumped storage concept and the use of bulb turbines.

Thermal and nuclear power plants, as well as open-flume hydroelectric turbines, operate most efficiently at relatively constant loads. Thus, during off-peak periods, if the plant can be operated at or near full load, the power in excess of the load requirement is comparatively low in cost. If a suitable location is available, this power can be used during low demand periods to pump water up to a storage reservoir. During periods of peak demand, the water can be used then to drive hydroelectric generators to supplement the power available from the base load stations. In effect, pumped storage is a method of storing low cost, off-peak power for use during peak demand periods.

Some pumped storage plants are equipped with both turbines and pumps. The generator is built as a combination motor/generator. During pumping operations, the turbine may be disconnected from the main shaft, or the turbine casing may be blown dry with compressed air so that the turbine operates without load. Similar arrangements are used with the pump. Quite a few new installations are equipped with reversible pump/turbines, and reversible motor/generators. There may be some sacrifice in efficiency with the combination pump/turbine. For example:

1. As a pump it may be less efficient than a unit designed as a pump alone.
2. As a turbine it may be less efficient than a unit designed as a turbine alone.

However, the lower cost of the combination machine generally offsets this loss of efficiency.

Bulb turbines (also called tubular turbines) are low head machines for what are referred to as ''flow-of-stream'' river applications. They can be used for relatively small, low cost installations that can be readily blended into the surrounding countryside. These machines have permitted the development of hydroelectric power in locations where installation of the older types of turbine would not be practical or economic.

I. TURBINE TYPES

Several types of hydraulic turbine are used. They can be considered to be impulse (Pelton) or reaction (pressure) types. Reaction turbines include the inward flow (Francis), diagonal flow (Deriaz), and propeller types. Bulb turbines are reaction turbines using a propeller-type runner. The choice of which type of unit to use in a particular application is a function of the pressure head and the quantity of flow available.

A. Impulse Turbines

In an impulse turbine, usually called a Pelton turbine, jets of water are directed by nozzles against shaped buckets on the rim of a wheel. The impulse force of the jets pushes the buckets on the rim of a wheel and causes the wheel to revolve. The buckets move in the same direction as the jets of water. For a turbine of this type to operate efficiently, the velocity of the jets must be high; thus, a high pressure head of water is required. Pelton turbines are usually designed for pressure heads in the range of about 500–3900 ft (150–1200 m). Single units with outputs up to 200 MW have been built.

Pelton turbines are built with the shaft either horizontal or vertical. Horizontal shaft machines are built with either one or two nozzles per runner. Usually, they are used for small to moderate-sized installations. A single runner may be connected directly to a generator, or two runners may be used, both on the same side of the generator or one on each side.

Vertical shaft Pelton turbines with four to six nozzles are now being used for larger installations. A cutaway of an installation with four nozzles is shown in Figure 13.1. In this machine, the nozzle tips are actuated by hydraulic pistons located inside the nozzle bodies. The deflectors are actuated by a ring and lever arrangement, which in turn is actuated by hydraulic servopistons. In many older machines, servopistons that operate the nozzle tips are located outside the nozzles, and a mechanical linkage is used to operated the nozzle tips.

Pelton turbines are used in a number of pumped storage applications. Both horizontal and vertical shaft machines are used for this purpose. If a horizontal shaft machine is used, the turbine is often mounted at one side of the generator/motor with the pump at the other side. Other arrangements can be used. In vertical shaft machines, the turbine

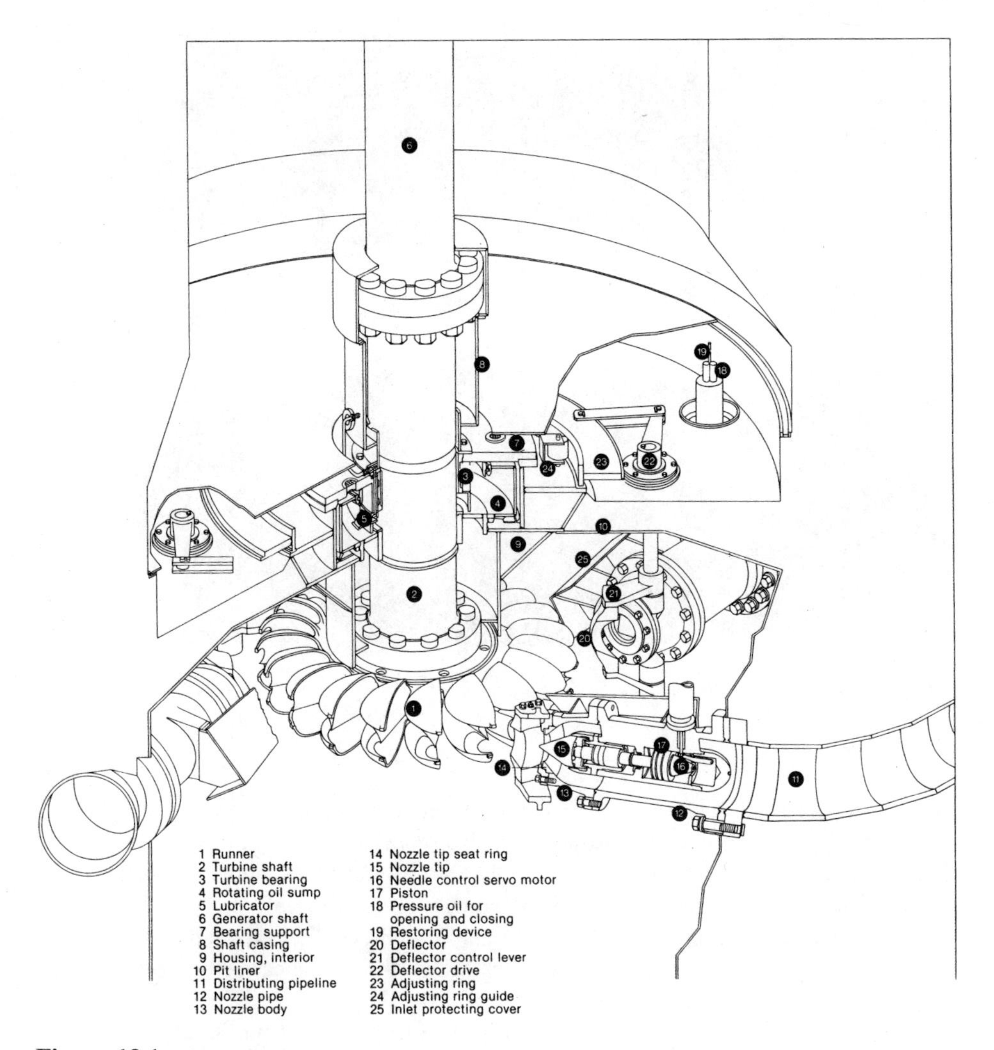

Figure 13.1 Cross-sectional view and components of a Pelton turbine.

normally is mounted above the pump so that the pump will operate with a positive pressure at the suction. The pump usually is coupled to the shaft through a clutch, which allows it to be disconnected and stopped during turbine operation. In some installations, the water is blown out of the pump, and it is left coupled to the turbine during turbine operation. The water is blown out of the turbine for pump operation. In some cases, the turbine is used to start the pump and motor and bring them up to speed. The water flow to the turbine is then shut off and the water is blown out of the casing.

B. Reaction Turbines

In reaction turbines, the flow of water impinges on a set of curved blades which, in effect, are the nozzles. The reaction of the water on the nozzles (blades) causes them to move in the opposite direction.

Figure 13.2 Francis turbine.

In a Francis turbine (Figure 13.2), the water flows radially inward from a volute casing and is turned through 90 degrees in the blades before flowing to the tail race outlet. In a Deriaz turbine (Figure 13.3 and 13.7) the direction of water flow is partially turned in the volute casing so that the flow is diagonally through the blades. The shaped boss then turns the water through the remaining angle to direct it into the tail race outlet. In propeller turbines, the direction of flow is controlled by a volute casing or a flume so that the water flows axially through the turbine.

Francis and Deriaz turbines are intermediate to high head machines. The various propeller turbines (fixed blade, Kaplan, and bulb) are low head machines.

1. Francis Turbines

Francis turbines are probably the most widely applied hydraulic turbine. They are now also widely used as reversible pump/turbines. Originally used for intermediate head installations, the range of use of the Francis turbine has been extended well up into the head range that was formerly the exclusive province of impulse turbines. Francis turbines are being used at heads ranging from about 65 to 1650 ft (20–500 m), and even higher head units are under development. Outputs in the 200–400 MW per unit range are common, and units rated at up to 750 MW are in service.

Small Francis turbines are sometimes built with horizontal shafts, but larger Francis turbines are nearly always vertical shaft machines. Water is brought in from the penstock through a spiral (volute) casing (Figure 13.4) from which it is directed into the turbine guide vanes by fixed vanes in the stay ring or speed ring (Figure 13.5). The movable guide vanes control the flow of water into the turbine to maintain the turbine speed constant as the load varies. They can be operated by a single regulating ring actuated by one or

Figure 13.3 Deriaz Turbine; also see cross-sectional view (Figure 13.7).

Figure 13.4 Spiral casing for Francis turbine.

Figure 13.5 Assembly of Francis turbine (servomotor at top right).

two hydraulic servomotors (Figure 13.5), or by individual servomotors (Figure 13.6). With a regulating ring, the individual vanes are connected to the ring through shear pins so that if one vane is jammed by a foreign object during closing, the remainder of the vanes can still be closed to shut off the turbine.

2. Diagonal Flow Turbines

The efficiency of Francis turbines drops off quite rapidly if the flow is less than the design value, or if the pressure head varies significantly. Where either of these conditions exists, a diagonal flow or Deriaz turbine can sometimes be used.

In the Deriaz turbine, both the guide vanes and the runner blades are adjustable (Figure 13.7). The runner blades are adjusted by a hydraulic servomotor located either in the turbine shaft or in the runner boss. The guide vanes are usually controlled by a regulating ring, similar to those used with Francis turbines. Movement of the guide vanes and runner blades are synchronized by the control system to maintain runner speed with changes in load, and keep the efficiency high as changes in the pressure head occur.

Deriaz turbines are presently built for pressure heads in the range of about 60–425 ft (18–130 m). Designs suitable for heads up to about 650 ft (200 m) are available. The

Figure 13.6 Guide vane adjustment with individual servomotors. This operating mechanism for a Kaplan turbine is equally applicable to Francis turbines.

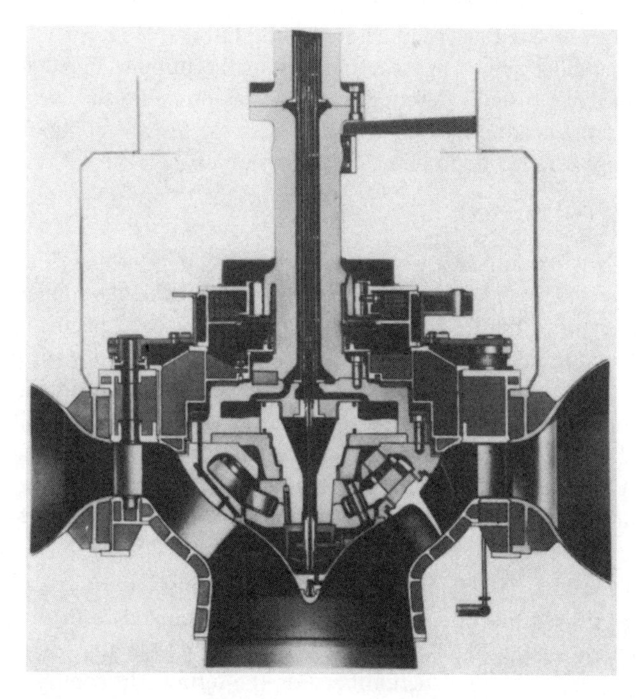

Figure 13.7 Cross section of Deriaz turbine.

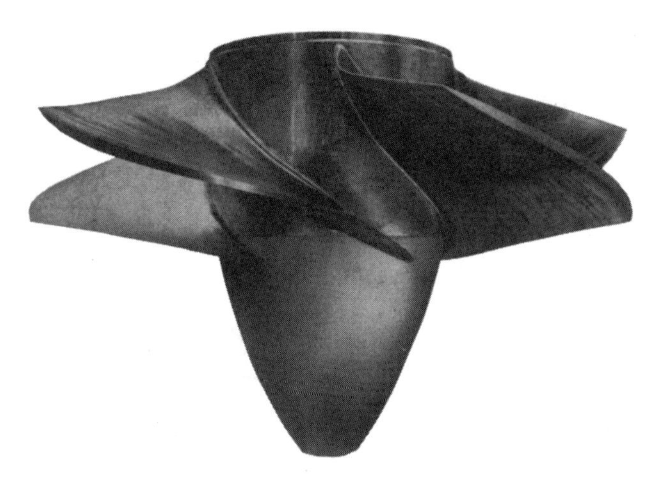

Figure 13.8 Fixed blade propeller turbine.

usual construction is with a vertical shaft. Single units with ratings up to about 150 MW are in service. The larger units are reversible pump/turbines.

3. Fixed Blade Propeller Turbines

Fixed blade propeller turbines are vertical shaft machines. The blades of the runner (Figure 13.8) are cast integrally, with the hub or welded to it. Because of the low heads at which propeller turbines operate, comparatively small changes in the head water or tail water level can make significant changes in the total head acting on the turbine. With a fixed blade turbine this may cause marked changes in the efficiency of the turbine. For that reason, fixed blade propeller turbines (usually referred to simply as propeller turbines) are used only in locations where the head is fairly constant. Relatively few installations of this type of turbine exist. Outputs range up to about 40 MW per unit.

4. Kaplan Turbines

The Kaplan turbine (Figure 13.9) is the largest class of low head hydraulic turbine. The runner blades are adjustable, operated by either a hydraulic servo-motor inside the runner hub or by a servomotor in the shaft with a mechanical linkage to the blade adjustment mechanism in the hub. The former arrangement is used for most new machines. As with Deriaz turbines, the controls for the guide vane adjustment and runner blade adjustment are synchronized to provide an optimum setting for each load and head condition. A typical Kaplan turbine installation is shown in cross section in Figure 13.10.

Kaplan turbines are built for heads from about 13–250 ft (4–75 m). Single units with power outputs from as little as 1 MW to units with outputs in the 175 MW range are in service.

(a) Bulb Turbines. The bulb turbine is actually a special application of the Kaplan turbine. As shown in Figure 13.11, the conventional Kaplan turbine is mounted vertically with a spiral casing carrying the water into the stay ring. From the turbine, a draft tube carries the water out to the tail race, creating a suction head on the turbine. In contrast, the bulb turbine has the turbine mounted in a bulb-shaped section of the flume with the

Figure 13.9 Kaplan turbine.

water flowing essentially straight through the turbine. The arrangement is much more compact and is less costly to construct.

The shaft of bulb turbines is either horizontal or angled slightly downward toward the turbine. With a horizontal shaft, the generator is usually mounted in a bulb-shaped casing inside the flume and is driven directly from the turbine (Figure 13.12). A mechanical drive to a generator mounted on the surface may also be used. With an angled shaft, the shaft may be extended out of the flume to drive a surface mounted generator directly, or the generator may be mounted inside the flume.

Bulb turbines for heads up to 75 ft (23 m) are in service. Unit outputs are usually less than about 10 MW, but units up to 50 MW are in service.

(b) S-Turbines. S-turbines (tubular turbines) are similar to bulb turbines except that the generator is not located in the bulb. The runner blade controls (where regulated) are contained in the bulb but a drive shaft extends out of the Kaplan runner and through a portion of the S-shaped draft casing into a generator room, where it is connected to a generator. The S-turbine operates with heads of approximately 49 ft (15 m) and power outputs up to 15 MW.

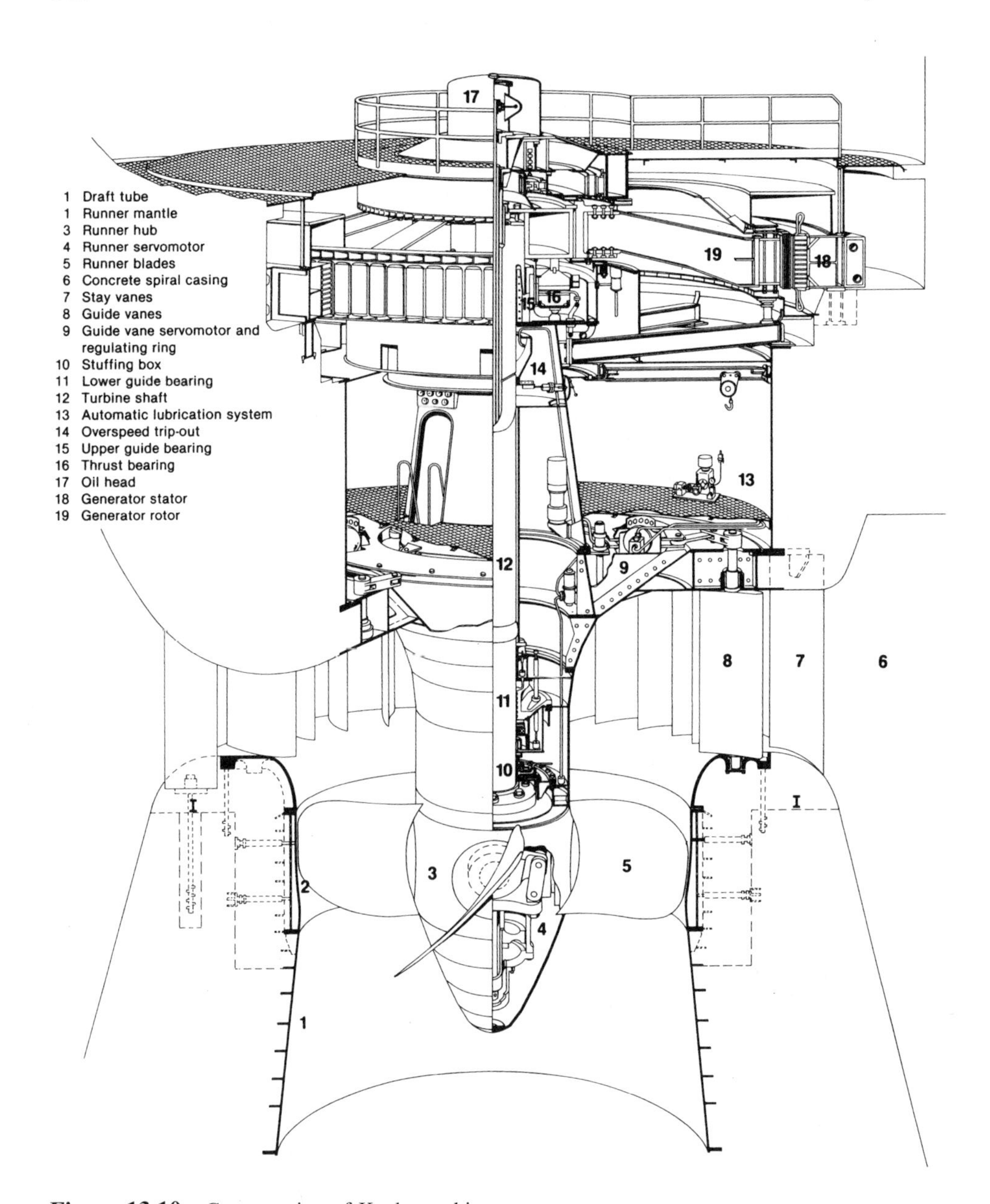

Figure 13.10 Cross section of Kaplan turbine.

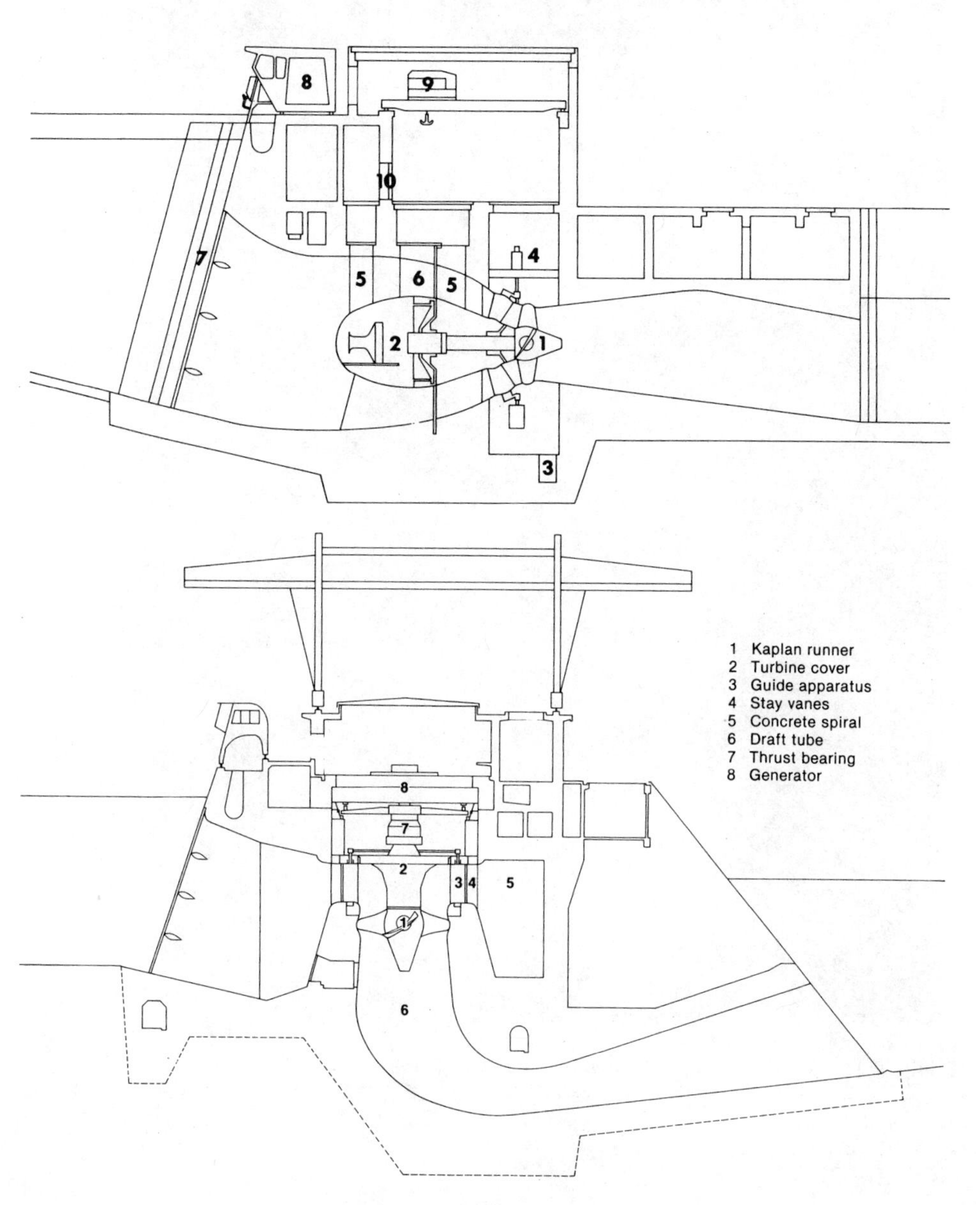

Figure 13.11 Comparison of bulb (top) and Kaplan (bottom) turbines.

Figure 13.12 Bulb turbine installation.

II. LUBRICATED PARTS

The main parts of hydraulic turbines requiring lubrication are the turbine and generator bearings, the guide vane bearings, the control valve, governor, and control system, and the compressors.

A. Turbine and Generator Bearings

Horizontal shaft machines require journal bearings to support the rotating parts, including the generator armature. With the exception of Pelton turbines, thrust bearings are also

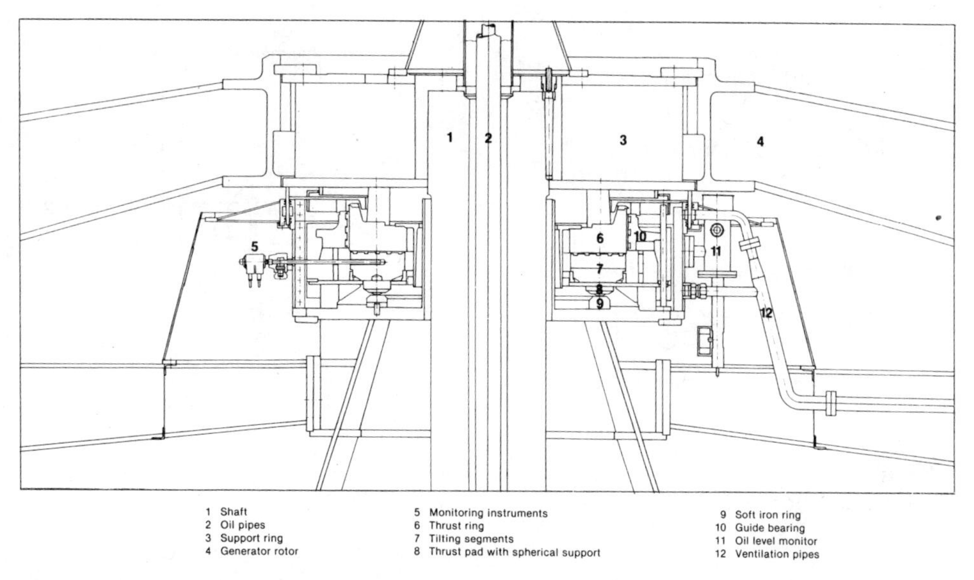

Figure 13.13 Umbrella-type bearing construction.

required to absorb the thrust of the water acting on the runner. Vertical shaft machines require guide bearings to keep the shaft centered and aligned, and thrust bearings to carry the weight of the rotating parts and absorb the thrust of the water acting on the runner.

Vertical shaft machines have a guide bearing above the turbine, and one or two guide bearings at the upper, or generator, end of the shaft. In some cases with long shafts, an additional guide bearing is installed about midway between the turbine guide bearing and the generator guide bearing. For reasons of accessibility, normally the thrust bearing is installed at the upper end of the shaft, either above the armature or just below it. A combination guide and thrust bearing located above the armature and a guide bearing below is sometimes referred to as a *two-bearing* arrangement. In the so-called *umbrella* type, a combination thrust and guide bearing (Figure 13.13) is located below the armature, and in the *semi umbrella* type separate guide and thrust bearings are located below the armature. Most current medium and large machines are of either the umbrella or semiumbrella type.

1. Journal and Guide Bearings

The bearings of horizontal shaft machines are of the fluid film type, with babbitt-lined, split shells. Oil lifts to assist starting may be used in the journal bearings when the rotating parts are extremely heavy. At stabilized operating conditions, oil-lubricated journal and guide bearings will operate in the range of 140°F (60°C).

Two general types of turbine guide bearing are used in vertical shaft machines. Many older turbines are equipped with water-lubricated rubber or composition bearings. This construction minimizes sealing requirements at the top of the turbine casing but is not too satisfactory when the water carries silt and other solids. As a result, most machines are built with a stuffing box at the top of the turbine, and a babbitt-lined guide bearing (Figure 13.10). Various types of seal are used in the stuffing boxes, including carbon ring packing and gland packing.

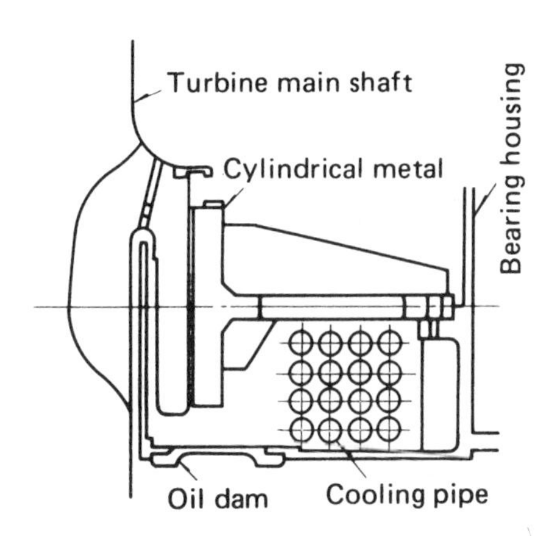

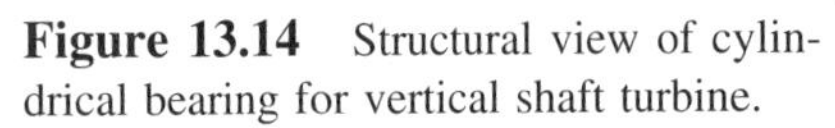

Figure 13.14 Structural view of cylindrical bearing for vertical shaft turbine.

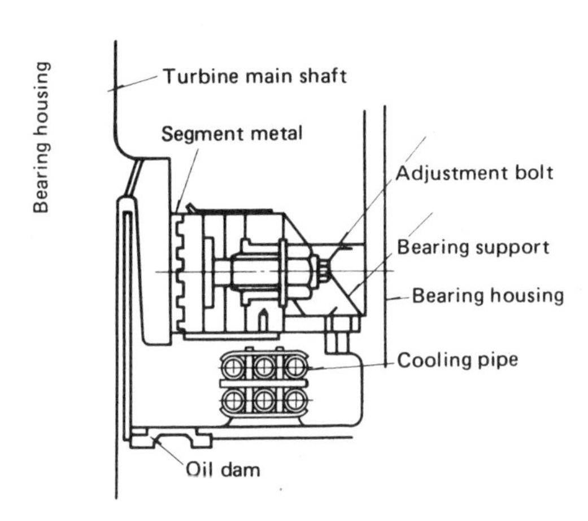

Figure 13.15 Cylindrical shell and segment-type guide bearings.

The upper guide bearings are either babbitt-lined, split cylindrical shell-type bearings (Figure 13.14) or segment bearings (Figure 13.15). The segment-type bearings are becoming increasingly popular because they permit easy adjustment of shaft alignment and bearing clearance. The segments may be crowned, that is, machined to a radius slightly larger than that of the journal plus the thickness of the oil film to permit easier formation of oil films. The construction shown in Figures 13.14 and 13.15, with the journal formed by an overhanging collar on the shaft and an oil dam extending up under the collar, is now common. It eliminates the need for an oil seal below the bearing. Bearings running directly on a machined journal on the shaft, with either an oil seal below the bearing or an oil reservoir fastened to the shaft and rotating with it, are also used.

2. Thrust Bearings

Thrust bearings of horizontal shaft machines are of the tilting pad or fixed pad type. For a reversible pump/turbine, tilting pad bearings designed for operation in either direction must be used. Thrust bearings for vertical shaft hydroelectric units are among the most highly developed forms of these bearings now in use. As pointed out earlier, the thrust bearing must support the weight of the rotating parts plus the hydraulic thrust of the water acting on the turbine runner. Single bearings capable of supporting loads in excess of 2000 tons are in service. Because of the high thrust loads thrust bearings operate at the highest temperatures. Operating temperatures generally are in the range of 212°F (100°C).

Tilting pad thrust bearings are used on all larger machines. Some older machines are equipped with tapered land bearings. Bearings of the Kingsbury and Michell type, in which the pads tilt on a pivot or on a rocking edge on the bottom of the pad, were used on many older machines. Most newer designs use flexible supports under the pads to permit the slight amount of tilt needed to form wedge oil films. Bearings with flexible pad supports, which can be run in either direction, are particularly suited to reversible pump/turbine units. Springs (Figure 13.16), elastomeric pads, interconnected oil pressure cylinders, and flexible metallic supports are all used. Spherical supports are also used. The bearing segments may be flat, or crowned slightly to aid in the formation of oil films.

Figure 13.16 Spring–supported tilting pad thrust bearing. Stationary portion showing oil lift slots.

In many of the larger machines, provision is made to pump up the bearings with high pressure oil to assist starting (Figure 13.16).

B. Methods of Lubricant Application

The bearings of hydroelectric sets are either self-lubricated or supplied with lubricant by a central circulation system. Circulation systems may be either unit systems (a separate system is used for each unit in a station) or station systems (all the units in a station are supplied from one system). In many cases, one or more of the bearings may be of the self-lubricating type, with a unit system supplying the other bearings.

In self-lubricated bearings, the oil is contained in a tank surrounding the shaft (Figures 13.13–13.15). Oil is lifted by grooves in the bearings, or by a ring pump on the shaft. Cooling coils can be located in the tank, or with a cylindrical shell bearing a cooling jacket may be located around the bearing shell. External cooling coils may also be used, but these are generally suitable only for relatively high speed machines, which generate sufficient pumping force to circulate the oil through the external circuit.

C. Governor and Control Systems

Older hydroelectric units were equipped with mechanical hydraulic control systems with a mechanical speed-sensitive device and a hydraulic system to actuate the guide vanes, and the runner blades if a Deriaz or Kaplan turbine were used. Newer machines are often equipped with electrical speed-sensitive devices and electronic systems.

Older hydraulic turbine hydraulic systems generally operated at 150 psi and used the same oil as the bearing oil system. Current units operate with pressures in the 1000 psi range but can go as high as 2000 psi. The hydraulic systems are now usually separate systems and require antiwear hydraulic fluids for the higher pressure systems. There is also a trend toward use of environmentally acceptable fluids for these applications. Hydraulic pumps are driven by electric motors. Air-charged accumulators (air over oil) are used to maintain system pressure and supply the large fluid flow necessary to adjust rapidly to meet sudden changes in load. They also provide a source of fluid under pressure to shut the turbine down in the event of a failure in the system. Emergency shutdown may also be assisted by the use of closing weights on the guide vane operating mechanism, or by designing the vanes that will be closed by water pressure if the hydraulic system fails.

D. Guide Vanes

The guide vanes, or wicket gates, are manufactured with an integral stem at each end that serves as the bearing journal. One bearing is used at the bottom and one or two bearings at the top. A thrust bearing may also be required. These bearings, as well as the bearings of the operating mechanism, are grease lubricated. Centralized application systems are now usually used to supply these bearings.

E. Control Valves

In some turbines, the guide vanes are arranged to close tightly and act as the shutoff valve for the turbine. In most installations, however, separate closing devices on the water inlet are used. In the case of pump turbines, closing devices on both the inlet and outlet are used.

Closing devices include sluice valves, rotary valves (Figure 13.17), butterfly valves, and spherical valves. All are designed for hydraulic operation. Closing weights may be used for emergency shutdown. Bearings are grease lubricated.

F. Compressors

In most hydroelectric plants, compressed air is required to maintain the pressure in the hydraulic accumulators. Compressed air is also used to blow out the draft tube and turbine casing when maintenance is to be performed. Compressed air also blows out the pump or turbine when the changeover from pump to turbine operation, or vice versa, is made in pump/turbine installations. Compressed air is also used in some impulse turbine installations to keep the tail water out of the turbine when the tail water level is high. Compressors operated in hydroelectric plants are critical pieces of equipment. Air compressors can be four-stage units and can operate with discharge pressures up to 1000 psi.

III. LUBRICANT RECOMMENDATIONS

The need for extreme reliability and long service life of hydroelectric plants generally dictates that premium, long life lubricants be used. Rust- and oxidation-inhibited premium oils are usually used for oil applications. Viscosities usually are of ISO viscosity grade 32, 46, 68, or 100, depending on bearing design, speeds, and operating temperatures. Oils with excellent water-separating characteristics are desirable. While start-up temperatures are rarely below freezing, the oils used must have adequate fluidity for proper circulation at those temperatures. Where oil lifts are not used for starting, oils with enhanced film strength may be desirable to provide additional protection during starting and stopping.

Hydraulic system requirements for older units were generally be met with oils of the same types used for bearing lubrication. More modern high pressure systems have been separated from the bearing oil systems and may require antiwear-type hydraulic fluids. Good air separation properties are desirable to ensure that air picked up in the accumulators separates readily in the reservoir.

Figure 13.17 Rotary closing valve.

Greases used in grease-lubricated bearings require good water resistance and rust protection. They should be suitable for use in centralized lubrication systems and should have good pumpability at the lowest water temperatures. Both lithium and calcium soap grease are used. NLGI no. 2 consistency greases are usually used, but in some extremely cold locations, NLGI no. 1 consistency greases are selected.

Compressors used in hydroelectric plants can be lubricated as outlined in Chapter 17.

14

Nuclear Reactors and Power Generation

Nuclear reactors fall into the following categories: zero-power research reactors, test reactors, special isotope production reactors, and power reactors. Basically, all nuclear reactors are similar in that they all utilize the fission chain reaction process to provide heat energy through the splitting (fission) of the heavy nuclei of fissionable materials. This reaction produces about 1×10^8 times the energy release of burning 1 carbon atom of fossil fuel plus the production of extra neutrons needed to sustain the chain reaction. The fuel used is generally ^{235}U (uranium-235), ^{233}U (uranium-233), or ^{239}Pu (plutonium-239). This chapter provides general information on reactors, with emphasis on those used in power generation.

I. REACTOR TYPES

The power reactor, whose main function is to furnish energy, consists broadly of a core containing nuclear fuel, a moderator (although this is eliminated in fast neutron reactors), a cooling system, a control system, and shielding. In practice, it is possible to design an almost endless number of different but basically similar reactor types by using various combinations of fuel, coolant, and moderator. It would seem that such a variety could lead to confusion, but in actuality, certain combinations are ruled out by unavailability of some of the components or by economics. For example, many areas must use natural uranium because of the lack of enrichment capabilities. This rules out certain reactors, such as the fast flux. Also, the use of natural uranium puts a limitation on the type of moderator, critical size, and power level, and although heavy water is a good moderator, especially for reactors using natural or low enrichment fuels, its cost has militated against its widespread use.

For these reasons, various countries throughout the world have pursued particular course of designs depending on the availability of materials for construction, moderator, and fuel. For example, most European nations and Canada based their first-generation reactor designs on the use of natural uranium because of a lack of enrichment facilities.

On the other hand, the United States, with its extensive system built for defense purposes, has concentrated its reactor designs on enriched fuels. Most countries using nuclear reactors currently have the ability to produce or obtain enriched fuel.

A. Basic Reactor Systems

Among the hundreds of combinations of fuel, coolant, moderator, and so on that are theoretically possible as reactor systems, six basic types have been studied in research stages and have resulted in demonstration or commercial reactors.

1. Pressurized-water reactor (PWR)
2. Boiling water reactor (BWR)
3. Sodium–graphite reactor (sometimes called light-water-cooled, graphite–moderated reactor: LGR)
4. Fast breeder reactor, including the liquid metal, fast breeder reactor (LMFBR)
5. Gas-cooled reactor (GCR)
6. High temperature, gas-cooled reactor (HTGR)

Figure 14.1 shows the schematics for each of these reactor designs, with Figure 14.1e representing both gas-cooled and high temperature gas-cooled reactors.

1. Pressurized Water Reactor

Fission heat is removed from the reactor core by water pressurized at approximately 2000 psi to prevent boiling. Steam is generated from secondary coolant in the heat exchanger. The major characteristics of this reactor are as follows.

Light water (H_2O) is the cheapest coolant and moderator.
Water is a well-documented heat transfer medium, and the cooling system is relatively simple.
High water pressure requires a costly reactor vessel and leakproof primary coolant system.
High pressure, high temperature water at rapid flow rates increases corrosion and erosion problems.
Steam is produced at relatively low temperatures and pressures (compared with fossil-fueled boilers) and may require superheating to achieve high plant efficiencies.
Containment requirements are extensive because of possible high energy release in the event of a primary coolant system failure.

2. Boiling Water Reactor

Fission heat is removed from the reactor by conversion of water to steam in the core. Such reactors have the following major characteristics.

Light water is the coolant, moderator, and heat exchange medium, as in a pressurized-water reactor.
Reactor vessel pressure is less than in the primary circuit of the pressurized reactor.
Steam pressures and temperatures are similar to those of pressurized-water reactors.
Heat exchangers, pumps, and auxiliary equipment requirements are reduced or eliminated.

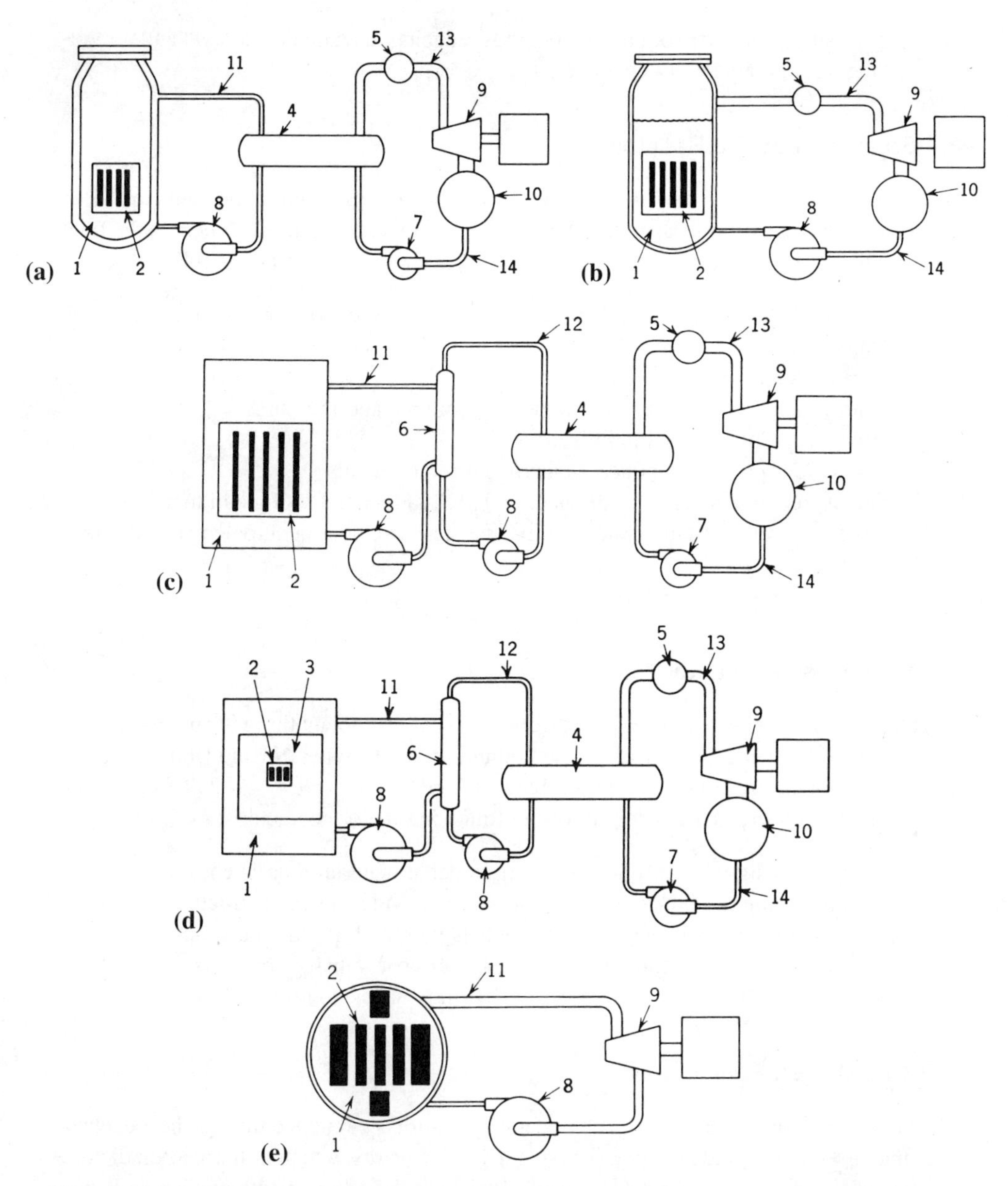

Figure 14.1 Common nuclear reactor schematics. (a) Pressurized-water reactor. (b) Boiling water reactor (direct cycle). (c) Sodium–graphite reactor. (d) Fast breeder reactor. (e) Gas-cooled and high temperature gas-cooled reactor. *Key:* 1, reactor; 2, core; 3, blanket; 4, boiler; 5, steam drier; 6, intermediate heat exchanger; 7, feed water pump; 8, circulating pump; 9, turbogenerator; 10, condenser; 11, primary coolant; 12, intermediate coolant; 13, steam 14, condensate; 15, circulating fuel.

Power surge causes a void formation, thus reducing the core power level and providing an inherent safety characteristic.

3. Sodium–Graphite Reactor

Molten sodium metal transfers high temperature heat from graphite-moderated core to an intermediate exchanger. Intermediate sodium–potassium coolant transfers heat to the final water in the boiler for steam generation. The major characteristics are as follows.

The high boiling point of liquid metal eliminates pressure on the reactor and primary systems.
High reactor temperatures are permitted.
Steam is generated at relatively high temperatures and pressures.
Corrosion problems are minimized.
Low coolant pressures reduce containment requirements.
Violent chemical reaction with water and high radioactivity of alkali metal requires a triple-cycle coolant system with dual heat exchange equipment to minimize hazards.
The core is relatively complex.

4. Fast Breeder Reactor

Heat from fission by fast neutrons is transferred by sodium coolant through an intermediate sodium cycle to steam boilers as in the sodium–graphite type. No moderator is used. Neutrons escaping from the core into a blanket breed fissionable^{239} Pu-239 from fertile ^{238}U blanket. Fast breeder reactors have the following major characteristics.

Reactor is designed to produce more fissionable material than is consumed.
Low absorption of high energy neutrons permits wide choice of structural materials.
Low neutron absorption by fission products permits high fuel burn-up.
A small core with a minimum area intensifies heat transfer problems.
Core physics, including short neutron lifetime, makes control difficult.

5. Gas-Cooled Reactors

Heat removed from the core by gas at moderate pressure is circulated through heat exchangers that produce low and high pressure steam. Such reactors, which utilize carbon dioxide gas, graphite moderator, and natural uranium fuel, have the following major characteristics.

Utilize natural uranium fuel and relatively available materials and construction.
Permit low pressure coolant and relatively high reactor temperatures.
Containment requirements are moderate and corrosion problems minimal at low temperatures.
Reactor size is relatively large because of natural fuel and graphite moderator.
Power density (kilowatt output per liter of core volume) is extremely low.
Poor heat transfer characteristics of gases require high pumping requirements.
Steam pressures and temperatures are low.
Carbon dioxide gas is relatively cheap, safe, and easy to handle.

6. High Temperature, Gas-Cooled Reactors

Heat from the reactor core is carried by inert helium to the heat exchanger for generation of steam or directly to a gas turbine. The gas returns to the reactor in a closed cycle. These reactors have the following major characteristics.

Good efficiency can be achieved in a dual cycle with a minimum gas temperature of 1400°F (760°C).
High fuel burn-up is possible and conversion of fertile material permits lower fuel costs.
Minimum corrosion of fuel elements will be caused by inert gas.
High temperature coolant minimizes the disadvantages of poor heat transfer characteristics of the gases.
Fuel element failure may cause contamination of turbine in direct cycle.
The design of fuel elements for long life is complicated by high temperatures.
The supply of helium worldwide is limited.
Graphite is combustible.

II. RADIATION EFFECTS ON PETROLEUM PRODUCTS

In general, radiation damage may be defined as any adverse change in the physical and chemical properties of a material as a result of exposure to radiation. Radiation damage is a relative term for the changes in a material that may have adverse effects on the operation of the nuclear plant. This is true of organic materials in particular; for example, the evolution of a gaseous hydrocarbon from a liquid organic material may result in an explosion hazard and an increase in liquid viscosity. Similarly, radiation of an organic fluid may result in unwanted increase in molecular size, with consequent thickening or solidification of the liquid or grease. In the study of radiation damage, we are concerned mainly with the adverse or undesirable changes in the lubricants that affect their ability to perform adequately in the machinery involved. It should be noted that that lubricants can still perform their lubrication function after reaching levels deemed unsatisfactory for continued use by conventional laboratory evaluations. This aspect is important in applications where equipment (reactor and other containment equipment) may not be accessible until such events as fuel rod changes, set up on 18- to 24-month cycles. If analysis of these lubricants indicates undesirable changes in their characteristics, it become necessary to decide whether the lubricant can be allowed to perform until the time for a scheduled outage arrives or whether other alternatives need to be considered.

Broadly speaking, there are two mechanisms of radiolysis that must be considered in a study of the damage to organic fluids. One is the primary electronic excitation and ionization of organic molecules caused by β particles, γ rays, and fast neutrons. The other is the capture of thermal neutrons and some fast neutrons by nuclei that would cause changes in the nuclei and the generation of secondary radiation that would result in further damage.

Two methods are utilized to measure radiation energy. One measure, the quantity of energy to which the materials exposed, is called the roentgen (R); the other, the amount of energy the material absorbs, is called the rad. For γ radiation, the exposure unit (roentgen) is defined as the quantity of electromagnetic radiation that imparts 83.8 ergs of energy to 1 gram of air.

The radiation dosage of a material is defined as an absorption of 100 ergs of energy by 1 gram of material from any type of radiation. Actually, absorbed energy will vary with the type of radiation, and the effect will depend on the material exposed. For γ radiation, however, one rad absorbed is approximately equivalent to 1.2 R of radiation dosage. The rad is useful for comparing the equivalent energy of mixed radiation fluxes but does not distinguish between types.

From a radiation damage standpoint, 1 rad of neutron flux causes 10 times more biological damage to tissue than an equivalent amount of absorbed energy of γ rays. For petroleum products, however, the dosage, as measured by such effects as viscosity increase, is almost equivalent for the two types. This is discussed in more detail later in this chapter. The general levels of radiation dosage are as follows:

Dosage (Rs)	Effect
200–800	Lethal to humans
<5 million	Negligible to petroleum products
5–10 million	Damaging to petroleum products
>10 million	Survived by only most resistant organic structures

Based on experimental work, the damage to petroleum products may be summarized in the following list.

1. Liquid products (fuels and oils) darken and acquire an acrid, oxidized odor.
2. Hydrogen content decreases and density increases.
3. Gases such as hydrogen and light hydrocarbons evolve.
4. Physical properties change, higher and lower molecular weight materials are formed, and olefin content increases.
5. Viscosity and viscosity index increase.
6. Polymerization to a solid state can occur.

It must be appreciated that the intensity of these effects or the incidence of one or more of them depends on the amount of absorbed energy, the exact composition of the specific petroleum material, and other environmental conditions such as temperature, pressure, and the gaseous composition of the atmosphere.

A. Mechanism of Radiation Damage

Organic compounds and covalent materials do not normally exist in an ionized state and therefore are highly susceptible to electronic excitation and ionization as the result of deposited energy. Covalent compounds, including the common gases, liquids, and organic materials, consist of molecules that are formed by a group of atoms held together by shared electron bonding, which yields strong exchange forces. The molecules are bound together by relatively weak van der Waals forces.

Conversely, ionic compounds, such as inorganic materials, which include salts and oxides, are already ionized (metals may be considered to be in an ionized state) and are not susceptible to further electronic excitation. Ionic compounds consist of highly electropositive and electronegative ions held together in a crystal lattice by electrostatic forces in accordance with Coulomb's law. There is no actual union of ions in the crystal to form molecules, although all crystals may be considered to be composed of large molecules of a size limited only by the capacity of the crystal to grow.

Therefore, the effect of radiation energy on nonionic compounds is to form ions, radicals, and excited species and thereby make the compounds more reactive with themselves or with the atmospheric environment. On the other hand, the effect of radiation on ionic compounds is to change the properties of the compound related to crystal structure.

B. Chemical Changes in Irradiated Materials

The physical and chemical properties of hydrocarbon fluids that make them important as lubricants change during irradiation to varying degrees based on chemical composition and the presence of additives. These changes may be traced to alteration of the chemical structure of the materials. Nuclear irradiation, either directly or by secondary radiation, deposits high level energy in the irradiated organic substance and causes ionization and molecular excitation. The ions are excited molecules that rapidly react to form free radicals, which further combine or condense (Figure 14.2).

The changes in chemical structure may be measured by various classical methods: for example, it is possible to determine the approximate number of free radicals formed by the use of scavengers such as iodine. In addition, either hydrogen or light petroleum fractions are evolved as gas. Investigations have shown that both carbon–hydrogen and carbon–carbon bonds can be broken by radiolysis. The dissociated or ionized molecules can condense, rearrange, and form olefins or other products, depending on the environment. At temperatures below 400°F (204°C), temperature effects do not seem to be significant.

Because most petroleum lubricants contain combinations of saturated and unsaturated aliphatic and aromatic compounds, the reactions of these principal hydrocarbon classes have been studied under the influence of ionizing radiation. These studies (Table 14.1) indicate, as would be suspected, that unsaturated hydrocarbons are most reactive and aromatics the least affected. Saturated compounds fall somewhere between the two

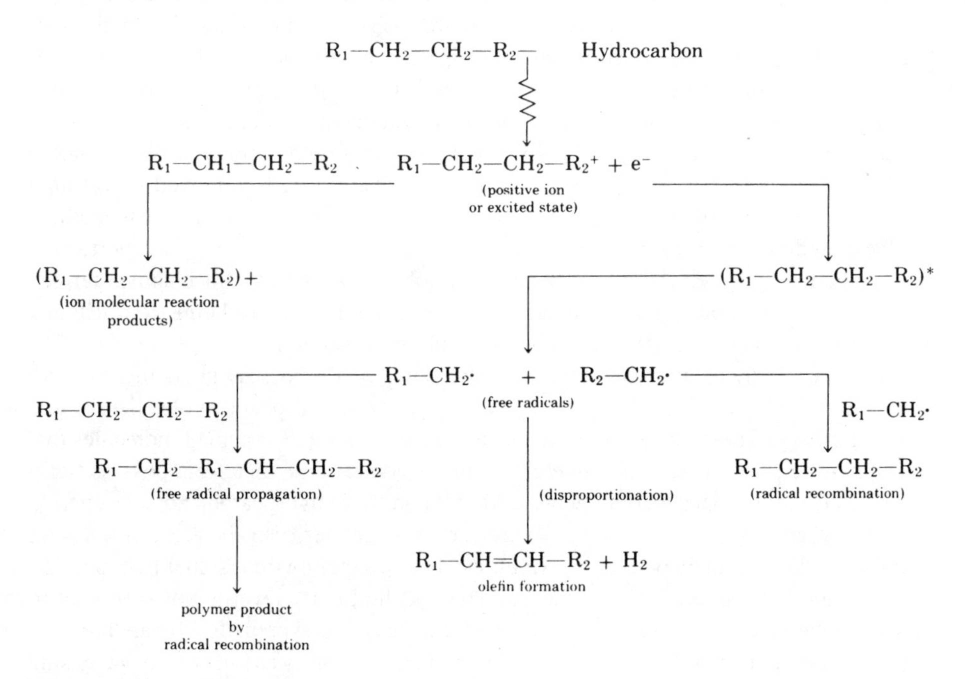

Figure 14.2 Radiolysis processes in hydrocarbons.

Table 14.1 Summary of Radiation Changes in Organic Compounds

	G values[a]		
Condition	Saturated hydrocarbons	Unsaturated hydrocarbons	Aromatic hydrocarbons
Polymerization	None	10–10,000	None
Cross-linking	~1	6–14	<1
H_2 evolution	2–6	1	0.04–0.4
CH_4 evolution	0.06–1	0.1–0.4	0.001–0.08
Destruction of irradiated material	4–9	6–2000	<1

[a] G = molecules produced, destroyed, or reacted per 100 eV of energy absorbed
Source: Nuclear Engineering Handbook. Used by permission of McGraw-Hill Book Co., Inc.

extremes. The results are expressed as G values: that is, the number of molecules reacting or produced for each 100 eV of ionizing radiation. For example, at least 6–14 molecules of unsaturated hydrocarbons react for each 100 eV. In certain instances, high G values (up to 10,000) result for free radical polymerization, whereas low G values ($G = 6$) are formed for random cross-linking, and very low values ($G < 1$) for methane formation.

Aromatic materials, with a G value for destruction of 1, are highest in radiation resistance. The principal reaction is cross-linking, from which very small amounts of gas evolve.

Additional studies were made on a range of petroleum oils representative of typical paraffinic, naphthenic, and aromatic materials varying in sulfur content. The viscosity was plotted against aromatic and sulfur content (Figure 14.3). The data show that as the aromatic content increases, the viscosity increase is reduced almost linearly. The effect of sulfur content is similar but more marked. Further, it was noted that radiation damage appears greatest for oils with the highest molecular weights or the highest initial viscosity.

Because it was found that naturally occurring compounds improved the radiation stability of petroleum oils and that these compounds were usually removed by refining procedures, the effect of using synthetic aromatic and sulfur as additives was studied. From these studies, it appears that a disulfide, or an alkyl selenide, provides good protection against radiation damage. The disulfides prevent polymerization by a mechanism termed free radical chain stopping. The disulfides have an advantage also of being good EP and antiwear agents but do not prevent oxidation or olefin formation.

It is well known, however, that aromatic compounds possess good thermal and radiation stability and in the latter case protect less stable aliphatic molecules by the transfer of energy. These compounds are usually characterized by complex molecules that resonate between a number of possible electronic structures and, therefore, possess fairly stable excited energy states. In other words, when a paraffinic hydrocarbon absorbs energy, it is raised to an unstable state in which the energy is greater than the electronic forces that constitute the chemical bonds. The result is a bond fracture with residual free radicals. In an aromatic with an equivalent absorbed energy, the higher level is not sufficient to sever the greater electronic binding forces, and the energy is eventually liberated as heat or light. The aromatic petroleum extracts exhibit radiation stability as follows, in decreasing order: polyglycols, paraffinic hydrocarbons, diesters, and silicones. Aromatic compounds

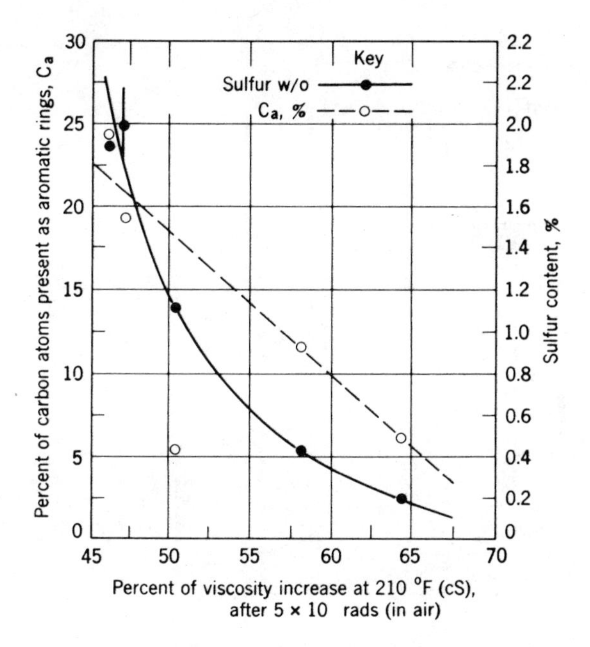

Figure 14.3 Radiation stability versus sulfur and aromatic content.

were studied to determine their effects both as pure synthetic fluids and as antiradiation additives to mineral oils. The results, given in Figures 14.4 and 14.5, show the following relationships.

1. The aromatics with bridging methylene groups between aromatic molecules are less efficient as protective agents than antiradiation additives with direct links between aromatic rings.
2. Long chain alkyl groups attached to the aromatic rings make less effective protective agents, probably because of a difference in stability of the compound and a lowering of the aromatic ring content.
3. Small amounts of a free radical inhibitor in addition to the aromatic additive substantially reduce the viscosity increase.
4. The protection afforded is not simply a direct function of aromatic content; in fact, it would appear that 40% of added aromatic material is a practical maximum. Beyond 40%, it is preferable to use a pure aromatic of suitable physical characteristics.

A study of the changes in properties and performance of conventional lube oils after irradiation shows the following.

1. Conventional antioxidant additives of the phenolic or amine type confer little radiation stability to base oils and are preferentially destroyed between 10^8 and 5×10^8 rads.
2. Didodecyl selenide, which is known to be an effective antioxidant, also has radiation-protective properties. The oxidation stability is effective after an irradiation of 10^9 rads.

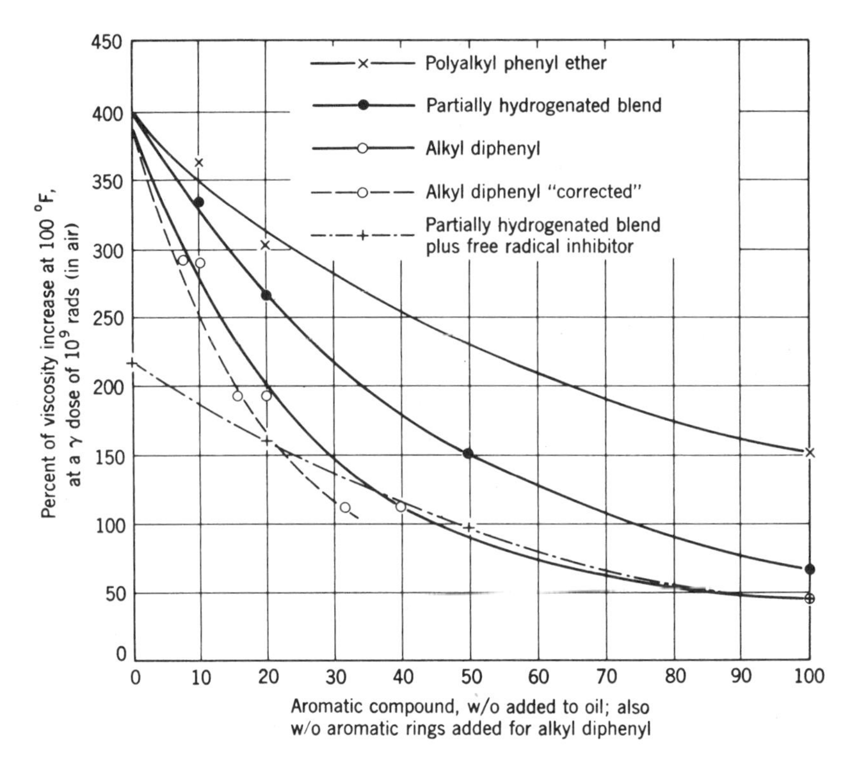

Figure 14.4 Radiation protection of synthetic aromatic additives.

3. Diester-based oils, phosphate esters (antiwear additives), and halogenated EP agents produce acids at a low radiation dose.
4. Polymers such as polybutenes and polymethacrylates cleave readily and thus lose their effect as VI improvers.
5. Silicone antifoam agents are destroyed at low radiation dose.
6. In most cases, the presence of air, compared with an inert atmosphere, increases radiation damage by a factor of 1.6–2.3 times, as indicated by viscosity increase.

In summary, high quality, conventional lubricating oils are suitable for radiation doses up to 10^8 rads. It should be noted that base oils that undergo severe hydrotreating or hydrocracking processes during refining may need to have sulfur- or aromatic-containing additives supplied to the finished product. These refining processes remove much of the content of sulfur and aromatic compounds. Further radiation resistance can be formulated into a good quality petroleum oil by use of antiradiation agents such as radical scavengers or aromatic structures. These formulated oils will protect up to exposures of 10^9 rads. Above these doses and at high temperatures, synthetic-type lubricants that use partially hydrogenated aromatics blended with aromatic polymers are required. The effect of temperature at high radiation dose (10^9 rads) has been studied under a nitrogen atmosphere for these fluids (Figure 14.5).

If, however, materials are to be satisfactory as lubricants, not only must they have good thermal and irradiation stability, but their wear characteristics must be satisfactory.

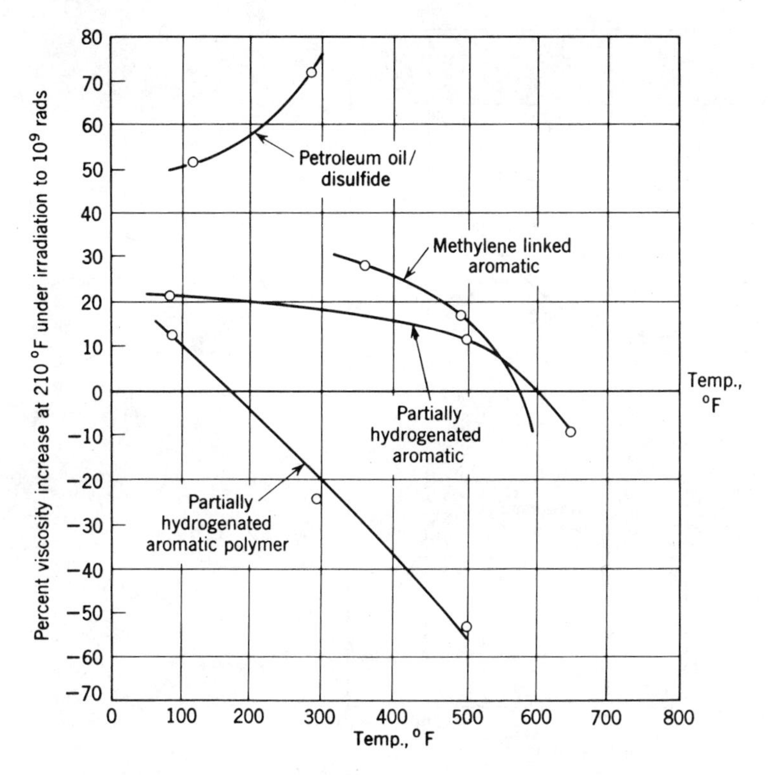

Figure 14.5 Radiation stability versus temperature; materials were studied under a nitrogen atmosphere.

This phase can use conventional laboratory and field testing for new lubricants. Additional special testing is also required to verify that the irradiated lubricants retain sufficient antiwear properties to protect the equipment over the expected service intervals.

1. Grease Irradiation

The damage caused by high energy radiation on greases is twofold. First, the radiation attacks the thickening structure and causes separation and fluidity. Following this, continued irradiation results in polymerization of the base oil, resulting in thickening and gradual solidification. The precise pattern of change is dependent on the type of thickener, the gel structure, and the radiation stability of the thickener and the base oil.

In general, greases have been evaluated for radiation damage by determining the worked penetration, following irradiation, and comparing it with the original value. These irradiation evaluations are usually of the static type, but testing under dynamic conditions during irradiation has yielded markedly different results. Typical greases that have organic soap components of alkali earth metals, although resistant to high amounts of radiation, break down at total doses of approximately 10^8 rads. Micrographs of soap structure showed a drastic change in the normal fiber structure of the gelling agent. On the other hand, greases made with nonsoap thickeners, such as carbon black, were less affected. Micrographs of a carbon black grease showed the same structure (agglomeration of carbon particles dispersed throughout the oil phase) both before and after irradiation.

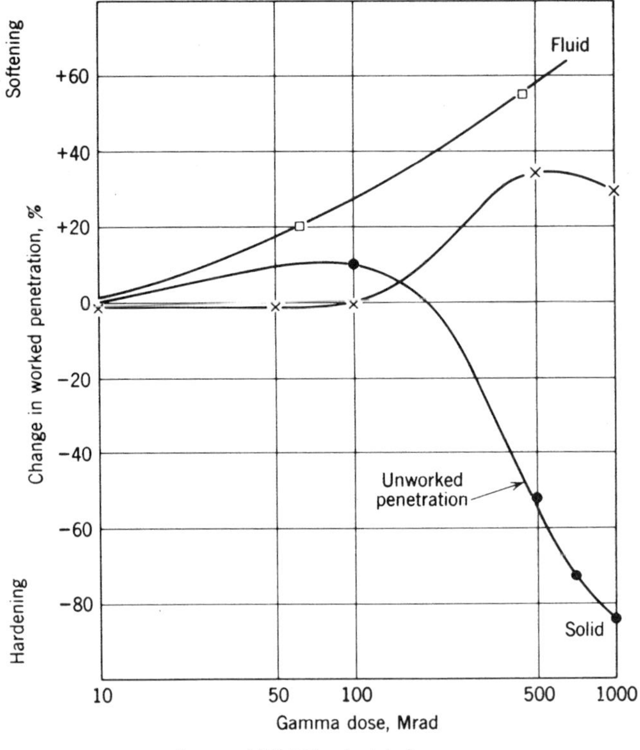

Figure 14.6 Effect of radiation on greases.

The stabilization of the thickening structure under irradiation solves the problem of softening or bleeding of the base oil but will not prevent the eventual solidification of the grease. This is a function of the base oil, and the solutions discussed in connection with lubricating oil (use of antiradiation additives or synthetic organics as base fluids) are valid.

The mechanism of change for three greases is shown in Figure 14.6. In one case, the grease had an unstable thickener and progressively softened to fluidity. Although such a grease might protect a bearing, the problem of leakage would be great, and incompatibility with reactor components would be a concern. The second grease gradually decreased in penetration (solidification) after an initial increase or softening. Such a grease would cause failure in the lubricated mechanism. The third grease showed good stability with a slight softening up to 10^9 rads.

2. Radiation Stability of Thickeners

The selection of the thickener or solid phase of a grease designed for nuclear applications requires consideration of compatibility as well as resistance to radiation, high temperatures, mechanical shear, and operating atmosphere.

Table 14.2 Elements on Which the UKAEA Places Restrictions[a] Are for Radiation-Resistant Lubricants Used in Reactors Employing Magnox Fuel Cans

None allowed	Mercury
0.1% allowed	Barium, bismuth, cadmium, gallium, indium, lead, lithium, sodium, thallium, tin, zinc
1% allowed	Aluminum, antimony, calcium, cerium, copper, nickel, praseodymium, silver, strontium

[a] These limits can be exceeded where it can be shown that the metals are present in stable compounded form and that practical compatibility tests are satisfied.

Certain elements are unsuitable because their presence within or close to the reactor core could seriously affect neutron economy or react with the fuel element cladding to cause destruction and possible release of fission products. Accordingly, the United Kingdom Atomic Energy Authority (UKAEA) has restricted lubricant composition (Table 14.2).

The effect of atmosphere can be illustrated by air, which has a serious oxidizing effect, especially when coupled with radiation and high temperatures. Conventional antioxidants are destroyed as noted earlier. Some of the organic-modified thickeners have an antioxidant effect and perform dual functions. Hot pressurized carbon dioxide can cause rapid degeneration of conventional soap-thickened greases, presumably by means of carbonate formation.

In selection of a thickener, the compatibility of the thickener and base fluid is of paramount importance. Even an exceptionally radiation-resistant thickener, when in combination with certain base fluids, may at best yield weak gels and soften easily. For example, a satisfactory grease structure is extremely difficult to obtain when an Indanthrene pigment is used with a paraffinic bright stock.

Various nonsoap thickeners that form good grease structure with both mineral oil and synthetic fluid bases are available. These thickeners may be grouped as follows.

1. *Modified clays and silicas.* Typical of the modified clays are Bentone and Baragel, which are formed by a cation exchange reaction between a montmorillonite clay and a quatenary ammonium salt. This reaction produces a hydrocarbon layer on the surface of the clay, which makes it oleophilic. Finely divided silicas may be treated with silicone to render them hydrophobic, or, as with Estersil, the silica may be esterified with *n*-butyl alcohol.

2. *Dye pigments.* Organic toners or dye pigments are utilized as grease thickeners (e.g., Indanthrene).

3. *Organic thickeners.* Typical of this type are the substituted aryl ureas characterized by the diamide–carbonyl linkage, which may be formed in situ by the reaction of diisocyanate with an aryl amine.

The behavior of these thickeners, when used in conjunction with a synthetic fluid, is shown in Figure 14.7.

As with fluid lubricants, antiradiation compounds may be added to the grease to increase its radiation stability.

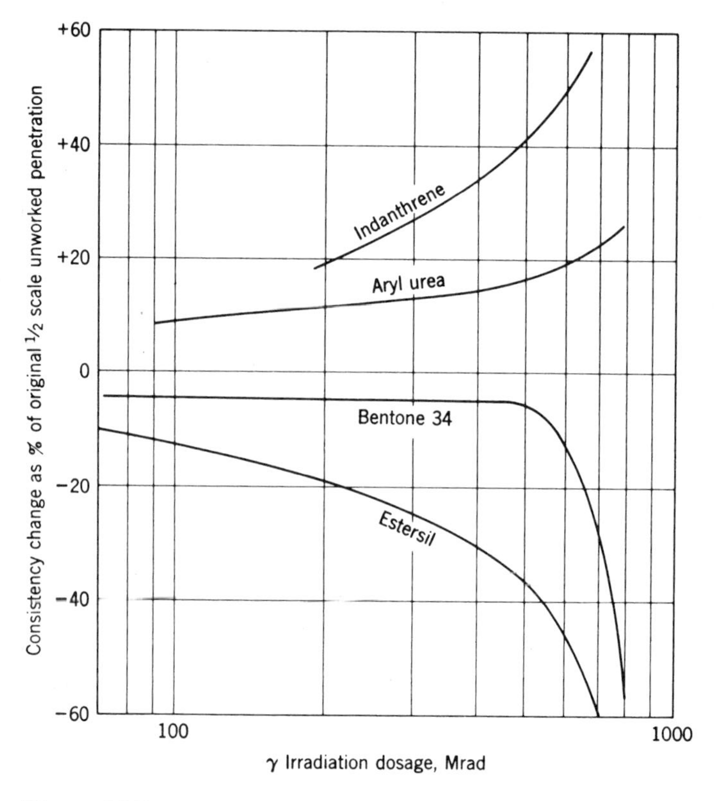

Figure 14.7 Effect of thickener on grease polymerization.

III. LUBRICATION RECOMMENDATIONS

The advent of nuclear energy has added a new dimension to the requirements of lubricants and other petroleum products in industrial applications. All equipment used in the nuclear industry—research and power reactors; fuel processing machinery; conveyors, manipulators, and cranes in irradiation facilities; viewing windows and shield doors in hot cells; heat exchange units—requires oils, greases, and other organic fluids to perform conventional and special functions in an irradiated atmosphere.

Nowhere are the operating conditions of radiation, temperature, and atmosphere more damaging than in a power reactor field. At the outset, equipment was designed to operate without conventional petroleum lubricants because little was known of the behavior of petroleum products under irradiation; moreover, the exact severity of the application was overestimated. This placed a design and economic burden on nuclear power generation. As experience was gained in the operation of these plants, the original position was reconsidered. First, specific operating parameters of radiation flux, temperature and so on, were obtained, which realistically established the requirements. Then research into the radiation resistance of petroleum materials showed that conventional lubricants can withstand doses up to 10^7–10^8 rads and still perform their lubrication function. New specialized lubricants were developed to withstand more than 10^9 rads. As a consequence of this knowledge and experience, proved engineering designs employing conventional petroleum greases and lubricating fluids have been adopted since the late 1950s.

Because the nuclear power industry is so complex and is under continuous scrutiny, patterns of design and operating conditions are changing. In addition, most equipment is unique, and each plant requires separate consideration before proper lubricants and lubrication schedules can be established. Little repetitiveness exists in equipment, especially in the reactor area, and in components associated with safety issues. Therefore, the best that can be accomplished in this chapter on lubrication recommendations is to furnish the background experience and establish guidelines so that lubrication engineers can, after a survey of the specific conditions, recommend the best lubricants for a particular application.

A. General Requirements

All lubrication engineers are familiar with the effect of factors such as speed, load, temperature, and time on the life of bearings and gears, and on the physical and chemical properties of oils and greases exposed to these environmental conditions.

In conventional applications, the effect of speed, load, and temperature are evaluated in making recommendations. Selecting the correct lubricant and service interval is determined by evaluating the lubricant's anticipated performance under the most critical of these conditions. For example, speed may be the determining factor with antifriction bearings, and the proper grease must resist excessive softening for the service period. In highly loaded applications, antiwear or extreme pressure properties may be the determining consideration. In most cases, however, temperature is the most critical factor. Operating severity may be determined by the degree of heat and the extent of exposure. The additional factor of radiation in nuclear applications affects lubricants in much the same manner as heat. Both are modes of energy and, as we saw earlier, lubricants, like all organic materials, undergo major structural changes when certain thresholds of absorbed energy are reached. We know that petroleum products undergo thermal cracking and polymerization at certain temperatures and, likewise, that cleavage and cross-linking occur at certain radiation dosage thresholds.

In nuclear applications, radiation energy is expressed as flux or dose rate. The particular type of radiation and the units have already been discussed. It is sufficient to state here that this dose rate, if applied over a specified interval, will yield the absorbed dose for an exposed material. If, for example, a grease can absorb a dose of 5×10^9 rads before suffering physical or chemical changes that will render it useless as a lubricant, this grease can be used for 5000 days if the dose rate is absorbed at a rate of 1 Mrad per day or for only 5 days if the dose rate absorbed is 1000 Mrad per day. As stated earlier, other factors such as temperature and atmosphere will increase or decrease these intervals.

B. Selection of Lubricants

Much has been published and extensive studies have been made that lead to the recommendations of correct lubricants. In this final analysis, however, the selection of the proper lubricant and its application to any particular equipment must be made by a lubrication engineer for each specific instance, based on operating conditions and the type of unit (bearing, gear, cylinder) requiring lubrication. Nowhere is this more pertinent than in the lubrication of the equipment in the reactor and containment areas of the nuclear power plant. Because of the uniqueness of the designs and the severity of the operating environment, each plant can be markedly different. Therefore the experience of the lubrication engineer is important, and blanket recommendations serve only as guidelines. The accumulated experience of lubrication engineers and equipment manufacturers has been helpful in numerous plants in the solution of lubrication problems and, in many cases, in the elimination of mechanical problems as well.

In selecting lubricants for a nuclear power plant, the engineer consulted in the design phases should be cognizant of development work at equipment manufacturers and should participate in practical evaluations of prototype units under the operating conditions. In surveying plant requirements, particular attention must be paid to the radiation flux profile that has been calculated for the various parts of the plant and compared with actual surveys during operation of similar plants. Some of this information is available in the form of a design basis event (DBE) developed for each plant. Extreme conservatism has been the rule in estimations of nuclear power plant requirements, often to the detriment of practical solutions.

15

Automotive Chassis Components

The engine and power train, generally considered to be part of the automotive chassis, were presented in detail in Chapter 11. In this chapter we are concerned only with suspension and steering linkages, steering systems, wheel bearings, and brake systems.

I. SUSPENSION AND STEERING LINKAGES

A suspension is an arrangement of linkages, resilient members (such as coil and leaf springs, struts, or torsion bars), and shock absorbers. These components attach the wheels to the frame or body of a vehicle. In front wheel drive and four-wheel-drive vehicles, the front suspension includes pivots that permit the front wheels to be turned for steering as well as providing the drive for the front wheels. Some off-highway and farm vehicles do not have steerable front wheels. In this type of equipment, steering is accomplished either by an articulated joint located between the front and rear sections of the machine or by disconnecting the power and applying braking to the wheels or tracks on the inside of the turn. Many self-propelled combines and some trucks and forklifts use steerable rear wheels.

Many suspension designs are used for passenger cars and trucks. Almost universally, passenger cars are equipped with "independent" front suspension, which permits the front wheels to move independently of each other with respect to the frame and body. Although this type of front suspension has been extended to trucks, many medium and heavy trucks are still equipped with rigid front axles. With the exception of some luxury and high performance cars, rear wheel drive vehicles are generally equipped with rigid rear suspension systems. Front wheel drive vehicles are generally equipped with independent rear suspension. The front wheel drive passenger cars are the predominant type of vehicle being marketed today.

In one rear wheel drive passenger car front suspension (Figure 15.1), the wheels are pivoted on a pair of ball joints (Figure 15.2) seated in the outer ends of the upper and lower control arms. These joints permit the wheels to move up and down with respect to the frame and to be turned for steering. Additional pivot points at the inner ends of the

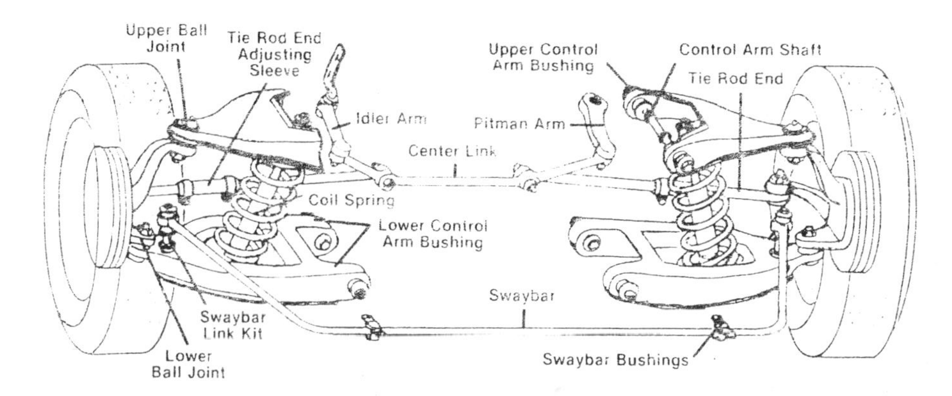

Figure 15.1 Passenger car front suspension (rear wheel drive). In this typical front suspension for rear wheel drive vehicles, the lubrication points are lower and upper ball joints and steering linkages, although many suspension and steering components may not require relubrication.

control arms (A-frame) permit them to swing up and down against the restraint of the coil springs and shock absorbers. Front suspension arrangements for front wheel drive vehicles use struts (Figure 15.3) in a similar fashion, with a ball joint on the lower portion connected to a swing arm. Instead of the swing arm being directly connected to the coil spring, the upper portion of the strut contains the spring and shock absorber.

Rigid front axles are currently used on heavy vehicles such as over-the-road trucks and buses but may be encountered in some older lightweight to medium-sized trucks. Where rigid front axles are used, usually the wheels are mounted on a yoke and kingpin (Figure 15.4) to permit turning the wheels for steering. Although some lightweight and medium-sized trucks have the ''rigid'' front axle separate for each wheel, they are still

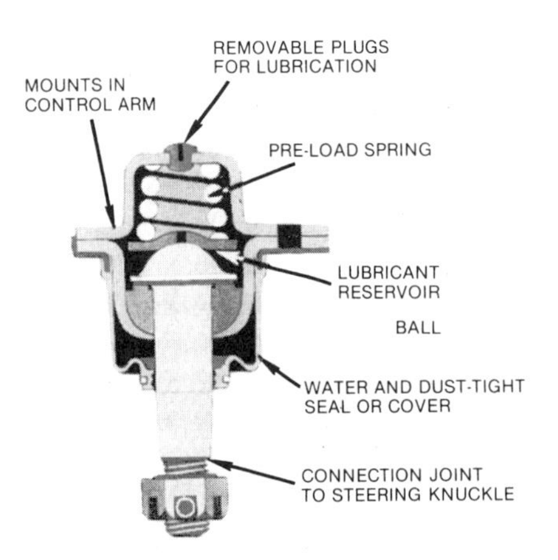

Figure 15.2 Suspension ball joint. This type of joint is designed for periodic relubrication, at low pressure, through the removable plug (some contain a grease fitting). Some similar joints have a vent on the seal so that they can be lubricated through a fitting.

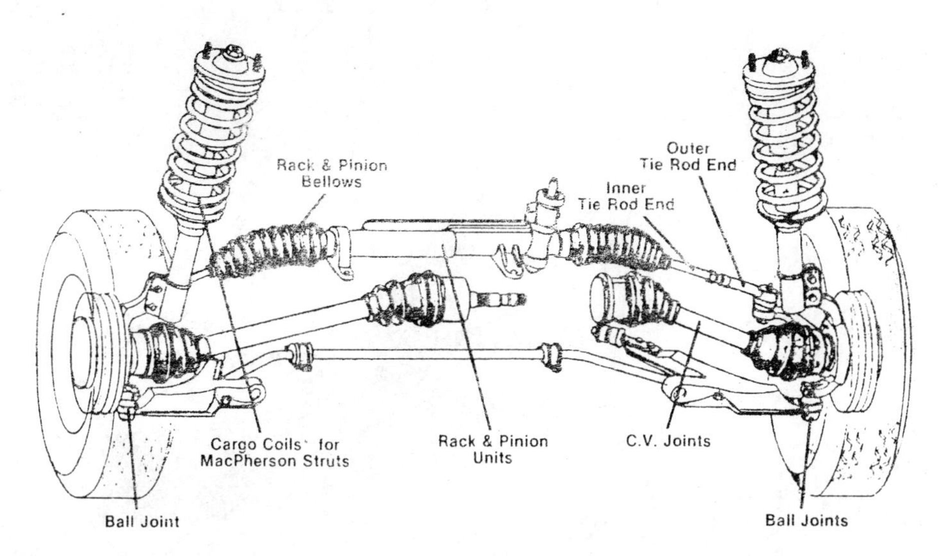

Figure 15.3 Front suspension for front wheel drive vehicles (struts). Generally only relubrication needed is at the ball joints in the lower portion of the strut and steering linkage; many suspension components may not require relubrication.

referred to as rigid. The usual rear suspension, because there is no requirements for steering, is simpler in design than the front suspension. Either leaf or coil springs may be used. Some additional complexity may be introduced with independent rear suspension to permit the wheels to move independently of each other while still transmitting the driving force to the wheels.

The most common type of steering linkage used on passenger vehicles is the parallelogram type (Figure 15.5). This term is derived from the parallelism maintained between the pitman and idler arm at all linkage positions. The pitman and idler arm is connected to an intermediate rod through pivot bearings, such as those shown at the right in Figure 15.6. All other connections are made through ball joints (sometimes referred to as tie-rod ends) such as those shown at the left of Figure 15.6. The pitman is operated by the steering gear, discussed in Section II.

Rack-and-pinion steering is now used on most light cars. In this design, no pitman or idler arm is used. The intermediate rod is replaced by a rack that meshes with a pinion on the end of the steering shaft. This arrangement results in a simple, direct-acting linkage.

Other linkage arrangements may be found on farm and off-highway equipment. For example, with the row crop tractors, no linkage is used. The wheels are mounted on a single vertical shaft that is rotated directly by the steering gear. Generally, where steerable front wheels are used, some type of linkage similar in principle to the parallelogram type is used. However, hydrostatic steering without linkages or gears, as such, is also employed.

A. Factors Affecting Lubrication

A number of factors contribute to the lubrication problems associated with suspension and steering linkage components. Since motion of the pivot points is oscillating, through relatively short arcs, fluid lubricating films cannot be formed, and lubricant films are

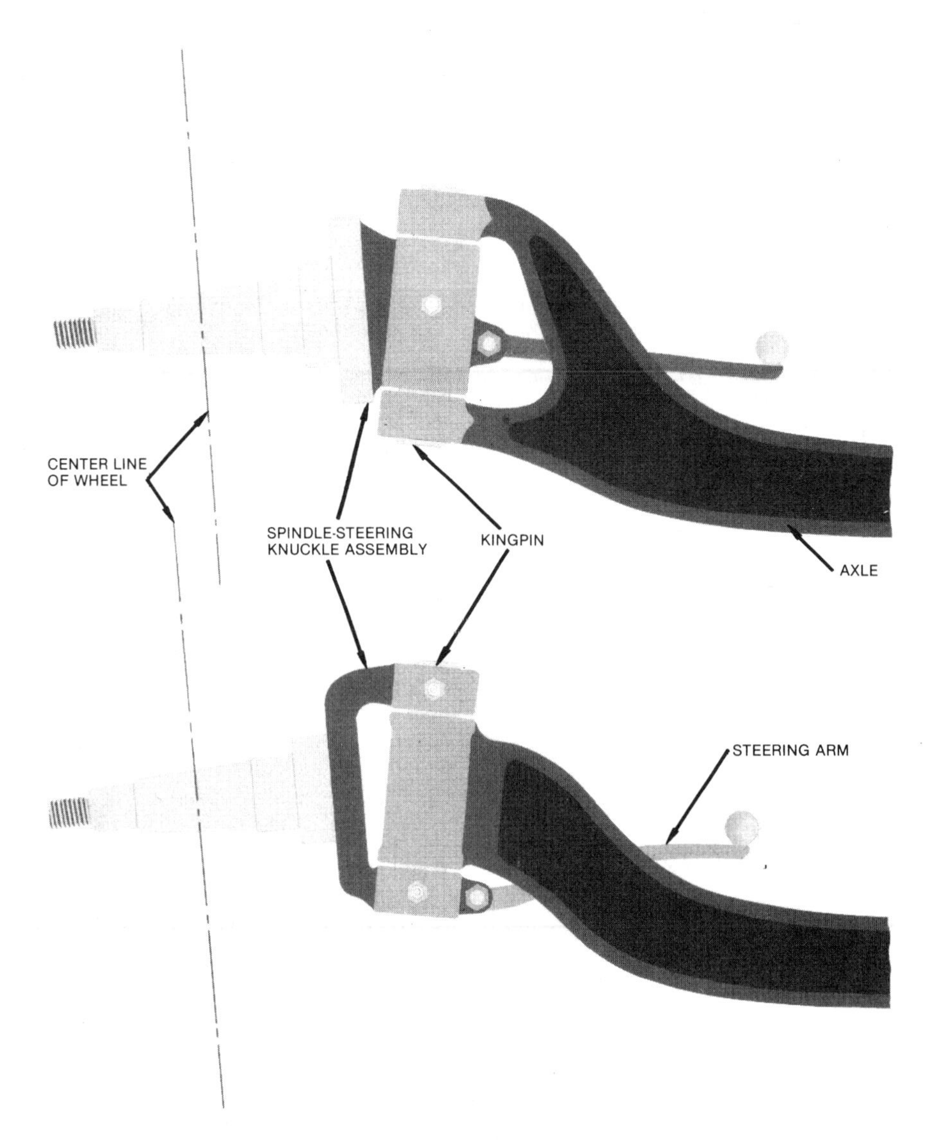

Figure 15.4 Rigid axle for front wheel mounting. In this design, the yoke is on the wheel spindle–steering knuckle assembly. In other designs, the yoke is on the axle. Relubrication is provided by fittings in the center of the upper illustration and fittings in the upper and lower portions of the spindle in the lower illustration.

wiped off. Since these components are located below the body, they are exposed to water (sometimes containing salt and sand used for melting ice) and dirt. Although effective sealing reduces the possibility of entry of these contaminants, seal failure or leakage is always possible. Loads are usually high, and severe shock loads are often present, especially in suspension members.

Over the past several years, procedures for lubrication of suspension and steering linkage components have changed markedly, particularly for passenger cars. Lubrication

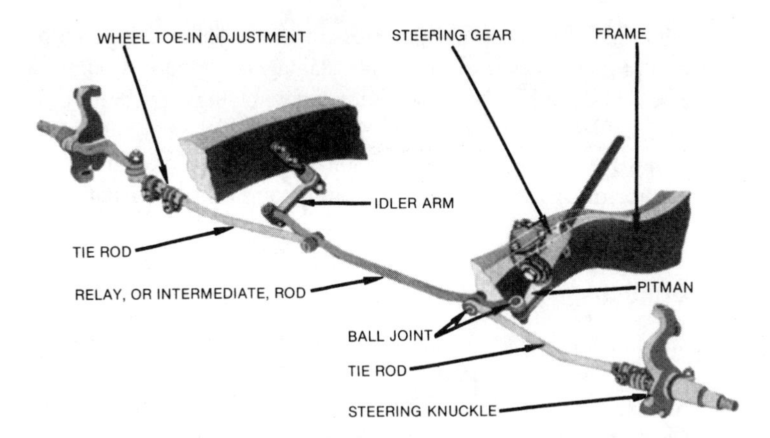

Figure 15.5 Typical steering linkage (nonpower steering). Some variation of this design is used on the majority of passenger cars today. Where rack-and-pinion steering gear is used, the rack replaces the intermediate rod. The pitman and idler arms are eliminated.

intervals have been extended with improved designs, particularly with respect to sealing, and by the use of better quality greases. A number of vehicle manufacturers have eliminated the requirement for periodic lubrication entirely, either by adopting designs that use flexible elastomeric bushings or by coating the rubbing surfaces with a wear-resistant, low friction coating. Nevertheless, most ball joints are equipped with some means of relubrication.

Two types of seal are used on ball joints designed for relubrication: umbrella seals and balloon seals. Umbrella seals permit excess lubricant to escape readily, while providing a degree of protection against contaminant entrance. Joints with umbrella seals can be

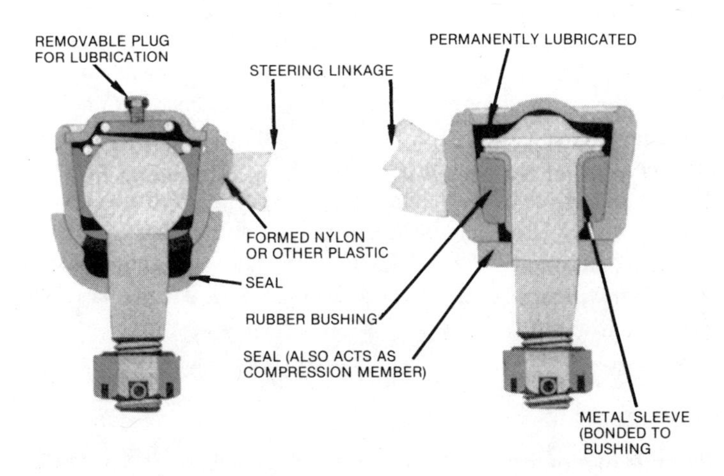

Figure 15.6 Typical steering linkage ball joints. The joint on the right is not truly a ball joint and is designed for rotation only in a single plane. It is suitable for connecting the intermediate link to the pitman and idler arm.

equipped with fittings and lubricated with usual high pressure grease-dispensing equipment. Balloon seals without a pressure-relief feature can be damaged if grease is pumped into the joint under high pressure. Thus, many joints are equipped with plugs. To relubricate the joint, the plugs must be removed and a fitting installed. Grease is then applied with a low pressure adapter to prevent damage to seal. Balloon seals with a pressure-relief feature have superseded this type of seal. With this design, lubrication fittings usually are factory-installed.

B. Lubricant Characteristics

Most suspension and steering linkage pivot points designed for lubrication are lubricated with grease. Fluid lubricants are used in some types of off-highway equipment. These applications may involve the use of central lubrication systems to replenish the lubricant automatically. Lubricants used include automotive gear lubricants of API GL-4 or GL-5 quality and semifluid greases. Rheopectic greases have also been used. These greases, which are semifluid as manufactured and dispensed, stiffen when subjected to mechanical shearing in bearings. If operating and ambient temperatures permit, greases may be preferred because of their better stay-put and sealing capabilities.

The grease used in suspension and steering linkage components, particularly when extended lubrication intervals are recommended, should provide the following.

1. Good oxidation resistance and mechanical stability
2. Protection against corrosion by both fresh water and salt water
3. Resistance to water washout
4. Good wear protection under the conditions of loading and motion involved
5. Some sealing against the entrance of dirt, water, and other contaminants
6. Resistance to pounding, leaking, and squeezing out
7. Friction-reducing capabilities to decrease steering effort and provide smoother riding characteristics

Several grease formulations are used and provide acceptable performance in these applications. Most of these formulations are similar to NLGI no. 2 in consistency and are made with an oil viscosity in the SAE 40–50 range (ISO viscosity grade 150–220). They are intended for multipurpose use. Usually, this oil viscosity is considered satisfactory for wheel bearing greases. In addition, special applications, such as wheel bearings on disk brakes, require greases that are resistant to high temperatures. Lithium complex based greases, which are used to a considerable extent, are modified by the addition of extreme pressure and antiwear additives to provide better load-carrying capabilities and lower wear rates. Colloidal or fine particle size solid lubricant materials such as molybdenum disulfide or polyethylene are often added, particularly when the grease is intended for use in passenger car suspensions and off-highway equipment pivot and hinge pins. These materials generally reduce both friction and wear.

II. STEERING GEAR

The steering gear that transmits motion from the steering wheel and steering shaft to the pitman arm of the steering linkage generally employs some type of worm gear mechanism acting against short levers or a gear segment connected to the pitman shaft. In cam and lever type steering gears, the worm is referred to as the cam. It meshes with pins on a

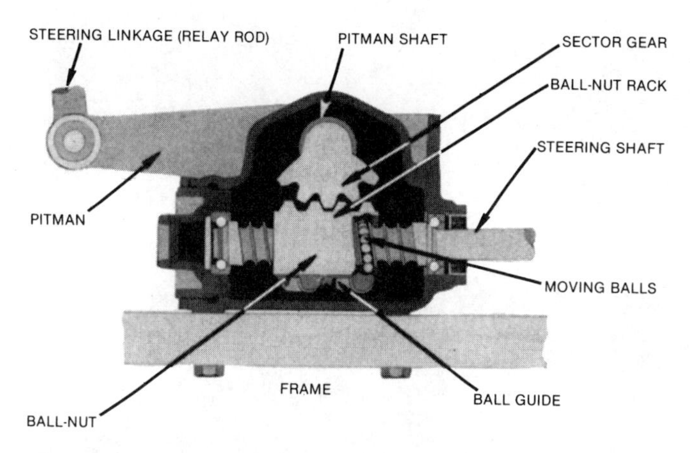

Figure 15.7 Recirculating ball type steering gear. As indicated, the recirculating balls act as a low friction mechanism for transmitting the motion of the steering shaft to the ball nut.

lever attached to the pitman shaft. The pins may be either solid or mounted on roller bearings to reduce friction. Other variations have a segment of a worm, or a complete worm wheel meshing with the worm, or a throated worm meshing with a roller in the shape of a short worm. Probably the most common steering gear design used in passenger cars is the recirculating ball type in which the worm drives a ball nut through a series of steel balls (Figure 15.7). The path of the balls includes an external guide tube that permits them to recirculate freely in either direction. This arrangement provides low friction between the steering shaft and ball nut, helping to keep steering effort low.

Most cars today are equipped with rack-and-pinion steering (Figure 15.8). In this design, a small pinion on the end of the steering shaft meshes with a rack mounted in a guide tube. The tie rods are connected directly to the rack through a mounting bracket. This simple linkage produces better ''road feel'' for the driver and may reduce steering effort.

Power steering, or power-assisted steering, is now widely applied to all types of automotive equipment. In addition to the rack-and-pinion steering systems, there are two other types generally used: the linkage type (Figure 15.9) and the integral type (Figure 15.10). Although the illustrations show passenger car applications, the principles are applicable to systems used on other equipment.

In linkage-type power steering, a double-acting hydraulic cylinder is attached between a point on the frame and the intermediate rod. A control valve is mounted so that it is operated by the pitman arm through a ball joint. Steering load deflects the spool valve, allowing power steering fluid to flow into the appropriate reaction chamber in the power cylinder. The amount of power assistance (pressure) is proportional to the spool valve deflection, which increases with the steering effort. High pressure hydraulic fluid is supplied from a small pump mounted on and driven by the engine. When the pitman moves in one direction, as a result of turning the steering wheel, the control valve allows fluid under pressure to flow into the correct end of the cylinder. This provides the hydraulic assist that forces the intermediate rod in that direction. When the steering wheel is centered, the valve is also centered and fluid can flow freely in and out of either end of the cylinder.

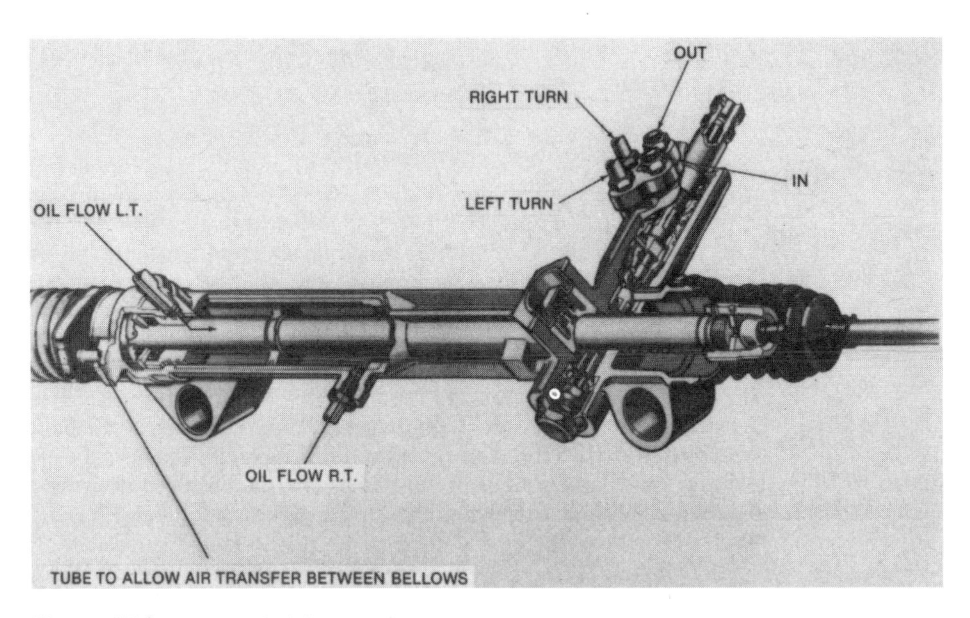

Figure 15.8 Rack-and-pinion steering gear.

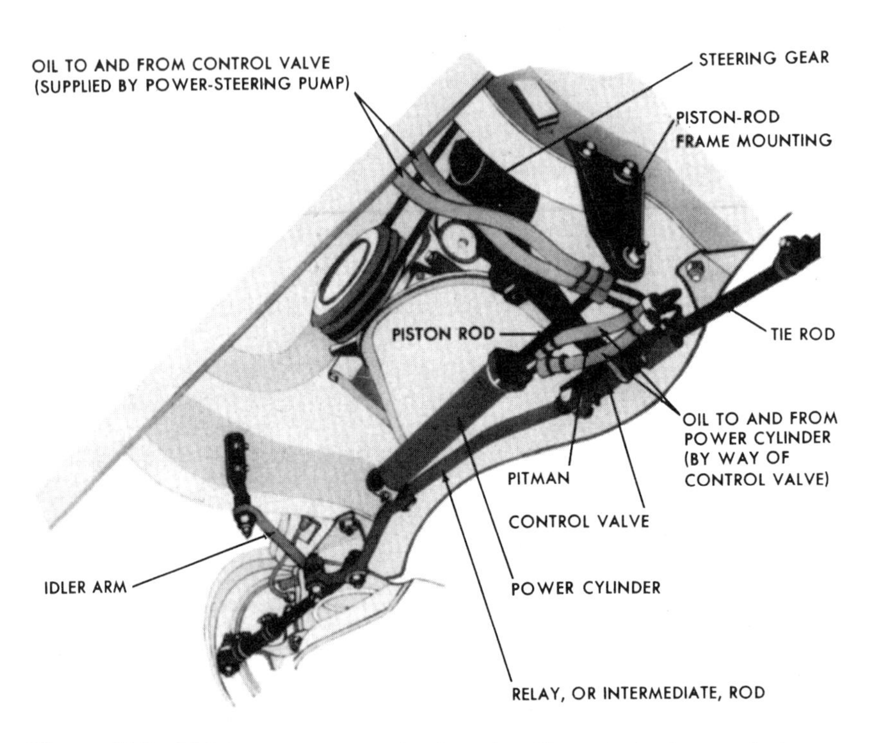

Figure 15.9 Linkage-type power-assisted steering. The control valve may be incorporated in the cylinder, rather than being a separate unit as shown. The valve opening is proportional to the force applied to the pitman, so the amount of hydraulic assist provided is proportional to the turning force applied to the steering wheel.

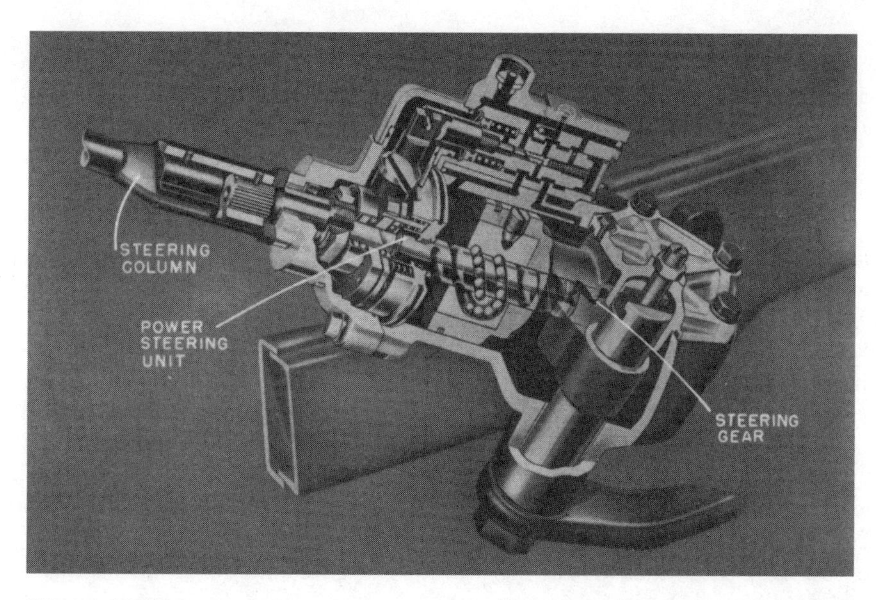

Figure 15.10 Integral-type power steering. Longitudinal forces from the worm act on reaction members to control the opening of the control valve and the amount of assist provided. In the other common design, the control valve is actuated by a torsion bar in which the amount of twist is proportional to the turning force applied to the steering wheel. With both designs, the vehicle can be steered if no power is available, since there is a mechanical connection between the steering shaft and the ball nut.

With this arrangement, the steering gear itself is a conventional type, although the ratio may be somewhat lower than is the case when the hydraulic assist is not installed. The steering gear is lubricated separately with a special lubricant. The power steering fluid only serves as the hydraulic medium.

Integral-type power steering systems are the most common type in use today. In these systems, when developed from a recirculating ball-type steering gear, the ball nut is also a double-acting piston. Hydraulic fluid under pressure is admitted to one end, or the other, by means of a control valve. The control valve can be actuated by reaction forces along the steering shaft, or by torsional forces acting on a torsion bar that rotates a spool valve. In either case, the amount the valve opens, like the power assist provided, is designed to be proportional to the amount of force applied to the steering wheel. In this way, the driver's feel of the road is retained.

With both types of power steering, the steering shaft is connected mechanically through steering gear. Thus, if fluid pressure is not available, either because the engine is not running or because a system failure has occurred, the vehicle can still be steered (with some difficulty).

In the usual automotive practice, hydraulic fluid is supplied from an integrated pump and reservoir mounted on the engine. Some use, such as in off-highway equipment, has been made of a central hydraulic system to supply a number of mechanisms such as power steering, brakes, transmissions, and equipment hydraulics. Power steering systems on farm and construction equipment may be supplied from a separate system, but generally, they are supplied from the main hydraulic system of the machine.

A. Factors Affecting Lubrication

The lubricant in the steering gear is required to lubricate the gears and bearings. It must withstand shock loads, which are transmitted to the steering gear from the wheels hitting bumps and obstructions, and it must resist the wiping action of gear teeth. At the lowest expected operating temperatures, it must not exhibit excessive resistance to motion; yet it must have enough viscosity to lubricate properly at the highest expected operating temperatures reached.

The fluid in integral power steering systems serves both as the hydraulic fluid and as the lubricant for the gears and bearings. Thus, it must perform all the functions of the lubricant in conventional steering gears with the additional requirement of being a highly stable, shear-resistant hydraulic fluid.

The conditions under which the power steering pump operates make the hydraulic fluid service relatively severe, particularly in passenger cars. The operating speed of the pump at maximum engine speed may be more than eight times what it is at idle speed; yet at idle speed, the output must be sufficient to provide whatever power assist is required. This means that at high speeds, the pump output far exceeds the requirements of the system and the excess must be recirculated internally through a flow control (or relief valve) and the reservoir. Since the quantity of fluid must be kept low to keep the system compact, fluid temperatures can be quite high, and severe shearing of the fluid may occur. Under these conditions, wear of the pump vanes or rollers may be a concern.

B. Lubricant Characteristics

Most U.S. manufacturers fill their manual or linkage-type steering gear units with semifluid grease to minimize leakage during installation and shipment. Some automotive units, however, are filled with a multipurpose gear lubricant of API GL-4 or GL-5 quality. For tractors and similar equipment, the lubricant may be a multipurpose fluid designed for the use in tractor hydraulic systems, transmissions, and final drives. Except in the latter case, the lubricant used for makeup in the field is usually a multipurpose gear lubricant of about SAE 80W-90 or 85W-140.

Automatic transmission fluids were originally used in passenger car power steering systems. However, in some cases these fluids permitted excessive pump wear, and in other cases seal compatibility problems were encountered. Special fluids, formulate to address these problems, are now recommended for most power steering systems. Generally, a small amount of automatic transmission fluid may be used as field makeup, but if a complete change-out is required, the special power steering fluid should be used. For makeup and refill, the car manufacturer's recommendations should be followed.

III. WHEEL BEARINGS

Wheel bearings for automotive equipment are of the rolling element type. Since the bearings of steerable wheels must carry considerable thrust loads in addition to the radial loads, they must be angular contact ball, taper, or spherical roller bearings. Usually, the bearings are installed in pairs so that each bearing is subjected to thrust loads in only one direction. Since bearings of nonsteerable wheels are subjected to lower thrust loads, deep groove ball bearings or cylindrical roller bearings may be used. In many cases, only a single bearing is required.

Most wheel bearings in passenger car applications are designed for grease lubrication. Most current production vehicles with front wheel drive are equipped with wheel bearings that are ''packed for life'' on assembly and generally do not require repacking during the life of the vehicle. Some wheel bearings on auxiliary automotive equipment have fittings for relubrication. The bearings of nonsteerable driving wheels that are connected with axles through differentials or final drives are generally lubricated with the gear lubricant from the drives and do not require relubrication from an external source. Oil lubrication of non–driving wheel bearings is also used to some extent on trucks, trailers, and off-highway equipment. Lubrication with oil requires careful attention to sealing to prevent leakage that might find its way onto the brakes, causing loss of braking effectiveness or failure. Use of a grease reduces the leakage tendencies and simplifies sealing.

Front wheel bearings of rear wheel drive cars, trucks, and buses are designed to be removed periodically for cleaning, inspection, and relubrication. Repacking can be done by hand or with a bearing packer. Generally, the latter is preferred because it is quicker and somewhat less wasteful of grease, and less skill is required to accomplish a satisfactory packing job. On assembly, the shaft and housing should be coated lightly with the same grease to prevent corrosion; the housing should never be filled with grease, however, as this may cause overheating due to churning. Only specially trained personnel should perform the entire wheel bearing repacking. The final adjustment of the wheel bearing, running clearance, and preload is critical, and specifications vary from one vehicle to another. Further, removal of the disk brake caliper when necessary requires mechanical expertise.

A. Lubricant Characteristics

Greases for wheel bearings are expected to provide acceptable performance over long intervals. They must resist softening and leakage, as well as hardening, which will cause increased rolling resistance and heating. They should also provide good rust and corrosion protection, as well as reduced friction and protection against wear.

In stop-and-go city driving, the use of disk brakes will result in higher operating temperatures for the wheel bearings than for drum brakes. As a result, greases with higher dropping points and better high temperature stability are generally required for vehicles equipped with disk brakes.

In the past, wheel bearing greases were specialty short fiber, sodium soap products. These single-purpose greases are disappearing from use and have been replaced with multipurpose automotive greases. Most of the types of multipurpose greases mentioned as suitable for suspension and steering linkage components are being used for wheel bearing lubrication. These are usually of the NLGI no. 2 or No. 3 consistency. Where higher temperature capability is required for vehicles equipped with disk brakes, lithium complex or comparable products using other thickeners with higher temperature capability may be used. For consolidation purposes and to eliminate possible cross-mixing of products, it might be advisable to standardize on one product that meets all the requirements for a given application.

Oil-lubricated wheel bearings usually are lubricated with a multipurpose gear lubricant of the SAE 90 or 140 (ISO 150 or 320) viscosity. Multiviscosity gear lubricants such as 80W-90 or 85W-140 may also be used.

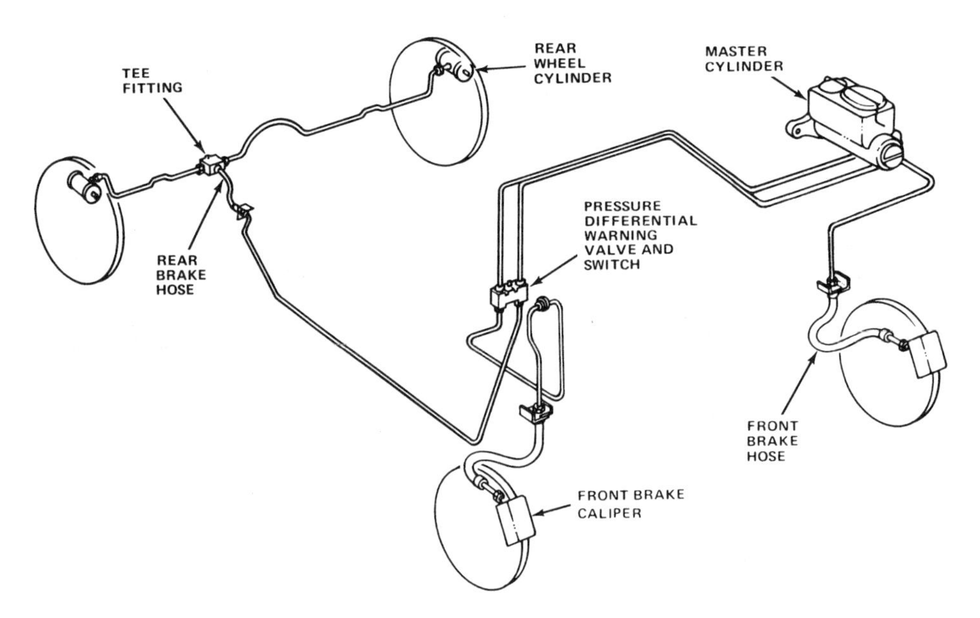

Figure 15.11 Conventional braking system.

IV. BRAKE SYSTEMS

Most service brake systems are hydraulically operated. There are two types generally in current service-conventional systems that use either disk brakes, drum brakes, or a combination of these (Figure 15.11) and antilock braking systems (ABS), which also use these components but are computer-controlled to avoid wheel lockup upon hard braking (Figure 15.12).

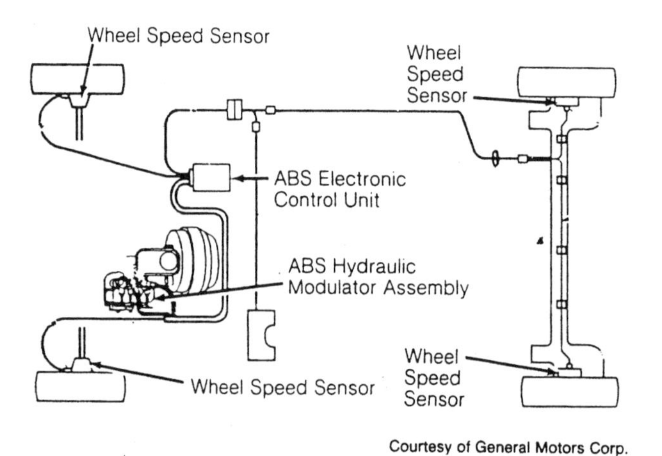

Courtesy of General Motors Corp.

Figure 15.12 Antilock braking system.

U.S. National Highway Traffic Safety Administration (NHTSA) regulations require a dual or two-sided system to ensure that some braking capability is retained if a failure occurs on either side of the system. Common practice is to operate the front wheel brakes from one side of the system and the rear wheel brakes from the other side. Also used are ''dual diagonal'' systems in which a front wheel and the diagonally opposite rear wheel are paired.

Operation of the brake pedal forces fluid from the master cylinder under considerable pressure to the wheel cylinders or disk brake calipers. In turn, the wheel cylinders act through connecting links to force the brake shoes out against the surfaces of the brake drums, or to force brake pads against the sides of the brake rotor (disk). A mechanical linkage to the rear wheel brakes usually is provided to serve as a parking brake, although some current production vehicles use a hydraulic interlock to actuate the parking brake system.

Disk brakes are widely used on front wheels of passenger cars and are being installed on all four wheels on a number of current production vehicles. Disk brakes are more resistant than drum brakes to brake fade or loss of effectiveness due to heat buildup with repeated application, and they also usually provide smoother stopping. Since higher application pressure is required, disk brakes are nearly always equipped with a power assist system to reduce the amount of pedal pressure required.

Generally, power assist systems involve a double-acting piston, or diaphragm, coupled to the master cylinder piston. Normally, both sides of the piston are under vacuum supplied by engine manifold vacuum through a reservoir. When the brake pedal is depressed, a valve is opened to allow atmospheric pressure to act on one side of the piston, assisting the pedal action in moving the master cylinder piston.

A. Antilock Braking Systems

There are some disadvantages to the conventional braking systems. If the wheels can lock up during hard stops, steering becomes more difficult and the stopping distances required are increased. ABS brakes address both these areas by regulating hydraulic pressures to each individual wheel to prevent lockup under hard braking. The ABS is designed to be activated when the vehicle is moving above a predetermined speed (generally >3 mph) and hard braking occurs. Under other conditions of low speed or normal braking, the response of the ABS is similar to that of conventional brake systems.

ABS components consist of a master cylinder, hydraulic lines, brakes, and booster similar to conventional brake systems. In addition are individual wheel speed sensors, hydraulic pressure modulators, and solenoid valves; at the heart of the system is the ABS control unit (Figure 15.12). The speed sensor for each wheel generates a voltage signal that is sent to the control module. The control module converts this signal to wheel rotational speed. When the control unit senses wheel lockup under hard braking, it activates the affected wheel's braking solenoid-operated valve, which then regulates the hydraulic pressure to control braking without the occurrence of lockup.

It is noteworthy to add that when the ABS is activated, the driver feels pulsations in the pedal due to regulation of the fluid needs and pressure within the system. A rapid clicking-type noise also is heard, signaling the operation of the solenoid valves, and there may be some vehicle body vibration due to the cycling of the individual wheel brakes (front to rear), which causes suspension movements as brake pressures are modulated.

The brake pedal on ABS should never be ''pumped'' because the control unit essentially does this function.

B. Other Braking Systems

Hydraulic retarders for dynamic braking are used on some heavy vehicles equipped with automatic transmissions. A torque converter does not transmit power well in the reverse direction, so with an automatic transmission, the engine cannot be used as effectively for dynamic braking. With a hydraulic retarder, which operates as a torque converter coupled in the opposite direction, energy from the wheel is converted into heat energy in the fluid in the retarder and then dissipated to the atmosphere.

Another type of assistance, called ''power boost'' or ''hydroboost,'' is a hydraulically operated power brake. The hydraulic booster consists of an open center spool valve and a hydraulic cylinder combined in a single housing. A dual master brake cylinder bolted to the booster is actuated by a push rod projecting from the booster cylinder. The power steering pump provides the hydraulic fluid under pressure to the booster cylinder. The master brake cylinder and the braking system use conventional brake fluid.

Oil-immersed or ''wet'' brakes are used on some crawler and wheeled tractors as well as on many other types of tractor and heavy construction and mining equipment. These brakes operate submerged in the lubricant for the final drive. Actuation is hydraulic, using fluid from the main hydraulic system on the machine. These systems are discussed in more detail in Chapter 16 (Section IX: Multipurpose Tractor Fluids).

C. Fluid Characteristics

Primarily, the fluid in brake systems is a hydraulic fluid. It must also, however, lubricate the elastomer seals in the wheel cylinders and calipers as well as protect system metal parts against rust and corrosion. In practice, it has been almost impossible to exclude moisture completely from brake systems. To prevent this moisture from collecting and freezing in cold weather or causing rust and corrosion, which can cause brake failure, the usual approach is to use brake fluids that are miscible with water. Petroleum-based fluids do not meet this requirement; so most brake fluids are based on glycols (alkylene glycol and alkylene glycol ethers).

In operation, considerable heat is transmitted from the friction surfaces of the brakes to the fluid. If this raises the temperature of the fluid sufficiently to cause vaporization of some of the fluid, braking effectiveness will be reduced because of the compressibility of the vapor. The problem may be more severe with disk brakes because the smaller areas and higher pressures of the friction pads may result in higher operating temperatures. With higher brake pad temperatures, more heat is transferred through the wheel calipers to the brake fluid.

The absorption of moisture by a brake fluid lowers the boiling point, since water boils at a lower temperature than the glycols used as a base for brake fluids. For this reason, brake fluid specifications include both an equilibrium reflux boiling point, which is the boiling point under reflux conditions of the pure fluid, and a wet equilibrium reflux boiling point, which is the boiling point of the fluid when it is contaminated with a specified amount of water. The amount that these boiling points can be raised is somewhat restricted because higher boiling point glycols also have higher viscosities at low temperature and may not provide satisfactory brake performance in cold weather.

The most widely accepted brake fluid specification is the U.S. Federal Motor Vehicle Safety Standard (FMVSS) No. 116 for grade DOT 3 fluids. Fluids meeting this U.S. Department of Transportation standard generally are suitable for all normal brake systems designed for nonpetroleum fluids. Some manufacturers specify a higher boiling point fluid for vehicles with disk brakes; and higher boiling fluids may also be required in certain types of severe service, such as mountain operations, particularly in hot climates, and for road racing.

The SAE Standard J1703 (Motor Vehicle Brake Fluid) is generally similar to the DOT 3 standard but somewhat less restrictive in its requirements. Most brake fluids will meet the DOT 3 and the SAE J1703 requirements.

Where mineral oil or silicon-based fluids are used, usually the fluids are developed specifically to meet requirements set out by the brake system manufacturer. These fluids must meet the requirements of FMVSS 116 and SAE J1703. Silicone brake fluids that meet the DOT 5 standard are being used in special application vehicles as well as in many motorcycle brake systems. The DOT 3 and DOT 5 brake fluids are not compatible and should never be mixed.

In service, it is important that every precaution be taken to keep brake fluids as clean and free of moisture as possible. Containers should be kept sealed when not in use and should never be reused. Master cylinder reservoirs should be kept filled to the proper level, since this minimizes the amount of breathing that can occur. Reservoir caps should be replaced properly after the fluid level has been checked or fluid added. A number of manufacturers use translucent plastic reservoirs to permit observation of fluid levels without removing the cap. This method reduces the possibility of introducing contaminants into the system.

V. MISCELLANEOUS COMPONENTS

The ''fifth wheel'' of tractor–trailer combinations requires effective lubrication to assure proper vehicle control. The grease must have good adhesion and water-resistant characteristics to resist wiping and removal by exposure to the elements. A variety of greases are used; however, the best results are obtained with greases containing antiwear and extreme pressure additives. The addition of molybdenum disulfide to the grease is generally beneficial.

16

Automotive Transmissions and Drive Trains

In automotive equipment, the power developed by the engine must be transmitted to the drive wheels to propel the vehicle. This is accomplished by the power train, the elements of which vary from application to application. In probably its simplest form, a bicycle equipped with an auxiliary engine, the power train may consist of only a belt drive with an idler pulley that can be actuated to engage or disengage the drive. At the other extreme, the power train may consist of some combination of a clutch or coupling, a transmission, transfer case, interaxle differential, constant velocity joints, drive shafts, front and rear differentials—possibly in tandem at the rear, driving axles—and planetary reducers at the wheel ends of the axles. Almost any arrangement between these two extremes will be encountered in most types of automotive equipment.

In addition to transmitting the power from the engine to the drive wheels, the power train performs several other functions. For example, the power train provides the following:

1. Mechanism for engaging and disengaging the drive so that the vehicle can be started and stopped with the engine running
2. Torque multiplication (speed reduction) so that sufficient torque is available to start from rest, accelerate, climb hills, and pull through soft ground
3. Mechanism for reversal of direction
4. Mechanism that allows one wheel to be driven at a higher or lower speed than the other when the vehicle is negotiating curves and turns
5. Change in direction of torque and power flow to couple a longitudinally mounted engine to the traverse drive axle (not needed in vehicles with traversely mounted engines)

The components of power trains can be considered conveniently starting from the engine and progressing toward the drive axle.

I. CLUTCHES

To engage and disengage the various gear ratios in a mechanical transmission, provision must be made to disconnect the engine from the power train. This is accomplished with a clutch.

Most road vehicles with mechanical transmissions are now equipped with a single-plate, dry disk clutch. A machined surface on the flywheel serves as the driving member. The driven member is usually a disk with a splined hub that is free to slide lengthwise along the splines of the transmission input shaft (clutch shaft) but drives the input shaft through these same splines. About 65% of the clutch plate area is faced on both sides with friction material. For example, a typical passenger car clutch plate with a 15 cm (10 in.) diameter would have a band of friction material about 6 cm (4 in.) wide along its outer portion. When the clutch is engaged, the pressure plate (driven member) is clamped against the driving member by a diaphragm-type spring or an arrangement of coil springs (Figure 16.1). The entire mechanism usually is called the clutch plate and cover assembly. To disengage the clutch, pressure is applied through mechanical linkage or a hydraulic system on the end of the throwout fork that pivots on a support in the clutch housing. When release pressure is applied, the fork transmits the force to the release levers in the cover assembly, which compresses the springs and retracts the pressure plate from the driven member.

On models with hydraulic clutch activator, the system reservoir should be kept filled to the correct level. Usually, conventional brake fluid is specified. Always follow manufacturer's recommendations. The only lubricated part of this type of clutch is the throwout bearing. In most cases, these bearings are "packed for life" on assembly and do not require periodic relubrication. In a few instances, these bearings are equipped with a fitting and require periodic lubrication, usually with multipurpose automotive grease. If fitted bearings are to be greased, care must be exercised to avoid getting grease on clutch faces for that would result in slippage and excessive heating. Also, various pivot points in the actuating linkage may require periodic lubrication, with a small amount of either engine oil or multipurpose automotive grease.

Multiple-plate clutches, usually of the oil-immersed or "wet" type, are used in some tractors and off-highway machines. With these clutches, the frictional properties of the fluid surrounding the clutch are extremely important if the clutch is to engage smoothly, resist slipping, and provide extended service life. These fluids are discussed in this chapter.

Single-plate, double-acting, dry disk clutches are used with some torque converter transmissions for buses and similar application. Two driven members are used, one on each side of the driving member. When engaged in one direction, the clutch connects the drive to the torque converter, which in turn drives the input shaft of the transmission. When engaged in the other direction, it connects the drive to a through shaft, which provides a mechanical drive to the transmission, to either the input or the output shaft, depending on the arrangement. Again, the only lubrication required is for the throwout bearing, which is located so that it is lubricated by the torque converter fluid.

II. TRANSMISSIONS

The transmission has three primary functions:

1. To provide a method of disconnecting the power train from the engine so that the vehicle can be started and stopped with the engine running

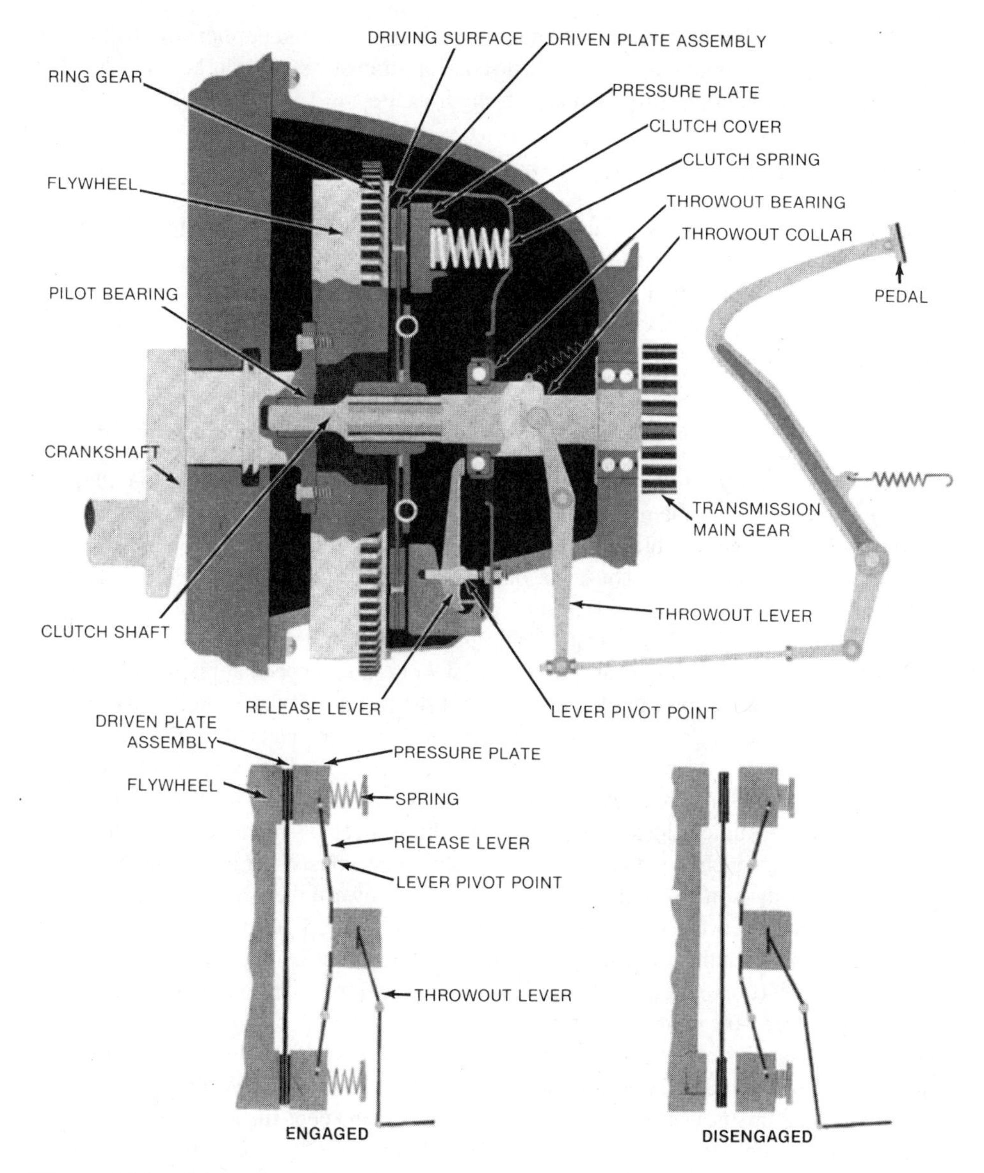

Figure 16.1 Single-plate, dry disk clutch.

2. To provide torque multiplication when greater driving torque is required at the wheels than is available from the engine
3. To provide a method of reversing the drive

Since torque multiplication is accompanied by speed reduction at the output end, the transmission also permits the operator to select different travel speeds for any given engine speed.

There is probably more variation in transmission design and application than in any other automotive component. For ease of discussion, transmissions can be considered to be mechanical, automatic, semiautomatic, or hydrostatic.

A. Mechanical Transmissions

A mechanical transmission is an arrangement of gears, shafts, and bearings in a closed housing such that the operator can select and engage sets of gears that give different speed ratios between the input and output shafts. In most cases, a set of gears that can be engaged to drive the output shaft in the opposite direction is also included. For a constant power, torque increases as speed is decreased; the transmission provides a series of steps of torque multiplication.

In an elementary sliding element transmission (Figure 16.2) one of each pair of gears is splined onto its shaft in such a way that it can be moved along it by a shift fork into and out of mesh with its mating gear. Drive is from the input shaft (also called the clutch shaft) through the main gear to the countershaft. The output shaft ends in a pilot bearing in the main gear, which is free to revolve at a speed or in a direction different from those of the input shaft. Thus, the output shaft is driven from the countershaft by whichever pair of gears is engaged, or directly from the main gear if the direct drive gear is engaged with the internal gear in the main gear.

Sliding element transmissions are used now only in low speed applications, such as tractors. In this application, the clutch must be disengaged and the vehicle must be at a complete stop before the gears can be engaged or the gear ratio changed. For other applications, syncromesh or synchronized transmissions are used. In this type of transmission, all gears are always in mesh, except the reverse gear. One of each pair of gears is free to revolve on its shaft unless locked to it by a clutching mechanism called a synchronizer. The synchronizer, which is keyed or splined to the shaft, consists of a friction clutch and a dog clutch. As the shift fork moves the synchronizer toward the gear, the friction cones make contact first to bring the shaft to the same rotational speed as the gear. The outer rim of the clutch gear then slides over its hub, causing a set of internal teeth to engage with a set of teeth (dogs) on the side of the gear. This then provides a positive mechanical connection between the gear and shaft.

Usually, synchronizers are equipped with a blocking (also called baulking) system to prevent engagement of the dog clutches until the gear and shaft speeds are fully synchronized. Generally, this is a spring-loaded mechanism, which keeps the teeth on the synchronizer from lining up with the teeth on the gear as long as there is any slip in the friction clutch.

Mechanical transmissions are built with up to about six gear ratios. If more ratios are required, as in the case of heavy trucks equipped with diesel engines, they are usually obtained by means of a two- or three-speed auxiliary transmission mounted behind the main transmission. With this arrangement, for each ratio in the main transmission there are two or three ratios in the auxiliary; for example, a four-speed main transmission with three-speed auxiliary becomes a 12-speed transmission. Heavy-duty transmissions are often built with twin countershafts to decrease gear tooth loading.

Sliding element transmissions are built with straight spur gears. Synchromesh transmissions for over-the-road vehicles are usually built with helical gears, both because they provide greater load-carrying capacity and because they operate more quietly. Transmissions for off-highway equipment may be built with either type of gearing.

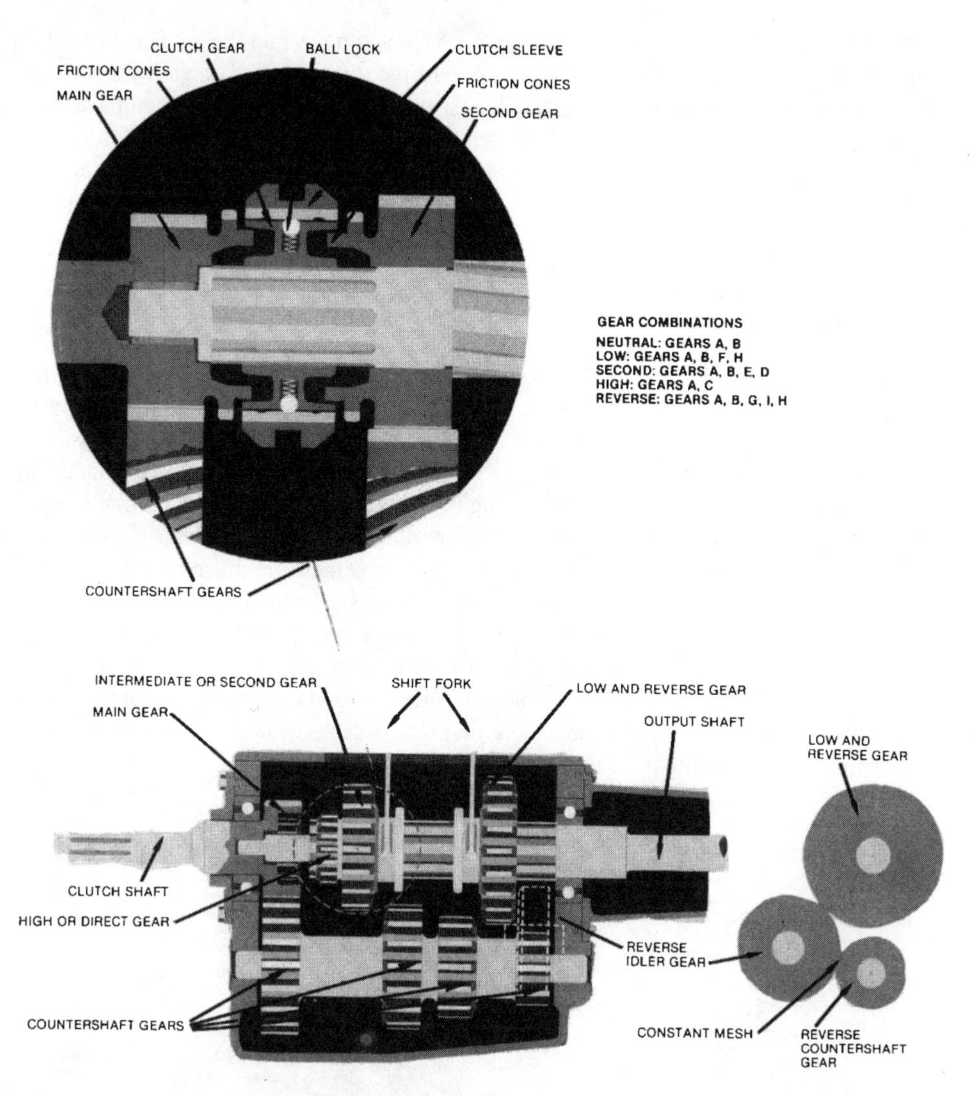

Figure 16.2 Elementary sliding element mechanical transmission. The shift forks move the gears into and out of mesh along with splined main shaft to change the output gear ratio.

B. Automatic Transmissions

Early passenger car automatic transmissions were built with a fluid coupling and a hydraulically operated power shift gearbox. The fluid coupling permitted enough slip with the engine idling to permit the vehicle to be stopped with the gears engaged. Power transfer efficiency was also good. However, since a fluid coupling does not multiply torque, the gearbox required four forward speeds to provide a smooth progression of gear ratios, and this resulted in considerable complexity. As efficient hydraulic torque converters were developed, they replaced the fluid couplings. In the late 1960s through the early 1980s, the gearboxes were standardized on an arrangement with three forward speeds and a reverse speed. Current production automotive vehicles have now been standardized on four forward speeds with an overdrive gear. Overdrive improves fuel economy and results in less engine noise at highway speeds. Truck automatic transmissions may have more forward speeds and may also have an arrangement to lock out or bypass the torque converter when the transmission is in any gear except first or reverse. Transit coach transmissions may have only a drive through the torque converter or direct drive. All these transmissions are sometimes referred to as ''hydrokinetic'' transmissions, since engine power is transmitted by kinetic energy of the fluid flowing in the torque converter.

1. Torque Converters

The simplest single-stage torque converter consists of three elements: a centrifugal pump, a set of reaction blades called a stator, and a hydraulic turbine (Figure 16.3). These three elements are installed inside a case filled with a hydraulic fluid. The pump is driven by the engine, and the turbine drives the input shaft of the gearbox. The pump blades are shaped so that they discharge the fluid at high speed and in the correct direction to drive the turbine. As the fluid flows out of the turbine, it strikes the fixed stator blades and is redirected into the inlet side of the pump, where any velocity it still retains is added to the velocity imparted to the fluid by the pump. With this arrangement, most of the power delivered to the pump is available to drive the turbine (some power is lost owing to fluid friction), and as long as the turbine is running at a lower speed than the pump, torque multiplication will occur. Most single-stage torque converters are designed for maximum torque multiplication ratio of slightly more than 2:1 which, at maximum load conditions, occurs in ''stall'' conditions when the turbine is stationary.

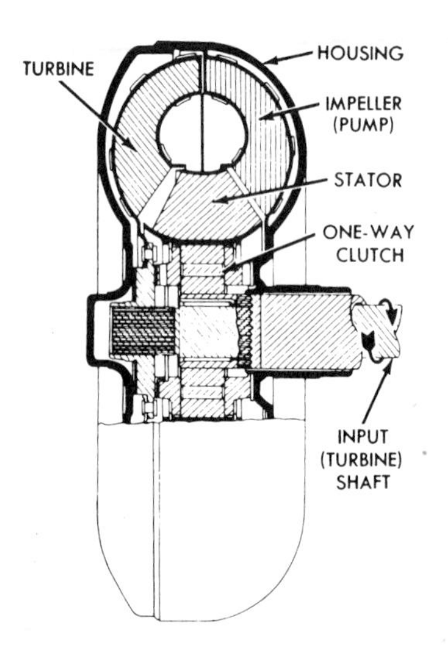

Figure 16.3 Three-element torque converter.

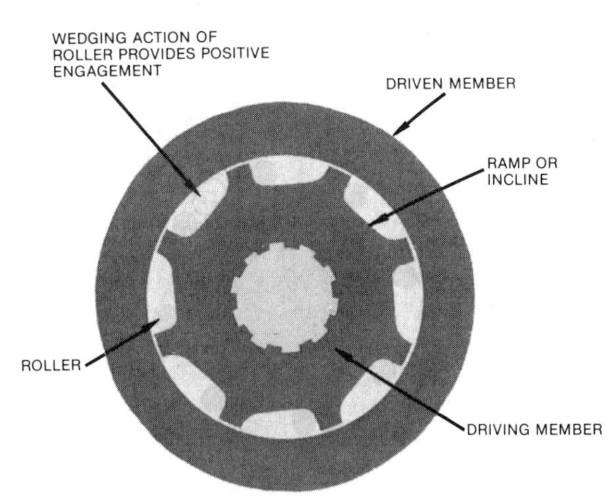

Figure 16.4 Overrunning clutch. When a clockwise force is applied to the movable member, the rollers wedge on the ramps to prevent rotation. When the force is released, the rollers move back down the ramps, permitting the movable member to rotate freely in the clockwise direction.

Torque converters can be built with more than one stage (i.e., additional pumps and stators in pairs) to give greater torque multiplication. However, this is done less frequently now, and the majority of units being built are single stage.

A torque converter does not transmit power very efficiently when the speed of the turbine reaches approximately the speed of the pump. To improve this power transfer efficiency, the stator is mounted on a one-way or overrunning clutch (Figure 16.4). With this addition, when the coupling stage is reached, the stator revolves (free wheels) with the turbine and the whole assembly performs as a fluid coupling.

The efficiency of power transfer through the converter when it is operating normally in the coupling phase is fairly satisfactory. However, to gain some percentage points in fuel economy, a number of car manufacturers have introduced torque converter ''lockup'' devices that eliminate all slippage in the coupling phase. The lockup mechanisms are designed to be effective when the transmissions are in direct drive and converter torque multiplication is not required.

At low engine speeds, the torque transmitted by a torque converter is so low that it either will not move the vehicle at all or will cause only a small amount of creep. This feature permits the vehicle to be stopped without disconnecting the engine from the power train.

2. Planetary Gears

Passenger car automatic transmissions are built with planetary and neutral gear sets to provide the additional torque multiplication, reversal of direction, and neutral. This type of gearing is also used in the automatic transmissions trucks and heavy equipment. Planetary gear-sets have the following advantages for these applications.

1. Ratio changes and reversal of direction can be accomplished through these constant mesh gears by locking or unlocking various elements of the gearset.
2. The gears are coaxial; thus they provide a compact arrangement.
3. Good load-carrying ability can be obtained from a relatively small gear set. The coaxial construction of planetary gears carries most of the operating loads. This allows the use of thin, lightweight aluminum die-cast housings because extreme mechanical loads will not be encountered. A simple planetary gear set is shown in Figure 16.5.

A single planetary gear set can provide direct drive, two stages of forward speed reduction, and reverse and can be operated in the overdrive phase, as well. By locking the sun gear and using the planet carrier as input, one can increase the annular gear output rotation speed. On earlier model cars, this overdrive effect for increasing fuel economy was very successful, and U.S. car companies now provide an automatic transmission with the overdrive feature. To provide the forward speeds used on most passenger car automatic transmissions today, generally two planetary gear sets or a compound planetary gear set with two sets of planet pinions and carriers are used. A typical three-speed transmission is shown in Figure 16.6.

3. Transmission

The gearbox of an automatic transmission is a ''power shift'' gearbox. That is, it can be engaged, or the gear ratio changed, while engine power transmitted by the torque converter is being applied continuously to the input shaft. These operations are performed by engaging and disengaging clutches in the drive lines to various planetary elements and by

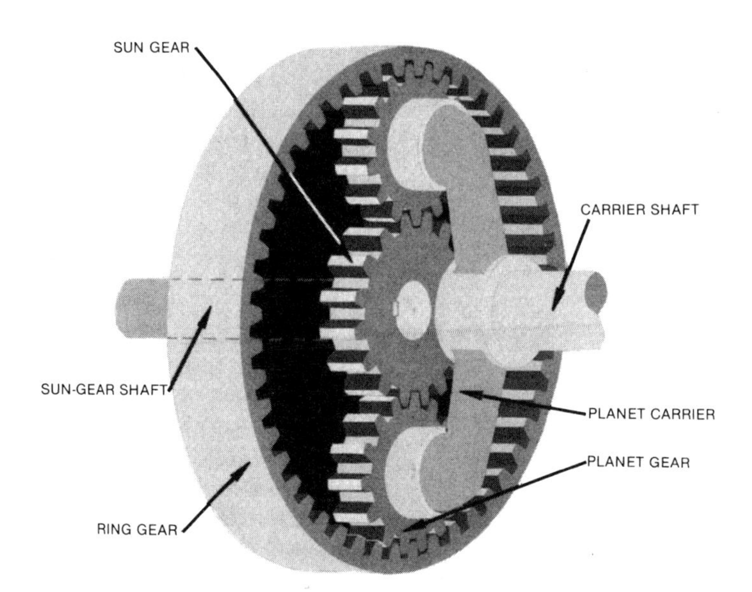

Figure 16.5 Planetary gear set. This arrangement uses only two planet gears. Automatic transmission gear sets commonly have either three or four planet gears.

applying and releasing brakes (sometimes called ''bands'') to lock or unlock elements. The clutches and brakes are operated by hydraulic servomechanisms, which are controlled by a complex valve arrangement. Hydraulic pressure to operate the servos is provided by an auxiliary pump, usually of the internal–external gear type, mounted in the front end of the gearbox.

In operation, the operator selects a driving range, and the gear ratio changes are made automatically within that range. Basically, speed-sensing and load-sensing devices on the output shaft of the gearbox determine these shifts. Throttle position and, in some models, engine vacuum are used to modulate the shift speeds. The farther the throttle is depressed, the higher the speed that shifts will occur. Normally, a forced downshift (sometimes referred to as a passing gear) is provided, permitting the operator to downshift the transmission for additional acceleration by ''flooring'' the accelerator pedal.

C. Semiautomatic Transmissions

A number of arrangements are used to reduce the operator effort required to select and engage gear ratios. Clutch operation may be made automatic, or the gearbox may be arranged so that a clutch is not needed. Since gear ratio selection and engagement are still performed by the operator, these arrangements are considered to be semiautomatic. Although used in the past, semiautomatic transmissions are not currently used in passenger car applications except in special applications.

In one type of semiautomatic transmission, a torque converter is installed ahead of an electropneumatic clutch and a three-speed, manually shifted gearbox. Electrical contacts in the shift lever knob are closed when slight downward pressure is applied to the knob. This completes the circuit to the vacuum valve, allowing vacuum from a vacuum storage

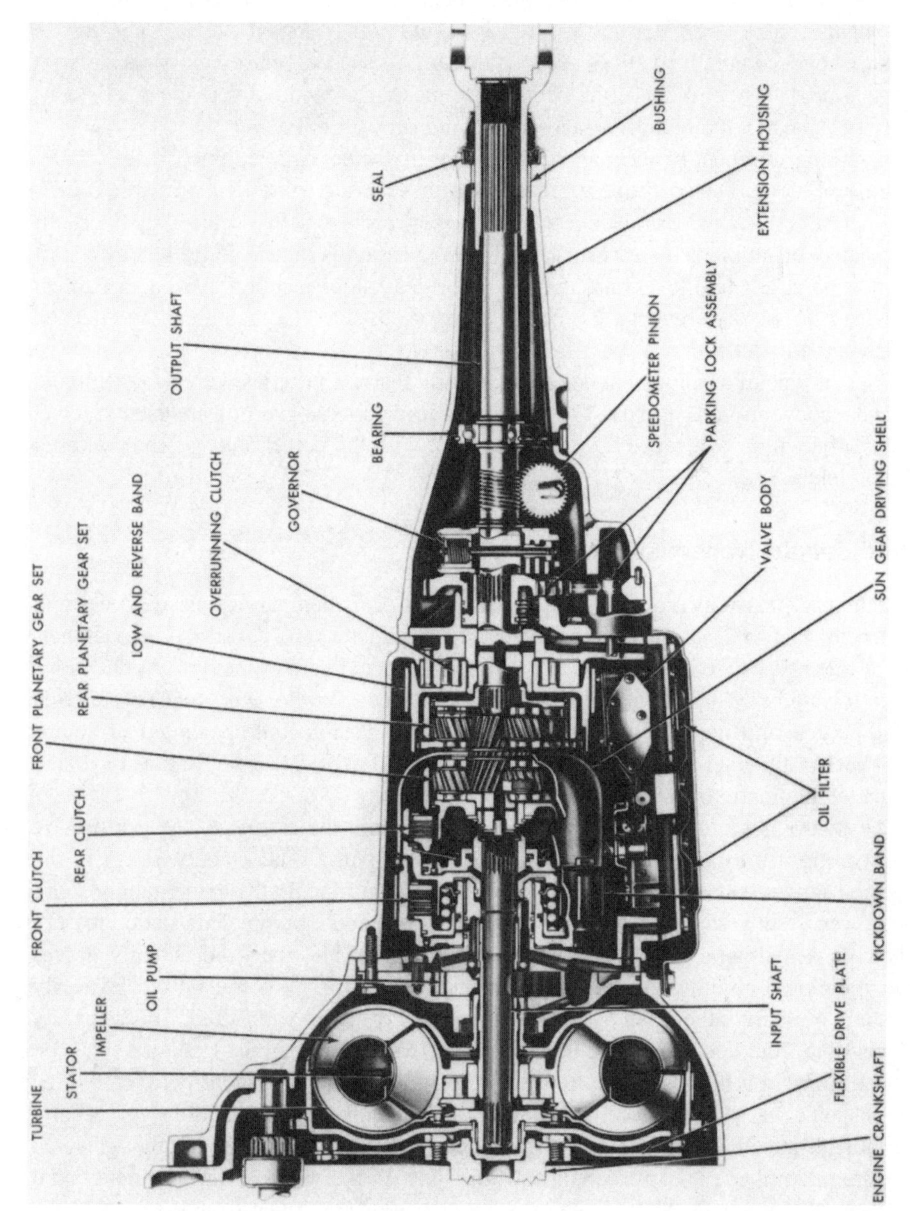

Figure 16.6 Cutaway of typical automatic transmission.

tank to act on the servo and disengage the clutch. Gearshifts can then be made in the normal manner, and when the knob is released, the clutch engages again. The torque converter provides enough multiplication so that a three-speed gearbox is adequate with a small engine. It also eliminates the need for some shifting and helps to cushion shocks that may occur when the clutch engages at the end of a shift.

Increasing numbers of farm and construction machines are equipped with ''power shift'' transmissions. The gearboxes of these transmissions are somewhat similar in principle to synchromesh transmissions, except that the synchronizers are replaced by hydraulically operated, oil-immersed clutches. The hydraulic circuit is in turn controlled by a shift lever. In some cases, two levers are used, one for gear ratio selection and one for direct shifting from forward to reverse. No clutch is required, since the gears can be engaged and disengaged under power.

Another type of arrangement has a power shift gearbox in series with a conventional clutch and a conventional gearbox. The clutch and conventional gearbox are used to select a range. Shifts within that range may then be made with the power shift gearbox without using the clutch.

D. Hydrostatic Transmissions

Hydrostatic transmissions are now used on many self-propelled harvesting machines and garden tractors, as well as significant numbers of large tractors and construction machines. Drives of these types are also used in many small lawn and garden tractors. Applications in trucks for highway operation are also being developed. In the sense that no clutch is used and no gear shifting is involved, this type of transmission could be called an ''automatic,'' but in all other respects the hydrostatic transmission has no similarity to the hydrokinetic automatic transmission.

The hydrokinetic transmission transfers power from the engine to the gearbox by first converting it into kinetic energy of a fluid in the pump. The kinetic energy in the fluid is then converted back to mechanical energy in the turbine. In the hydrostatic system, engine power is converted into static pressure of a fluid in the pump. This static pressure then acts on a hydraulic motor to produce the output. While the fluid actually moves through the closed circuit between the pump and motor, energy is transferred primarily by the static pressure rather than by the kinetic energy of the moving fluid. The relatively incompressible fluid acts much like a solid link between the pump and motor.

The pump in a hydrostatic system is of the positive displacement type. It may be either constant or variable displacement, but for mobile equipment applications, it is usually is a variable displacement type. Axial piston pumps are the most common, although some radial piston pumps are used for small transmissions. In the variable displacement, axial piston pump (Figure 16.7), the cylinder block and pistons are driven from the input shaft. Piston stroke, pump displacement, and direction of fluid flow are controlled by the reversible swash plate, which in this case is moved by a pair of balanced, opposed servopistons. The servopistons are, in turn, controlled by a speed control lever. On smaller units, where the forces acting on the swash plate are not as great, the speed control lever has direct control over swash plate position. With radial piston pumps, a movable guide ring is used to control piston stroke instead of a swash plate.

When the speed control lever of the pump in Figure 16.7 is in neutral, the swash plate is perpendicular to the pistons and no pumping occurs. As the speed control lever is moved in one direction, the swash plate is tilted, piston stroke is gradually increased,

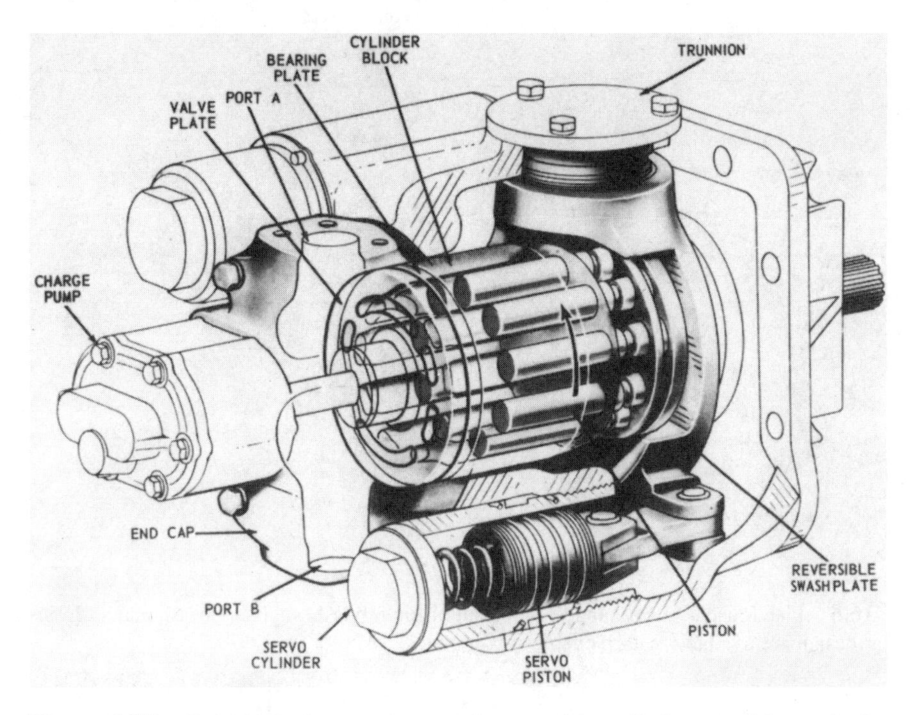

Figure 16.7 Variable displacement pump. The servopistons tilt the reversible swash plate on the trunnion to vary the displacement from maximum in one direction through zero to maximum in the other direction. In many cases, motion in the reverse direction is limited, with the result that the maximum reverse speed is half the maximum forward speed or less.

and fluid is pumped from one of the outlet ports. If the speed control lever is moved in the opposite direction, the swash plate is tilted in the opposite direction, and the piston stroke is moved 180 degrees around the case. Fluid is then pumped from the other outlet port. In combination with a reversing motor, this permits a continuously variable range of speeds from full forward to full reverse with only the single lever for control.

The motor in a hydrostatic system can be any type of positive displacement hydraulic motor. Axial piston motors usually are used for larger drives, and are also used for some smaller drives. Both gear motors and radial piston motors are used for low power drives. The motor is usually of the fixed displacement type (Figure 16.8) but may be a variable displacement unit. As noted, the motor is reversible with the direction of rotation dependent on the direction of flow in the closed-loop circuit to the pump.

In addition to the pump and motor, connecting lines, relief valves, and a charge pump are required. If the pump and the motor are in the same housing the connecting lines may be passages or, if the motor is mounted away from the pump, hoses may be used. The charge pump provides initial pressurization of the motor and replaces any fluid lost as a result of internal leakage. On small tractors it may also be used to supply fluid for remote hydraulic cylinders. A typical hydrostatic transmission schematic diagram was shown in Chapter 7 (Figure 7.26).

Various arrangements of the pump and motors are used. A variable displacement pump with a variable displacement motor may be used. Thus, the swash plates must be

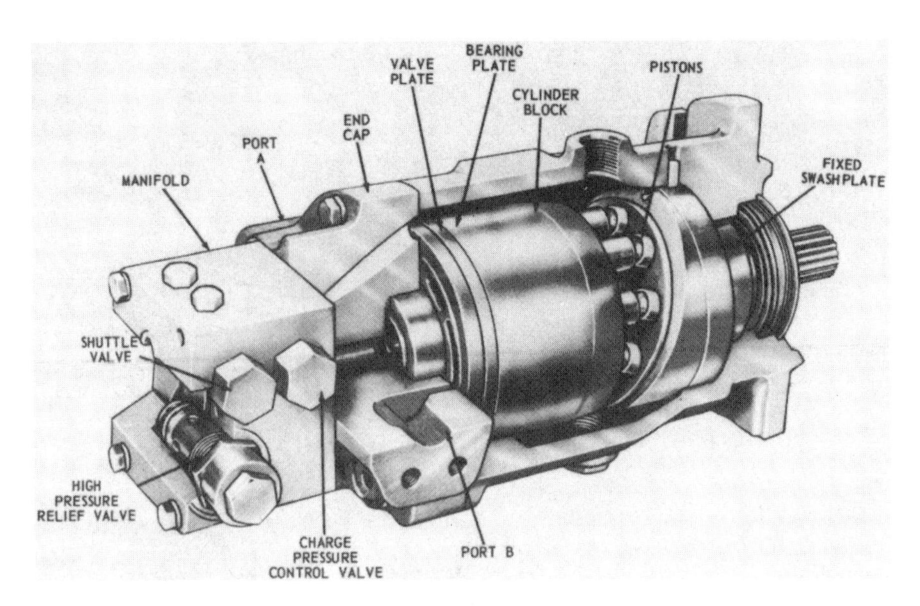

Figure 16.8 Fixed displacement pump. This axial piston motor has a fixed swash plate. Motors of similar design are available with movable swash plates.

linked and synchronized so that the motor displacement decreases as the pump displacement is increased. Because motor displacement is maximum when pump displacement is low, motor speed will be low and torque will be high. Conversely, motor displacement will be minimum when pump displacement is maximum, so the maximum speed will be high. The arrangement gives high starting torque and the widest range of speeds for any given size of pump and motor.

Another variation has a variable displacement pump and a two-piston swash plate or guide ring on the motor. The latter is controlled by a range lever. In the low range, motor displacement is greater, thus starting torque is higher and maximum speed lower.

Most drives in mobile-type equipment have a variable displacement pump in combination with a fixed displacement motor. This type of circuit gives a constant torque output, with the power output increasing as the pump displacement is increased.

The fact that the pump and motor do not need to be connected directly together permits considerable flexibility in arrangement. The pump may be connected directly to the engine output shaft and motors located at the driving wheels. In another arrangement, two pumps and two motors may be used, with each pump and motor driving one wheel. One wheel can then be driven forward with the other in neutral or reverse for spin turns.

One of the main disadvantages of hydrostatic drives is that they permit the operator to select any travel speed up to the maximum without varying the engine speed. The engine can be operated at governed speed to provide proper operating speed for elements such as the threshing section of a combine, but a full range of travel speeds is available to adjust to terrain or crop conditions. Operation is also greatly simplified.

E. Factors Affecting Lubrication

The differences in lubrication requirements of the various types of transmission require separate consideration of the factors affecting lubrication. Transaxle units are discussed later (Section IV).

1. Mechanical Transmissions

The elements in mechanical transmissions requiring lubrication are the bearings, gears, and sliding elements in the synchronizers. Bearings may be either plain or rolling element. As noted, gears are usually either straight spur or helical; gear loads are moderate to heavy. Normally, lubrication is by bath and splash, but some large transmissions may have integral pumps to circulate the lubricant.

Most mechanical transmissions are designed to be lubricated by fluid products. Soft or semifluid greases may be used in small units, such as the transmissions of some lawn and garden equipment and scooters.

Generally, the lubricant in a mechanical transmission is expected to remain in service for an extended period of time; normally, many passenger car manufacturers do not recommend periodic draining and refilling. Thus, the lubricant must have the chemical stability to resist oxidation and thickening under conditions of agitation and mixing with air. Operating temperatures may also be quite high. Plain bearings, thrust bearings, and synchronizer components are often of bronze or other copper alloys. Thermal degradation of the lubricant can result in formation of materials that are corrosive to these components. Severe agitation also occurs; therefore, the lubricant must have good resistance to foaming.

A lubricant selected for mechanical transmissions must have adequate fluidity to permit immediate circulation and easy shifting when a vehicle is started in cold weather. At the same time, the lubricant's viscosity at operating temperature must be high enough to maintain lubricating films and to cushion the gears so that operation is acceptably quiet.

A variety of lubricants are recommended by mechanical transmission manufacturers. Straight mineral gear lubricants suitable for API Service GL-1 (see discussion of automotive gear lubricants in Section VII) are recommended by a number of manufacturers. Most manufacturers will accept multipurpose gear lubricants, but only of API Service GL-4 quality, while others will accept either GL-4 or GL-5 quality lubricants. At least one manufacturer recommends DEXRON® (General Motors Company registered trademark) Automatic Transmission Fluid, but permits the use of SAE 80W-90 or SAE 85W-140 gear lubricants if operating on the DEXRON fluid results in objectionable noise. Manufacturers of farm and construction machines frequently install the transmission in a common sump with the final drive; the sump may also serve as the reservoir for the central hydraulic system on the machine. Special fluids designed for service as combination heavy-duty gear lubricants and hydraulic fluids are usually required for these applications. It is important to check manufacturers' recommendations to assure adherence to specific requirements.

2. Automatic Transmissions

In some installations the torque converter is located in a separate housing with its own supply of hydraulic fluid. However, in most of the passenger car automatic transmissions, the torque converter and the gearbox operate from a common fluid reservoir.

In a torque converter, the fluid serves mainly as a power transfer fluid. It also lubricates the bearings and transfers heat resulting from fluid friction and power losses to a cooler or to the transmission case for dissipation in the atmosphere. Power transfer efficiency increases with decreasing viscosity. Heat transfer efficiency also generally increases with decreasing viscosity. These factors dictate the lowest viscosity that is practical for a torque converter fluid. On the other hand, high operating temperatures and the need for long service life of the fluid require oxidation resistance properties, as well.

Where the torque converter operates from the same fluid supply as the gearbox, the lubrication requirements of the gearbox cannot be met unless the physical characteristics of the fluid are compromised.

The fluid in the gearbox portion of an automatic transmission performs several functions.

It lubricates the gears and bearings of the planetary gear sets.
It serves as a hydraulic fluid in the control systems.
It controls the frictional characteristics of the oil-immersed clutches and brakes.
It provides a degree of cooling.

These functions must be performed under a variety of operating conditions that tend to make the service severe.

Automatic transmissions are expected to engage and shift properly at low temperatures when a vehicle is started in cold weather. In operation, temperatures in the order of 250–300°F (121–149°C) may be reached. Gear loads are relatively heavy, and the fluid is exposed to severe mechanical shearing both in the gears and in the hydraulic circuit. Changes in temperature inside the unit produce some breathing of air, this tends to promote oxidation of the fluid, particularly when operating temperatures are high. Where the gearbox operates on the same fluid as the torque converter, the severe churning in the torque converter tends to cause foaming. Seal compatibility of the fluid is also an important consideration.

To satisfy these requirements for automatic transmission fluids, various highly specialized products have been developed. They are discussed in detail in this chapter.

3. Semiautomatic Transmissions

Since semiautomatic transmissions of the type discussed comprise a combination of torque converter and a mechanical transmission, the lubrication requirements given earlier for these units apply.

Power shift transmissions used in heavy equipment have lubrication requirements not unlike the gearbox section of automatic transmissions. Because of the higher torques transmitted, somewhat higher pressure may be required in the hydraulic system to obtain proper engagement of the clutches. This in turn may apply somewhat higher mechanical shear stresses to the fluid. Again, frictional characteristics of the fluid, and its compatibility with the clutch materials, are critical if the clutches are to engage smoothly and firmly.

4. Hydrostatic Transmissions

Since a hydrostatic drive is a high pressure hydraulic system, the basic fluid requirements correspond closely with those of industrial hydraulic systems. Good oxidation stability is required, as well as good resistance to foaming and good entrained air release. Antiwear properties are also required since, operating pressures are usually in excess of 2500 psi (17.2 MPa).

In addition, the requirement that hydrostatic drives operate over a wide range of temperatures generally dictates the use of very high viscosity index (VI) fluids with good low temperature fluidity. Since the fluid is a major factor in proper sealing of the pump pistons, the high temperature viscosity of the fluid is important, and excellent shear stability is required to maintain this viscosity in spite of the severe shearing that occurs in the pump and motor.

Many hydrostatic drives are operated from a common reservoir with the differential or final drive. In these cases, the fluid used must also provide satisfactory lubrication of the gears and bearings.

The hydrostatic drives used on garden tractors usually are designed to operate on automatic transmission fluids (ATFs). These fluids are readily available and generally provide the combination of performance characteristics required. ATFs may also be recommended for hydrostatic drives on larger machines. Engine oils, often in SAE 10W-30 viscosity, may also be recommended. For a hydrostatic drive on a larger machine that is operated from a common reservoir with other drive elements, the fluid recommended is usually one of the special fluids discussed in Section VIII: Multipurpose Tractor Fluids.

III. DRIVE SHAFTS AND UNIVERSAL JOINTS

Road vehicles have the wheels connected to the body and chassis through springs, but the engine and transmission are mounted directly on the chassis. Where a rigid (live) axle is used and the springs flex, the position of the axle with respect to the engine and transmission changes. Thus, there must be provision in the power connection between the transmission and drive axle to accommodate these changes in angular contact. This is accomplished in the drive (propeller) shaft and universal joints.

Typically, a drive shaft consists of a tubular shaft with a universal joint at each end. The universal joints allow for angular changes and a slip joint at one end allows for changes in length. Some long drive shafts are made in two parts with a center support bearing to minimize whip and vibration. Three universal joints are then used.

Universal joints are usually of the cross or cardan type (Figure 16.9). Ball-and-trunnion universal joints were used to some extent in the past. If there is any angular misalignment between the driving shaft and driven shaft, joints of both these types will transmit rotation with fluctuating angular velocity. The amount of fluctuation increases with increasing misalignment, rising from about 7% at 15° misalignment to over 50% at 40°. Since this fluctuation in velocity may be accompanied by vibration, the drive shafts, in which these joints are used, are designed for minimum misalignment. Another approach is to use the so-called constant velocity (CV) joints.

One type of constant velocity joint consists of two cross-type joints in a tandem assembly. Several other designs are available. Constant velocity joints are now being used

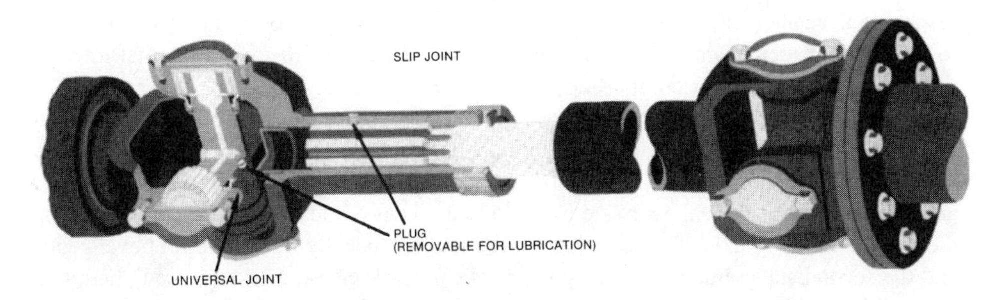

Figure 16.9 Cross-type universal joint. This type of joint, which is often called a cardan joint, may have needle roller bearings or plain bearings at the ends of the cross. Fittings may be used for relubrication. Since the weight of the fitting can cause imbalance, however, a more common arrangement is to use special flush fittings, or plugs that must be removed and replaced with fittings during lubrication.

to some extent in propeller shafts and are used as the drive axles of front engine, front wheel drive or rear engine, rear wheel drive vehicles, and in conventional arrangement vehicles with independent rear suspension. In some cases, rather than constant velocity joints, cross-type universal joints may be used at each end of the axle shaft, positioned so that the changes in angular velocity cancel out.

A. Lubrication

In some designs, the transmission lubricant lubricates the universal joint and the slip joint at the transmission end of the drive shaft. In other designs, the joints are lubricated with grease. Many joints are now ''packed for life'' on assembly and require servicing only if other repairs are being made. Some joints require periodic disassembly and repacking, while some are equipped with a fitting or plug for periodic relubrication. The plug can be replaced with a special fitting while relubrication is being performed. In most cases, the grease used is a multipurpose automotive grease, sometimes with the addition of molybdenum disulfide.

B. Drive Axles

The drive axle usually contains one or more stages of gear reduction such as the gears in the differential, which enable the wheels to be driven at different speeds. Also, in vehicles with a longitudinally mounted engine, the drive axle provides the gears with the capability to produce a 90° change in direction of power flow to couple the transverse axle shafts to the longitudinal transmission output shaft. In passenger cars and most trucks, the gear reduction in the drive axle is the final stage of gear reduction in the power train. In heavy trucks, and farm and construction equipment, additional stages of speed reduction, usually called final drives, may be used at the wheel ends of the drive axle.

In the most common passenger car and light truck arrangement (Figure 16.10), the drive shaft couples through a universal joint to the front end of a pinion shaft of the differential. The pinion gear at the rear of this shaft meshes with the ring gear, which is bolted or riveted solidly to the differential case. The differential, in turn, drives the half-axle shafts.

In most drive axles of this type, hypoid gears are used for the reduction stage. This type of gear design has high load-carrying capacity in proportion to the size of the gears and operates quietly. In addition, the offset position of the centerline of the pinion, with respect to the centerline of the gear, permits the drive shaft to be located lower. This helps to lower the center of gravity of the vehicle and reduces the size of the tunnel through the floor of the passenger compartment that covers the drive shaft.

With front engine, front wheel drive or rear engine, rear wheel drive cars, spiral bevel gears are usually used for this reduction state. If the engine is mounted transversely with either of these arrangements, the 90° change in direction of power flow is not required and straight spur or helical gears are used. In many heavy trucks, two stages of reduction are used in the drive axle. The first stage of reduction is usually a set of spiral bevel gears, and the second stage either straight spur or helical gears. Some trucks are equipped with worm gears, which require a large total reduction.

C. Differential Action

As a vehicle turns, the wheels on the outside of the turn follow a longer path than those on the inside of the turn. To compensate for this and other differences in rolling distances between the driving wheels, a differential is used.

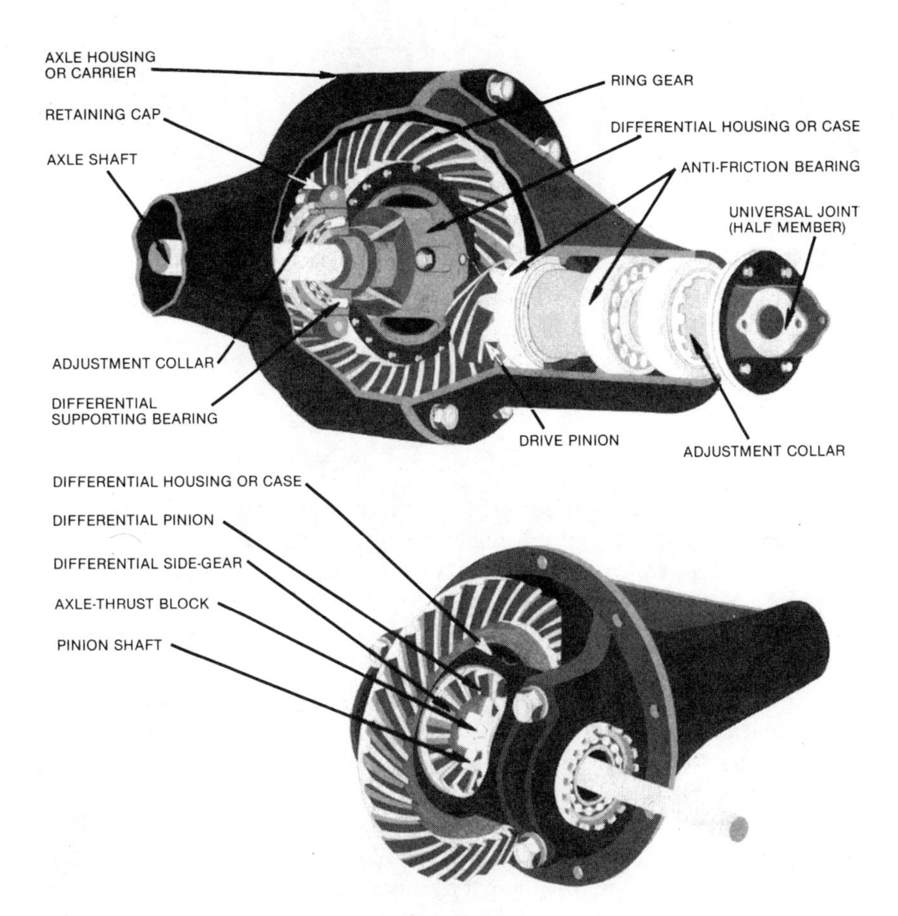

Figure 16.10 Hypoid-type drive axle. In the lower view, the differential case is cut away to show one of the side gears.

The principle of operations of a differential is shown in Figure 16.11, where the arm represents the differential case, which is bolted to the ring gear. The differential pinion is free to rotate on its shaft, and it meshes with the side gears, which drive the axle shafts. When driving resistance is equal at both wheels, the pinion acts as a simple lever to drive the axle shafts at the same speed as the ring gear. If greater rolling resistance is encountered at one wheel than at the other, the unbalanced reaction forces acting on the pinion will cause it to rotate on its shaft. The wheel encountering the least resistance will then be driven faster and the wheel encountering the most resistance will be driven slower. The increase in speed of one wheel will be exactly equal to the decrease in speed of the other.

Differentials are built with either two or four pinions. Two pinions are mounted on a shaft, which runs across the case and is mounted in a bearing at each end. Four pinions are mounted on a cross-shaped member called a spider, and the case is split to permit assembly. Normally, straight bevel gears are used for the pinions and side gears.

D. Limited-Slip Differential

Conventional differentials have a major drawback in that exactly the same torque is delivered to both wheels regardless of traction conditions. Thus, if one wheel is on a surface

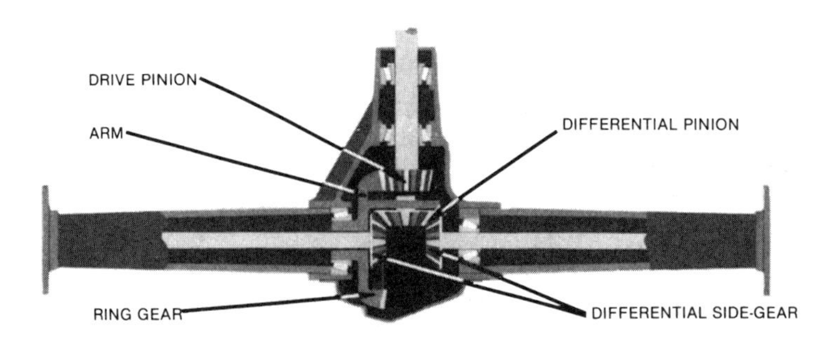

Figure 16.11 Elementary differential. In an actual differential, at least two pinions are used and the arm is replaced by a case that more or less completely encloses the pinions and side gears.

with low enough traction for the applied torque to exceed the traction, that wheel will break loose and increase in speed until it is revolving at twice the speed of the ring gear, whereupon the other wheel will stop revolving. All the power will then be delivered to the spinning wheel, and no power will be delivered to the wheel with traction. Limited slip, or torque biasing, and locking-type differentials have been developed to overcome this problem.

The limited-slip differentials used in passenger cars are all similar in principle. Clutches are inserted between the side gears and the case. When these clutches are engaged, they lock the side gears to the case and prevent differential action. Either plate-or cone-type clutches may be used. A typical unit using cone-type clutches is shown in Figure 16.12. Initial engagement pressure for the clutches is provided by the springs. As torque is applied to the unit, normal gear reaction forces tend to separate the side gears, which apply more pressure to the clutches. The more torque is applied, the more closely the unit approaches a solid axle. When differential action is required, the changes in torque reaction at the wheels tend to reduce the pressure on the clutches, permitting them to slip. Coil springs, dished springs, and Belleville springs are all used to provide the initial engagement pressure.

In a variation of the unit shown in Figure 16.12, the cones are reversed, with the result that increasing torque input reduces the engagement pressure on the clutches. This is referred to as an ''unloading cone,'' spin-resistant, differential. It has been found useful for the interaxle differential of four-wheel-drive vehicles and some high performance cars. Both torque biasing and locking differentials are used for trucks and off-highway equipment. Some locking differentials lock and unlock automatically, while others are arranged so the operator can lock them when full traction at both driving wheels is needed. Because of the higher torque inputs involved with these machines, more positive locking arrangements than the clutches used in passenger cars are required for the torque biasing differentials. One type uses cam rings and a set of blunt-nosed wedges that operate much in the manner of an overrunning clutch. Other types use special tooth profiles on the pinions such that a torque bias in favor of the wheel with the best traction is always provided.

E. Factors Affecting Lubrication

The hypoid gears used in drive axles are among the most difficult lubricant applications in automotive equipment. The high rate of side sliding between the gear teeth tends to

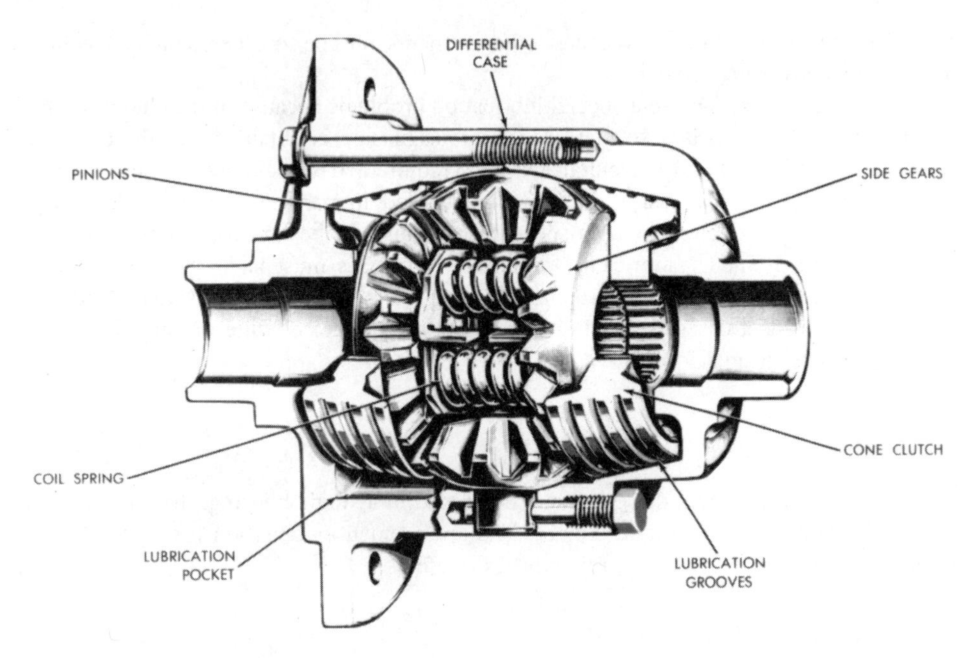

Figure 16.12 Limited-slip differential. A typical limited-slip differential using cone-type clutches.

wipe lubricant films from the tooth surfaces, and the amount of sliding in proportion to rolling increases as the offset between the shaft center lines is increased. The gears must transmit high torques, and shock loads are often present. The drive axle is normally not sprung; thus, to keep the unsprung weight low, it is desirable to make the gears as compact and lightweight as possible. This has resulted in the use of high strength, hardened steel for these gears, and tooth loading is usually high.

Where spiral bevel gears are used in the drive axle, conditions are more favorable to the formation of lubricant films and tooth loading is generally lower. Worm gear axles present special problems because of the high rate of sliding between the teeth and because of the metallurgy that must be employed. Where spur gears are used, lubrication conditions are generally not severe.

Many manufacturers no longer recommend periodic lubricant changes for passenger car drive axles. The lubricant for these axles must be suitable for extended service, often at high temperatures.

Breathing of moisture into drive axles frequently occurs. Combined with the severe churning action of the gears, this can contribute to foaming of the gear lubricant and can also promote rust and corrosion.

An important consideration for the drive axles of all vehicles operated in cold weather is the ability of the lubricant to flow to the teeth to provide lubrication of the gears, and to be carried up by the gears in sufficient quantities to lubricate the pinion shaft bearings.

In farm and construction equipment, the drive axle and the differential are often installed in a common sump with the transmission, which may also serve as the reservoir for the hydraulic system. Oil-immersed clutches and brakes may also be involved. In these cases, the lubricant requirements of these other elements must be considered in the selection of the lubricant for the drive axle.

Limited-slip axles present special lubrication problems because of the clutches. The clutches must engage firmly so that proper torque biasing is obtained, but the clutches must slip smoothly when differential action is required. If the clutches do not release properly, or if they stick and slip, chatter will result and, in extreme cases, one wheel may be forced to break traction in high-speed turns. This can be an unsafe condition. To minimize these problems, lubricants for limited-slip axles must have special frictional properties, which may conflict with their ability to protect the gears against wear, scuffing, and scoring. In most cases, a small amount of chatter or noise while turning corners is considered to be normal for these units.

IV. TRANSAXLES

When a transmission and a drive axle are combined in a single housing, the unit may be referred to as a transaxle (Figure 16.13). The arrangement is common for front engine, front wheel drive or rear engine, rear wheel drive cars.

A. Factors Affecting Lubrication

The drive axle reduction gears used in a transaxle are either spiral bevel or, with a transverse engine, spur or helical gears. As a result, the lubrication requirements are not as severe as with hypoid gears. At the same time, the lubricant must meet the requirements of the transmission portion of the transaxle, which often has synchronizer elements that are

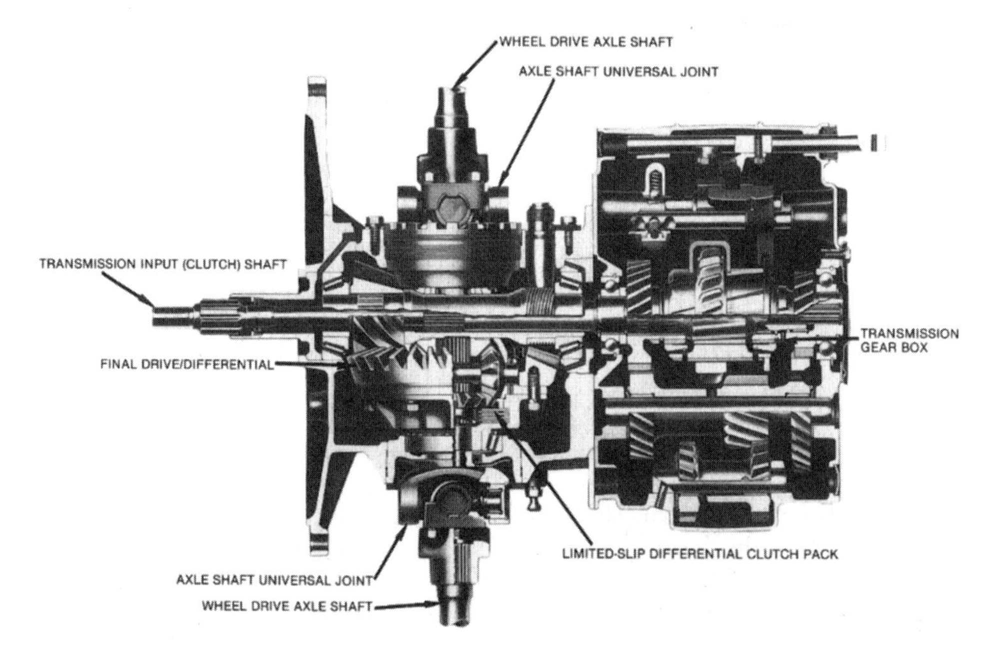

Figure 16.13 Typical transaxle cutaway with manual transmission.

sensitive to active extreme pressure agents designed for use in hypoid axle gears. Most builders of transaxles now recommend either specially formulated lubricants, lubricants for API Service GL-4 or GL-5, or multiviscosity engine oils.

Some engine/transaxle combinations with automatic transmission, whether at the front or rear of the vehicles, are usually equipped with separate compartments for the transmission and final drive. The transmission compartment contains automatic transmission fluid; the final drive compartment contains suitable gear lubricant for the spiral bevel gears. The current trend is to use front wheel drive transaxles with transverse engines. This design is usually provided with a common compartment that contains automatic transmission fluid for both the transmission and final drive.

V. OTHER GEAR CASES

A number of other gear cases may be used in various types of automotive equipment. Some of the more important are discussed in Sections V.A–VI.D.

A. Auxiliary Transmissions

Auxiliary transmissions are used with mechanical transmissions to provide a larger choice of reduction ratios; see Section V.E, Factors Affecting Lubrication. These are usually two- or three-speed gearboxes. Shifting is by means of a range control lever, which, on trucks, usually controls a hydraulic circuit to perform the actual shifting. On low speed equipment, shifting may be by means of a mechanical linkage. In some cases, the auxiliary is built into the same housing as the main transmission; in other cases it is a separate unit. Gears in auxiliary transmissions may be either spur or helical type. Bearings may be either plain or rolling element.

B. Transfer Cases

A transfer case is required with most four-wheel-drive vehicles to provide a second output shaft to drive the second axle. A transfer case may also provide a power takeoff to drive accessory equipment such as a hoist. In some heavy equipment, a transfer case is not required because the main transmission is provided with both front and rear outputs.

With the conventional four-wheel-drive arrangement, operation at highway speeds can cause the buildup of stresses in the drive line due to the different distances traveled by the front and rear wheels. If these stresses become excessive, they may cause one wheel to break traction and ''hop.'' To prevent this, the drive to the front wheels normally is disconnected for highway travel. A recent approach is to use a third differential in place of the transfer case. Differentiation in this unit prevents the buildup of stresses, which in turn makes it possible for the front wheel drive to be left engaged all the time.

C. Overdrives

An overdrive is an arrangement that drives the transmission output shaft at a higher speed than the input shaft. At cruising speeds this reduces engine rpm and improves fuel economy. Two general approaches are used.

Many of the five-speed mechanical transmissions used in small cars have an overdrive fifth gear. Fourth gear is made with a ratio only slightly greater than 1 : 1 and fifth gear has a ratio slightly lower. The normal progression of ratios in the gearbox is maintained, but in fifth gear the engine rpm will be somewhat below the rpm of the output shaft of the transmission.

With three-speed transmissions an auxiliary unit is added to the rear of the main transmission. This unit has an arrangement of planetary gears that provides a step-up ratio. The unit may be controlled electrically or hydraulically. When not engaged, it acts as a solid coupling. When the operator moves the control lever to the engaged position, the overdrive is activated but does not engage until a predetermined cut-in speed is reached and the operator momentarily releases the accelerator pedal. The overdrive will then remain engaged until car speed drops below the cut-in speed, or until the operator forces disengagement by fully depressing the accelerator pedal. Separate overdrive units are usually bolted to the rear of the main transmission, and provision is made for lubricant to flow from one case to the other. Separate drain plugs are usually provided, and special care in filling may be necessary to ensure that both units are properly filled.

D. Final Drives

Several types of drive unit are used at the wheel ends of the drive axles to obtain additional reduction or to rotate the wheels. Planetary reducers are used on many large off-highway trucks. They are also used on many tractors and on some self-propelled harvesting machines. Planetary speed increasers are used at the front wheels of conventional tractors equipped with power front wheel drive to match the travel speed of the smaller front wheels to that of the larger rear wheels. Chain drives of various types are used on self-propelled harvesting machines. Chains or gears are also used to couple a single drive axle to tandem driving wheels on certain types of heavy construction machine. Drop housings, which may or may not involve speed changes, are used on farm tractors to increase the clearance under the tractor for row crop work.

In many cases, these final drives have a separate lubricant supply. In other cases, they are lubricated from the drive axle or a common sump that supplies other units.

E. Factors Affecting Lubrication

Auxiliary transmissions, transfer cases, and overdrives are generally similar to mechanical transmissions in their lubricant requirements. As noted, auxiliary transmissions and overdrives frequently are coupled to the main transmission so that the same lubricant supply serves both units. Transfer cases are usually independent, but do not present any special lubrication problems. If an interaxle differential of the limited-slip type is used instead of a transfer case, the frictional characteristics of the lubricant are extremely important for the proper operation of the clutches.

Final drives present a range of lubrication problems because of the diversity in design of these units. Some are lubricated from the drive axle, thus, they are designed to operate on the type of lubricant that is suitable for the axle. To simplify lubrication, many final drives with independent lubricant reservoirs are also designed to operate on one of the lubricants required for other parts of the machine. Chain drives may present special problems. Some are fully enclosed and run in a bath of oil, but others are enclosed in relatively loose fitting dust shields. In the latter case, hand oiling or a drop feed oiler may be used. In extremely dusty conditions; it may be necessary to let the chains run dry. Also, it may be desirable to remove the chains periodically and soak them in a bath of oil so that some lubrication will be present inside the rollers on the pins.

VI. AUTOMOTIVE GEAR LUBRICANTS

In automotive gear units, gears and bearings of different designs and materials are employed under a variety of service conditions. The selection of the lubricant involves careful consideration of these factors and their relationship to performance characteristics of the lubricant. Some of the more important performance characteristics of automotive gear lubricants are discussed next. The reader is also referred to Chapter 3's discussions of additives, physical and chemical characteristics, performance and evaluation tests of lubricants (Sections I, II, and III, respectively).

A. Load-Carrying Capacity

One of the most important performance characteristics of a gear lubricant is its load-carrying capacity, that is, its ability to prevent or minimize wear, scuffing, or scoring of gear tooth surfaces. The load-carrying capacity of straight mineral oil is adequate for the conditions under which some gears are operated; however, most gears require lubricants with higher load-carrying capacity. This higher capacity is provided through the use of additives. Lubricants of this type are generally referred to as extreme pressure (EP) lubricants.

To provide differentiation between automotive gear lubricants with different levels of EP properties, the American Petroleum Institute (API) prepared a series of five lubricant service designations for automotive manual transmissions and axles. These service designations describe the service in which the various types of lubricant are expected to perform satisfactorily. In addition, for the two service designations intended for use in hypoid axles, antiscore protection must be equal to or better than that of certain reference gear oils. And the lubricants must have been subjected to the test procedures and must provide the performance levels described in ASTM STP-512 (Laboratory Performance Tests for Automotive Gear Lubricants Intended for API GL-4 and GL-5). The ASTM standard describes tests for performance characteristics in addition to those for load-carrying capacity.

B. API Lubricant Service Designations

The gear lubricant service designations are as follows:

API GL-1: designates service characteristics for automotive spiral bevel and some truck manually operated transmissions that have components sensitive to additive materials such as EP additives. These transmissions are designed for operation under conditions of low unit pressures and sliding velocities so mild that straight mineral oil can be used satisfactorily. Oxidation and corrosion inhibitors, defoamants, and pour depressants may be utilized to improve the characteristics of lubricants for this service. Frictional modifiers and extreme pressure agents shall not be utilized.

API GL-2: designates service characteristics for automotive-type worm gear axles operating under conditions of load, temperature, and sliding velocities that cannot be accommodated by lubricants satisfactory for API-GL-1 service. The API-GL-2 gear oils generally contain fatty-type additives, making them satisfactory for worm gear and other types of industrial gearing.

API GL-3: designates service characteristics for manual transmissions and spiral bevel axles operating under moderately severe conditions of speed and load. These service conditions require a lubricant having load-carrying capacities greater than those that will satisfy API GL-1 service, but below the requirements of lubricants satisfying API GL-4 service.

API GL-4: a classification still used commercially to describe lubricants for differentials and transmissions operating under moderate to severe conditions such as hypoid gears. These oils may be used in manual transmissions and transaxles where EP oils are acceptable. Limited-slip differentials generally have special lubrication requirements. The supplier should be consulted regarding the suitability of a given lubricant for such differentials. Information helpful in evaluating lubricants for this type of service may be found in ASTM STP-512.

API GL-5: designates service characteristics for gears, particularly hypoid, in passenger cars and other automotive equipment operated under conditions of high speed and shock load, high speed and low torque, and low speed and high torque.

API MT-1: designates lubricants for manual transmissions that do not contain synchronizers. These oils are formulated to provide levels of oxidation and thermal stability higher than those obtainable from API GL-1, GL-4, and GL-5 category oils.

MIL-PRF-2105E: the military specification for automotive gear lubricants. It combines the requirements of API GL-5 and MT-1.

The foregoing lubricant service designations have generally provided satisfactory levels of performance for current automotive gear units. Some difficulties have been experienced in low temperature service because operating temperatures are not always high enough to fully activate the various EP agents. This disadvantage is usually overcome by using a higher dosage level of additive in lubricants intended specifically for arctic-type service.

C. Viscosity

The viscosity of a gear lubricant has some effect on load-carrying capacity, leakage, and gear noise. At low temperatures, it also determines ease of gear shifting and has considerable influence on flow to gear tooth surfaces and bearings.

The viscosities of automotive gear lubricants are usually reported according to SAE Recommended Practice J306 (Axle and Manual Transmission Lubricant Viscosity Classification: see Chapter 3, Section II.J.2). This classification system provides for measurement of low temperature viscosities according to ASTM D 2983 (Standard Method of Test for Apparent Viscosity of Gear Oils at Low Temperatures Using the Brookfield Viscometer). The 150,000 cP (150 Pa·S) value selected for the definition of low temperature viscosity in this revision was based on a series of tests in a specific axle design, which showed that pinion bearing failures could occur if the lubricant viscosity exceeded this value. Since, however, other axle designs may operate safely at higher viscosities or fail at lower viscosities, it is the responsibility of the axle manufacturer to determine the viscosity required by any particular axle design under low temperature conditions.

Other gear applications may have different limiting viscosities. For example, for satisfactory ease of shifting, many manual transmissions require a lubricant viscosity not exceeding 20,000 cP at the shifting temperature. However, this does not necessarily mean that a lubricant with a viscosity exceeding 20,000 cP at the lowest expected starting temperature cannot be used satisfactorily. At low temperatures, it is usually necessary to idle the engine for a short period before driving away. During this period, the main gear and countershaft in the transmission will be revolving if the clutch is engaged. As long as the lubricant does not channel, some of it will be picked up and circulated by the gears. The resulting fluid friction may warm the lubricant, and the shear and churning conditions may increase the lubricant's fluidity sufficiently for satisfactory shifting. Gear or bearing failures usually do not occur during this period because the loads are relatively low, being those resulting from fluid friction in the lubricant only. In a drive axle, on the other hand, as soon as the drive is engaged, loads are relatively high and adequate flow of lubricant is critical.

Under this system, conventional gear lubricant based stocks can be used to formulate multigrade lubricants such as 80W-90 and 85W-140. Lubricants in these viscosity grades are generally suitable for a wide range of operating temperatures in automotive gears, and formulation can be achieved fairly readily. Multigrade lubricants covering an ever wider range, such as 75W-90 or 80W-140, are also formulated, but these usually require either special high VI base oils such as some of the synthetics, or the use of VI improvers. VI improvers must be selected with extreme care because under the severe mechanical shearing that exists in gears, the use of an unsuitable material can result in the rapid loss of the contribution of the VI improver to high temperature viscosity.

D. Channeling Characteristics

Under low temperature conditions, a lubricant may "channel"; that is, the lubricant may become so solid that when gear teeth cut a channel through it, it does not flow back rapidly enough to provide fresh material for the gear teeth to pick up.

The channeling temperature of a lubricant is somewhat related to its pour point, and also to its low temperature viscosity. However, the pressure head tending to cause flow into a channel cut in a lubricant may be greater than in the pour point test, and the conditions in the Brookfield viscometer are sufficiently different that it can provide a viscosity measurement for a lubricant that might not flow under other conditions. In effect, the Brookfield viscometer will indicate that a lubricant can be carried up by the gears and distributed, but it does not necessarily indicate whether the lubricant will flow to the gear teeth to be picked up.

E. Storage Stability

In extended storage, particularly if the storage temperature is excessively high or low, or if moisture is present, some of the additive materials may separate from some highly additized gear lubricants, or reactions that can change the properties of these materials may occur. With some of the older additive systems, such instability was occasionally a severe problem, but the newer additive systems generally exhibit improved oil solubility and a reduced tendency for the components to react with each other. Proper storage and control of inventory to prevent excessively long storage are, however, still important; see Chapter 18: Handling, Storing, and Dispensing Lubricants.

F. Oxidation Resistance

Operating temperatures of drive axles and manual transmissions in normal passenger car service are usually moderate, and thus the oxidation resistance of the lubricant may not be a major consideration. In heavy-duty service, such as trailer towing and the use of many commercial vehicles, high operating temperatures may be encountered. In these applications, to avoid the development of sludges that could restrict oil flow, the lubricant must have adequate oxidation resistance to prevent excessive thickening. Oxidation may also result in the development of materials that are corrosive to some of the metals used.

G. Foaming

The churning in gear sets, combined with contaminants such as moisture that enter the case through breathers, tends to promote foaming of the gear lubricant. Foaming may cause overflow with loss of lubricant and may interfere with lubricant circulation and the load-carrying ability of lubricant films.

Defoamants are used in gear lubricants to reduce the foaming tendency. Some of the commonly used defoamants are not soluble in the oil. As a result, defoamants are carried in suspension. If the density of the defoamer is significantly greater than that of the oil, the defoamer can settle out during extended storage. As a check on this, gear lubricants submitted for U.S. military approval must pass a foam test both when freshly prepared and after 6 months of storage. To meet this requirement, organic defoamants rather than silicone defoamants are now generally used in gear lubricants.

There is considerable evidence that gear lubricants that pass the laboratory foam test are not always satisfactory in service. Some form of full-scale testing in automotive gear sets is usually used to check on this aspect of the performance of new gear lubricant additive systems and gear lubricant formulations.

H. Chemical Activity or Corrosion

Extreme pressure agents function by reacting chemically with metal surfaces. This reaction normally is initiated when local overheating results from the rubbing of surface asperities through the oil film. If the reactivity of the extreme pressure agents is too high, some chemical reaction may occur at normal temperatures, resulting in corrosion and metal loss. Copper and its alloys are particularly susceptible to this type of corrosion and, although copper alloys are not normally used in drive axles, they are frequently used in manual transmission components.

I. Rust Protection

Some moisture enters gear sets through normal breathing, and heavy water contamination may occur in certain types of off-highway equipment. The gear lubricant must provide rust protection adequate to prevent rusting that might result in gear or bearing damage, particularly during shutdown periods.

J. Seal Compatibility

To maintain control of leakage, gear lubricants must be formulated to be compatible with the elastomeric materials used as seals. This means that the lubricant must not cause excessive swelling, or shrinkage and hardening. A slight amount of swelling is usually considered desirable to help keep the seals tight.

K. Frictional Properties

As pointed out earlier, the frictional properties of the lubricant are critical in limited-slip drive axles, particularly the types that use clutches to accomplish lockup. It is generally considered that for slip in the clutches to be initiated smoothly, the lubricant must have a lower coefficient of friction at low sliding speeds than it has at high sliding speeds. This requires the addition of special friction modifiers to the lubricant, or formulation of additive system with one or more components that have these special frictional properties. However, the strong polar attraction of the friction modifiers to metal surfaces may serve to reduce access to these surfaces by the extreme pressure agents. This in turn may reduce the load-carrying ability of the lubricant. Also, the frictional properties of the friction modifiers may change relatively rapidly, requiring limited-slip additive replenishment or a complete fluid change-out.

To minimize these difficulties with limited-slip axles, some manufacturers apply special coatings to their axle gears to reduce the severity of the service on the gear lubricant. Regular lubricant changes are also recommended by some manufacturers.

Friction modification of the lubricant may also offer benefits in conventional axles, particularly during break-in or during severe service. The reduction in sliding friction between the gear tooth surfaces reduces the amount of heat generated at these surfaces. Since less heat is generated, less heat will be rejected to the lubricant, and the operating temperature of the lubricant will be lower. With lower operating temperatures, oxidation and thermal degradation of the lubricant may be less serious. During break-in, surface temperatures high enough to affect the metallurgy of the gear tooth surfaces have been experienced in some axle designs. The reduction in friction from the use of friction modifiers can provide some relief from this problem too.

L. Identification

In addition to the API lubricant service designations and the SAE viscosity classification, automotive gear lubricants are frequently identified by U.S. military specifications. Some manufacturer specifications also exist, but products meeting them are usually used only for factory fill of new machines, or for sale through the service parts organization of the sponsoring company.

M. U.S. Military Specifications

Only one gear lubricant specification is now current. (MIL-PRF-2105E) Others may be encountered as performance references:

U.S. Military Specification MIL-L-2105C—This obsolete specification requires performance essentially equivalent to API-GL-5. It covers SAE 75W, SAE 80W-90, and SAE 85W-140 viscosity grades. All grades must pass the gear performance tests. The SAE 75W grades replaces the former MIL-L-10324A (sub zero) specification, and the 80W-90 and 85W-140 grades replace the 80, 90, and 140 grades of MIL-L-2105B.

U.S. Military Specification MIL-L-2105B—This obsolete specification differed from MIL-L-2105C in that it covered viscosity grades corresponding to an earlier SAE viscosity classification. Generally, only the SAE 90 grade was fully tested in the qualification program.

U.S. Military Specification MIL-L-2105—This obsolete specification described lubricants essentially equivalent to API-GL-4.

U.S. Military Specification MIL-L-10324A—This obsolete specification described one low viscosity grade intended for use in arctic regions. These requirements are now supplied with SAE 75W products meeting MIL-L-2015C.

VII. TORQUE CONVERTER AND AUTOMATIC TRANSMISSION FLUIDS

The specialized fluids used in passenger car, truck, bus, and off-highway automatic and semiautomatic transmissions are all manufactured to meet specifications developed by equipment manufacturers such as General Motors Corporation, Ford Motor Company, DaimlerChrysler, and Caterpillar.

A. Torque Converter Fluids

Separately housed torque converters are used with some limited passenger car automatic transmission arrangements, but the main application is in transit coach drives and some construction equipment drives. Separately housed torque converters for passenger cars are operated on automatic transmission fluids. Some early torque converters for vehicles with diesel engines were designed to operate on diesel fuel. Since the fuel was recirculated from the vehicle tank to the torque converter, the bulk fuel in the tank served as a heat sink to absorb heat resulting from power losses in the torque converter. This arrangement was not very successful because thermal degradation of the fuel sometimes resulted in sludge and deposits in the torque converter.

To provide better oxidation stability, low viscosity mineral oils were adopted. These oils were about 60 SUS (10.5 cSt) at 100°F (38°C). Again, some problems with thermal degradation were encountered, particularly in extended service and in newer, larger vehicles with higher operating temperatures. Some quantities of these fluids are still used in older equipment, mainly outside the United States; but the most commonly used products now are oxidation-inhibited fluids of about 100 SUS (20.5cSt) at 100°F (38°C).

Two original equipment manufacturer specifications covering these 100 SUS fluids are generally recognized: the General Motors Corporation Truck & Coach Division Specification No. C-67-I-23 for Type 2 V-Drive Fluids, and the Dana M-2003 specification for torque converter fluid. These specifications have the same physical requirements, and the performance requirements are similar.

B. Automatic Transmission Fluids

All passenger car automatic transmission fluids commercially available in the United States are developed to meet either General Motors, Ford, or DaimlerChrysler specifications.

Most other automobile manufacturers accept fluids of one or the other of these types, although some require additional testing to ensure compatibility with the requirements of their transmissions. For heavy truck and off-highway automatic and semiautomatic transmissions, Allison C-4 or Caterpillar TO-4 type fluids may be recommended.

Passenger vehicle ATFs have evolved over the years as car weights have increased and decreased and operating conditions have subjected the fluids to more severe service. Other factors such as extension or elimination of drains have dictated a need for products that are more resistant to deterioration in service. Several automobile manufacturers have advertised 100,000 miles before the first major service interval, recommending transmission fluid changes at this interval when the vehicle is driven under "normal service."

The various versions of the ATF specifications are outlined briefly in Sections VII.B.1–VII.B.9.

1. General Motors Type A Fluid

The specification and qualification procedure for products that came to be known as ''type A'' fluids was introduced in 1949. These products provided satisfactory performance in the transmissions of that time and were recommended by a number of manufacturers in addition to General Motors. In 1956 the type A specification was revised to require improved quality. Fluids meeting the new requirements were suitable for use in transmissions for which type A fluids had been originally recommended.

2. General Motors DEXRON Fluid

In 1967 General Motors issued the DEXRON specification, representing a further improvement in quality, mainly in the areas of high temperature stability, low temperature fluidity, and retention of frictional properties in service. Fluids were intended to be suitable where type A or type A, suffix A fluids had been recommended originally, but, some European manufacturers continued to show preference for the older type A, suffix A fluids. In addition to the General Motors applications, DEXRON fluids were at that time acceptable to Chrysler, American Motors, and some Borg-Warner affiliates.

3. General Motors DEXRON II Fluid

The DEXRON II specification was intended originally as a multipurpose fluid suitable for passenger car automatic transmissions, for the lubrication of rotary engines, and as a replacement for the Allison C-2 fluids recommended by the Detroit Diesel Allison Division of General Motors for automatic and semiautomatic transmissions used in trucks and off-highway equipment. These fluids, however, proved to be deficient in corrosion protection for the brass alloys used in certain transmission fluid coolers. The specification was revised to include a brass alloy corrosion test, and the rotary engine lubrication requirements were dropped. Fluids qualified under this version of the specification are generally suitable for applications for which any of the earlier General Motors approved fluids were recommended originally.

The DEXRON II fluids have somewhat lower low temperature viscosity than the DEXRON fluids, and the oxidation stability is improved. Oxidation stability for the DEXRON II fluids was evaluated in a three-speed automatic transmission, probably a more severe test than the older two-speed transmission part of the DEXRON test. Additionally, the rate of airflow into the transmission during the test is twice as great in the DEXRON II test. A humidity cabinet rust test is also required in the DEXRON II specification.

DEXRON IId and IIe were interim specification requirement changes to enhance specific properties of the DEXRON II. The DEXRON IIe, introduced in late 1990 and still often referred to as a viable specification, was replaced by the DEXRON III specification in 1993. Both the DEXRON IIe and III specifications call for improved low temperature performance, improved shear stability, frictional characteristic durability, and improved oxidation and thermal stability. The DEXRON III provides the longest service interval capability to satisfy the requirements to meet the fill-for-life or substantially extended service intervals advertised by some of the automobile manufacturers. The DEXRON III is back-serviceable to all previous vehicle transmissions that specified DEXRON IIe, IId, DEXRON, type A-suffix A, and type A fluids.

4. Ford Fluids

In the past, specifications for the Ford automatic transmission fluids were issued under the designation ESW-M2C33, with a suffix letter indicating different revisions. They are usually referred to as ''type F'' fluids. M2C33-E/F was issued in 1959 and is suitable for use in all Ford automatic transmissions (except the C-6) manufactured prior to 1980. A number of other manufacturers, including some of the Borg-Warner affiliates, also recommended this type of fluid.

Early in 1980, Ford issued specifications M2C138-CJ and M2C166-H, referred to as types ''CJ'' and ''H'' respectively. Only type CJ fluid was recommended for the C-6 and ''JATCO'' automatic transmissions, the former is the largest Ford passenger car automatic transmission manufactured. Types CJ and H differ from the earlier fluids in that they contain friction modifiers and are not recommended where type F fluids are recommended.

In 1988 Ford introduced its MERCON® specification designed to meet the requirements of the Ford transmissions manufactured since that date and through the mid-1990s. These fluids back-serviceable to Ford transmissions manufactured in 1981 and later. The MERCON V specification, introduced in 1997, requires a normal service interval of 30,000 miles except for some of the Ford models that do not require change-outs, as mentioned earlier. These fluids have stricter requirements for oxidation stability and wear protection. Although compatible, the MERCON fluids should not be used for change outs (if required) if MERCON V is specified. The MERCON V fluids have limited back-serviceability, so manufacturers' recommendations should be followed on vehicles produced before 1997.

5. Fluid Comparison

The major difference between the General Motors fluids and the M2C33 Ford fluids is in the frictional properties. The Ford fluids have a higher coefficient of friction at low sliding speeds than at high sliding speeds, while the General Motors fluids are the opposite (Figure 16.14). Thus the Ford type F fluids provide faster and more positive lockup of the clutches and bands at low sliding speeds. The Ford type F fluids are physically compatible with the General Motors fluids, but the two should not be mixed. Any mixing changes

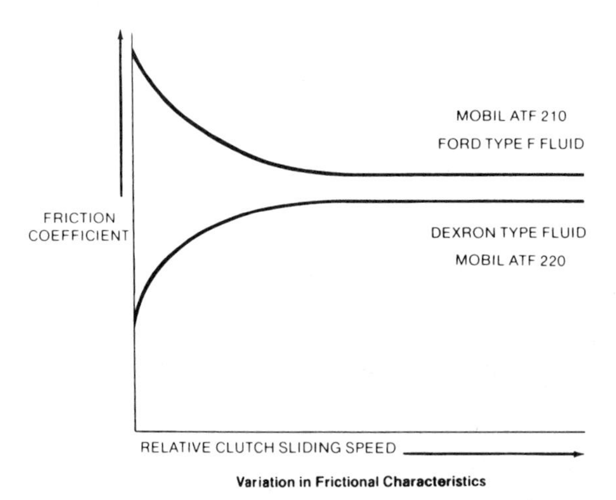

Figure 16.14 Automatic transmission fluid frictional characteristics. As shown, the coefficient of friction of the Ford fluids decreases with increasing sliding speeds, while that of the DEXRON fluids increases.

the frictional properties and may result in poor transmission shifting or, in extreme cases, failure of the transmission to engage.

6. DaimlerChrysler Corporation

DaimlerChrysler has historically allowed the use of DEXRON fluids in its transmissions. The company introduced its own specification, Chrysler MS 7176 ATF + 3®, to be used for complete change-outs but still allows the use of DEXRON III for topoff. The ATF + 3 is not always readily available in the marketplace but can be obtained at Chrysler dealers or some transmission repair shops. Recently, DaimlerChrysler introduced a new specification, MS 9602 ATF + 4®. The ATF + 3 had recommended service intervals of 30,000 miles, while the ATF + 4 never needs changing even under severe service conditions.

***Caution:* Care and good judgment must be exercised to assure long transmission service life. Transmission fluids can sometimes be subjected to temperatures approaching 300°F. If exposure to such very high temperatures occurs for extended intervals, the transmission fluid may deteriorate so much that changing the fluid becomes advisable. Fluid should be checked periodically for severe darkening and burnt odor. If these conditions exist, the advice of a dealer or service provider should be sought.**

---

7. Multipurpose ATFs

Most of the products on the market today are formulated to meet the requirements of most of the transmission manufacturers. For example, one product may be qualified against DEXRON III, MERCON V, Ford types CJ and H, and Allison C-4. Part of the testing for these qualifications requires that candidates also pass vane and piston hydraulic pump tests, making them suitable for use where hydraulic fluids passing the Denison HF-0 (piston and vane), Vickers 35VQ25 vane pump, and Sundstrand piston pump tests are recommended. Generally, these multipurpose fluids will not be qualified against the Chrysler ATF + 3 or ATF + 4 requirements.

8. Allison Transmission Division of General Motors: Type C Fluids

The General Motors Allison Division offers both fully automatic and semiautomatic transmissions for trucks and off-highway equipment. Early designs were intended to operate on SAE 10W heavy-duty engine oils, but variations in the performance of some of these products led to the issuance of the type C-1 specification. Type C-1 oils were SAE 10W heavy-duty engine oils that had performed satisfactorily in Allison transmissions. In 1967 the type C-2 specification was issued to replace the type C-1. Type C-2 was intended only as a transmission fluid and included tests such as a transmission oxidation test (operated at a somewhat lower temperature than the test used in the DEXRON II program) and a power steering pump wear test. Type C-3 incorporated several viscosity grades as long as they complied with the SAE J300 viscosity grade requirements. Type C-4 was recently introduced and, if labeled as an ATF, such an oil must meet both DEXRON III and MERCON requirements. Some heavy-duty diesel engine oils met the C-4, and part of the specification required that a U.S. military standard (Mil-L-2104E) be met for any oil also classified as an engine oil. Some DEXRON fluids and a number of special fluids were developed specifically to meet it.

Allison automatic transmissions for on-highway vehicles are designed to operate on either DEXRON III or MERCON fluids that meet the Allison C-4 requirements.

9. Caterpillar®

The current TO-4 specification of Caterpillar Tractor Company is designed to meet the requirements of the firm's oil-cooled friction compartments such as transmissions, final drives, and planetary gears, including wet braking systems. The TO-4 specification does not include any requirements to meet any engine oil performance criteria. The TO-4 specification includes three monoviscosity grade products: SAE 10W, 30, and 50. The TO-4 fluids are a replacement for the earlier version, CD/TO-2 oils, which were API Service CD (Mil-L-2104C) engine oils that had passed the Caterpillar TO-2 test. The TO-2, API CD, and Mil-L-2104C are all obsolete tests. The tests for TO-2 qualification were designed to evaluate the retention of frictional properties of the oil after repeated engagement of oil-immersed clutches. The TO-4 tests include additional tests for high temperature /high shear, and lower temperature requirements and testing in both gear (FZG) and hydraulic pumps (Vickers vane) to assure performance in hydraulic and gear systems.

VIII. MULTIPURPOSE TRACTOR FLUIDS

Manufacturer specifications for fluids for the drive (transmissions, differentials, final drives) and hydraulic systems of farm and industrial tractors and self-propelled farm equipment have been extensively developed and used. At least 25 different fluid specifications are used by the equipment manufacturers, although several are now considered obsolete. Several equipment manufacturers market many of these fluids through their parts departments, and in many cases these fluids, or an equivalent product that meets the same specifications and performance standards, must be used during the warranty period. This requirement can cause considerable complexity in oil inventories and applications for the operator with several different makes of equipment. The risk of misapplication may also be high under these conditions.

Many reasons can be given for the large number of these fluids and their variations in characteristics. Some of the differences among the products result from differences in design philosophies and construction material among the manufacturers. The number and types of machine elements served by the fluid may be extensive, and only minor differences in materials of construction, applied loads or operating characteristics can have a marked effect on the lubricant properties required for satisfactory performance. When these systems were first being developed, there were no commercially available fluids that provided the type of performance desired by the equipment manufacturers. Thus, these manufacturers tried to develop fluids that were specific to their needs. Table 16.1 lists many of these OEM tractor fluid specifications.

There are continuing attempts to classify these fluids with a view to achieving some simplification, and this has been done to some extent with the UTTOs (universal tractor transmission oils) and the STOU (super tractor oils, universal). The STOU oils are formulated for use in engines, transmissions, wet brakes, and hydraulics. The UTTOs are more popular in North America and the STOUs in Europe. The success of these fluids depends

Table 16.1 OEM Tractor Fluid Specifications

OEM	Specification
AGCO	
Deutz-Allis	PF821XL
Massey-Ferguson	M1110[a]
	M1127A[a]
	M1127B[a]
	M1129A[a]
White Farm	M1135
	M1139
	M1141
	Q1826
John Deere	J14C[a]
	J20C
	J20D
	J21A[a]
	J27
Case	MS 1204[a]
	MS 1205[a]
	MS 1206[a]
	MS 1207
	MS 1210
	Hy-Tran[a]
New Holland	M2C41B
	M2C48B
	M2C48C
	M2C86B
	M2C134D
	M2C159-B1/C1
	M2C159-B2/C2
	M2C159-B3/C3
Kubota	UDT

[a] Obsolete specification.

on their ability to satisfy all the needs of the various manufacturers and their availability to the markets where needed.

Multipurpose fluids must satisfy the lubrication requirements of hydraulic systems, gears, and bearings. They must also work with wet braking systems, and they must provide the proper frictional characteristics for clutches in transmissions and power takeoffs. In addition, they must be compatible with all system components and capable of performing under a wide range of operating conditions.

A. Fluid Characteristics

In addition to the usual physical and chemical requirements, most multipurpose tractor fluids have requirements for seal and other component compatibility. The tests for this characteristic vary widely both in the types of seal elastomer and friction material used for testing and in the test conditions. Various other bench-type tests for such properties as wear sensitivity or tolerance, filterability, compatibility with other similar fluids, and low temperature performance are also used in many of the specifications. Many of the specifications also include performance tests, including such tests as automatic transmission oxidation tests, transmission durability tests, cold engagement tests, various hydraulic pump tests, and gear wear tests, as well as wet brake, clutch chatter, and durability tests, friction tests, frictional property retention tests, and dynamometer or full-scale field tests.

Some of the more important properties of the multipurpose tractor fluids are discussed briefly in Sections VIII.A.I–VIII.A.5.

1. Viscosity and Viscosity Index

The type of equipment in which these fluids are used is often operated on a year-round basis. This means that the fluids must have adequate low temperature fluidity to flow to parts requiring lubrication, and to the inlet of hydraulic pumps, at normal winter starting temperatures. All the specifications prescribe a maximum allowable viscosity at 0°F (−18°C), and some also limit the viscosity at even lower temperatures. Since these limits are determined by the equipment manufacturers on the basis of tests on their own equipment, it is probable that the fluids will give satisfactory low temperature performance in the equipment for which they are recommended. To meet a desired viscosity at high temperature, however, there may be some compromises in the maximum allowable low temperature viscosity.

In high temperature operation, the viscosity of the fluid should be sufficient to permit adequate maintenance of hydraulic pressure and to protect gears and bearings from excessive wear. While many of the fluids are similar in viscosity at 100°C (212°F), there are some divergences that may result from different attitudes toward VI improvers. The degradation products of some VI improvers are reported to cause glazing of certain types of facing material used on oil-immersed (''wet'') clutches and brakes. This condition can cause slippage and other problems. Therefore, some equipment manufacturers do not permit the use of VI impovers in their fluids, with the result that the viscosity characteristics of those fluids are limited to what can be obtained from conventional high VI mineral oil bases.

In general, VI-improved fluids are desirable to maintain adequate high temperature viscosity while maintaining the low temperature viscosity in a range that will permit satisfactory operation at the lowest ambient temperatures encountered in service. Where such additives are used, careful selection of shear-stable VI improvers is necessary to obtain materials that will resist the mechanical shearing in hydraulic pumps and gears. Many of the specifications include shear stability tests such as high temperature/high shear (HT/HS) requirements.

2. Foam and Air Entrainment Control

Churning and pumping through the system can promote foaming, which can seriously degrade the performance of hydraulic components and may interfere with the formation of proper lubricating films. Since the rest time of the fluid in the reservoir is low, the fluid not only must resist foaming, it must also permit rapid collapse of any foam that does form. Excessive entrainment of air in the fluid can also cause difficulties in hydraulic components, so the fluid must release entrained air as rapidly as possible. Careful selection of the components of the fluid and use of an effective defoamer are necessary to meet these requirements.

Some breathing of moisture into these systems always occurs. Since moisture can promote foaming, some of the manufacturers require that their fluids have good foam resistance when they are contaminated with a small amount of water.

3. Rust and Corrosion Protection

Much construction and farm equipment operates at least part of the time under wet or humid conditions. Breathing of this moisture into the lubrication systems can cause rust and corrosion of ferrous parts, particularly above the oil level during shutdown periods. Nearly all the specifications have some requirement for protection against rust and corrosion, with some type of humidity cabinet test being the most common. Quite effective rust inhibitors are necessary to pass this type of requirement.

As with gear lubricants, the extreme pressure agents used in multipurpose tractor fluids function by reacting chemically with the metal surfaces. If the reactivity is too high, corrosion may occur. Nearly all the specifications contain a copper strip corrosion test, although the time and temperature vary considerably. Since oxidation of a fluid in service can increase its corrosivity, a combination oxidation/corrosion requirement must be met.

4. Oxidation and Thermal Stability

Operating temperatures often are high in multipurpose tractor systems as a result of heavy loads and continuous operation. Thermal and oxidative degradation of the fluid can result in thickening, which can reduce hydraulic pump capacity and promote the formation of varnish and sludge. The latter materials can plug filters and leave deposits on clutch facings, where they can cause slippage, or in hydraulic control valves, where they can control to be erratic.

Some type of oxidation test is used in nearly all the specifications. These tests can be interpreted in various ways. Excessive evaporation loss usually results from the use of volatile, low viscosity components in the base oil blend (NOACK volatility tests are incorporated to evaluate volatility). The amount of viscosity increase is somewhat indicative of the amount of oxidation of the fluid. Sludge or sediment may indicate thermal decomposition of additive components, some form of chemical reaction between additive components, at high temperatures.

Various performance tests are also used in many of the specifications to evaluate the stability of fluids in long-term service or under simulated service conditions. These tests may be field or dynamometer tests or specialized tests such as transmission oxidation tests.

5. Frictional Characteristics

The growing use of oil-immersed clutches and wet brakes has led to increased emphasis on the frictional characteristics of the fluid and the retention of these characteristics in service. Generally, friction modifiers must be included in the fluid if the clutches and brakes are to engage smoothly and resist chatter and squawk. Detergent dispersants may be desirable to help keep the facing surfaces clean. Good oxidation stability is important to reduce the amount of varnish-type materials that might deposit on the friction surfaces. The additive materials must also be carefully selected, since some are reported to cause rapid deterioration of some types of friction material. The overall problem is complicated by the varying levels of friction performance required in different systems and the many types of friction material involved. It is also very important that the fluids retain their frictional properties after extended service.

B. Extreme Pressure and Antiwear Properties

Multipurpose tractor fluids must provide adequate protection against wear and gear tooth surface distress on a variety of gearing that operates under conditions of loading varying from mild to severe. Shock loads are often present. In hydraulic systems, protection against wear of pumps operating at pressures that may be in the order of 4000 psi (28 MPa) or higher must be provided. Further complicating the situation, some of the specifications permit, or require, zinc dithiophosphates, which give good antiwear performance in hydraulic pumps and certain types of gear application, while other specifications prohibit the use of these materials.

In addition to a considerable number of bench tests for EP properties, several specialized gear tests are used by the equipment manufacturers. Differences in gear types and operating conditions make it difficult to compare the gear loading-carrying capacity of the various products.

Since the fluids must satisfy the lubrication requirements of hydraulic systems, gears and bearings, they are subjected to a range of tests such as the Vickers vane pump test (ASTM D 2882) and the FZG test for gearing. Additional pump testing such as the Denison HF-0 (piston and vane pumps) or the Sundstrand piston pump test may be also required.

17
Compressors

Compressors are manufactured in several types and for a variety of purposes. In addition to being used to compress gas, many compressors serve as blowers or can be used as vacuum pumps. Lubrication requirements vary considerably, depending not only on the type of compressor but also on the gas (including any contaminants) being compressed. In general, air and gas compressors are mechanically similar. Thus the main difference is in the effect of the gas on the lubricant and the compressor components. The lubricant plays roles in preventing wear, achieving sealing, minimizing viscosity dilution and additive reactions with the gas, and preventing corrosion. Refrigeration and air conditioning compressors require special consideration because of the recirculation of the refrigerant and mixing of the lubricant with it.

Compressors are classified as either positive displacement or dynamic. The positive displacement class includes reciprocating (piston) types, several rotary types, and diaphragm types. Dynamic compressors are either of the centrifugal or axial flow type, although mixed flow machines that combine some elements of both types are used.

Excluding refrigeration and air conditioning applications, in terms in the number of machines, more compressors are used to compress air for utility use than for any other purpose. Although many applications require high pressures, the vast majority of pneumatic equipment is designed for pressures between 90 and 100 psig, and therefore most compressed air systems are designed to operate between 100 and 125 psig. These requirements are met by both portable compressors used on construction projects, in mining, and in other outdoor applications and by stationary compressors used to provide plant air in applications ranging from service stations to industrial plant types. In the past, these were largely the province of reciprocating compressors, but large numbers of compressors of other types are now being used for reasons involving various factors such as design and metallurgical improvements, the need to increase speed capabilities, and the need to achieve size reduction.

In plant air applications, the vibration associated with medium-sized and larger reciprocating compressors generally necessitates either heavy vibration-absorbing mount-

Table 17.1 Effect of Staging on Discharge Temperatures[a]

Discharge pressure gage		Discharge temperature					
		One stage		Two stages		Three stages	
psi	kPa	°F	°C	°F	°C	°F	°C
70	483	398	203	209	98	—	—
80	552	426	219	219	104	—	—
90	621	452	233	226	109	—	—
100	689	476	247	238	114	—	—
110	758	499	256	246	119	—	—
120	827	519	271	254	122	182	83
250	1724	—	—	326	163	225	108
500	3447	—	—	404	207	269	132

[a] Temperatures based on adiabatic compression (actual temperatures will be lower due to heat losses within system).

ings or special isolating mountings. The pulsation in air delivery may require pulsation dampers. Rotary and centrifugal compressors, often supplied as packaged units, can be installed with minimal mounting requirements from the vibration absorption point of view. In comparison to reciprocating units, these packaged rotary units are more compact, offer nonpulsating flow, have fewer wearing parts, and require less maintenance.

Centrifugal and axial flow compressors deliver oil-free air. Positive displacement compressors are also available in nonlube designs, although these are not recommended for severe operating conditions such as high pressures (oil helps seal), high temperatures, or wet gas, or when there are corrosive elements in the gas. It should be recognized that many users of nonlube positive displacement compressors will use some lubricant, particularly in reciprocating cylinders, to provide longer life of components and improved sealing. Certain reactive gases, such as oxygen, require special consideration, and petroleum-based lubricants should not be used for gases of these types.

The improvements in the design of helical lobe (screw) compressors have resulted in higher pressure capability, with efficiencies approaching those of reciprocating compressors. This advantage, combined with the lower noise levels associated with rotary machines, also has increased the use of rotaries.

When one is considering compressor lubrication, it is necessary to recognize that compressing a gas causes its temperature to rise. The more gas is compressed, the higher will be its final temperature. When high discharge pressures are required, compression is carried out in two or more stages, with the gas being cooled between stages to limit temperatures to reasonable levels. This measure also improves compressor efficiency and reduces power consumption for the range of temperatures that can be reached—see Table 17.1, which is based on adiabatic* compression of air with an intake pressure of 14.7 psia (101.3 kPa abs), a temperature of 60°F (15.6°C), and intercooling between stages to the

* Adiabatic compression assumes that all the mechanical work done during compression is converted to heat in the gas. That is, the cylinder is assumed to be perfectly insulated, ensuring that no heat is lost from it.

same temperature. These temperatures, although higher than those reached in actual practice since some of the heat is removed by cooling the cylinder walls, are indicative of the temperatures that must be considered when selecting compressor lubricants.

I. RECIPROCATING AIR AND GAS COMPRESSORS

Reciprocating compressors are used for many different purposes involving mild conditions to extremes of pressure and volume requirements. As a result, a great variety of designs are commercially available. Most reciprocating compressors are of the single- or two-stage type, with smaller numbers of multistage machines—three, four, or more stages such as shown in Figure 17.1. From the lubrication point of view, single- and two-stage machines generally are similar, while multistage units may have somewhat different requirements, depending on pressures, temperatures, gas conditions, and the size and speeds of the pistons.

The principal parts common to all reciprocating compressors are pistons, piston rings, cylinders, valves, crankshafts, connecting rods, main and connecting rod bearings (crankpin bearings), and suitable frames that generally contain the lubrication system. Double-acting machines (which compress on both faces of the pistons—refer to Figure 17.2) require piston rods, packing glands, crossheads, and crosshead guides; the connecting rods are connected to the crossheads by crosshead pins. Crossheads and associated parts are also used in some multistage, single-acting compressors, but the majority of single-acting compressors are of the trunk piston type, with the connecting rods connected directly to the pistons by piston pins (wrist pins). For lubrication purposes, all parts associated with the cylinders, including pistons, rings, valves, and rod packing (on double-acting machines), are considered to be cylinder parts (Figure 17.3). All parts associated with the driving end, including main, connecting rod, crosshead pin or wrist pin bearings, and crankshaft and crosshead guides, are considered to be running parts or running gear. In many applications, lubricant requirements differ so substantially that there are two lube systems to separate the cylinder lubrication from the running gear lubrication.

Reciprocating compressors are provided with cooling facilities to limit the final discharge temperature to a reasonable value and to minimize power requirements. The cylinder walls and heads are cooled, and in the case of two-stage and multistage machines,

Figure 17.1 Multistage reciprocating compressors. (Courtesy of Ariel Corporation.)

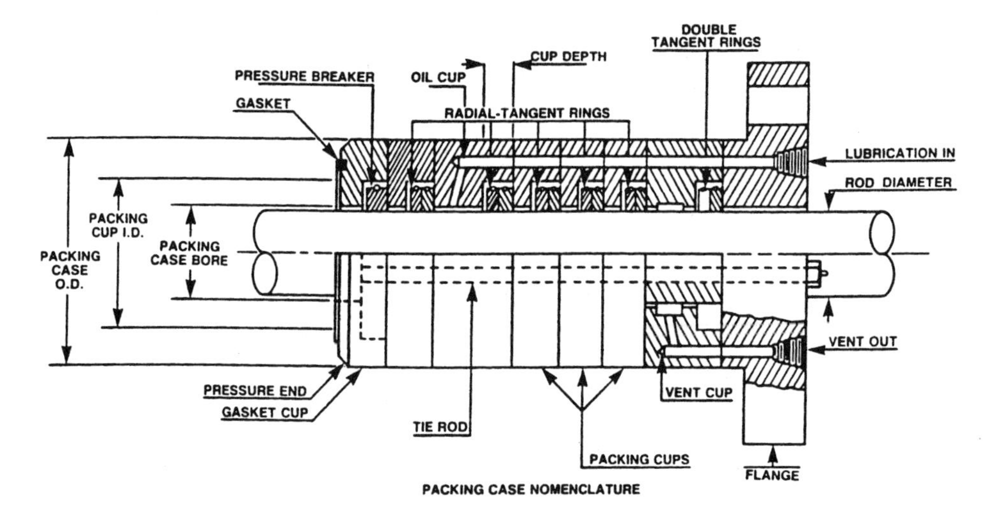

Figure 17.2 Mechanical rod packing design and operation. (Courtesy of Morris Compressor Supply, Inc.)

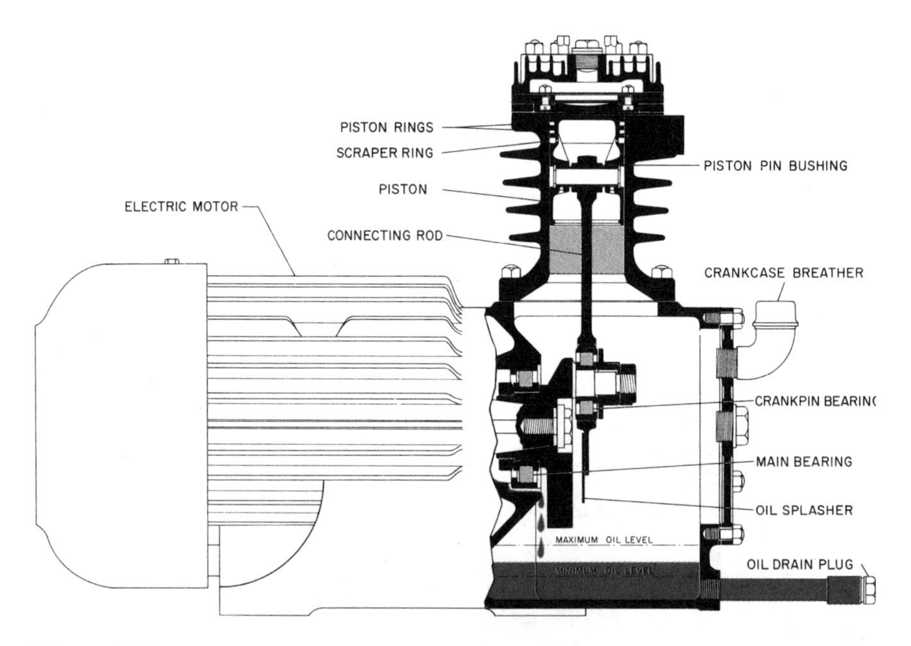

Figure 17.3 Vertical single-acting compressor. In this simple 3 hp machine designed for a discharge pressure of 100 psig (690 kPa), the motor and compressor are combined in a single compact unit. Lubrication is by splash system.

the gas being compressed is cooled between stages in intercoolers. Cooling can be by air or water, but in the larger machines, water cooling is usually required. In captive cooling water systems, glycol and inhibitors are used to minimize corrosion and any potential to freeze in cold operations or during shutdown periods. Frequently, the gas being cooled is further cooled in aftercoolers. In the case of air, this helps remove water and thus prevents or minimizes condensation of moisture in the air distribution system. Aftercoolers also act as separators to assist in removing oil that may be carried over from the cylinders. Air receivers are used in most large industrial systems, not only to provide a reserve to accommodate varying supply demands but also to reduce compressor pressure pulsations, add radiant cooling capacity, and allow further separation of moisture and oil carryover.

A considerable amount of moisture can be condensed in intercoolers. For example, in a two-stage compressor taking in air at atmospheric pressure, 70°F (21°C) and 75% relative humidity, and discharging at 120 psig (828 kPa), about 3.75 gallons (14 liters) of water per hour will be condensed in the intercooler for each 1000 ft^3/min (1700 m^3/h) of free air compressed. This moisture content is based on saturated air at the second stage suction at a pressure of 50 psig and 80°F. This condensation behavior has an influence on the lubrication of subsequent stages. Figure 17.4 can be used to calculate moisture levels condensed in intercoolers and aftercoolers for the various pressures, temperatures and humidity conditions.

A. Methods of Lubricant Application

In reciprocating compressors, the cylinders and running gear may be lubricated from the same oil supply, or the cylinders may be lubricated separately.

1. Cylinder Lubrication

Except when cylinders are open to the crankcase, oil is generally fed directly to the cylinder walls at one or more points by means of a mechanical force-feed lubricator. In a few cases, main oil feed to the cylinders is supplemented by an additional feed to the suction valve chambers (pockets). For some small-diameter, high pressure cylinders of multistage machines, oil is fed only to the suction valve chambers. Essentially, all the oil fed to cylinders, which are not open to the crankcase, is carried out of the cylinders by the discharging gas and collects in the discharge passages, piping, and other system components such as receivers.

Cylinders that are open to the crankcase are lubricated by oil thrown from the reservoir by means of scoops or other projections on the connecting rods and cranks. When this splash lubrication method is used, the pistons are provided with oil control rings similar to those used in automotive engines, which are designed to prevent excessive oil feeds to the cylinders.

Compressor valves require very little lubrication. Usually the small amount of oil required spreads to the valves from the adjacent cylinder walls or is brought to them in atomized form by the stream of air or gas. However, when air compressors operate under extremely moist suction conditions, it is sometimes necessary to provide supplementary lubrication by means of force-feed lubricator connections to the suction valve chambers.

Valve operators, such as used for unloader valves, may hold valves open or closed for certain types of pressure regulation system. These generally require very little lubrication. As with suction and discharge valves, a small amount of oil is carried over from adjacent cylinder walls or is brought to the valve operators in atomized form by the gas stream.

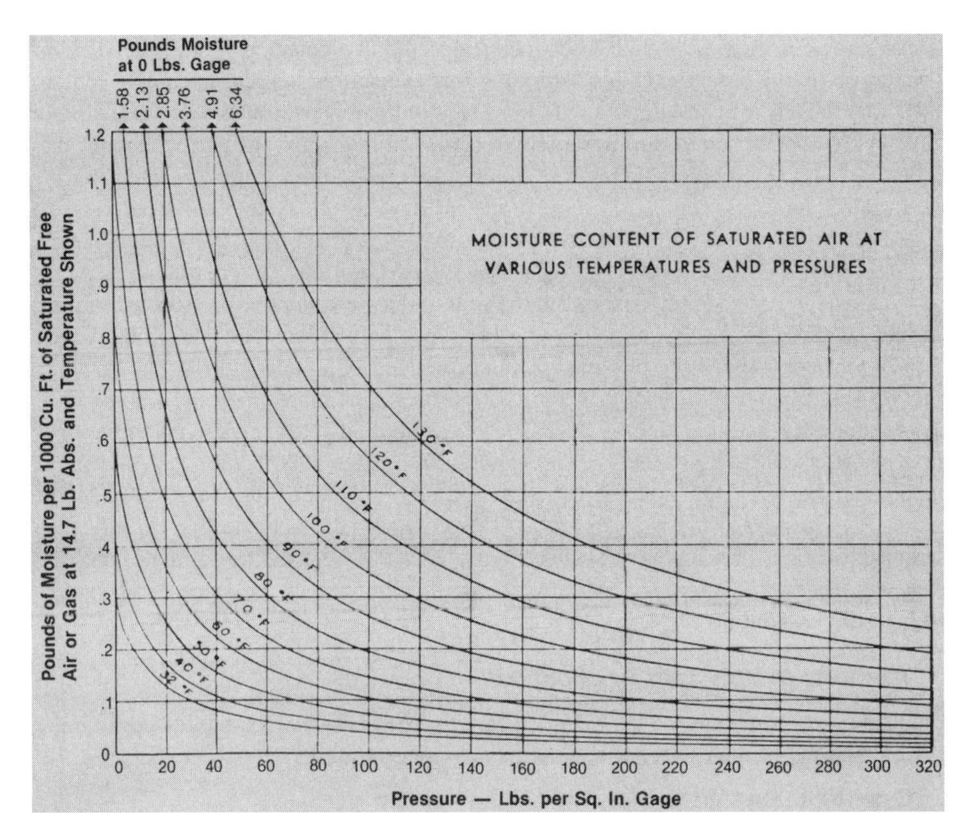

Figure 17.4 Moisture condensed in intercoolers and aftercoolers: example. Air at 0 psig, 80 F, and 70% relative humidity (RH) is taken into a single-stage compressor, compressed to 80 psig, and cooled in an aftercooler to 80°F. If the intake air had been saturated, it would have contained 1.58 lb of moisture per 1000 ft^3, as indicated at the left end the 80-degree line. At 70% RH, it actually contained 1.58 × 0.70, or 1.11 lb of moisture per 1000 ft^3. After compression and cooling, the air is saturated and contains 0.24 lb of moisture per 1000 ft^3 (of free air), as read on the vertical scale by projecting to the left of the intersection of the 80°F line and the 80 psig line. The rest of the moisture (1.11 − 0.24, or 0.87 lb/1000 ft^3) has been condensed in the aftercooler.

When metallic piston rod packing is used on double-acting machines, stuffing boxes are lubricated by means of force-feed lubricators. Usually, when nonmetallic packing is used, stuffing boxes are adequately lubricated by oil from the compressor cylinders. In some cases, however, mechanical force-feed lubricators (or drop feed cups in older compressors) are used.

2. Bearing (Running Gear) Lubrication

In practically all reciprocating compressors, the oil for lubrication of the running gear is contained in a reservoir in the base of the compressor. Oil from the bearings, crosshead, or any cylinder open to the crankcase, drains back to the reservoir by gravity. However, a variety of methods or combinations of methods are used to deliver oil from the reservoir to the lubricated parts.

Oil may be delivered to the lubricated parts entirely by splash. If this is done, a portion of, or projection from, one or more crankshafts or connecting rods dips into the oil and throws it up in a spray or mist that reaches all internal parts.

Many horizontal compressors have a flood system for bearing and crosshead lubrication (see Chapter 9, Figure 9.18). Oil is lifted from the reservoir by disks on the crankshaft and is removed by scrapers. The oil is then directed to the bearings by passages, or allowed to cascade down over the crosshead bearing surfaces.

A full-pressure circulation system is often used for running gear lubrication. A positive displacement pump draws oil from the reservoir and delivers it under pressure to the main bearings (if plain) and connecting rod bearings, then through drilled passages to the wrist pin bearings (bushings) and crosshead (if used). Where rolling element main bearings are used, the small controlled feed of oil required for them is commonly supplied by a drip or spray from the cylinder walls or rotating parts. Sometimes a jet of oil is directed toward these bearings. In some compressors oil is supplied under pressure to the connecting rod bearings from which it is thrown by centrifugal force to the cylinder walls. The wrist pins are then lubricated by oil scraped from the cylinder walls and directed to the pin bearing surfaces by drilled passages in the pistons.

B. Single- and Two-Stage Compressors

Industrial single- and two-stage compressors range in size from fractional-horsepower units to large machines of 20,000 hp (14,900 kW) or more. The smallest compressors are of the vertical single-acting type, usually air cooled (see Figure 17.3). Larger compressors are built in a variety of arrangements, including vertical single-acting (air- or water-cooled), vertical double-acting water-cooled (Figure 17.5), and horizontal balanced opposed piston compressor (Figure 17.6). The largest machines are of the latter type. Machines may be single cylinder or multi cylinder with cylinders in tandem, opposed, or in a V (Figure 17.7) or a W configuration (Figure 17.8). Assuming atmospheric conditions at the compressor suction, single-stage machines are available for pressures up to about 150 psig (1030 kPa). Two-stage machines are available for pressures up to about 1000 psig (6.9 MPa), although applications with pressures this high will generally use three or four compressor stages. Most two-stage compressors are designed for discharge pressures in the range of 80–125 psig (550–860 kPa).

1. Factors Affecting Cylinder Lubrication

The operating temperature in compressor cylinders is an important factor because of its effect on oil viscosity and oil oxidation, and on the formation of deposits. Since oil viscosity is reduced at high temperatures, when operating temperature is high, higher viscosity oils are required to maintain adequate lubrication films.

The thin films of oil on discharge valves, valve chambers (valve pockets), and piping are heated by contact with hot metal surfaces and are continually swept by the heated gas as it leaves the cylinders after compression. This is a severe oxidizing condition, and all compressor oils oxidize to an extent that depends on the conditions to which they are exposed as well as on their ability to resist this chemical change. Oil oxidation is progressive. At first, the oxidation products formed are soluble in the oil, but as oxidation progresses, these materials become insoluble and are deposited, mainly on the discharge valves and in the discharge piping, which are the hottest parts. After further baking, these deposits are converted to materials that are high in carbon content. These materials, along with contaminants that adhere to and bond with them, are commonly called carbon.

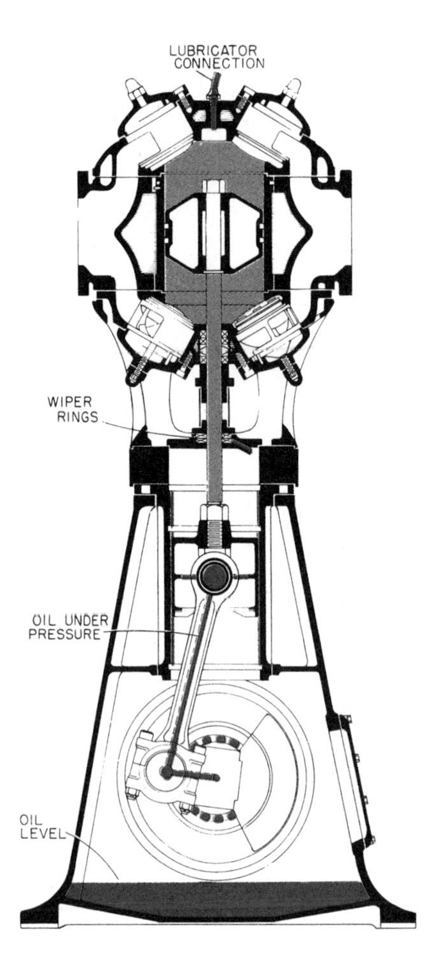

Figure 17.5 Vertical double-acting compressor. This single-stage machine uses rolling element main bearings. A pressure circulation system supplies oil to the connecting rod bearings and through-drilled passages in the connecting rods to the crosshead guides. Oil draining back from the crosshead area is broken up into a fine mist, which lubricates the main bearings. A mechanical force-feed lubricator supplies oil to the cylinders and wiper rings.

Deposits on discharge valves may interfere with valve seating and permit leakage of hot, high pressure gas back into the cylinders. This high temperature gas heats the gas taken in on the suction stroke, causing the temperature at the start of compression to be raised, and also the final discharge temperature. This so called recompression results in efficiency losses as well as reduced flow capacity. Since the action is repeated on each stroke, the effect is cumulative, although for a constant rate of leakage, there is a tendency for the temperature to level off at some higher value. If leakage increases, there is a further increase in temperature.

Deposits also restrict the discharge passages and cause cylinder discharge pressures to increase and overcome the pressure drop caused by the restriction. Again, an increased discharge temperature accompanies the increased discharge pressure. Abnormally high

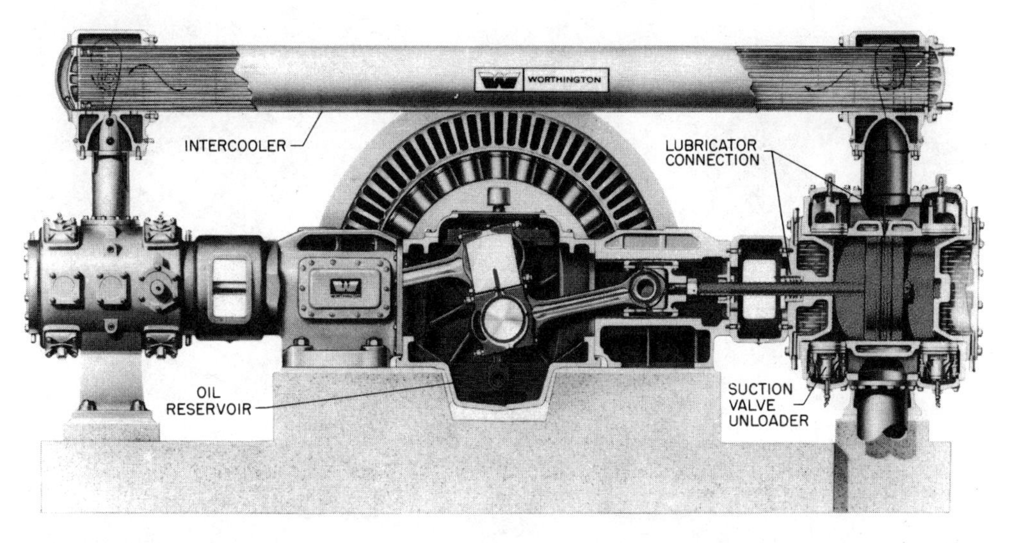

Figure 17.6 Horizontal balanced opposed piston compressor. This construction is used for very large capacities. There may be two to eight cylinders, each pair being arranged as shown. Oil is supplied to the cylinders and packings by a mechanical force-feed lubricator (not shown) and to the running parts by a pressure circulation system. A gear pump, driven from the crankshaft or a separate motor, delivers oil to a main distribution line and from there to the precision insert main and connecting rod bearings, the bronze crosshead pin bushings, and the crosshead slippers.

discharge temperatures due to these effects result in more rapid oil oxidation and, therefore, contribute to further accumulation of deposits and temperature rise. If not properly addressed, this cycle may eventually result in a fire or explosion.

A variety of contaminants are often present in the gas being compressed. Hard particles abrade cylinder surfaces and may interfere with ring and valve seating. Some contaminants promote oil oxidation by catalytic action, and some entrained chemicals may react directly with the oil to form deposits. Solid deposits adhere to oil-wetted surfaces and contribute to deposit buildup on discharge valves and in discharge passages. In many cases, the deposits found in compressors are largely composed of contaminants with a relatively small proportion of carbonaceous materials from oil oxidation. If contaminants of these types are found, gas filtration on the suction side of the compressor should be improved.

All the oil fed to compressor cylinders is subjected to oxidizing conditions. Since most of the oil fed to cylinders eventually leaves through the discharge valves where temperatures are highest, keeping the amount of oil fed to a minimum helps to minimize the formation of deposits on these areas as well as to reduce excessive oil carryover to downstream equipment. Recommended base rates of oil feed to cylinders are shown in Table 17.2. Manufacturers of gas compressors generally recommend that oil be supplied to compressor cylinders at the rate of 1 pint for each 2 million ft^2 of area swept by the piston. The Compressed Air Institute suggests the use of 1 pint of oil for each 6 million ft^2 of area swept by the piston in air compressors. These are general recommended starting points and may need to be adjusted (up or down) based on gas and operating conditions in the compressor.

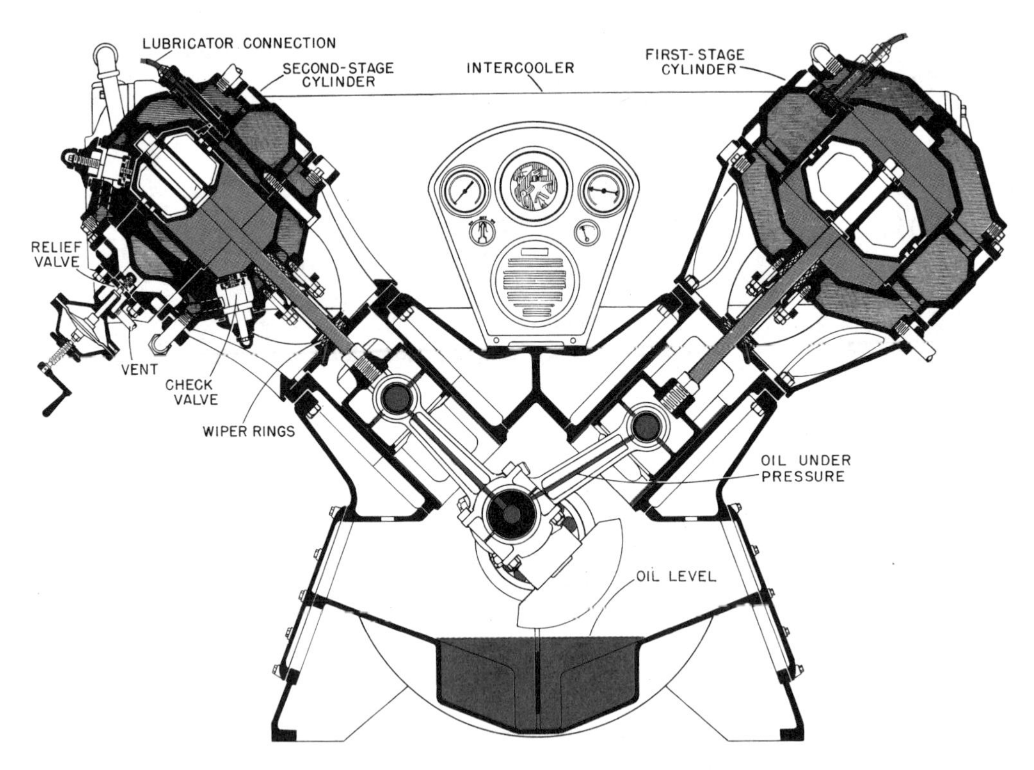

Figure 17.7 V-type two-stage double-acting compressor. The cylinders are lubricated by means of a mechanical force-feed lubricator that is geared to the crankshaft. Oil is delivered to the bearings through drilled passages by a pressure circulation system.

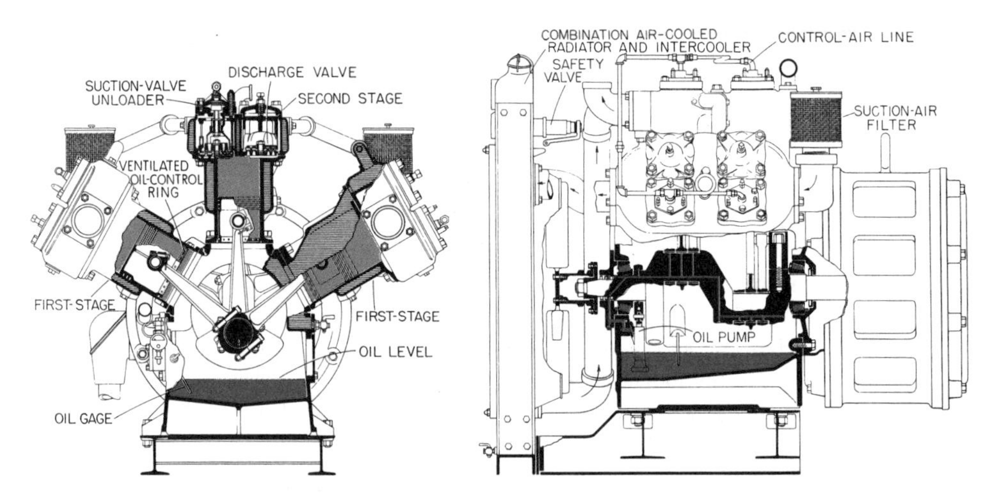

Figure 17.8 W-type two-stage single-acting compressor. The first stage has two cylinders in each bank, while the second stage has only one cylinder for each low pressure bank. Two stages of six cylinders form the complete compressor. The radiator contains two sections: an air-cooled intercooler and a cooler for the cylinder jacket water. The connecting rod bearings are lubricated by pressure from a plunger pump. The tapered roller main bearings and cylinder walls are lubricated by oil thrown from the connecting rod bearings.

Table 17.2 Reciprocating Air Compressor Lubrication Guide (Dry Air)[a]

Cylinder diameter (in.)	Piston displacement (ft^3/min)	Rubbing surface (sfm)	Oil feed per cylinder drops/min[b]	Oil feed per cylinder pints/day
Up to 6	Up to 65	Up to 500	0.66	0.12
6–8	65–125	500–750	1	0.18
8–10	125–225	750–1100	1.33	0.24
10–12	225–350	1100–1500	1–2	0.34
12–15	350–600	1500–2000	2–3	0.48
15–18	600–1000	2000–2600	3–4	0.62
18–24	1000–1800	2600–3600	4–5	0.86
24–30	1800–3000	3600–4800	5–6	1.15
30–36	3000–4500	4800–6000	6–8	1.44
36–42	4500–6500	6000–7500	8–10	1.80
42–48	6500–9000	7500–9000	10–12	2.16

[a] Oil feed to cylinders is in drops per minute and pints per day based on 8000 drops per pint at 75°F (24°C). To use this table for cylinder feed rates for gases other than air, multiply the feed rates shown times 3, which will provide the equivalent 1 pint for each 2 million square feet of swept cylinder surface. These are base starting points and may need adjustments according to gas conditions and operating parameters.
[b] The oil feed rates given are for water-filled gravity and vacuum-type site-feed lubricators. For glycerine-filled sight-feed lubricators, multiply the feed rates in drops per minute by 3 to achieve the listed pints per day.

Moisture is a factor in single- and multistage air compressors principally because of condensation that occurs in the cylinders during idle periods when cylinders cool below the dew point of the air remaining in them. The water formed tends to displace the oil films, exposing the metal surfaces to rusting. The amount of rust formed during any single idle period may be minor and indeed may be scuffed off as soon as the compressor is started again. In time, however, the process will result in excessive wear. In addition, rust tends to promote oil oxidation, and the rust particles contribute to accelerated formation of deposits. If this potential exists, the use of oils that have superior rust-inhibiting qualities and are fortified with effective additives that will adhere to metal surfaces should be considered. Oils of this type will help reduce the potential of the moisture and other liquids from contacting the metal surfaces during idle periods. These oils will also help during operation.

2. Factor Affecting Running Gear Lubrication

In general, the factors that affect compressor bearing lubrication—loads, speed, temperatures, and the presence of water and other contaminants—are moderate. The main requirement for adequate running gear lubrication is that the oil be of suitable viscosity at operating temperature.

Much of the oil in circulation in compressor crankcases is broken up into a fine spray or mist by splash or oil thrown from the rotating parts. Thus, a large surface of the oil is exposed to the oxidizing influence of warm air, and oil oxidation will occur at a rate and to an extent depending on the operating condition and the ability of the oil to resist this chemical change. Conditions that promote oil oxidation in crankcases are mild in comparison to the oxidizing conditions in compressor cylinders, discharge valves, and discharge piping. However, the lubricant in the crankcase may remain in service for thousands of hours, as opposed to cylinder lubricant, which is continually replenished.

C. Multistage Compressors

Three-stage compressors for continuous duty are available up to 2500 psig (17 MPa). Four-, five-, six-, or seven-stage machines are available up to 60,000 psig (414 MPa) or higher. High discharge temperatures may or may not accompany these high discharge pressures. When possible, the machines are designed for compression ratios of 2.5 : 1 to 4.0 : 1 per stage, with 60°F (15.6°C) suction temperature and adequate cooling. This practice limits discharge temperatures from each stage to less than 375°F (190°C). However, compression ratios as high as 6 : 1 are often employed, and with 60°F (15.6°C) suction temperature, the adiabatic discharge temperature is well over 400°F (204°C). Further, with a suction temperature of 110°F (43°C), which is not uncommon for the later stages, the discharge temperature may approach 500°F (260°C). Even higher compression ratios and discharge temperatures may be encountered in compressors designed for intermittent duty.

1. Factors Affecting Lubrication

The same factors discussed in connection with single-and two-stage compressors—operating temperatures, contaminants, and oil feeds—also affect the lubrication of cylinders in multistage compressors. In addition, the lubrication of multistage compressors is often affected by the entrainment of water and oil in the suction gas carried to the higher pressure stages, and by the high cylinder pressures.

Water in the form of droplets is often entrained in the air leaving the intercoolers and carried into the higher pressure stages of multistage air compressors. This water, moving at high velocity into relatively small cylinders, tends to wash the lubricant film from the cylinder walls. If an unsuitable oil is in service, this results in inadequate lubrication and excessive wear, incomplete sealing against leakage, and exposure of metal surfaces to rust and corrosion.

Some of the oil carried into the intercoolers usually is entrained with the water and carried into the high pressure cylinders. While this oil contributes to lubrication of the succeeding stages, the results are not necessarily beneficial. This oil has already been exposed to oxidizing conditions in the lower pressure stages, and thus the total exposure is increased two or more times. As a result, these conditions could contribute to accelerated buildup of deposits even in the absence of excessively high temperatures.

High cylinder pressures acting behind the piston rings increase the rubbing pressure between the rings and cylinder walls. In addition, in trunk piston compressors, the connecting rod produces a thrust force against the cylinder wall with considerable pressure that increases with increasing compression pressure. The rubbing surfaces involved are parallel; movement between them does not tend to form thick oil films; and as pressure increases, there is a greater tendency to wipe away the thin films that are formed.

The lubrication of the running gear of multistage compressors presents essentially the same problems as single- and two-stage machines.

2. Lubricating Oil Recommendation

As mentioned at the outset, the lubrication of compressors is a function not only of the type of compressor but also of the gas being compressed. In general, gases can be considered to be of four types: air, inert gases, hydrocarbon gases, and chemically active gases.

There are marked differences between compressor cylinder and bearing lubrication, but, in many cases, it is possible to use a single oil for the lubrication of both. In some cases, the oil required to meet the cylinder lubrication requirements may not be suitable for bearing lubrication. For example, it may be too high in viscosity, may contain special compounding that is not compatible with materials used in some bearings, or may be of a special type that is not suitable for the extended service expected of bearing lubricating oils. Except in the case of air compressors, the following discussion pertains to the characteristics of oils for cylinder lubrication. The oils used for bearing lubrication of air compressors are usually suitable for the lubrication of bearings of compressors handling other gases. Where a problem might exist, the crankcase lubrication system is usually adequately isolated from contamination by the gas being compressed or the cylinder lubricant.

(a) Air Compressors. The oils recommended for, and used in, air compressors vary considerably. Such factors as discharge pressure and temperature, ambient temperature, moistness or dryness of the air, and design characteristics of the machine must all be considered in the selection.

Single- and two-stage trunk piston type stationary compressors operating at moderate pressures and temperatures with dry air are generally lubricated with premium rust- and oxidation-inhibited oils. In portable service, these compressors may be lubricated with detergent-dispersant engine oils, typically oils for API Service SH, SJ, CE, CF, CG-4, or CH-4. The engine oils are also being used in many stationary compressors in which the air is moist, or deposits or wear problems have been experienced with circulation or turbine-type oils. The viscosity grade used is frequently ISO VG 100 (SAE 30), but both lower and higher viscosity oils are used, depending on ambient temperatures and machine requirements.

Under mild conditions, rust- and oxidation-inhibited turbine oils are also used as cylinder lubricants for crosshead-type compressors. Under wet conditions, compounded oils are used. The compounding may be either a fatty oil or synthetic materials. During the break-in period, either for a new or rebuilt compressor, higher viscosity oils are used and oil feed rates are increased.

When high discharge temperatures have resulted in rapid buildup of deposits on valves and in receivers with conventional lubricating oils, synthetic oils (usually diester or polyglycol based) are often used. The straight synthetics offer better resistance to thermal degradation, as well as the ability to dissolve and suspend potential deposit-forming materials. These lubricants are usually of ISO VG 100 or 150, although higher viscosity blends are used for extremely high pressure cylinders or for severe moisture conditions.

In a number of cases in which receiver fires have been experienced, fire-resistant compressor oils are used. Phosphate ester based oils are one example of a fire-resistant oil that is used in compressors as well as other applications for which fire-resistant oils are desirable. For larger machines and higher pressures, or with moist air, the viscosity is usually ISO VG 100 (SAE 30) or ISO VG 150 (SAE 40). Lower viscosity grades may be used under moderate conditions with dry air.

When there is a need for oil-free air, reciprocating compressors, which operate without cylinder lubrication, are used. Nonlube rotaries and dynamic compressors can also be used for applications calling for oil-free air. The oil-free reciprocating compressors have polytetrafluoroethylene (PTFE), carbon, or filled composition rings and rider bands, which do not require lubricant. In some newer designs, the pistons may also be filled plastic composition or contain composition buttons in the piston skirts to prevent contact of the piston with the cylinder surfaces.

The running gear of crosshead compressors is lubricated normally with premium rust-and oxidation-inhibited circulation oils. In some cases in which synthetic oil is being used as the cylinder lubricant, the same oil is recommended for the running gear, offering the potential of extending crankcase oil drain intervals.

(b) Inert Gas Compressors. Inert gases do not react with the lubricating oils and do not condense on cylinder walls at the highest pressures reached during compression. Examples are nitrogen, carbon dioxide, carbon monoxide, helium, hydrogen, and neon. Ammonia is relatively inert, but some special considerations apply to it.

The inert gases generally do not introduce any special problems, and they can be handled suitably by the lubricants used for air compressors. However, carbon dioxide is slightly soluble in oil and tends to reduce the viscosity of the oil. If moisture is present, carbonic acid, which is slightly corrosive, will form. To minimize the formation of carbonic acid, the system should be kept as dry as is practical. To counteract the dilution effect, higher viscosity oils than those normally used in air compressors are desirable.

Ammonia is usually compressed in dynamic compressors, but occasionally it may be compressed in positive displacement compressors. In the presence of moisture, it can react with some oil additives and oxidation products to form soaps. Ammonia is not compatible with antiwear compounds such as zinc dialkyl dithiophosphate (ZDDP), and oils containing additives of these types should not be used. Automotive engine oils and most antiwear-type hydraulic oils contain ZDDP. Ammonia may also dissolve in the oil to some extent, resulting in viscosity reduction. Highly refined straight mineral oils are usually used. Synthesized hydrocarbon-based lubricating oils are also used because of their low solubility for ammonia.

When gases are compressed for human consumption, such as carbon dioxide for use in carbonated beverages, carryover of conventional lubricating oils is undesirable. Generally, medicinal white oils are required for cylinder lubrication under these circumstances. This does not apply when air is being compressed for the manufacture of oxygen for human consumption. The subsequent liquefaction of the air and distillation to separate the oxygen will leave the oxygen free of lubricating oil carryover. To minimize ''burnt oil'' odor that can be carried over, however, care is required in selecting the lubricant, controlling the rate of oil feed, and keeping the system clean. If conventional lubricants are used for this purpose, their effect on any catalysts should be evaluated, and steps should be taken to ensure that the oil does not contact the oxygen.

Petroleum hydrocarbons in the lungs can cause suffocation and possible pulmonary disease. Therefore, air compressors for scuba diving equipment (breathing air) should use nonlube compressors.

Under some circumstances, conventional petroleum oils cannot be used in inert gas compressors. This is the case in some process work, where traces of hydrocarbons cannot be tolerated in the process gas or some constituents of lubricating oils might poison catalysts used in later processes. Compressors similar to those used to produce oil-free air or systems equipped with sophisticated filtration and conditioning equipment are used where hydrocarbon carryover cannot be tolerated. Both special low sulfur, straight mineral oils and polybutenes are used where carryover of conventional oils might poison catalysts. Another example would be in paint booth applications, where carryover of silicon-type additives would create problems on surfaces to be painted (fish eyes).

In some cases, a compressor may be used to compress alternately an inert gas and a chemically active gas: for example, hydrogen and oxygen. Petroleum products form an explosive combination with oxygen. Therefore, they must not be used where oxygen is being compressed. The lubricant or lubrication system used for the compression of oxygen must also be used with the inert gas.

(c) Hydrocarbon Gas Compressors. More horsepower is consumed in compressing natural gas than any other gas except air. When the volumes of other hydrocarbons that are compressed for the chemical and process industries are considered, the total horsepower consumed in compressing hydrocarbons is extremely large. Dynamic compressors usually are used if the hydrocarbons must be kept free of lubricating oil contamination, but if high pressures are required, reciprocating compressors are used. With improved technology and the ability of some rotary compressors to achieve higher pressure and volume capacities, there is also a trend toward the use of rotaries in hydrocarbon compression.

While natural gas is mainly methane, other gases usually are present in small portions. These include ethane, carbon dioxide, nitrogen, and heavier hydrocarbon gases. The heavier hydrocarbon gases are similar in many respects to the hydrocarbons that are compressed for process purposes. Occasionally these heavier hydrocarbons are in liquid form, which complicates the lubricant selection process.

The temperature at which a material will condense from the gaseous state to the liquid state (also the temperature at which it will pass from its liquid state to the gaseous state, i.e., its boiling point) increases with increasing pressure. With the higher boiling point, heavier hydrocarbons, the condensation temperature may be above the cylinder wall temperature at the pressure in the cylinders. The condensate formed under this condition will tend to wash the lubricant from the cylinder walls and dissolve in the lubricating oil, resulting in viscosity reduction. Using an oil that is somewhat higher in viscosity than would be used for air under the same operating conditions can generally compensate for the dilution effect. Generally, compounded oils help to resist washing where condensed liquids are present in the cylinders. It is usually advisable also to operate with somewhat higher than normal cooling jacket temperatures, to minimize condensation. This also requires the use of higher viscosity oils.

Natural gas that contains sulfur compounds as it comes from the well is referred to as ''sour'' gas. Compressors handling sour gas are usually lubricated with detergent-dispersant engine oils—automotive or natural gas engine oils. These oils provide better protection against the corrosive effects of sulfur. The viscosities used most frequently are ISO VG 100 and 150 (SAE 30 and 40); but if the gas is wet (i.e., carrying entrained liquids), heavier oils may be used. The compressor of integral engine–compressor units is usually lubricated with the same oil used in the engine. But depending on the contaminants contained in the gas, compressor cylinders may require a lubricant different from that used in the engine crankcase.

(d) Chemically Active Gas Compressors. Among the chemically active gases that must be considered most frequently are oxygen, chlorine, hydrogen chloride, sulfur dioxide, and hydrogen sulfide.

Petroleum oils should not be used with oxygen because they form explosive combinations. Oxygen compressors with metallic rings have been lubricated with soap solutions. Compressors with composition rings of some types have been lubricated with water. Compressors designed to run without lubrication are also being used. Some of the inert synthetic lubricants, such as the chlorofluorocarbons or fluorinated oil, can be used safely and provide good lubrication. Dry-type solid lubricants such as Teflon or graphite can be used to minimize metal-to-metal contact in this service.

Petroleum oils should not be used for the lubrication of chlorine and hydrogen chloride compressors. These gases react with the oil to form gummy sludges and deposits. If the cylinders are open to remove these deposits, rapid corrosion takes place. Compressors designed to run without lubrication are used. Diaphragm and nonlube rotary compressors are also used for the corrosive and reactive gases.

Sulfur dioxide dissolves in petroleum oils, reducing the viscosity. It may also form sludges by reacting with the additives in the presence of moisture, or by selective solvent action. The system must be kept dry to prevent the formation of acids. Highly refined straight mineral oils or white oils from which the sludge-forming materials have been removed, either by acid treating or severe hydroprocessing of the base stocks, are often chosen. Oil feed rates should be kept to a minimum.

Hydrogen sulfide compressors must be kept as dry as possible because hydrogen sulfide is corrosive in the presence of moisture. Compounded oils are usually used, and rust and oxidation inhibitors are considered to be desirable.

II. ROTARY COMPRESSORS

The five main types of rotary positive displacement compressor are straight lobe, rotary lobe, helical lobe (more commonly referred to as screw compressors), rotary vane, and liquid piston. There are many design variations available for each of these types, based on application requirements. They can be single- or multiple-stage units that are designed for low pressure/high flow to relatively high pressure. The lubricant coming in contact with the gas being compressed for rotary screw and rotary lobe compressors can either be dry (nonlube) or flooded lubrication. Dry lubrication is in reference to compressor rotors only. Rotary vane compressors almost always are flooded, while liquid piston units almost always are nonlube.

A. Straight-Lobed Compressors

Straight-lobed compressors are built with identical two- or three-lobed impellers that rotate in opposite directions inside a closely fitted casing (Figure 17.9). Timing gears outside the case drive the impellers, and these gears maintain the relative positions of the impellers. The impellers do not touch each other or the casing, and no internal lubrication is required. No compression occurs within the case, since the impellers simply move the gas through. Compression occurs because of backpressure from the discharge side. Compression ratios are low, and for this reason these machines are often referred to as blowers rather than compressors. Straight-lobed compressors are available in capacities up to about 30,000 ft^3/min (51,000 m^3/h) and for single-stage discharge pressures up to about 25 psig (172 kPa).

1. Lubricated Parts

The principal parts of straight-lobed compressor requiring lubrication are the timing gears and the shaft bearings. The bearings may be either plain or rolling element (usually roller) type. Timing gears are precision spur or helical type. The bearings and gears are generously proportioned to minimize unit loads and wear, since small clearances between impellers and casings must be maintained for efficient operation. For most bearings and gearing, lubrication is by means of an integral circulation system. Some rolling element bearings at the end opposite the drive may be grease lubricated.

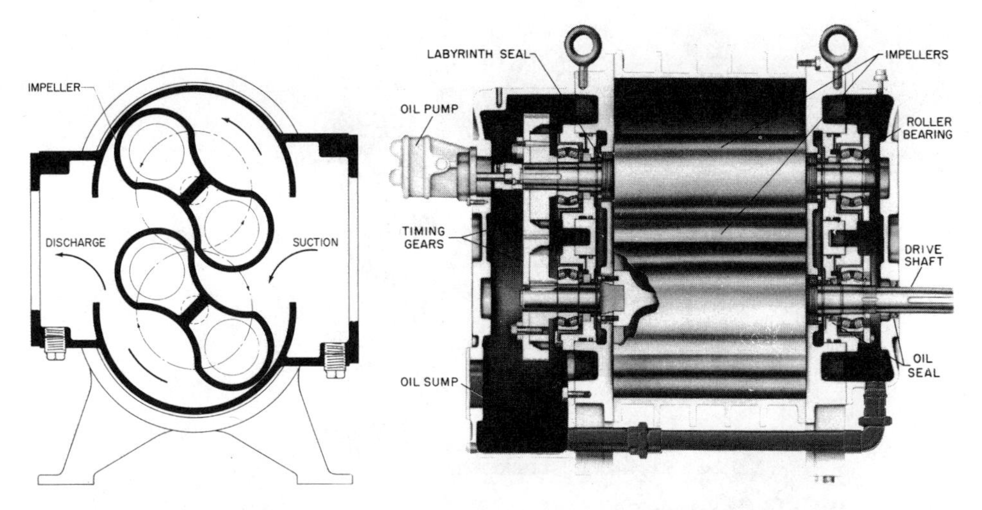

Figure 17.9 Rotary straight-lobed compressor. As the impellers rotate, air (or gas) is drawn in at the suction opening, trapped between the impellers and the casing, then forced into the discharge area against the pressure of the system. Lubricating oil is supplied under pressure to the bearings and timing gears by means of an integral circulation system.

2. Lubricating Oil Recommendations

Lubricating oils used in straight-lobed compressors are either mineral turbine/circulating oil quality containing rust and oxidation inhibitors or synthetic lubricants (where temperature extremes are encountered and long life is desired). Since the lubricant does not contact the gas being compressed in most applications, the oil generally does not require highly additized or compounded oils. For high temperatures or with heavy loads, antiwear oils may be used to provide additional protection for the gears. Viscosities usually recommended are ISO VG 150 for normal ambient and operating temperature and ISO VG 220 for high temperatures. Synthetic lubricants are often used when extreme low temperatures are involved.

B. Rotary Lobe Compressors

Most rotary lobe compressors are designed for discharge pressures up to a maximum of 200 psig (1380 kPa) and outputs up to about 1100 ft^3/min. They are commonly used in applications similar to straight-lobed compressors such as blowers or when small amounts of compressed air may be required. The rotors (Figure 17.10) are supported by antifriction bearings, and their position relative to each other is controlled by timing gears. The rotary lobe compressors are available in both nonlube and flooded-lube designs.

1. Lubricated Parts

The main components of a rotary lobe compressor that require lubrication are the bearings and the timing gears. The lubrication is essential to prevent wear in the gears and bearings, since wear will result in loss of efficiency and possibly contact of the lobes with each other or with the cases. In the compressors of flooded type, the oil helps seal the clearances

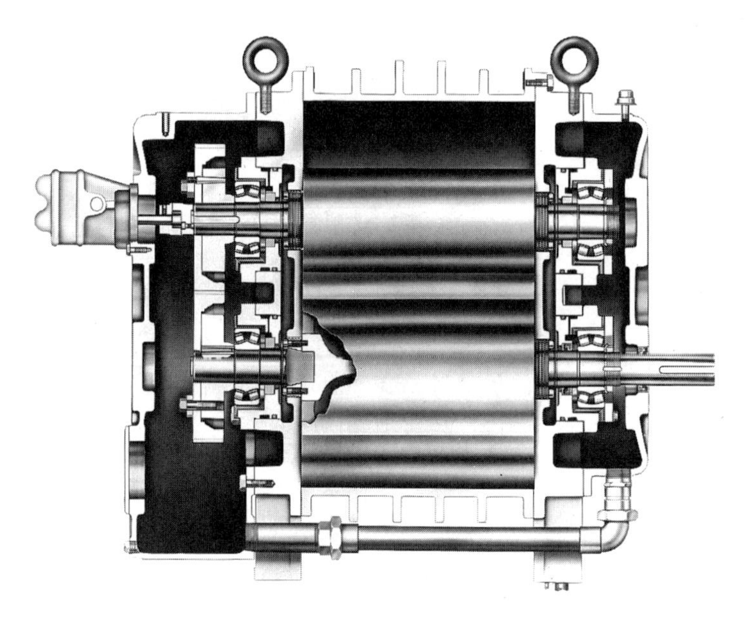

Figure 17.10 Rotary lobe compressor.

between the lobes and the cases, allowing greater pressure capability, helps cool the system, and also serves as a corrosion inhibitor in the presence of gases that contain moisture or other contaminants. In flooded applications, the lubricant introduced to the lobes goes out with the discharge gas and must be recycled back to the compressor. Generally, one common system is used for both the gears and bearings and the compressor lobes.

2. Lubricating Oil Recommendation

The lubrication requirements for the gears and bearings will generally be satisfied with an ISO VG 46 or 68 synthetic turbine or circulating-type oil. Depending on manufacturer and application, a different viscosity oil may be required. Oil selection will be based on the operating parameters of the compressor (nonlube or flooded) and the condition of the gas being compressed.

C. Rotary Screw Compressors

Also called helical lobe compressors, rotary screw compressors are available in single-impeller and the more common two-impeller (rotor) designs. In the two-impeller types, one common design uses a four-lobed male rotor meshing with a six-lobed female rotor (Figure 17.11). Timing gears may individually drive the rotors, or the male rotor may drive the female rotor. Gas is compressed by the action of the two meshing rotors as illustrated in Figure 17.12. The machines come in single- and multiple-stage units.

Two variations of two-impeller screw compressors are used. In the ''flood-lubricated'' type, the oil is injected into the cylinder to absorb heat from the air or gas as it is being compressed. The oil also functions as a seal between the rotors. Since oil is available in the cylinder (casing) to lubricate the rotors, these machines are now usually built without timing gears. They require an external circulation system to control the

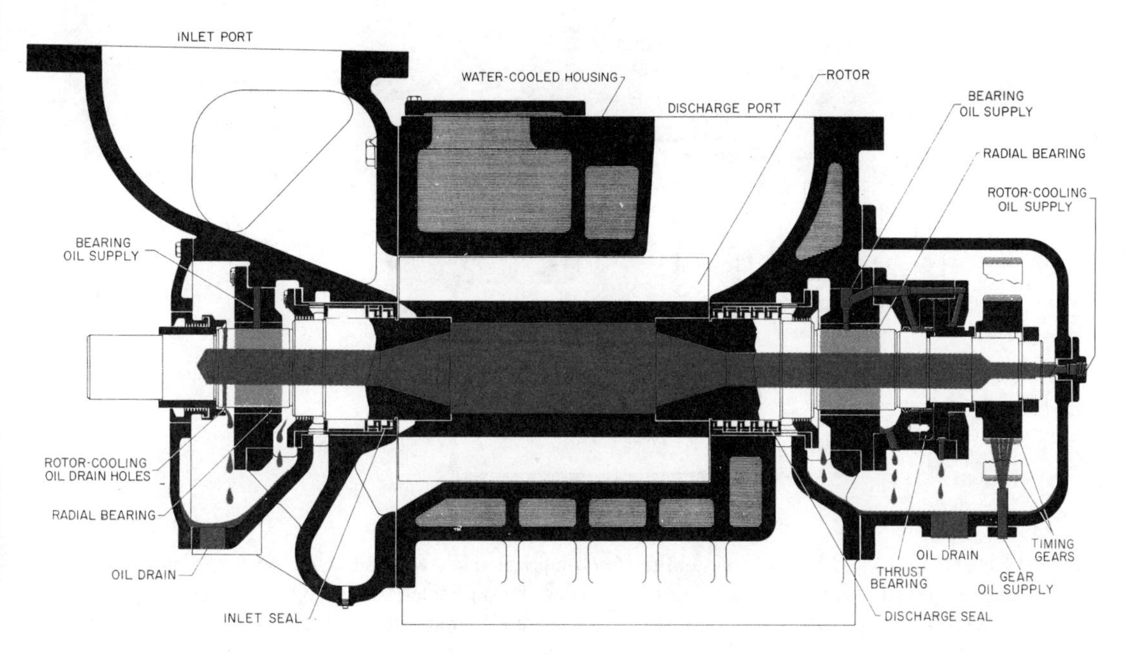

Figure 17.11 Two-impeller rotary screw compressor. This large stationary machine has timing gears, and no oil is used inside the casing. The machine is water cooled, with oil-cooled impeller shafts. Multiple carbon ring shaft seals are shown. Shaft bearings are plain, and a tilting pad thrust bearing is used.

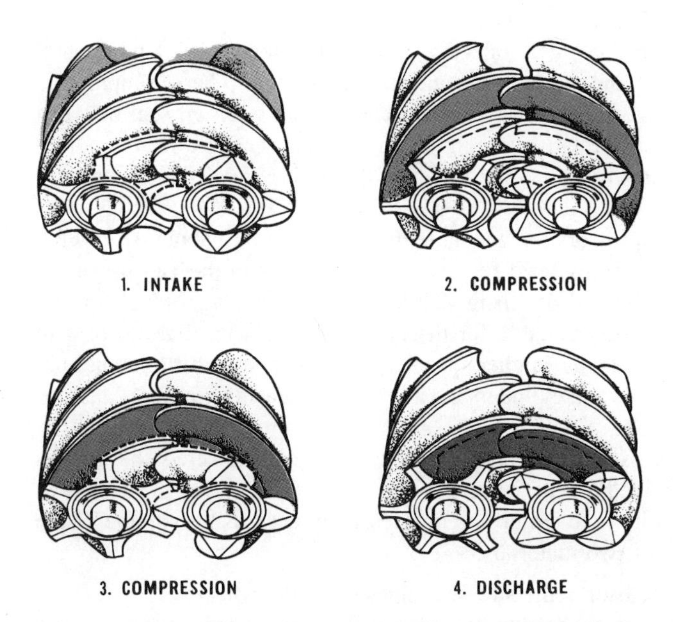

Figure 17.12 Compression in a screw compressor. As the screws unmesh at the rear and below the centerline of the view labeled ''intake,'' the expanding space created draws gas in through the inlet, filling the grooves between the screws. As the screws advance, they seal off this gas and carry it forward, compressing it against the discharge end head plate.

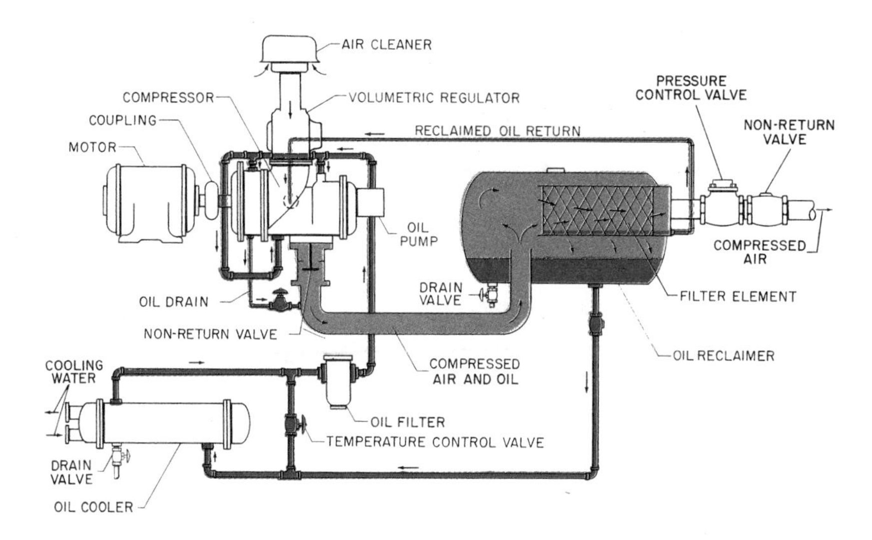

Figure 17.13 Lubrication system for flood-lubricated screw compressor. When the oil is cold, the temperature control valve is open, bypassing the oil cooler. As the oil warms up, the valve gradually closes, being completely closed when the oil temperature reaches about 180°F (82°C). All the oil then passes through the oil cooler. The arrangement provides rapid warm-up of the oil to minimize condensation but limits the maximum oil temperature to provide long oil life.

temperature of the oil and an oil removal system to remove most of the oil from the discharge air or gas (Figure 17.13).

In dry-screw compressors, no oil is injected. Since the rotors are not lubricated, timing gears are required to keep the rotors from contacting each other. These compressors can be used to deliver oil-free air or gas. However, because there is no oil seal between the rotors, operating speed must be relatively high to minimize gas leakage. Water cooling of the cylinder and such features as oil-cooled impeller shafts (Figure 17.11) are usually required.

Two-impeller screw compressors are available in capacities up to about 41,200 ft^3/min (70,000 m^3/h) and are commonly used for discharge pressures in the 125 psig (860 kPa) range for single-stage units and 350 psig (2408 kPa) for two-stage units. Special application screw compressors are available for discharge pressures up to 1500 psig in compound- and multiple-stage designs. Higher speed capabilities and closer tolerances in the clearance areas between rotors and cylinders are allowing increasing pressure capabilities and improved efficiencies.

The single-screw compressor (Figure 17.14) has a driven conical screw that rotates between toothed wheels. The teeth of the wheels sweep the thread cavities of the screw, compressing the air or gas as it is moved up the progressively decreasing volume of the cavity. The wheels are made of two materials:

A metal backing, which absorbs the stresses imposed on the wheels
A wider plastic facing, which contacts the sides of the thread cavities

Oil is injected into the case for lubrication, cooling, and sealing. As with flood-lubricated, two-impeller screw compressors, equipment for oil cooling and oil removal is required as part of the compressor running gear.

Figure 17.14 Single-screw compressor. The gear wheels are shown resting in the screw to indicate the manner in which they sweep the thread.

1. Lubricated Parts

The lubricated parts of dry-screw compressors are the gears and bearings. Shaft bearings may be either plain or rolling element, and thrust bearings may be either tilting pad or angular contact rolling element. Gears may be either precision-cut spur or helical type.

In flood-lubricated machines, in addition to the gears and bearings, the oil lubricates the contacting surfaces of the rotors. In both types, oil-lubricated seals may be used where gas leakage must be kept to a minimum.

2. Lubricant Recommendations

Flood-lubricated screw compressors in portable and plant air applications are lubricated by oils of ISO VG 32 (SAE 10) to ISO VG 68 (SAE 20). Heavy-duty engine oils suitable for API service classification SH, SJ, CE, CF, CG-4, or CH-4 are used widely in mobile screw compressors, as are automatic transmission fluids. Engine oils generally contain high levels of detergents and corrosion inhibitors, which tend to pick up moisture. This characteristic may affect selection based on the need to separate water from the oil. Premium quality rust- and oxidation-inhibited oil with good demulsibility characteristics should be considered if low ambient temperatures, cyclic operation, or very humid conditions are anticipated. Circulation and turbine-type oils offer improved water-separating characteristics.

Even with engine oils and automatic transmission fluids, problems with varnish and sludge can be encountered in severe operations. Polyalphaolefin-based synthetic lubricants are now the most commonly used oils in these applications. Other types of synthetic lubricant, such as diester, polyglycol, and synthetic blends, can also be used. These materials also allow the extension of drain intervals, which helps offset their higher initial costs.

Flood-lubricated screw compressors are being used for compressing gases other than air. One example is their expanded use in the field gathering of natural gas. Owing to improvements in design that permit greater pressure and volume capabilities, as well as efficiencies closely approximating those of reciprocating compressors, they are seeing increased usage with the compressing of hydrocarbon gases. Because of the intimate mixing of the oil with the gas, oil selection is even more critical than in reciprocating compressors. In handling hydrocarbon gases, polyglycol-based synthetic lubricants are used. The polyglycols have low solubility for hydrocarbons, and thus dilution of the lubricant is minimized and good separation of the oil and gas can be obtained. Synthesized hydrocarbon-based lubricants also provide an option for lubrication of screw compressors handling hydrocarbon gases.

Dry-screw compressors require lubrication of the gears and bearings only. Either synthetic or premium rust- and oxidation-inhibited circulating and turbine-type oils of a viscosity suitable for the gears are usually used. In some cases, oils with enhanced antiwear characteristics may be desirable for added protection of the gears.

D. Sliding Vane Compressors

Sliding vane compressors are positive displacement machines in which the vanes are free to move in slots in the rotor mounted eccentrically in a casing (cylinder) (Figure 17.15). Rotation of the rotor causes the vanes to move in and out of the rotor slots, creating pockets that increase and decrease in volume. The vanes can be held out against the case by coupling pins as in Figure 17.15, by springs, or by centrifugal force alone. The casing is not circular if pins are used, since all diameters through the axis of the rotor must be equal.

Sliding vane compressors are available in capacities exceeding 6000 ft^3/min (10,200 m^3/h) and discharge pressures exceeding 100 psig (690 kPa) for single-stage and 125 psig (860 kPa) for two-stage units.

Two general types of cooling are used for rotary vane compressors. Large stationary machines are water cooled. Portable and plant air machines may be air or water cooled, with the addition of oil injection into the cylinder. As with screw compressors, the latter machines may be referred to as flood lubricated, or by such terms as oil injection cooled

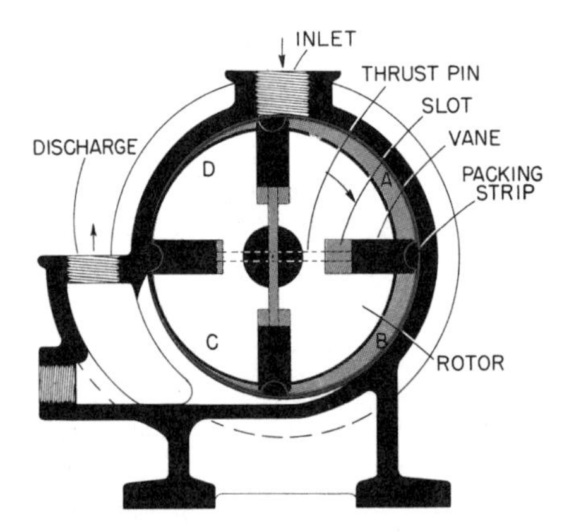

Figure 17.15 Rotary sliding vane compressor. The ports are located so that air (or gas) is drawn into pockets of increasing volume *A* and discharged from pockets of decreasing volume *B*. In this design the bore of the cylinder is not circular, since the total length of two vanes plus the pin, measured through the axis of the rotor, is constant.

or direct oil cooled. As with flood-lubricated screw compressors, an external system is required to remove the oil from the discharge gas and cool it.

1. Lubricated Parts

All the sliding surfaces in the cylinder require oil lubrication to minimize friction and wear. The lubricating oil also aids in protecting internal surfaces from rust and corrosion, adds a certain degree of cooling, and helps seal clearances between vanes, rotors, cylinder walls, and heads. Shaft bearings of either plain or rolling element type are designed for oil lubrication. Where the shaft passes through the head, packing glands or seals are used, and these are supplied with oil to minimize wear and friction and to assist in preventing leakage of high pressure air or gas.

2. Lubricant Recommendations

In sliding vane compressors, some cylinder surfaces are subjected to heavy rubbing pressures as a result of the action of the vanes in their slots. The forces acting on the extended vanes tend to tilt the vanes backward, causing heavy pressure at the inner leading edges of the vanes and the trailing edges of the slots. This combination results in high resistance to the inward motion of the vanes and, consequently, heavy pressure between the vane tips and the cylinder wall. Such pressure tends to rupture the oil films and cause wear and roughening of the surfaces involved. This process gradually diminishes efficiency. Inattention to good maintenance practices or selection of poorer quality oils could also contribute to deposit formation on vanes and in rotor slots, further restricting the free movement of the vanes in the slots.

Flood-lubricated rotary vane compressors are usually lubricated with oils of the same types recommended for flood-lubricated screw compressors. These oils give good protection against wear under the conditions encountered in the cylinders.

Large stationary rotary vane compressors differ considerably in their lubrication requirements from the flood-lubricated type. Water jackets are used to cool these machines. Cylinder lubricant is fed by force-feed lubricators. Feed rates, in relation to compressor capacity, must be much higher than for reciprocating machines. These machines tend to run cool at the inner cylinder walls, and thus condensation is a problem. Heavy-duty engine oils of SAE 30 or 40 grade (ISO VG 100 or 150) are widely used. Where discharge temperature is above 300°F (149°C), higher viscosity oil or the use of synthetic compressor oils is recommended. Compounded compressor oils of comparable viscosity are also used.

E. Liquid Piston Compressors

In a liquid piston compressor (Figure 17.16), rotation of a multibladed rotor causes a liquid, usually water, to be thrown outward, forming an annulus that rotates at rotor speed and maintains a seal between the rotor blades and the housing. As the annulus rotates, the pockets formed between the rotor blades and the annulus increase or decrease in volume as in a sliding vane compressor. Because of the annulus, the suction and discharge ports must be located in the cylinder heads, rather than on the outer perimeter of the cylinder. The liquid annulus eliminates the need for sliding, sealing surfaces, so there are essentially no wearing surfaces inside the cylinder.

Liquid piston compressors are available in capacities exceeding 16,000 ft^3/min (27,200 m^3/h) and for pressures up to 100 psig (690 kPa).

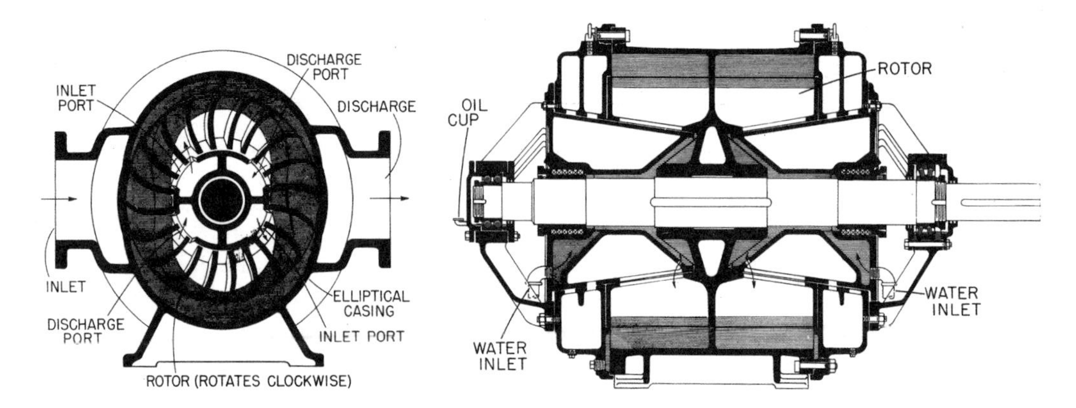

Figure 17.16 Rotary liquid piston compressor. The water annulus, in combination with the oval casing, forms pockets between the rotor blades that increase and decrease in volume much like those in sliding vane compressors. However, since there is no contact between the rotor blades and the casing, no internal lubrication is required. Water or other liquid medium is fed continuously to replace the liquid phase carried out with the compressed gas.

1. Lubricated Parts

The only parts of liquid piston compressors requiring lubrication are the shaft bearings. All models have rolling element bearings, which may be lubricated by oil bath or grease.

2. Lubricant Recommendations

Oil-lubricated bearings of liquid piston compressors are usually lubricated with premium rust- and oxidation-inhibited circulating and turbine quality oils. Viscosities are selected according to bearing requirements, as discussed in Chapter 8. Grease-lubricated bearings are lubricated with premium ball and roller bearing greases. NLGI grade 2 multipurpose greases are most often satisfactory.

F. Diaphragm Compressors

Diaphragm compressors have been around for many years in low pressure/low flow applications. Many of the small units consist of a simple diaphragm that is moved back and forth in a reciprocating motion either by a rod mounted to a rotating crank or by means of an eccentric cam. More recently, with improved designs, these units are being used for very high pressures with moderate flow capacities. A common usage is in refinery operations in the production of white oils, where pressures can exceed 50,000 psig. Compression ratios can exceed 20:1. Diaphragm compressors are also used in handling the corrosive or reactive gases found in many process industries, since exotic corrosion-resistant materials such as Inconel, Monel, and Hastelloy can be used for one diaphragms and casings. The vast majority of diaphragm compressor components are made of carbon steel or stainless steel, which also can handle many of the chemically reactive gases.

1. Lubricated Parts

The lubricated parts of a diaphragm compressor consist essentially of a hydraulic system. The lubricant acts as a hydraulic oil to provide the reciprocating motion of the ram that operates the diaphragm. This ram operates in a range of 200–400 cycles per minute, and hydraulic pressures can exceed 2000 psig (13.8 MPa). Depending on the medium being compressed and the discharge temperatures, the oil also functions to remove heat from the diaphragm. In severe operating conditions, oil coolers may be required.

2. Lubricant Recommendations

Typically, the oil used for diaphragm-type compressors is an ISO VG 68 or 100 antiwear-type hydraulic oil. Since the oil does not come in direct contact with the gas being compressed, it is not necessary to use highly additized or compounded oils for these systems. If there is any potential for contacting the gas being compressed, special considerations apply. If system oil temperatures are consistently high, synthesized hydrocarbon-type hydraulic oils are often used.

III. DYNAMIC COMPRESSORS

In contrast to positive displacement compressors, in which successive volumes of gas are confined in an enclosed space, which is then decreased in size to accomplish compression, dynamic compressors first convert energy from the prime mover into kinetic energy in the gas. This kinetic energy is then converted into pressure. The rotors of dynamic compressors do not form a tight seal with the casings; thus, some internal leakage occurs past the rotors. These clearances are getting much tighter and speeds much higher to allow greater efficiencies and higher pressures. The two main types of dynamic compressor are centrifugal and axial flow. Quite often, both these designs are incorporated into one single-shaft unit.

A. Centrifugal Compressors

In a centrifugal compressor, a multibladed impeller rotates at high speed in a casing (Figure 17.17). Air or gas trapped between the impeller blades is accelerated and thrown outward and forward in the direction of rotation. The air leaves the blade tips with increased pressure and very high velocity and enters a diffuser ring. In the diffuser ring, increasing area in the direction of flow causes a reduction in velocity and a substantial increase in pressure. The air or gas then enters a volute casing, where again increasing area in the direction of flow causes a further reduction in velocity and an increase in pressure. The outward flow of air through the impeller creates a reduced pressure at the inlet (or eye, of the impeller), causing air to be drawn into the compressor.

Centrifugal compressors are particularly adapted to supplying large volumes of air (or gas) at a relatively small increase in pressure. They are inherently suited for high speed operation—up to 60,000 rpm—and can, therefore, be direct-connected or geared to high speed driving motors or turbines. Speed increasing drives may also be used. Centrifugal compressors are built with capacities exceeding 650,000 ft^3/min (1.1 million m^3/h).

Compression ratios range up to about 5:1 with one impeller stage. A single-stage machine operating with atmospheric intake may be designed for discharge pressures as high as 60 psig (414 kPa). Multistage compressors with up to 10 stages are built. For example, a 10-stage internally geared compressor will have the final-stage pinion rotating at about 50,000 rpm and will be capable of discharge pressures as high as 2900 psig [200 bar(–20 MPa)].

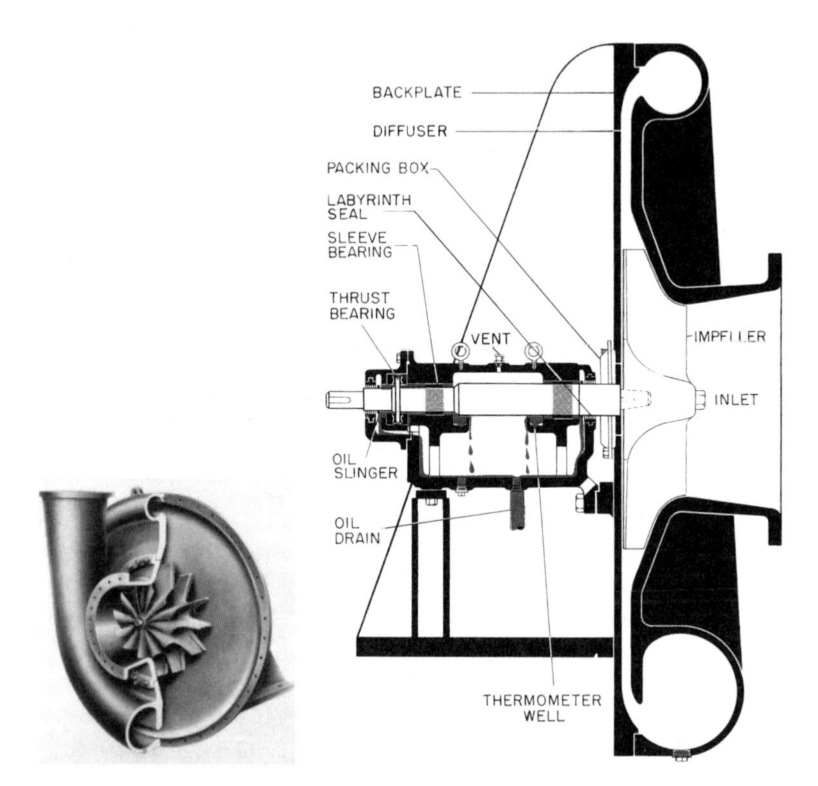

Figure 17.17 Single-stage centrifugal compressor. In this so-called pedestal machine, the impeller is overhung on a short shaft supported by two bearings mounted on a pedestal. The thrust bearing is of the fixed shoe design.

Centrifugal compressors are used extensively as boosters in petroleum refineries, in chemical process plants, and, to a lesser extent, in natural gas pipeline operations. In booster service, air or gas is taken in at or above atmospheric pressure and discharged at a still higher pressure. If suction pressures are high, casing and shaft seals must be designed to withstand them.

Centrifugal compressors are also built for plant air applications. These units are generally designed with capacities from 100 to 15,000 ft^3/m (1700–25,500 m^3/h) at discharge pressures generally in the range of 125 psig (860 kPa). The best performance from a pressure ratio standpoint is about 4 or 5 : 1 in a single-stage centrifugal compressor. From two to five stages are used, with three or four stages being the most common. Pinions mounted around a large bull gear drive the impellers. In this way, each impeller is driven at its optimum speed. If a turbine drive is used, the bull gear may be driven directly from the turbine. With an electric motor drive, the bull gear may be driven through a speed increaser. Intercooling between stages is normally used.

Centrifugal compressors discharging at low pressure usually have no provision for cooling. Higher pressure machines are cooled by water passages in the diaphragms between stages, or by injecting a volatile liquid into the gas stream. It should be noted that because of the very high velocities encountered in dynamic compressors, any liquid (water or other) or solid contaminants in the gas stream will cause erosion of the impellers and blading.

Where the shaft passes through the casing, seals are provided. Several types are used: soft braided packing, labyrinth seals, carbon ring seals, oil film seals, and mechanical (end face) seals. The mechanical seals are similar to those used for reciprocating compressors but designed for a rotating shaft. The type of sealing used depends on such factors as pressure, speed, gas being compressed, and the amount of leakage that can be tolerated.

1. Lubricated Parts

The bearings and gears (if used) of centrifugal compressors require lubrication, but there is no need for lubrication inside the impeller casing. Oil film seals and contact as well as noncontact (gas-pressurized, close-clearance labyrinth) seals are used on high pressure compressors. Contact-type seals must be supplied with oil to provide lubrication, and cooling, and to enhance sealing.

With a single impeller, or group of impellers all facing in the same direction, there is an axial thrust due to the backward force exerted by the discharging gas on the rim of the impeller. This thrust force can be balanced internally by means of a balancing drum or dummy piston, but a thrust bearing must be used to support all or part of the thrust load and to keep the impeller accurately located in the casing. Thrust bearings are generally angular contact ball, collar, fixed shoe, or tilting shoe type.

Shaft bearings of high speed units are generally plain (sleeve type), but rolling element bearings are used in some units. Plain bearings, including collar, fixed shoe, and tilting pad shaft bearings, are oil lubricated. Tilting pad shaft bearings are used on high speed dynamic compressors because of two desirable characteristics: they are self-aligning, and they exhibit good dampening. Rolling element bearings may be either oil- or grease-lubricated, depending on the operating parameters of the compressors and other component requirements, such as gears.

For compressor driven by a step-up gear set, the drive gears and one or more of the bearings may be lubricated from the same system. In the case of packaged plant air units, the bull gear and bearings are contained in an oil-tight casing, and oil is circulated to them under pressure from a shaft-driven oil pump. If a speed increaser is used, it is lubricated from the system that supplies the bearings.

2. Lubricant Recommendations

Owing to the high speed of most centrifugal units, the oil that is most commonly used is an ISO VG 32 synthetic or turbine quality oil containing rust and oxidation inhibitors. The high pitch line velocities on some of the gearing, particularly on the internally geared high pressure machines, will dictate the limitations on viscosity to avoid inlet shear heating in the gear mesh. Lower or higher viscosity oil may also be used, depending on compressor design and operation. Additional comments on lubrication that apply to both centrifugal and axial flow machines are given Section III.B.2 (lubrication recommendations for axial flow compressors).

B. Axial Flow Compressors

An axial flow compressor contains alternating rows of moving and fixed blades. High velocity is imparted by the moving blades to the air (or gas) being compressed. As the

Figure 17.18 Six-stage axial flow compressor. This compressor is equipped with plain journal bearings and a tilting shoe thrust bearing. Oil is supplied to the bearings by a circulation system.

air flows through the expanding passages between the fixed and moving blades, the velocity is reduced and transformed into static head, that is, pressure. A stage consists of one row of moving blades and one row of stationary blades (see Figure 11.8) and a relatively large number of stages—in some applications, more than 20 stages in one casing can be used. Axial flow compressors (Figure 17.18) are compact, high speed machines suitable for handling large volumes of air at compression ratios as high as 30:1 in a single cylinder or casing. Machines capable of handling 1 million ft^3/min (1.7 million m^3/h) or more have been built. They may be driven by electric motors, steam turbines, or gas turbines, the latter being fairly common in the chemical process industries, where large volumes of gas are available to drive the turbine, and large volumes of compressed air or gas are required. Axial flow compressors are also used in all except the smaller sizes of gas turbines (see Chapter 11: Stationary Gas Turbines), including industrial or aircraft-type that are used for power generation and ship propulsion.

1. Lubricated Parts

As with centrifugal compressors, the parts of axial flow compressors requiring lubrication are the shaft bearings, the thrust bearings that are used to take axial thrusts and maintain axial rotor position, and any oil-lubricated seals that may be used.

2. Lubricant Recommendations

Since the lubricants used in centrifugal and axial flow compressors are not exposed to the conditions inside the cylinder or casing, lubricant selection is based on the requirements of the bearings. If a speed increaser is supplied from the same lubricating oil system, the oil must be selected to meet the gear requirements, which are generally more severe than those of the bearings. Premium rust- and oxidation-inhibited circulating and turbine-type oils are used most commonly. If only bearings require lubrication, oils of ISO VG 32 are usually used. If a speed increaser is used, somewhat higher viscosity oils, typically ISO VG 46 or 68, may be required, depending on the gear pitch line velocities and other requirements of the gearing. In some limited cases, lower viscosity oils may be required.

In packaged plant air units, there is considerable tendency for air to be whipped into the oil, possibly resulting in foaming. Oils with good air release properties and foam resistance are preferred. The premium long life turbine oils in ISO VG 32 are usually recommended.

If a dynamic compressor is driven by a gas turbine, both machines may be on the same lubrication system. The lubricant selection is then based on the requirements of the gas turbine. Both specially developed mineral oil based and synthesized hydrocarbon based lubricants are used. Generally, the viscosity used is ISO VG 32, with some manufacturers requiring ISO VG 46 for their turbines.

Some difficulties have been experienced with dynamic compressors handling ammonia or ''syngas'' for ammonia production. As pointed out in connection with reciprocating compressors (Section I), ammonia can react with some inhibitors and oxidation products to form insoluble soaps and sludges. Although it is uncommon for the ammonia gas to contact the lubricating oil directly in dynamic compressors, there could possibly be some contamination via seal oil systems or faulty seals. If the soaps and sludges formed should be thrown out in the coupling connecting the speed increaser to the compressor, coupling imbalance could result, or eventual binding, leading to increased vibration. The deposits can also end up in oil coolers, resulting in poorer heat transfer characteristics. Deposits may also form in the speed increaser. Highly stable oils with inhibitors that are not affected by ammonia have been effective in reducing the severity of these problems. Both mineral oil and synthesized hydrocarbon based products are used. The viscosity used is usually ISO VG 32 or 46.

IV. REFRIGERATION AND AIR CONDITIONING COMPRESSORS

The basic principles of the refrigeration compression cycle are shown in Figure 17.19. The five essential parts basic to every system are shown: evaporator, compressor, condenser, receiver, and expansion valve (or capillary). Liquid refrigerant flows from the receiver under pressure through the expansion valve to the evaporator coils, where it evaporates, absorbing heat and resulting in a cooling action. The vapor is then drawn into the compressor, where its pressure and temperature are raised. At the higher pressure in the discharge

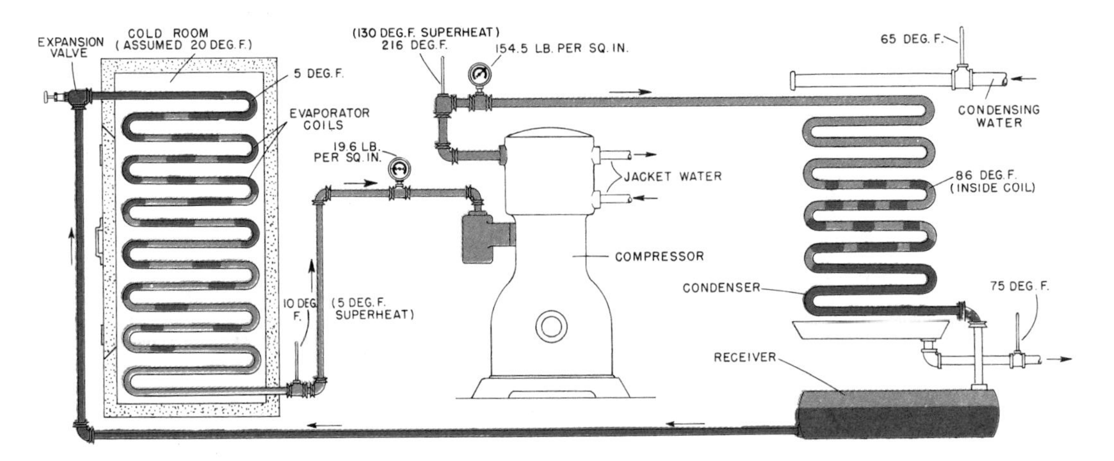

Figure 17.19 Basic single-stage compression refrigeration system. The elements shown are common to all compression refrigeration systems, whether refrigeration or air conditioning.

of the compressor, the condensing temperature of the refrigerant is higher than it would be at atmospheric pressure. When the hot, high pressure vapor flows from the compressor to the condenser, the cooling water (air in some applications) removes enough heat from it to condense it. The heat removed from the refrigerant in the condenser is equal to the amount of heat removed from the cold room (cooling action) plus the heat resulting from the mechanical work done on the refrigerant in the compressor that is not removed by the jacket cooling of the compressor. In many commercial installations, the evaporator cools a heat transfer fluid such as brine, which is then pumped through the area to be cooled. Smaller units, such as home refrigerators and freezers, room air conditioners, and automotive air conditioners, have air-cooled rather than water-cooled condensers.

In commercial installations, two or three stages of compression may also be used. If system pressures or cooling capacities dictate the use of two stages of compression, two-stage compressors are used, or a combination of separate single-stage compressors. Rotary sliding vane, scroll, or rotary screw compressors are sometimes used at low to moderate pressures or for booster purposes. Multistage reciprocating compressors are used for large air conditioning installations, with a trend toward the use of more scroll compressors. Reciprocating compressors are commonly used for refrigeration systems, with a trend toward the use of rotary vane. Centrifugal compressors are also used on some commercial refrigeration systems as well as in chillers. Reciprocating, sliding vane, and scroll compressors are used for automotive air conditioning systems, with some screw and axial piston compressors also used. Some very small units such as dehumidifiers may be equipped with diaphragm-type compressors. Reciprocating compressors are used in most other applications.

Most reciprocating compressors for commercial installations are of the single-acting, trunk piston type and have closed crankcases. As a result of refrigerant leakage past the pistons, the crankcases are filled with a refrigerant atmosphere. The same is true of axial piston units used for automobile air conditioning. Crosshead and double-acting compressors have open crankcases. The majority of small to medium-sized electric motor driven refrigeration and air conditioning units are hermetically sealed, with all the operating parts, including the electric motor, inside the sealed unit. Evaporators may operate either dry or flooded. In dry evaporators, only refrigerant vapor is present, while flooded evaporators have both liquid and vapor present.

A. Factors in the Compressor Affecting Lubrication

1. Cylinder Conditions

The oil film on the cylinder walls of a reciprocating refrigeration compressor is subjected to low temperatures at the suction ports and to moderately high temperatures near the cylinder head. Since viscosity decreases with temperature, the oil near the suction ports will have considerably higher viscosity than the oil near the cylinder head. Nevertheless, the oil must spread in a thin film over the entire working surface. Spreading is accomplished by the piston rings (or the piston itself in small compressors without piston rings) as the pistons move back and forth. The oil must distribute rapidly, but to do this it must not be too high in viscosity. On the other hand, an oil too low in viscosity will not protect against wear.

Oil carried out of the cylinders to the valves and discharge piping is subjected to the temperature of the discharging refrigerant. Ordinarily the temperature of the discharging refrigerant is not high; for example, the discharge temperature of a single-stage ammonia compressor with a compression ratio of about 5 : 1 should not be much in excess of 250°F (121°C). Some single-stage units operate at higher ratios, and higher discharge temperatures; but in most small compressors, the valve temperatures remain moderate because of the relatively large cooling area in proportion to cylinder volume. The discharge temperatures of compressors operating on fluorocarbon refrigerants are lower than those of equivalent machines operating on ammonia, although the compressors of automobile air conditioning systems may operate at quite high discharge temperatures.

When two or more stages of compression are used, the operating temperature in each stage usually is lower than in single-stage machines. In rotary compressors, the discharge temperatures are also usually moderate because of low compression ratios.

2. Oxidation

In compressors with enclosed crankcases, temperatures are normally moderate and the entire machine is filled with refrigerant vapor. Very little, if any, air is present. Under these conditions, oxidation in the usual sense does not occur, although it is doubtful that it can be avoided entirely. Limited oxidation does not impair the lubricating value of an oil because the initial oxidation products formed are soluble in the oil. If oxidation progresses too far, eventually some of the soluble oxidation products become insoluble when the oil is cooled. These products could plug or restrict capillary tubing or orifices inside the system.

3. Bearing System Conditions

The general requirements of the bearing systems of refrigeration compressors are similar to those of other comparable compressors. However, some special factors must be considered. In the compression of air or gases such as hydrocarbon gases, it is desirable that the oil not be miscible with the gases, whereas in closed refrigeration systems, the oil must be somewhat miscible with the oil to be able to circulate throughout the system and get to all the components in need of lubrication.

In compressors with closed crankcases, there is very little exposure to oxygen, and thus oxidation stability of the oil is not a major concern. If the same oil is used for both bearings and cylinders, as in many small units, however, the oil must have oxidation stability adequate for the cylinder conditions.

When ammonia is used as the refrigerant in compressors with closed crankcases, any additives used in the oil must be types that are not affected by ammonia. A refrigerant that is soluble or partially soluble in the oil, as are the majority of the fluorocarbon refrigerants, will dilute the oil and reduce its viscosity, a sequence of events that must be considered in the selection of the oil viscosity.

The motor in a hermetically sealed unit is completely surrounded by a mixture of refrigerant and oil. Thus the oil must have good dielectric properties, must not affect the motor insulation, and must not react with the copper motor windings or other system materials at elevated temperatures. Since most such units are operated on fluorocarbon refrigerants, the dilution effect of the refrigerant on the viscosity must be considered.

When the crankcase and cylinders are completely isolated from each other, as in compressors having crosshead construction, the oil in the crankcase is exposed to air and there is intimate mixing of the warm oil with air. These conditions are favorable to oxidation and require a chemically stable oil to resist oxidation.

B. Factors in the Refrigeration System Affecting Lubrication

If the oil carried out of the compressor cylinder forms gummy deposits in the condenser, or congeals or forms waxlike deposits in the evaporator, capillary tube, or expansion valve, there may be serious reduction in the heat exchange capacity. Heat insulating deposits in the evaporator make it necessary to carry a lower evaporator temperature to produce the required refrigeration effect. This in turn requires a lower evaporator pressure and increases the power required by the compressor for a given refrigeration duty, owing to the increased pressure range through which the gas must be compressed. In addition, at the lower suction pressure, the vapor density is lower, forcing the compressor to handle a greater volume of vapor and thus reducing refrigeration capacity. Heat insulating deposits in the condenser increase the temperature difference between the cooling medium (water or air) and the condensing refrigerant. The resulting higher condensing temperature makes higher compression necessary and increases power consumption.

Whether heat insulating deposits will be formed depends on the properties of the lubricating oil, the refrigerant in use, the evaporator temperature, and the equipment used in the system. The effects of some of the common refrigerants are considered separately.

1. Fluororocarbons

Chlorofluorocarbons (CFCs) are being phased out for use in air conditioning and refrigeration systems because of their potential negative effects on the ozone layer. As a result, more environmentally friendly non-CFC refrigerants are being developed. Several alternative refrigerants have been around for many years, such as ammonia, hydrocarbons, carbon dioxide, methyl chloride, and others that do not pose problems from the ozone depletion standpoint. These will be continued to be used in many applications. Non-CFC fluorocarbon refrigerants such as R-134a, R-123, and blends such as R-404A, R-407C, and R-410A are replacements for the CFCs. The use of these alternative refrigerants is increasing rapidly.

Each of the alternative materials has specific properties and operating characteristics that must be understood and handled appropriately to ensure maximum system performance as well as the safety of the people working with them and the public potentially exposed to them. In many systems, CFCs are and will remain in service. The Montreal Protocol banned the production of CFCs as of January 1, 1996, and hydrochlorofluorocarbons (HCFCs) were limited to production levels as of the same date, with a cease production date in the year 2030.

Air conditioners in older automobiles, as well as, many home refrigerators and air conditioners, were filled with CFCs, and many of these units are still in service. When systems containing CFCs need servicing, they must be refilled with CFCs manufacture before January 1, 1996, use reclaimed CFCs from older systems, or retrofit the systems to accept R-134a or one of the alternative environmentally friendly refrigerants. Gradually, all the CFCs and HCFCs will be replaced by alternative HFC materials, as well as by other gases such as isobutane, propane, and ammonia.

With the refrigerants that are miscible or partly miscible with oil, enough of the refrigerant dissolves in the oil to depress the pour point of the oil sufficiently to prevent congealing of the oil on evaporator surfaces in most cases. However, there is a temperature at which a heavy, flocculent precipitate first appears when a mixture of Freon 12 and 10% of the oil is chilled. The temperature at which this occurs depends on the refrigerant, the percentage of oil in the refrigerant, and on the oil. Refrigeration systems using fluorocarbon refrigerants are often designed to ensure that approximately 10% oil is present in the evaporator. In some cases, the evaporator is actually charged with this amount of oil. Under these conditions, the floc point of the oil (also known as the critical separation temperature) represents the lowest temperature that can be used with that oil.

The waxy materials that precipitate from these oil–refrigerant mixtures can also clog expansion valves and capillary control tubes, preventing their proper functioning. However, the concentration of oil in the refrigerant at the expansion valve is usually lower than in the evaporator, so the floc point is depressed below what it would be at a 10% concentration. As a result, if the oil selected has a low enough floc point for conditions in the evaporator, it usually will not cause difficulties in the expansion valve or capillaries. Difficulties in these areas attributed to mineral oils are frequently due to ice crystals formed by minute quantities of water in the system.

Oil selection can go a long way to minimizing problems related to lubrication in systems using fluorocarbons. The use of highly refined naphthenic or paraffinic mineral oils works satisfactorily with both the CFCc and HCFCs. The base stocks for these oils are usually severely hydroprocessed or acid-treated to remove wax and other materials undesirable from a refrigeration oil standpoint. For HCFCs, alkyl benzene synthetic lubricants provide excellent performance, as does mineral oil. Widely used products for HFCs are polyol esters, polyalkylene glycol, and polyvinyl ether.

2. Ammonia, Carbon Dioxide

Oil is slightly miscible in anhydrous ammonia and carbon dioxide. Generally, not enough of the gas dissolves in the oil to have a significant effect on the pour point of the oil. Thus, if the pour point of the oil is above the evaporator temperature, oil will congeal on the evaporator surfaces and form an insulating film that interferes with heat flow and efficient performance of the system. To remove the oil, the evaporator must be periodically warmed, liquefying the oil so that it will drain from the surfaces to a location from which it can be removed. With flooded evaporators, refrigerant flow may be so rapid that there is little or no opportunity for the oil to collect on evaporator surfaces, and the pour point of the oil may not be a major concern. Ammonia is not compatible with copper or brass and cannot be used in systems containing these metals. As with CFCs and HCFCs, ammonia works well with highly refined mineral oil. CFCs and HCFCs also can use polyalphaolefins (synthesized hydrocarbons), polyalkylene glycols, and polyol esters.

3. Hydrocarbon Refrigerants

Isobutane and propane gases are being used as replacements for CFC refrigerants in some applications. These gases are primarily used in smaller units such as hermetic household refrigerators.

4. Sulfur Dioxide

Sulfur dioxide has a selective solvent action that with conventional lubricating oils results in sludge. It, therefore, requires the use of highly refined white oils or group III base stocks with low levels of additive.

5. Lubricating Oil Recommendations

Table 17.3 shows general lubricant recommendations by refrigerant type. The lubricants are classified according to base type. The requirements of oils for refrigeration systems can be summarized as follows.

1. The oil should be of proper viscosity to distribute readily at the system's lowest temperatures yet provide adequate films to protect against wear in the cylinders and crankcases.
2. The oil should have chemical stability adequate to resist oxidation and the formation of deposits in crankcases open to the atmosphere, and to resist the deteriorating influence of high temperatures at compressor discharge.
3. In closed systems without oil separators, the oil should be miscible with the refrigerant, to ensure that the oil will circulate through the system and return to lubricate the compressor. In closed systems with separators, it is desirable that the oil not be miscible with the refrigerant, to facilitate separation. In open-

Table 17.3 Lubricating Oil Recommendations Based On Refrigerants

	Lubricating oil	
Refrigerant	Mineral oil[a]	Synthetic[b]
Fluorocarbons		
CFC- 11, 12, 113, 114, 500, 502	Yes	PAO, POE
HCFC- 22, 123, 125, 408A(blend)	Yes	PAO, AB
HFC- 134a, 143a	No	POE, PAG, PVE
Blends 404A, 407C, 410A	Yes	POE, PVE
Ammonia	Yes	PAO, PAG, POE
Carbon dioxide	Yes	PAO

[a] Mineral oils are to be highly refined paraffinic or naphthenic. White oils or severely hydroprocessed base stocks should be used.
[b] PAO, Polyalphaolefin; POE, polyolester; AB, alkylbenzene; PAG, polyalkylene glycol; PVE, polyvinyl ether.

crankcase systems, it is desirable that the oil not be soluble or miscible with the refrigerant, to minimize dilution.

4. The oil should be able to withstand system temperatures without breakdown, and it should not inhibit the heat transfer characteristics of the refrigerant.
5. The oil must be chemically stable and must not react with the refrigerant or system components. Some additives in the oil can react with the refrigerant to form deposits or sludges.
6. The oil must reduce friction and minimize wear.
7. The oil must keep the system clean and stay in service for extended intervals.

Oil viscosities recommended vary from as low as ISO VG 7 to as high as ISO VG 150.

18

Handling, Storing, and Dispensing Lubricants

ExxonMobil lubricants are quality products made to exacting standards. Their quality levels are designed to provide effective and economical performance when they are used as recommended in the applications for which they are intended. Every practical precaution in product storage and handling is taken by ExxonMobil to ensure that the products are maintained on specification through delivery to the customer. After delivery, it becomes the responsibility of the customer to exercise proper care in handling, storing, and dispensing the lubricants. This is necessary both to protect the quality built into each lubricant so that it can deliver optimum performance in the use for which it is intended and to maintain product identification and any precautionary labeling that may exist. A good lubrication program should also include appropriate steps to ensure proper handling and disposal of used lubricants.

The first steps in obtaining optimum performance from correctly selected proper quality oils and greases are proper handling, storing, and dispensing. These are necessary for two primary reasons: first to preserve the integrity of the products, and second to preserve identification and any precautionary labeling. It is poor practice to buy high quality lubricants and then permit degradation through contamination or deterioration before the products are used. Poor storage and handling practices also increase the risk of misapplication because the identification on the containers may become illegible through improper handling or the products may have been transferred to inadequately or improperly marked containers. Proper handling, storing, and dispensing are also important for plant and personnel safety, to protect against health hazards, and to minimize the risk of environmental contamination. Most petroleum products are combustible and require protection against sources of ignition. They are not generally health hazards, but it must be recognized that excessive exposure to them can be undesirable and should be avoided. Good hygienic practices should be exercised when contact has occurred. Finally, contamination or leakage produces waste that must be disposed of, aggravating the disposal problems and environmental concerns.

If simple precautions are not observed, contamination of lubricants with subsequent damage to machines can occur during storage or during transfer of oil or grease from the original container to the dispensing equipment, or to the equipment to be lubricated. Pumps, oil cans, grease guns, measures, funnels, and other dispensing equipment must be kept clean at all times and covered when not in use. Where operating conditions justify them, centralized dispensing or lubrication systems that keep the lubricants in closed systems and, therefore, protected against contamination, are highly recommended. Systems of this type are available to handle oils and greases of many types. There are other advantages: lubrication servicing generally can be performed faster, which results in less waste; integral metering devices can supply important consumption data; and bulk deliveries with economies in the purchase cost of lubricants frequently can be obtained.

Deterioration of lubricants can result from exposure to heat or cold, intermixing of brands or types, oxidation, prolonged storage, chemical reaction with fumes or vapor, entrance of dust and abrasive particles, and water contamination.

The observance of relatively simple precautions and procedures in the handling, storing, and dispensing of lubricants and associated petroleum products can achieve significant economic and operating benefits.

Economic benefits that can be obtained are based mainly on the elimination of waste due to the following preventable factors:

1. Leakage or spills from damaged or improperly closed containers
2. Contamination due to exposure of lubricants to dust, metal particles, fumes, and moisture
3. Deterioration caused by storage in excessively hot or cold locations
4. Deterioration due to prolonged storage
5. Residual oil or grease left in containers at the time of disposal or return
6. Mixing of incompatible brands or types of lubricant
7. Leaks, spills, and drips when a reservoir is being charged or a machine lubricated

Operating benefits, which also are reflected in dollar savings, include the following:

1. Reduction of machine problems attributable to the lubricant, resulting in fewer downtime occurrences.
2. Reduced material handling time. It has been estimated that labor costs for lubricant application can be as high as eight times the cost of the lubricant applied.
3. Better housekeeping. Oil or grease spilled on floors is a major safety and fire hazard.

While the information contained in this chapter provides general suggestions on good practices, it is the responsibility of the purchaser to identify and adhere to federal, state, and local regulations. Areas such as plant safety, handling and storing of flammable or combustible materials, any special precautions to address health and safety issues, fire prevention and protection, ventilation, and disposal of wastes must be integral parts of the lubrication program. The recommendations and suggestions in this chapter are believed to be consistent with the standards of the Occupational Safety and Health Administration (OSHA) of the federal Department of Labor, issued as of the date of publication. But it must be recognized that in many cases more stringent state or local standards may apply. These should always be checked to ensure the establishment and maintenance of conformity with all applicable standards.

I. HANDLING

In the sense used here, ''handling'' includes operations involved in the receipt of the supply of lubricants and the transfer of these materials to in-plant storage. The type of handling depends on the form of receipt of the lubricants: in packages, in mini-bins, or in bulk.

A. Packaged Products

All shipments of oils, greases, and associated petroleum products in containers up to and including 55-gallon (U.S.) oil drums and 400 lb grease drums are considered to be packaged products.

Figure 18.1 shows the size, shape, and weight of the most popular standard lubricant containers. To aid in planning storage space, racks, or shelves, the dimensions of the containers illustrated and cartons of smaller packages are shown as well. Most containers are made of sheet steel, in thicknesses and constructions suitable for their contents, the intended service, and the applicable freight regulations. Exceptions are grease gun cartridges and the nonmetal lubricant containers typically used for smaller packages that are made of plastic or spirally wound fiberboard with an oil- or grease-proof liner and metal ends.

As delivered, each container is sealed with appropriate covers, gaskets, lids, bungs, caps, and seals. In addition, each container is stamped, embossed, stenciled, lithographed, or labeled with the brand name of the lubricant it contains and appropriate safety information.

Most packaged lubricants are delivered to the user by truck or freight car. The 55 gallon-oil drum or 400 lb grease drum is the most common container, since it is usually the most economical way for the small user to purchase packaged lubricating oil, greases, or other petroleum products. For medium or large users, mini-bins and bulk deliveries, discussed later, may offer still greater savings.

Drums can be unloaded without damage from trucks or freight cars that do not have a hydraulic tailgate by sliding them down wood or metal skids (Figure 18.2). Skids are commercially available in 16 in. (0.41 m) widths and in lengths from 6 to 12 ft (1.8–3.6 m). Before unloading, the brakes of the truck should be set firmly, and if any possibility of movement exists, the wheels to the truck or freight car bed should be blocked.

Caution.* *A full oil or grease drum weighs about 450 lb (204 kg). To minimize the risk of bodily injury, it is good practice to have drums handled by two people unless mechanical aids such as forklift trucks, drum handlers, or chain hoists are available.

Most trucks used in the regular delivery of drums and similar heavy packages are equipped with hydraulic tailgates (Figure 18.3), which are used to lower drums from the truck bed level to the ground or loading platform level. Hand winch lifts (Figure 18.4) can do the same job. Special jaws enclose and grip drums about the diameter under the upper rolling hoops for direct movement from the truck. Under no conditions should drums or pails be dropped to the ground or onto a cushion such as a rubber tire. Doing so may burst seams, with consequent leakage losses and possible contamination of the contents. Leakage also may constitute an environmental, safety, or fire hazard. The empty drums may also have a refundable deposit or resale value if they are not damaged. Increasing

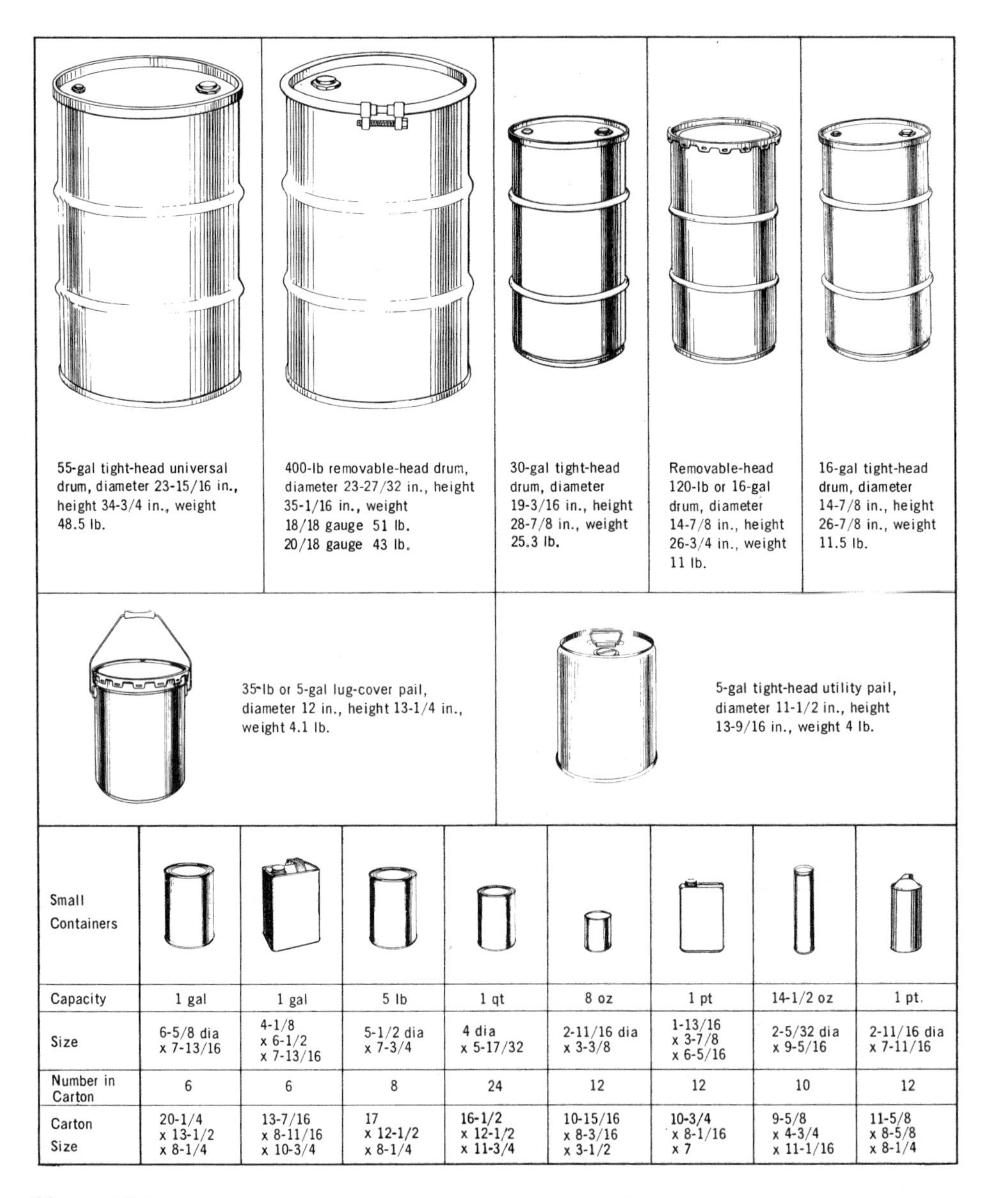

Small Containers								
Capacity	1 gal	1 gal	5 lb	1 qt	8 oz	1 pt	14-1/2 oz	1 pt.
Size	6-5/8 dia x 7-13/16	4-1/8 x 6-1/2 x 7-13/16	5-1/2 dia x 7-3/4	4 dia x 5-17/32	2-11/16 dia x 3-3/8	1-13/16 x 3-7/8 x 6-5/16	2-5/32 dia x 9-5/16	2-11/16 dia x 7-11/16
Number in Carton	6	6	8	24	12	12	10	12
Carton Size	20-1/4 x 13-1/2 x 8-1/4	13-7/16 x 8-11/16 x 10-3/4	17 x 12-1/2 x 8-1/4	16-1/2 x 12-1/2 x 11-3/4	10-15/16 x 8-3/16 x 3-1/2	10-3/4 x 8-1/16 x 7	9-5/8 x 4-3/4 x 11-1/16	11-5/8 x 8-5/8 x 8-1/4

Figure 18.1 ExxonMobil lubricant containers (not to scale). Standard containers commonly used for Mobil products; outside dimensions shown in inches. Carton sizes for smaller packages are also given. Note that a 55-gallon or 400 lb drum on end requires about 4 ft^2 of floor space, and on its side, about 6.5 ft^2. The 14.5 oz package is a grease cartridge for loading hand guns. The 1-pint container at the lower right is a spray can (aerosol) for ease of application of specific products.

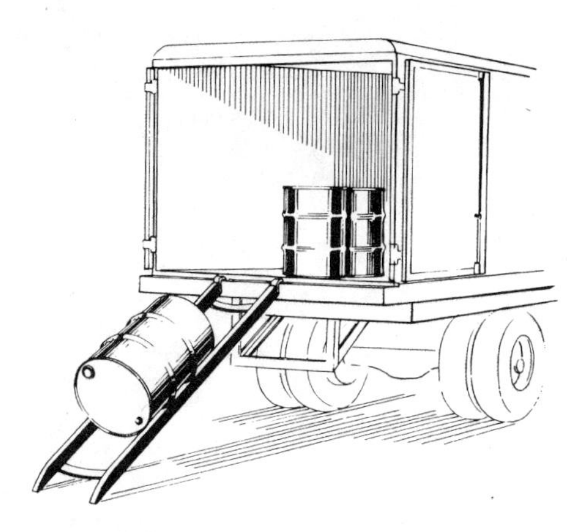

Figure 18.2 Drum skid. Drums can be unloaded from trucks or freight cars by sliding them down a wooden or metal skid. The skid must be securely fastened to the bed of the truck or freight car.

Figure 18.3 Hydraulic lift gate. Many trucks used for delivering packaged products are now equipped with hydraulic lift gates, which can be used to lower the packages to the ground or a loading dock.

amounts of lubricants in both drums and smaller packages are being delivered to customers on pallets. These can be unloaded with a forklift truck and transported directly to storage. Generally, this unloading must be accomplished at a loading dock because it is usually necessary to drive the lift vehicle onto the truck or freight car to complete the unloading. Where a lift truck is to be driven onto a truck or freight car, applicable OSHA, state, or local regulations relative to such operations must be adhered to. These include regulations regarding the method for reaching the loading dock and may include other requirements.

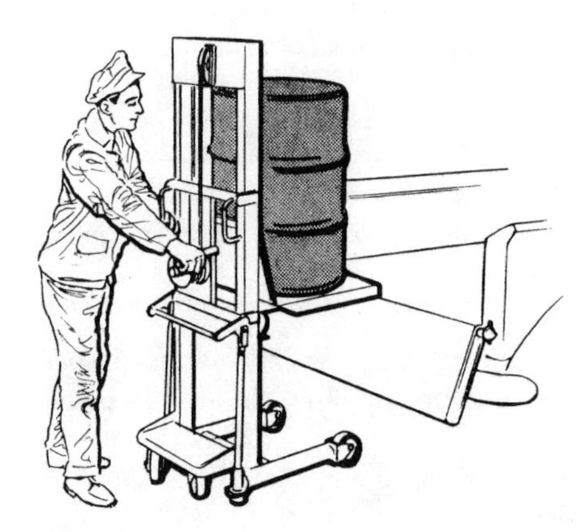

Figure 18.4 Hand winch lift. A simple lift mechanism of this type can be used to raise or lower a drum to ground level from a truck bed. A floor lock holds the lift in position while the drum is being loaded or lowered, and the casters permit moving for short distances.

Figure 18.5 Lift truck for handling drums. The hydraulically actuated arms on this truck will clamp and lift four drums. Other models are available that will clamp one or two drums at a time, and rotators can be added to permit tilting or inverting the drums.

1. Moving to Storage

After unloading, drums can be moved safely to the storage area by properly equipped forklift trucks, either on pallets or held in specially equipped fork jaws (Figure 18.5). If forklift trucks are not available, the drums can be handled and moved with barrel trucks or drum handlers (Figure 18.6). Drums should be strapped or hooked to the frame of a barrel truck and unloaded carefully. In some cases, drums can be rolled directly on the plant floor to the storage area, provided the distance is short. The rolling hoops will protect the shell from damage. For long distance transfer by rolling, metal tracks should be laid on the floor.

When drums are rolled, care should be taken to avoid hard objects that are higher than the drum body clearance, which might puncture the shell and result in leakage or contamination of the contents as well as potential injury to personnel handling the heavy drums.

Caution. When rolling drums, maintain firm control to prevent the possibility of a drum running away and injuring personnel, damaging machinery, or itself becoming damaged.

A hand-drawn or powered lift truck and skids can be used to move drums or pails along hard surfaces for long distances. An inclined ramp simplifies the task of getting drums onto pallets or skids.

Five-gallon oil or 35 lb grease pails are usually shipped on pallets. They should be handled with the same care as the larger containers.

Smaller containers of lubricants are usually packed in plastic or fiberboard cartons. These should be unloaded and moved to storage with care. The cartons should be left sealed until the product is required.

Figure 18.6 Manual drum handler. Hand-operated hydraulic systems clamp the drum, then lift it for transporting.

B. Bulk Products

The advantages of bulk delivery and storage of lubricants have resulted in the increased use of this method of operation wherever the quantities of lubricants used are sufficient to justify the installation costs. The term ''bulk'' in this context refers not only to deliveries in tank cars, tank trucks, tank wagons, and special grease transporters, but also to deliveries in any containers substantially larger than the conventional 55-gallon or 400 lb grease drums. In this latter category are a considerable number of special bulk bins or jumbo drums (mini-bins) that will transport on the order of 275 gallons (1041 liters) or more of oil or 3680 lb (1670 kg) of grease. Bulk bins usually are off-loaded and left on the customer's premises while the product is being used, but the lubricant also may be pumped off into on-site bulk storage tanks or supply tanks for central lubrication or dispensing systems (Figure 18.7).

Bulk handling and storage systems offer both economic and operating benefits to the lubricant user. Economic benefits include the following:

1. **Reduced handling costs** When delivery is made directly into permanent storage tanks, handling costs are reduced markedly over those for drums or smaller packages.

2. **Reduced floor space requirements** Permanent tanks occupy much less floor space than is required for an equal volume of oil in drums.

Figure 18.7 Bin for grease or oil. This type of bin, shown stacked ready to be moved into a plant, is used most widely by ExxonMobil, but other designs with different capacities may be used in some areas.

3. **Reduced contamination hazards** Bulk storage tanks usually are filled directly from the tank car or tank truck through tight fill connections. The exposure of the lubricant to contamination is greatly reduced in comparison to handling and dispensing the same volume of packaged product.

4. **Reduced residual waste** The user is charged only for the amount of product delivered into a bulk tank. Practice varies with bulk bins, but in many cases the user is allowed credit for residual material in the bin, or is charged for only the amount added to the bin when it is refilled. Since the amount of residual oil left in drums in usually up to one gallon, and the amount of grease left in drums may range up to 20 lb (9 kg), the amount of lost lubricant attributable to residual waste with these packages can be significant.

5. **Simplification of inventory control** The use of bulk storage facilities sharply reduces the number of individual items to be ordered, tallied, routed, and processed for return or salvage.

6. **Reduced container deposit losses** Damaged drums or containers that cannot be returned for other reasons such as excessive or unacceptable materials left in them result in lost deposit value and/or additional disposal costs.

The operating benefits to be achieved through the use of bulk facilities include the following.

1. **Simplification of handling** Less handing time means that personnel are freed for other tasks.
2. **Constant availability of needed lubricants** Time spent waiting for placement and opening of new drums and installation of pumps is minimized.
3. **Reduced contamination hazards** Minimizing contamination means fewer lubrication problems on the manufacturing floor.

4. **Lubricants can be made available readily at strategic points** Pumps and pipelines can be used to move lubricants from permanent tanks to locations at which repetitive lubrication operations are performed.
5. **Improved personnel and facility safety** Any reduction in handling operations reduces exposure to conditions that could result in injuries to personnel, plant, or environment.

These benefits can be achieved to their fullest extent only when the bulk system is properly designed and installed, and cooperation between the supplier and user results in the delivery, in timely fashion, of uncontaminated products. In determining the feasibility of installing bulk capabilities, other factors, such as compliance with federal, state, and local requirements, need to be considered. Insurance requirements are another consideration.

1. Unloading

Prior to the receipt of bulk deliveries of lubricants, whether they are in tank cars, tank wagons, or special grease transporters, certain precautions should always be observed. The storage tanks should be gaged to ensure that there is sufficient capacity available for the scheduled delivery. Empty tanks should be inspected, and flushed or cleaned if necessary. When large tanks must be entered for manual cleaning, applicable safety rules should always be observed. One person should never enter a tank alone; a second person should be available outside the tank to provide assistance if an emergency arises. Breathing apparatus suitable for oxygen-deficient atmospheres or other toxic atmospheres may be necessary, and suitable protective clothing should be worn. The tank should be certified for admittance by a qualified safety professional.

Before the delivery is unloaded, a check should be made to be sure that the correct fill pipe is being used, the valves are set correctly, and any crossover valves between storage tanks are locked shut. It is also advisable to take a midtank sample from the delivery container for a quality assurance reference. Each type of bulk delivery requires some additional special precautions. Some of the more important of these are outlined in the following sections.

2. Tank Cars and Tank Wagons

Only trained employees should unload flammable or combustible liquids from tank cars. The car brakes should be set, wheels blocked, and ''stop'' signs set out. Before attaching and unloading connection, the dome cover should be opened and the bottom outlet valve checked for leakage. If the valve is not leaking, its cap will usually be loose. All hoses, pumps, valves, and other connections should be checked for cleanliness. When unloading is complete, the hose should be disconnected, the dome cover closed, and the valve closed immediately. Caps or other protective covers should be reinstalled on hoses and other connections.

3. Special Bulk Grease Vehicles

While tank trucks have been used successfully to carry bulk grease, cleaning problems are considerable and tend to discourage the use of this type of vehicle. A more satisfactory approach has been the use of specially designed and constructed bulk grease vehicles. The

Figure 18.8 Bulk grease transporter.

current vehicles (Figure 18.8) are capable of carrying 38,900 lb (17,645 kg) of grease in two 19,000 lb (8618 kg) compartments. The tanks are fully insulated for heat retention in long hauls. For excessively long hauls or for deliveries made in cold temperatures, bulk grease vehicle tanks should be equipped with heating capabilities to facilitate off-loading. Two power pods, each equipped with its own engine and pumping unit, can unload the grease at a rate of up to 1000 lb (454 kg) per minute under proper conditions. The two power pods are interchangeable to provide maximum dependability.

Bulk grease vehicles are designed primarily for use in refilling bulk bins or the main storage tanks in centralized lubricant dispensing systems in plants having large volume requirements for specific lubricants. They usually are used for transporting pumpable greases in the consistency range up to and including NLGI grade 1. Other greases of higher consistency can be handled in bulk by additional heating; however, each application requires careful study and engineering to assure satisfactory operation.

II. STORING

Proper storage of lubricants and associated products requires that they be protected from sources of contamination and from degradation due to excessive heat or cold, and that product identification be maintained. Also important are the ease with which the products can be moved in and out of storage, and the ability to operate on a ''first in, first out'' basis. Increasingly important in the selection, location, and operation of petroleum product storage facilities are applicable fire, safety, and insurance requirements.

The National Fire Protection Association in its flammable liquids code NFPA 30* defines a flammable liquid as follows: ''Any liquid having a flash point below 100°F (37.8°C) and having a vapor pressure not exceeding 40 psia (2068 mmHg) at 100°F (37.8°C) shall be known as a Class I liquid.'' Flammable liquids (Class I) are further subdivided as follows.

---

* The definitions contained here are those of NFPA 30 (revision dated 1999).

Class IA consists of liquids having flash points below 73°F (22.8°C) and a boiling point below 100°F (37.8°C).
Class IB consists of liquids having flash points below 73°F (22.8°C) and a boiling point at or above 100°F (37.8°C).
Class IC consists of liquids having flash points at or above 73°F (22.8°C) and below 100°F (37.8°C).

A *combustible* liquid is defined as any liquid having a flash point at or above 100°F (37.8°C). Combustible liquids are divided into two classes, as follow:

Class II liquids consists of those with flash points at or above 100°F (37.8°C) and below 140°F (60°C).
Class III liquids consists of those with flash points at or above 140°F (60°C). They are divided into two subclasses, as follows:
Class IIIA liquids consists of those with flash points at or above 140°F (60°C) and below 200°F (93.3°C).
Class IIIB liquids consists of those with flash points at or above 200°F (93.3°C). This class of liquids, because of their lower flammability, does not require the special handling precautions that apply to the more flammable materials.

When a combustible liquid is heated for use near its flash point, it must be handled in accordance with the next lower class of liquids.†

Under OSHA regulations, most lubricants, because of their relatively high flash points, are class IIIB combustible liquids. Petroleum solvents that are commonly used for cleaning parts and equipment are class II combustible liquids. However, depending on the conditions of storage and use, they may require handling as class IC flammable liquids.

Suggestions included later in connection with oil house size and arrangement (Section III.A.3) are believed to be consistent with the OSHA standards in effect as of the date of publication. These standards, in turn, are similar to those of the National Fire Protection Association, but it must be remembered that more stringent state or local regulations could apply.

A. Packaged Products

Packaged lubricants and associated products can be stored outdoors, in a warehouse, or in an oil house. Outdoor storage should be avoided whenever possible. The best storage area for lubricants is a well-arranged, properly constructed, and conveniently located oil house. Warehouse storage is desirable when the oil house lacks the space needed to stock the complete lubricant inventory required.

1. Outdoor Storage

Storing lubricant drums or other containers out-of-doors is a poor practice. The hazards of this type of storage include the following.

1. Identifying drum markings may fade and become unreadable under the combined action of rain, sun, wind, and airborne dirt. Rusting can obliterate drum markings.

† OSHA regulations, which apply to any workplace having employees, require that when a combustible liquid is heated for use to within 30°F (16.7°C) of its flash point, it must be handled in accordance with the requirements of the next lower class of liquids.

When the markings become so deteriorated that the drum contents cannot be identified, it may be necessary to discard the material or take a sample and perform a laboratory analysis to determine the nature of the contents. This is a costly and time-consuming procedure. On the other hand, accidental use of the wrong lubricant as a result of incorrect identification can cause severe damage to machines or other equipment.

2. Seams of containers may be weakened by exposure to alternating periods of heat and cold, which causes the metal to expand and contract cyclically. The net result may be loss of the contents by leakage or contamination if the drums are subsequently subjected to rough handling or improper storage.

3. Water may get into the drum around the bungs, contaminating the contents. A drum standing on end with the bungs up can collect rainwater or condensed atmospheric moisture inside the chime (Figure 18.9). This water can gradually be drawn in around the bungs by the breathing of the drum as the ambient temperature rises and falls. This can occur even with the bungs drawn tight and the tamperproof seals in place.

4. Dirt and rust that accumulate inside the chime and around the bungs may contaminate the contents when the drum is finally opened for use.

5. Extremes of heat or cold can change the physical properties of some products and render them useless. Water that separates from soluble oils, invert emulsions, or fatty oils may congeal in compounded oils. Emulsifiers, in wax emulsions, may separate and cause irreversible degradation. Cold lubricants dispense slowly and consequently must be warmed up before use, with the possibility of further damage due to overheating. Cold lubricants dumped into operating systems may cause temporary operating problems owing to substantial difference to viscosity and thermal effects.

6. Contaminating rust can develop inside a container if water leaks in from any source.

Of course, if any of these conditions results in irreparable damage to the drums, these vessels may become a scrap disposal problem instead of a source of resale value. To minimize the harmful effects of unavoidable outdoor storage, a few simple precautions and procedures can be very helpful.

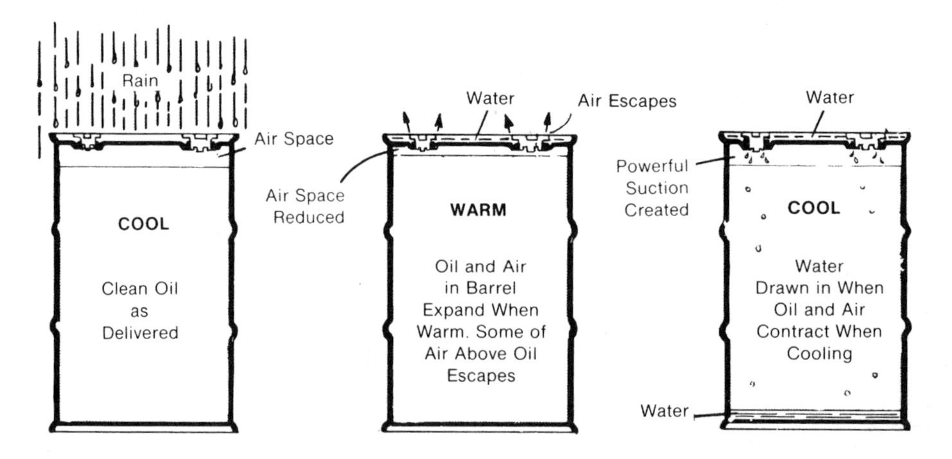

Figure 18.9 Entrance of moisture due to expansion and contraction (breathing) by upright drum.

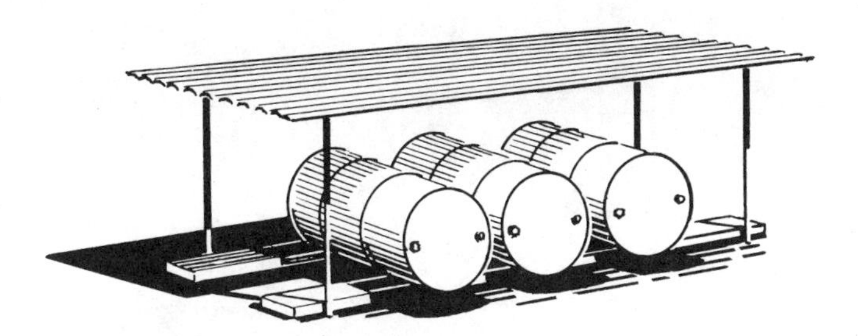

Figure 18.10 Outdoor drum shelter.

As a general rule, lubricants in drum containers should never be stored out of doors. When drums must be stored outside, a temporary shelter (Figure 18.10) or lean-to, or a waterproof tarpaulin, will protect them from rain or snow. Drums should be laid on their sides with the bungs approximately horizontal, as shown. In this position, the bungs are below the level of the contents, thus greatly reducing breathing of water or moisture and preventing water from collecting inside the chime. For maximum protection, the drums should be stood on end (as long as product identification is visible) with the bung ends down on a well-drained surface. Where all these approaches are impractical, drum covers such as those shown in Figure 18.11 can be used. These are available in both metal and plastic. Drums that have a bung on the side should be stored either on end or on the side with the bung down. Regardless of the position of storage, drums should always be placed

Figure 18.11 Metal drum covers. Weights have been placed on these covers to reduce the possibility that the covers will be blown off by high winds.

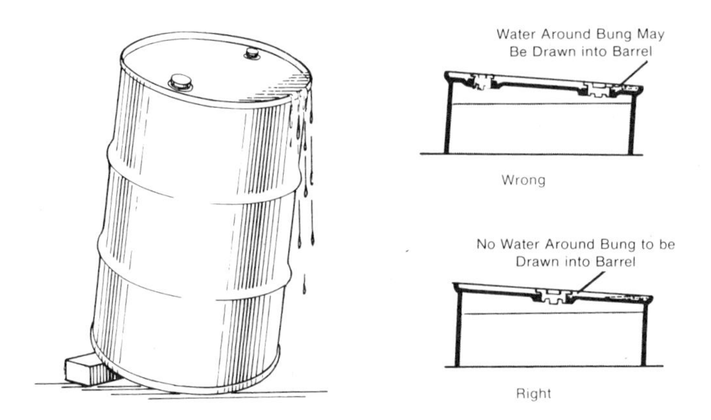

Figure 18.12 Correct method of blocking up a drum.

on blocks (Figure 18.12) or racks several inches above the ground to avoid moisture damage.

Drums that must be stored outdoors with the bung end up prior to use should be cleaned carefully to eliminate the hazard of collected rust, scale, or dirt falling into the drum contents.

The drum storage area should be kept clean and free of debris that might present a fire hazard.

2. Warehouse Storage

Mechanical handling equipment is needed for efficient movement of drums into and out of a plant warehouse. Hand- or power-operated forklift trucks or stackers (Figure 18.13) are widely used for this purpose. They offer the advantage of a single handling operation from warehouse storage to the oil house or point of use. A chain block or trolley with a proper drum sling mounted on an I-beam bridge (Figure 18.14) can be used to move drums in and out of storage. This type of equipment does not require the aisle space needed to maneuver a forklift. Mechanical handling equipment is closely regulated by OSHA. A thorough study is required to assure that the system and equipment meet all requirements.

Racks and shelving should provide adequate protection for all containers. Aisle space should be adequate for maneuvering whatever type of mechanical handling equipment is used. To maintain a ''first in, first out'' policy, no lubricant should be stocked in a way that blocks access to the older stacks. Care in this regard will minimize the hazard of deterioration due to excessive storage time. A wide variety of racks and shelving (Figure 18.15), both assembled and in components for assembly on-site, is commercially available.

When lubricants and other petroleum products are to be stored in a plant warehouse, location and rack construction should be considered with respect to applicable insurance, fire, and safety regulations. The material storage location should be considered also on the basis of receiving and dispensing convenience.

B. Oil Houses

1. Function

The function of an oil house in an industrial plant is to provide a central point for the intermediate storage and day-to-day dispensing of lubricants, cutting fluids, and other

Figure 18.13 Drum stacker. The lifting arms slip under a drum resting on the floor. The drum can then be lifted and moved into the storage rack. This unit is battery powered, but manual models are also available.

Figure 18.14 Chain block and trolley. A hand- or power-operated chain block mounted on an I-beam bridge can be used to move drums into and out of storage racks. The racks shown are constructed of 2 in. pipe and standard fittings. The pairs of flat strips supporting the upper drums can be removed to permit access to the drums below.

Figure 18.15 Storage racks. Racks of this type can be shop-built or obtained commercially in various heights and widths. Drums can be conveniently moved into and out of this rack with the stacker shown in Figure 18.13.

related materials needed to lubricate and maintain the plant's production equipment. If soluble cutting fluids are used, emulsions are generally prepared in the oil house, which in many cases is equipped to purify or recondition used or contaminated oils.

The oil house should be equipped to properly maintain and clean all the equipment used in daily lubrication work. This includes pumps, grease guns, oil cans, portable dispensing units, strainers, filter elements, and grease gun fillers. In addition, storage space should be provided for small containers of products, guns, cans, and other equipment necessary to properly dispense and apply lubricants.

The administrative function of the oil house includes the maintaining of machine lubrication charts and inventory records, conformity with established lubrication service schedules, and observance of good housekeeping practices.

2. Facilities

The simplicity or complexity of oil house facilities will depend largely on the size of the plant and the comprehensiveness of its lubrication program. Speaking generally, a well-equipped oil house will contain adequate stocks of the following items.

1. Drum racks, either of the rocker type or tiered
2. Oil and grease transfer pumps
3. Drum faucets
4. Grease gun fillers

5. Grease guns
6. Oil cans
7. Portable equipment such as oil wagons, lubrication carts, sump drainers, catch pans, and power grease guns
8. Maintenance supplies such as wiping rags, cleaning materials, wicks, replacement screens and filters, and spare grease fittings
9. Containers of absorbent materials for cleaning up oil spills
10. Grounding straps for use with combustible materials

In addition, batch purification equipment to recondition used oils may be desirable, soluble oil mixing equipment may be required, and solvent tanks for cleaning parts and lubrication equipment generally are required.

When plant usage of a specific oil, grease, or fluid warrants, a permanent bulk tank may be installed in the oil house.

3. Size and Arrangement

The size and arrangement of stock in an oil house are also determined largely by plant size and lubrication service requirements. The total rack storage needed can be calculated by determining the quantity of each type of product (oil, grease, solvent, etc.) that will need to be stocked in the oil house and then converting this to rack dimensions by means of the container dimensions shown in Figure 18.1. The amount of space needed for miscellaneous equipment (solvent tanks, mixing equipment, cabinets for lubrication service personnel, portable dispensing equipment, etc.) must also be considered.

Individual lockers for lubrication service personnel can contribute to good housekeeping and general oil house efficiency. Lockers of steel construction are durable and easy to keep clean. They should be big enough to store normal individual equipment, with adjustable shelving to permit maximum space utilization (Figure 18.16). This type of storage cabinet should be ventilated to prevent accumulation of vapors or fumes. If safety cans of cleaning solvents are to be stored in them, the lockers must be of an approved design.* Used wiping rags or waste should never be stored in these lockers, but should be placed in an approved disposal container.

Adequate cleaning capabilities are essential. Dispensing equipment (grease guns and the like) must be cleaned regularly for proper functioning. Airborne dirt and dust collect quickly on oil-wetted surfaces, and oil containers, faucets, and similar equipment should be cleaned regularly to remove these contaminants. Solvent tanks should be large enough to handle the largest equipment used. In general, baths of nontoxic safety solvents are permitted by safety regulations, provided the tanks have properly designed covers of the automatic self-closing type, and ventilation is adequate. Two tanks, one for cleaning and one for rinsing, will assure proper cleaning of equipment. As shown in Figure 18.17, a metal grating over a drain leading to the central trap prevents spilled solvent from collecting on the floor to form a safety or fire hazard and from finding its way to the sewers to contaminate the plant effluent water. The approved safety drip cans under the drain faucets collect any leakage from these sources.

---

* Reference to an ''approved design'' or ''approved type'' of container, locker, or other unit indicates that has been tested by and carries the label of one of the recognized testing organizations such as Underwriters Laboratories (UL) or Factory Mutual Laboratory (F/M).

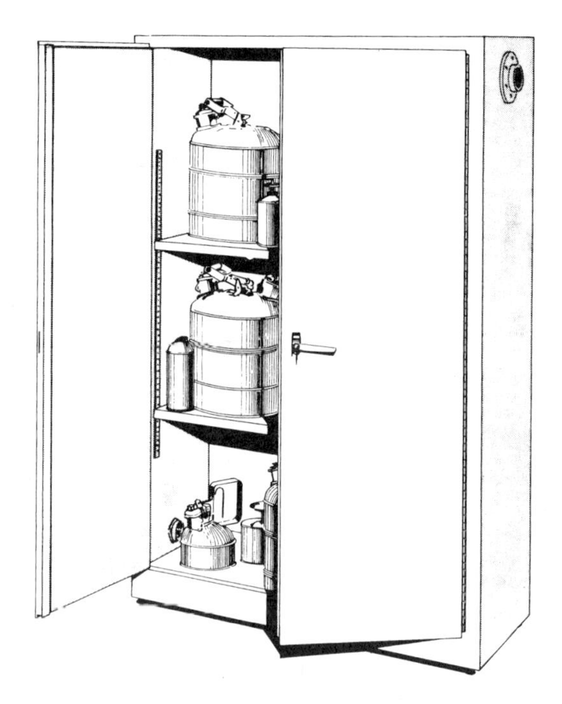

Figure 18.16 Individual locker. Adequate space is provided for the lubrication servicer's equipment. The locker should be provided with top and bottom ventilation to dissipate fumes and vapors. If flammables are to be stored in the lockers, a safety locker of approved design must be used.

One oil house layout (Figure 18.18) is offered as a guide to planning. Note that the sliding doors (1 and 6) are wide enough to admit the largest expected forklift load. The doors shown in this layout are self-closing fire doors, which may not be required by local fire regulations but generally offer the advantage of increased safety. The portable equipment storage area (13) provides ample maneuvering space for lift trucks to remove and replace the oil and grease drums (2 and 12). The open layout permits easy, unimpeded access from any part of the oil house to the door (6) leading to the plant. Federal, state, or local codes may require provisions for secondary containment of spills (diking).

4. Location for Optimum Utilization of Manpower

An oil house should be located as centrally as possible with respect to the lubrication service activity. A study of service requirements based on total travel of lubrication service personnel to their work from the oil house can help to determine the most economical and efficient location for the oil house. Travel distances from the warehouse, unloading dock, or other storage facilities should also be taken into account. In multibuilding plants it may be advantageous to build a separate oil house in a central location within the plant area. In such cases, an oil warehouse and the oil house may be combined in one building, with consequent savings in handling and storage costs.

5. Housekeeping

Orderliness and cleanliness in the oil house are essential. In an orderly storage arrangement, the chances of mistaken identification of products or applications are greatly reduced.

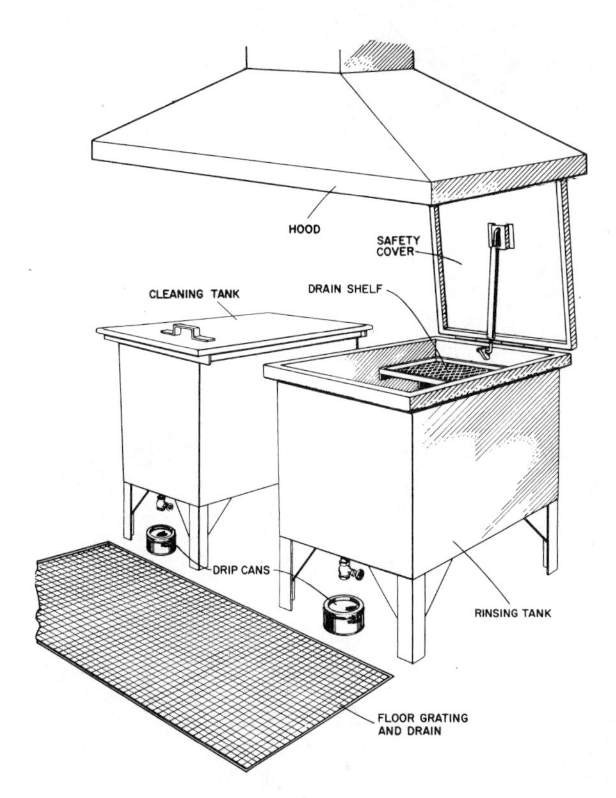

Figure 18.17 Cleaning tanks. While the overhead hood with forced ventilation is not mandatory in all cases, it is generally good practice. Self-closing covers, with a fusible link to permit automatic closing in the event of fire, are now required by OSHA regulations.

Regular cleaning schedules should be set up and maintained. Each container or piece of equipment should bear a label showing clearly the product for which it is used. This label should provide enough information to enable personnel to correctly identify the desired product in the original containers. For example, identifying an oil container as "Mobil DTE Oil" would not be adequate identification because the same plant might use several Mobil DTE oils, such as Mobil DTE 24 and 25 for hydraulic systems, Mobil DTE Oil 105 for compressors, and Mobil DTE Oil Heavy Medium for bearing lubrication. Color coding is frequently suggested for this identification, but it must be remembered that a considerable number of people are color-blind. Labels should be renewed as frequently as necessary to maintain legibility.

Every piece of equipment used in the oil house should have a space reserved for it and should be in its place when not actually in use. A chart or list of these locations posted near the product storage racks will facilitate location of needed items.

Observance of cleanliness and an orderly routine will be reflected in the attitude and efficiency of the lubrication personnel. Their increased sense of pride and responsibility will have a direct bearing on the lubrication service in any plant. It will also be a contributing factor to optimum safety against fire and personal injuries. In this respect,

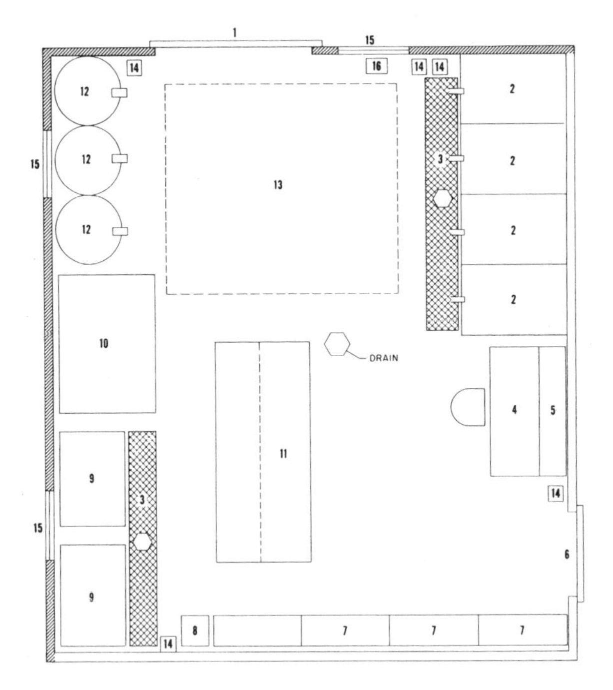

Figure 18.18 Oil house layout. Some of the features of this layout are as follows: 1 and 6, self-closing fire doors wide enough for passage of lift trucks or other material handling equipment; 2, drum racks; 3, grating and drain; 4, desk; 5, filing and record racks; 7, individual lockers or storage cabinets; 8, waste disposal container; 9, solvent cleaning tanks; 10, purification equipment, soluble oil mixing equipment, or other special equipment; 11, cabinets and racks for equipment, supplies, and small containers; 12, grease drums with pumps; 13, parking area (for oil wagons, etc.); 14, fire extinguishers; 15, ventilators; 16, container of sawdust or other absorbent.

cleanliness is accorded such importance that it receives major emphasis from compliance officers and some of the regulatory bodies.

***Note:* While comparatively few lubricants and associated products require precautionary labeling, in the few cases that call for it, similar precautionary labels should be affixed to all equipment used for those products. Dispensing and application equipment used for products requiring precautionary labeling should not be used for any other products until it has been thoroughly cleaned.**

---

6. Safety and Fire Prevention

Warning signs should be posted in every oil house to alert personnel to the presence of combustible materials. If flammables are used or stored, or class II combustibles are used in such a way that they must be treated as flammables, the applicable signs warning for the hazards associated with these materials should also be posted. Standard warning placards complying with OSHA or other applicable safety regulations relative to flammables or combustibles are generally available. ''No smoking'' signs, in red, should be prominently displayed and the no-smoking rule rigorously enforced.

Hand fire extinguishers and automatic systems are essential for oil house safety. Hand-operated fire extinguishers should be mounted at strategic points throughout the oil house, particularly near the cleaning tanks. They should be inspected periodically (at least as frequently as required by applicable fire regulations) to be sure that they are in satisfactory operating condition.

Fire extinguishing methods for flammable or combustible liquid fires include the following.

Suppression of vapor by foam
Cooling below the flash point by water spray or fog (not a direct stream, which could spread the fire!)
Excluding oxygen, or reducing it with carbon dioxide (CO_2) to a level insufficient to support combustion
Interrupting the chemical chain reaction of the flame with dry chemical agents or a liquified gas agent

All personnel employed in the oil house should be thoroughly instructed in the location and use of the fire extinguishing equipment.

Rags, paper, or other solid materials that have been soaked in flammable or combustible liquids should be placed in an approved type of disposal container with a self-closing cover. The container should be emptied at the end of each shift and the contents removed to a safe location for reconditioning or incineration. All spills should be cleaned up promptly. If an absorbent is used, it should be swept up promptly, placed in an approved disposal container, and removed to a suitable disposal area.

Ventilation equipment should be kept in first-class operating condition at all times. Without good ventilation, vapors can collect within the oil house, constituting a safety and health, as well as a fire hazard. The solvents used for cleaning parts and equipment in an oil house may present fire and health hazards. A few simple precautions can minimize these hazards.

1. Use only nontoxic solvents such as mineral spirits with flash points above 100°F (37.8°C).

2. Static electricity generated by the flow of products such as solvents can cause sparking, which can be a source of ignition. This is not generally a problem with products with flash points above 100°F (37.8°C), but simple precautions will eliminate the hazard. Drums containing solvents should preferably be grounded during opening and while installing faucets or pumps. If the drum is not resting on a grounded, conducting surface, an adequate ground can be obtained by clipping a wire between the drum chime and a water pipe. When small containers are filled from a solvent drum, a bond between the drum and container should be made. If both drum and container are resting on the same conducting surface, adequate bonding will be obtained if the faucet or nozzle is kept in contact with the container, otherwise a bond wire clipped between the container and the drum should be used (Figure 18.19).

3. Use only approved-type safety containers (Figure 18.20) for transferring solvents from drums or bulk storage to the cleaning tanks or point of use.

4. Prolonged inhalation of solvent fumes can cause headaches, dizziness, or nausea. Good ventilation around areas where solvents are used should be provided, preferably in the form of an exhaust system with hoods.

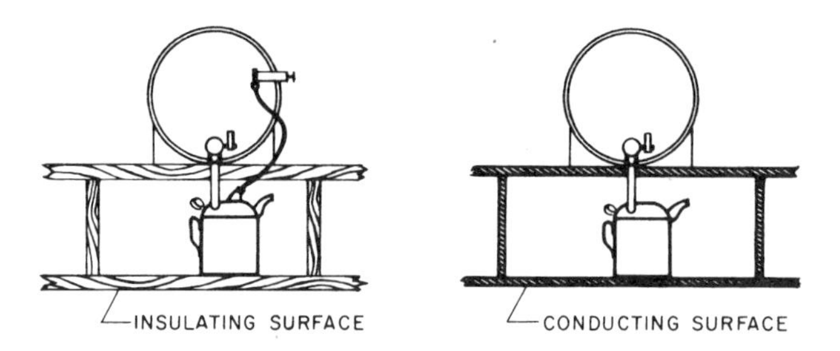

Figure 18.19 Bonding for control of static electricity. Left: if either container is on an insulating surface, a bond wire should be used. Right: if both are on the same conducting surface, an adequate bond can be obtained by keeping the faucet or nozzle in contact with the container being filled.

5. Keep covers on the solvent tanks when they are not in use. If an exhaust system with hoods is used over the solvent tanks, make sure that the exhaust system is operating before the covers of the tanks are opened.

If ventilation is inadequate (i.e., does not meet safety regulations), or problems of disposing of dirty solvents make it impossible to clean with petroleum solvents inside the oil house, arrangements should be made for cleaning in an area where the requirements can be met, such as a painting department or general cleaning area.

Figure 18.20 Safety container. The container is self-closing with a liquid-tight seal. The container must have a UL and/or F/M label.

7. Security

Experience indicates that the oil house should be kept closed, and sometimes locked, to other than lubrication personnel. This is necessary to prevent the confusion that may arise when unauthorized or uninstructed personnel are able to select lubricants from an area of open containers. Inadvertent choice of an improper lubricant may lead to damage to machinery or manufactured products.

C. Lubricant Deterioration in Storage

Lubricants can deteriorate in storage, usually as a result of one of the following causes:

Contamination, most frequently with water
Exposure to excessively high temperatures
Exposure to low temperatures
Long-term storage

Some contaminated or deteriorated products can be reconditioned for use, while others must be degraded to inferior uses, destroyed, or otherwise disposed of. In addition, portions of some contaminated products can be salvaged for use. The course of action to follow depends on such factors as the amount of product involved and its value. This information needs to be evaluated relative to the cost of reconditioning or salvaging, the type and amount of contaminants present, the degree of deterioration that has occurred, and the effect of the contamination or deterioration on the functional characteristics of the product as well as environmental, health and safety issues. Some of these considerations are discussed at more length in the subsections that follow.

1. Water Contamination

In many cases, water contamination is relatively easy to identify, although a quantitative indication of the amount present usually requires analysis from a laboratory or via an on-site water test kit. Frequently a quantitative determination is not required, and the presence of water can be detected by haze or suspended water. When doubt exists, a simple ''crackle'' test, in which a small amount of the oil is placed in a shallow dish and heated to the boiling point of water on a hot plate, provides a fairly positive identification of water. When water contamination has been identified, the steps that can be taken depend on the type of product and its intended use.

2. High Temperature Deterioration

(a) Greases. Some oil separation (bleeding) may occur in some greases when they are stored for prolonged periods, particularly in a high temperature environment. Slight bleeding may be normal. The separated oil will accumulate on the surface of the product. The amount of oil separated should be considered in deciding whether the grease can be used satisfactorily. If the amount of oil separated is relatively small, it can be poured or skimmed off and the remaining grease used satisfactorily. If excessive oil separation has occurred, it is probable that the bulk grease has changed sufficiently to be unsuitable for its intended use and must be disposed of.

Evaporation of the original water content of a water-stabilized calcium soap grease can cause separation of the product into oil and soap phases. However, this usually occurs only at quite high storage temperatures. As a rule, a grease that has separated is unfit for service.

(b) Lubricating Oils. Most premium grades of hydraulic, process, circulation, and engine oils are not affected by even prolonged storage at temperatures below 150°F (66°C). Prolonged storage at higher temperatures (direct container contact with furnaces, steam lines, etc.) may cause some darkening due to oxidation. When one is in doubt about the ability of the oil to perform satisfactorily, a sample should be sent to the laboratory for analysis and evaluation. This assumes that the amount of product involved is sufficient to justify the cost of the analysis before using the product.

(c) Products with Volatile or Aqueous Components This group of products includes diluent-type open gear lubricants, solvents, naphthas, conventional soluble oils, wax emulsions, emulsion-type fire-resistant fluids, and water–glycol fire-resistant fluids.

Long storage at elevated temperatures of diluent-type open gear lubricants can cause evaporation of the diluent. If a significant proportion of the diluent is lost by evaporation, the product will be difficult to apply.

Caution: The diluent used in some types of open gear lubricant may be classified as a health hazard if inhaled, and precautionary labeling is required. Containers of such materials should be kept tightly closed when not in use, and stored in a cool place.

Volatile products such as solvents and naphthas may suffer loss by evaporation. Usually the quality of the remaining product is not affected other than in ease of application. If the products are in sealed containers, high temperatures can cause distortion or bursting.

Some products that contain significant amounts of water, such as fire-resistant water-in-oil (invert) emulsions and water–glycol fluids, are usually unharmed by small amounts of water evaporation. Since invert emulsions lose viscosity with loss of water, and fire-resistant characteristics are dependent on the water content, caution needs to be exercised to assure appropriate water content in these fluids. Losses of water can be replaced by adding water (following the manufacturer's instructions) as needed. Agitation is usually necessary when water additions are made to these products.

Wax emulsions stored at over 100°F (37.8°C) will lose water content owing to evaporation. Minor losses will cause the formation of a surface skin on the product. This skin can be removed, leaving the remainder of the emulsion fit for use. However, before use, the product should be passed through a 40-mesh screen to remove any remaining undesirable materials. Excessive water loss may upset the chemical balance of the product and cause separation of the wax and water, producing a curdlike appearance. Under these conditions, the product cannot be reconditioned readily and usually must be disposed of.

3. Low Temperature Deterioration

Below-freezing temperatures normally will not affect the quality of most fuels, solvents, naphthas, and conventional lubricating oils and greases. The major difficulty associated with storage outdoors or in unheated areas during cold weather is with dispensing of products that are not intended for low temperature service. When possible, containers should be brought indoors and warmed before an attempt is made to dispense the product.

Products that contain significant amounts of water should not be exposed to temperatures below 40°F (4°C). Freezing of wax emulsions will cause separation of the wax and water phases, giving the product a lumpy, curdlike appearance. Under these circumstances, wax emulsions normally must be disposed of, since the products usually cannot be restored to their original condition. Certain fire-resistant invert emulsions are formulated to resist physical and chemical changes brought about by a small number of freeze–thaw cycles, but if the stress is excessive, the invert emulsions can sustain damage.

Repeated freezing or long-term exposure to freezing temperatures may destroy the emulsification properties of conventional soluble oils. Usually there is no change in appearance of the product, although it may have a cloudy cast. The product must be disposed of, however, if it has last its emulsifiability.

Some oils, when subjected to repeated fluctuations of a few degrees above and below the pour point, under go an increase in their pour point (pour point reversion) of 15–30°F (8–17°C). As a result, dispensing may be extremely difficult even when the ambient temperature is above the specified pour point of the product. Oils that contain pour point depressants or have relatively high wax content (e.g., steam cylinder oils) are the most prone to pour point reversion. Although this problem occurs relatively infrequently, products that may exhibit this phenomenon should be stored at temperatures above their pour points.

A product that has undergone pour point reversion may return to its normal pour point when stored for a time at normal room temperature. Cylinder oils may require 100°F (37.8°C) or higher temperature storage for reconditioning.

4. Long-Term Storage

Long-term storage at moderate temperatures has little effect on most premium lubricating oils, hydraulic fluids, process oils, and waxes. However, some products may deteriorate and become unsuitable for use if stored longer than 3 months to a year from the date of manufacture and packaging. An approximate guide to the maximum recommended storage times for some products is shown in Table 18.1. Products stored in excess of the times listed may still be of acceptable quality for use, but a precautionary lab test should be performed to assure that the chemical and physical characteristics of the original products are retained. This testing can be used to requalify the product for an additional storage period. If storage time and conditions for a product are extremely critical, this information is generally indicated on the package.

Table 18.1 Typical Product Storage Life

Product	Maximum recommended storage time (months)[a]
Diesel fuels	6
Defoamants	6
Gasoline	6
Greases	12–36
Calcium complex greases	6
Emulsion-type fire-resistant fluids	6
Soluble oils	6
Custom-blended soluble oils	3
Wax emulsions	6

[a] These times are only guidelines. Products may be requalified for an additional storage period through laboratory testing.

III. DISPENSING

''Dispensing'' in the sense used here includes withdrawal of the products from the oil house or other storage, transfer to the point of use, and application at the point of use. The point of use is defined as the location (bearing, sump, reservoir, friction surface, system, etc.) at which the product will perform a lubrication function of mechanical components. Packaged products may be dispensed directly from the original containers, or they may be transferred to dispensing or application units of various types. Bulk products may be transferred by means of portable equipment to the point of use, but frequently they are dispensed directly from the tanks or bins through permanently installed piping systems to automatic application devices, or to automatic or manually operated dispensing points. When lubricants and related products are dispensed by methods other than completely closed systems, the basic requirements for effective dispensing are as follows.

1. Containers or devices used to move lubricants and related products should be kept clean at all times.
2. Each container or device should be clearly labeled for a particular product and used only for that product.
3. The device used for introduction of a product to a point of final use should be carefully cleaned before the filling operation is started; this includes grease fittings, filler pipes and the area around them, filler screens, filler holes, and quick-disconnect fittings. Sumps and reservoirs should be thoroughly cleaned and flushed before filling for the first time, and they should be checked when being refilled and cleaned as necessary.

The most common dispensing equipment used in industrial manufacturing or process plants includes the following, in sizes, types, and quantities suitable to specific needs:

- In the oil house
  - Faucets
  - Grease gun fillers
  - Grease paddles
  - Highboys
- From oil house to machine
  - Lubrication carts and wagons
  - Oil cans
  - Safety cans
  - Portable greasing equipment
  - Portable grease gun fillers (where grease cartridges are not used)
  - Oil wagons
  - Sump pumps
  - Catch pans
  - Rags and funnels
  - Tools to gain access to reservoirs and/or drain systems

A. In the Oil House

1. Faucets

Faucets (Figure 18.21) are used to dispense oils, solvents, and other fluids from storage tanks or drums to containers for use in the manufacturing area. They are available in

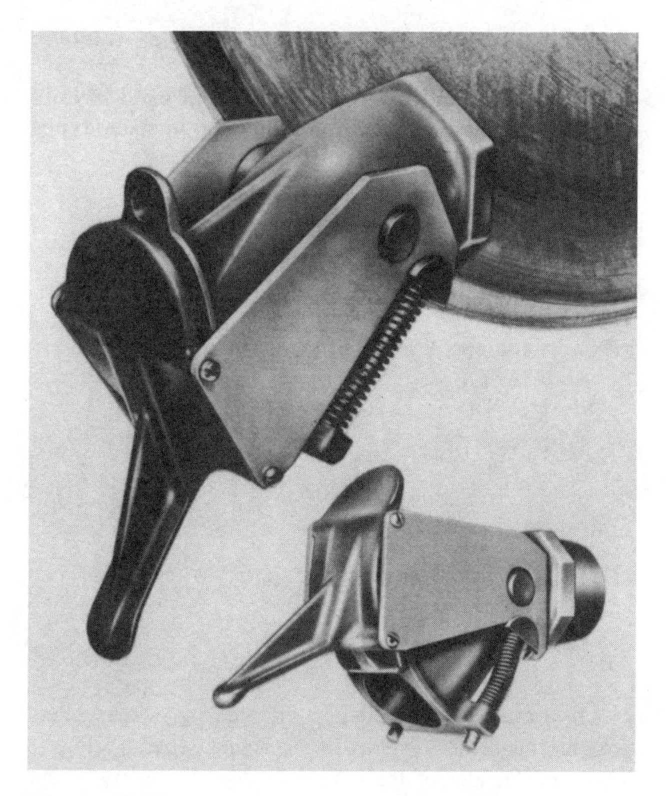

Figure 18.21 Faucet. This faucet has a desirable self-closing feature.

different sizes for fast or slow flowing fluids and to fit either the 2 in. or 3/4 in. drum opening. The faucet can be inserted in the opening and the drum lifted by a crane, lift truck, or chain hoist to a rack that will hold it in a horizontal position while the fluid is dispensed. An alternate method is to use one of the many rocker drum racks (Figure 18.22) to tip the drum onto its side and support it during dispensing. Racks of this type with casters permit easy positioning of the drum on the rack.

Drum faucets are available commercially in steel, brass, stainless steel, and plastic to suit specific fluid needs (noncorrosive, nonsparking, etc.). Some types provide a padlock hasp to prevent inadvertent or unauthorized opening. Faucets with a self-closing feature are generally preferable because they minimize spillage if the faucet is accidentally opened. Faucets equipped with flame arresters are available for use with flammable liquids.

Flexible metal hose extensions may be attached to some faucets to contact the lip or edge of the container being filled to reduce the risk of static electricity discharge.

2. Transfer Pumps

Pumps are used to transfer oils, greases, solvents, coolants, and other fluids from drums or tanks to other containers. In the manufacturing area, they may be used to fill reservoirs or lubrication equipment directly from drums or lubrication wagons.

A pump that can be inserted in the bung opening of a drum is desirable for dispensing oil. These pumps are of positive displacement design and can be obtained to deliver

Figure 18.22 Rocker-type drum rack. The hook on the handle engages the top of the drum chime, while the pads slip under the bottom chime.

measured quantities of lubricants. The simplest types are operated manually. An excellent type of hand-operated pump has a closable return (Figure 18.23) actuated by a spring, to return drippage to the drum without the danger of the contamination that would occur through an exposed drain. Air-operated drum pumps (Figure 18.24) and electrically operated pumps are also available.

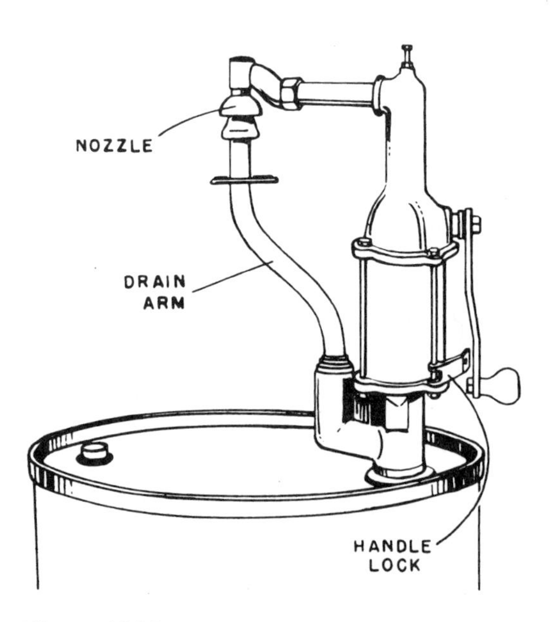

Figure 18.23 Drum pump with return.

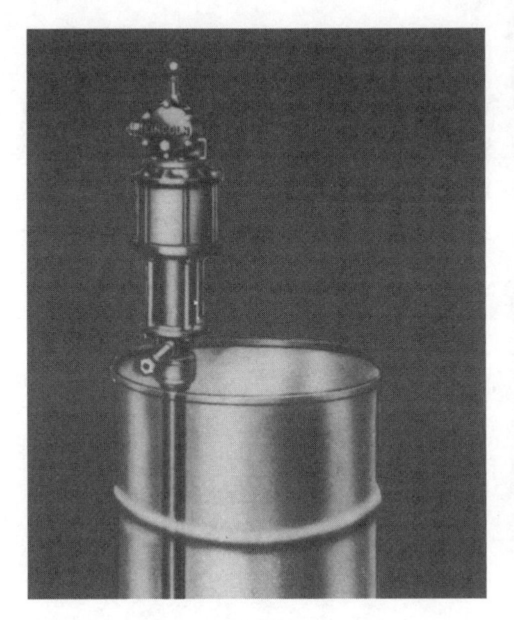

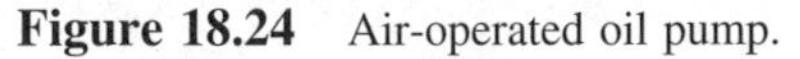

Figure 18.24 Air-operated oil pump.

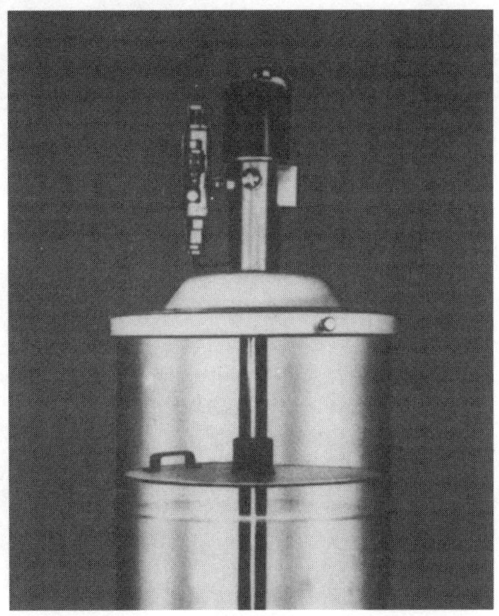

Figure 18.25 Air-operated grease pump.

Drum pumps are also available for dispensing semifluid or soft greases that are packed in closed-head drums. These pumps can be fitted into the 2 in. bung opening; they deliver 20 lb (9 kg) or more of grease per minute and are capable of high pressure discharge. They are used for transferring grease to smaller servicing equipment such as grease gun fillers and portable greasing equipment.

Stiffer greases are packed in open-head drums, and pumps for these products usually are mounted in the center of a drum cover that clamps on the top of the drum (Figure 18.25). Such pump arrangements are generally equipped with follower plates that push the grease toward the suction and keep the grease from adhering to the sides of the drums (Figure 18.26). Figure 18.26 also illustrates a hoist arrangement that can be used to remove and install pumps rapidly and easily with a minimum of lubricant contamination. The usual pumps can handle greases of the NLGI grades 2 and 3 consistency, depending on the type of grease.

3. Grease Gun Fillers

When grease cartridge guns are not used, filling is generally done with grease gun fillers. These can be used to fill directly from the original container or from a reservoir in the gun filler. Models are available to fit 35 lb pails and 120 and 400 lb open-head drums. They may be operated by hand (Figure 18.27), by air, or electrically.

If grease guns, fillers, and other grease-dispensing equipment are to be filled with a grease that is too stiff to pump with a transfer pump, a grease paddle normally must be used. Preferably, the paddle should be made of metal or plastic to prevent contamination that might occur with a wooden paddle. When filling is completed, the cover of the grease container should be replaced securely to prevent contamination.

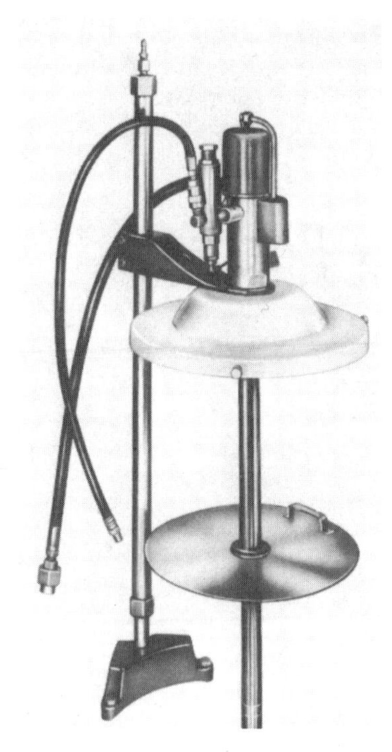

Figure 18.26 Air-operated pump and hoist. The pump is equipped with a follower plate to ensure feeding of relatively stiff greases and to minimize waste.

Figure 18.27 Hand-operated grease gun filler. This unit fits a 120 lb openhead drum and is equipped with a follower plate. Similar units fitting 35 lb pails or 400 lb drums are available.

4. Highboys

At one time, highboys with permanently mounted pumps were used to some extent for oil storage and dispensing. They have the advantage of neat appearance, compactness, and sturdiness, but they are difficult to clean and require an additional product transfer—with risk of contamination—to fill them. The usual highboy design has no drain for cleaning, and foreign matter tends to collect in the bottom of the tank. When new oil is added to the tank, the resulting agitation may cause this foreign material to be suspended in the oil. As a result, oil drawn off before settling has occurred may contain harmful contaminants.

When highboys must be used because of space limitations, they should be equipped to permit easy draining and cleaning.

B. From Oil House to Machine

Moving lubricants from the oil house to machinery where they are to be used is a critical phase of dispensing that justifies the same care as handling in the oil house. The problem, again, is one of preventing contamination and confusion of products. The task is usually further complicated by the necessity for transporting a variety of lubricants in different types of container or lubrication equipment. This phase of dispensing problems is essentially a matter of choosing containers or equipment that can be handled economically with a minimum risk of contamination and confusion. As a general rule, the most desirable containers or pieces of equipment are those that can be filled in the oil house and then emptied directly into the machine, or used to lubricate the machine. Each product should have its own container or piece of dispensing equipment, which should be clearly marked for that product. Such units should not be considered to be interchangeable unless emptied and thoroughly cleaned before being used for another product. It is also a good practice to consult a lubricant supplier to generate a list of compatible products that can be intermixed for noncritical applications and products that cannot be mixed because of incompatibility or potential changes in critical physical and chemical balances.

Caution.* *Galvanized containers or piping should not be used for transporting oil. Many of the industrial oils used today contain additives that can react with the zinc to form metallic soaps. These soaps may thicken the fluid and clog small orifices, oil passages, wicks, and so on. The zinc may carry over into circulating oils and cause interpretation problems when oil analysis programs are run.

1. Oil Cans

Although being outmoded rapidly and outnumbered by automatic oiling devices ranging from simple bottle oilers to complex centralized oiling systems, the common oil can is still widely used to carry oil to machines. Its chief advantages are traditional acceptance and low cost, combined with the need for an easy method of applying small quantities of oil to open bearings. Its chief disadvantages are high labor costs per unit of oil dispensed, the increased hazard of lubricant contamination compared with closed automatic systems, and the generally inferior performance of bearings designed for hand oiling. When hand oil cans must be used, thorough precautions should be taken to maintain lubricant cleanliness.

The simplest hand oil can (sometimes called a pistol oiler) is the diaphragm type. A more practical and desirable oil can is the positive delivery type that delivers a definite quantity of oil from any position (Figure 18.28). The trigger actuates a simple pump incorporated in the can, and an adjustment permits the quantity of oil delivered with each stroke to be varied.

If larger quantities of oil than can be delivered with an oil can are required, special containers are used. Open pails and cans invite contamination. A practical container for quantities of oil that can be hand-carried is the safety can (Figure 18.29). Various styles are available. They can be obtained in many capacities with self-closing spouts and fill covers. Some are provided with removable spouts of flexible metal (or plastic) hose to permit filling of less accessible reservoirs. Such cans are sturdy, easily cleaned, and closed against contamination. Self-closing safety cans of approved design should always be used for transporting flammable materials.

Plastic containers of oils in 1-pint to 5-gallon sizes are being used in some plants. Lightness, low cost, and freedom from rust or corrosion are advantages. Since, however, some plastics are affected by oils or by the additives in some oils, it is always necessary to check with the manufacturer to be sure that the plastic composition is compatible with the fluid to be carried.

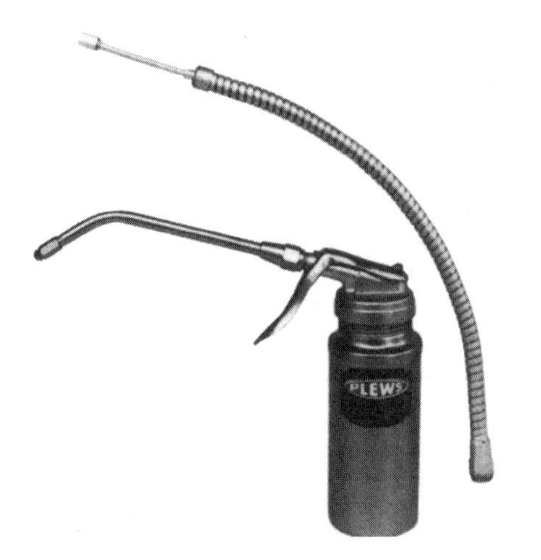

Figure 18.28 Pistol oiler. The flexible spout can be used to service hard-to-reach application areas.

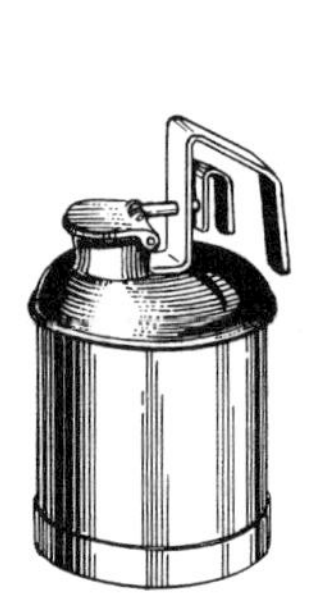

Figure 18.29 Large oil containers.

2. Portable Oil Dispensers

A wide variety of equipment is available to transfer quantities of oil from the oil house to the point of use. Included are bucket pumps (Figure 18.30) that can be used to transport a few gallons of liquid lubricant and discharge it directly into the machine sump or reservoir. If larger quantities must be transferred, wheeled dollies (Figure 18.31), which will transport a standard 16- or 55-gallon drum, are available. The drum can be equipped with a transfer pump that is operated manually, by air, or electrically, to allow the lubricant to be pumped directly into the machine. Even larger quantities can be transported in special oil carts (Figure 18.32), which also may be equipped with a manual or power-operated pump.

3. Portable Grease Equipment

Small quantities of grease (3–24 oz, 85–680 g) can be dispensed from hand grease guns. These are available in push, screw, lever (Figure 18.33), and air-operated types. Some of the lever and air-operated guns can be loaded by suction, with a gun filler, by cartridge, or by any of the three. When small quantities of grease at high pressure are required, special lever-operated guns, sometimes called booster guns, are available.

Couplers and coupling adapters can be fitted to grease guns to permit their use with many various types of fitting. Extension hoses can be attached to many guns to facilitate lubrication of hard-to-reach fittings.

Before a grease gun is loaded, it should be checked for cleanliness and cleaned if necessary. When a grease gun filler is used, the fittings on both the filler and the gun should be wiped clean with a lint-free cloth. Hand loading should be avoided whenever possible. If it is necessary to load from an open grease container, the surface of the grease should be checked for dirt and other contamination and, if necessary, the top layer scraped off to remove any contaminants and hardened grease. Personnel should make sure that the grease is of the correct thickener type to assure compatibility and performance. A

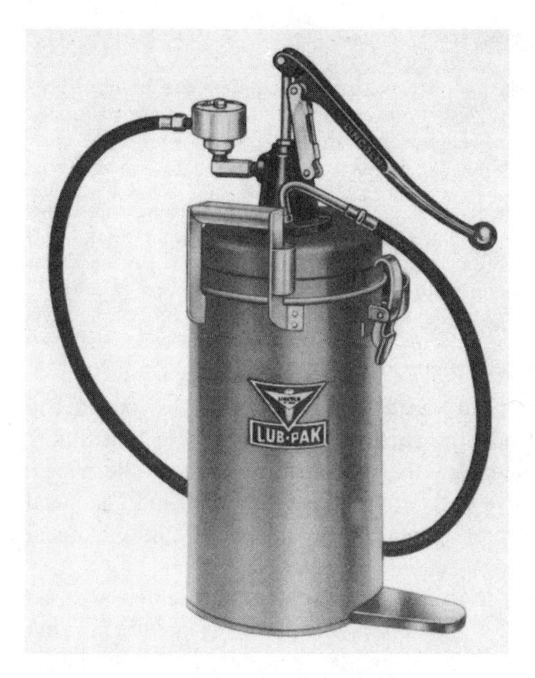

Figure 18.30 Lever-operated bucket pump. This unit holds 30 lb of a lubricant such as a gear lubricant. The extension on the base is to step on and hold the unit firmly while it is being operated.

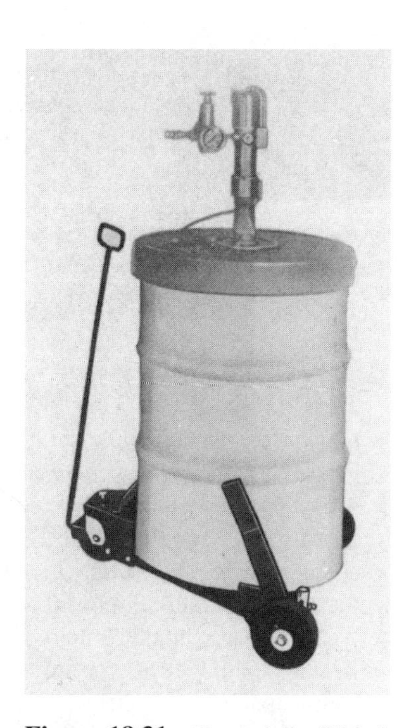

Figure 18.31 Drum dolly. With the wheels raised, a drum can be rolled onto the platform of the dolly. The wheels can then be lowered and the drum moved to the point of use, where product can be dispensed with a transfer pump.

Figure 18.32 Oil cart. The drain on the right facilitates cleaning the tank.

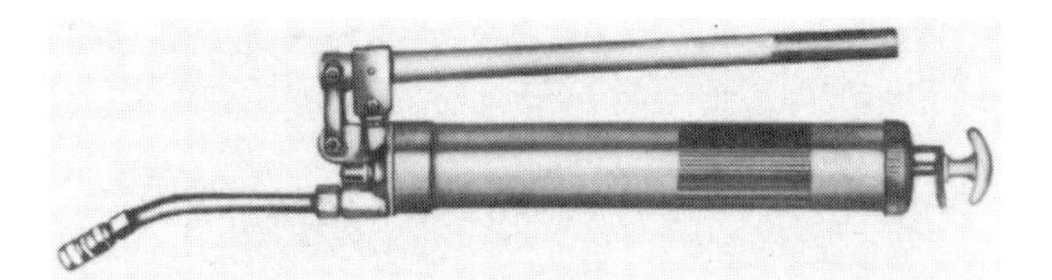

Figure 18.33 Lever-operated grease gun. Models are available to deliver various pressures. This 24 oz model can be loaded by suction, by paddle, or with a grease gun filler. Similar guns of smaller size are available for cartridge loading.

clean metal or plastic paddle should be used to transfer the grease from the container to the gun; alternatively, the gun may be loaded by suction filling. Before suction filling, the surfaces of the gun that will be submerged in the grease should be carefully wiped. Grease cartridges should be wiped to remove dust before loading into the gun. The metal top of the cartridge should be pierced or torn off carefully to avoid the possibility of metal or other materials falling into the grease.

Grease guns should be marked clearly to indicate which type of grease they are used for. A gun should be used for only one type of grease to avoid problems that might result from incompatibility of different brands or types of grease. If it is necessary to change the type of grease to be used, the gun should be thoroughly cleaned in solvent and dried before refilling. After filling, grease should be pumped through the nozzle to flush out any remainder of the previously used product and any traces of solvent.

If more of a particular grease is required than can be provided by a single filling of a gun, several alternatives are available. A gun filler on a pail or drum of grease, or a portable gun filler (Figure 18.34) can be taken to the location at which lubrication is being performed. Manual bucket pumps on standard 35 lb pails can be used. A variety of air-powered and electric guns are also available. Most of this equipment will handle semifluid and soft greases without difficulty. Some fibrous and firmer greases may require a follower plate arrangement to ensure that the grease slumps to the pump inlet. Very firm greases may require an arrangement that provides positive priming, as shown in Figure 18.35, to ensure feeding to the pump inlet.

4. Lubrication Carts and Wagons

The lubrication program of a plant may be so organized that personnel are regularly assigned to service comparatively distant parts of the plant, where they must apply a number of different oils and/or greases to various types of machinery. It is important in these cases to supply each person with a practical means for transporting the necessary supplies. Such equipment may be either purchased or built. It may be elaborate or simple, depending on the variety of lubricants used and on the machines to which the lubricants are to be applied. A simple, practical lubrication wagon may be nothing more than a shop cart with sufficient room for the needed supplies-oil containers, grease guns, hand oil cans, grease pails or portable gun fillers, and miscellaneous necessities such as spare grease fittings, tools, replacement cartridges, and clean lint-free cloths for cleaning lubrication equipment and application points. Lubrication instructions for the machinery to be lubricated should also be carried. Such a cart is flexible and permits a quick change in the types of lubricant carried.

Figure 18.34 Portable gun filler. The 30 lb tank on this unit can be filled in the oil house. The unit can then be taken to the plant to refill guns as required.

Figure 18.35 Electric power gun and cutaway. The cutaway view shows the helix arm and worm gear, which ensure feeding of firmer greases to the pump inlet.

More elaborate lubrication carts can be obtained from suppliers of lubrication equipment. The type shown in Figure 18.36 provides space for portable grease gun fillers, grease guns, and a number of containers of fluid lubricants. The types of cart shown in Figure 18.36 handles mainly fluid lubricants, but space is available to carry miscellaneous supplies. If there are provisions for practical and easy cleaning of the tanks, these carts are very satisfactory.

C. Closed-System Dispensing

A number of types of lubricant dispensing and application system can be considered to be essentially closed systems in which lubricant is exposed to the minimum possibility of contamination. These systems are not truly closed in that the supply tank or reservoir must be charged with lubricant periodically. The lubricant supplied to these applications can be dispensed at the point of use through a quick-disconnect fitting or metering device, reducing the risk of contamination greatly in comparison to the dispensing methods already discussed. One of the more common types of closed systems is discussed in Section III.D.

D. Central Dispensing Systems

While lubricants can be dispensed from bulk bins or bulk tanks into containers, oil wagons, grease gun fillers, and so on, transfer by some form of piping system often is justified. Such systems are usually custom-designed and built to handle large volumes of lubricants, but smaller systems may be assembled from equipment modules available from lubrication

Figure 18.36 Commercial lubrication cart. Space is provided for gun fillers, fluid lubricants, grease guns, and miscellaneous supplies.

equipment suppliers. Electric, hydraulic, or air-operated transfer pumps, similar to those discussed earlier for dispensing from drums that will handle greases or oils, are available to fit tanks or bulk bins. If lines extend beyond the capacity of the tank-mounted pump, booster pumps can be installed at suitable points along the lines leading to the dispensing stations. When distribution lines must be run outdoors or through cold areas, steam or electric tracers on the lines may be required to maintain pumpability of the lubricant.

At the remote dispensing stations, delivery can be made to grease gun fillers, power grease guns, reservoirs of centralized lubrication systems, or to any type of oil dispensing unit. Delivery can go directly to the point of final use, such as the reservoirs of large lubrication systems or units being lubricated on an assembly line. An example of lubricant being dispensed directly to the point of use is shown in Figure 18.37. The plant in which this diesel engine is installed is equipped with a 5000 gallon (19,000-liter) underground tank for cylinder oil, and a 3000-gallon (11,400-liter) tank for crankcase oil. Cylinder oil is pumped to the overhead hose reel and then dispensed through a metering nozzle to the cylinder lubricators. Crankcase oil is pumped to a 275-gallon (1040-liter) secondary tank indoors for preuse warming, then pumped directly to the crankcase. The pump for the later operation is controlled from the engine location, to permit the addition of the correct amount of makeup.

Dispensing nozzles of the type shown in Figure 18.38 can be used to control dispensing in systems of this type, or in assembly line filling of equipment such as gear units. The dial can be preset to deliver from 1 to 60 quarts and to shut off automatically. The meter also records the total amount of oil dispensed, up to 9999 gallons.

Comparatively little maintenance is required for central dispensing systems. Proper cleanliness precautions should be observed when storage tanks or reservoirs are filled. An adequate quantity of lubricant should always be maintained in the tank to ensure that the

Figure 18.37 Dispensing system for a large diesel engine. The overhead hose reel (arrow) is used to dispense oil directly to the cylinder lubricators through a metering nozzle. Crankcase oil is brought from the main storage tank to a secondary tank indoors, then pumped directly to the crankcase.

Figure 18.38 Metering control nozzle. The dial can be preset to deliver from 1 to 60 quarts of oil, then shut off flow. A counter provides a record of the total amount of lubricant dispensed.

transfer pump will not suck air that might cause binding or locking. Pump pressures should be checked periodically, and lines should be inspected for leaks or damage.

Again, as stated earlier, any industrial plant lubrication program should meet federal, state, and local codes and requirements, which includes health and safety factors as well as addressing environmental issues.

BIBLIOGRAPHY

Mobil Technical Bulletin

Handling, Storing, and Dispensing Industrial Lubricants

Technical and Regulatory References

National Fire Prevention Association, NFPA 30
Occupational Safety and Health Administration, OSHA 29 CFR.

19

In-Plant Handling and Purification for Lubricant Conservation

Conservation of natural resources and the protection of the environment have become common goals in society today. Depending on current plant practices, these goals can be accomplished generally with minimal cost impact in the areas of lubricant use and disposal. Control of lubricants inside the plant to prevent the generation of oily wastes is just one such operation. Any lubricant that becomes a waste increases operating costs from the standpoint of material purchases, waste disposal, and environmental protection issues. A plant lubricant that becomes a waste product may require reclamation or disposal under proper control to prevent pollution for the following reasons.

1. Contamination with water, foreign matter, dust, process materials, or wear metals (from the lubricated equipment), or dilution
2. Degradation during use due to depletion of additives, increases in total acid number, formation of oxidation products, change in viscosity, or loss of lubricity
3. Escape from system through leaks, spills, line breakage, faulty gaskets, or excessive foaming; overlubrication in all-loss systems; carry-off on products in process

One key to preventing any lubricant from becoming a waste product lies in the selection, storage, and handling of these products in the plant, from receipt as new materials to disposal of used materials, as discussed in Chapter 18.

With the increased emphasis on the subject of pollution of ground and surface waters, strict regulations have been enacted covering the composition of industrial effluent to both surface waters and municipal sewage plants. Regulations covering plant effluent and stream water quality vary from area to area, and depending on the jurisdictional control of the particular body of surface water, groundwater systems or sewage plant authority may be under the control of federal, state, municipal, or regional groups. The first step for anyone concerned with the problem of preventing pollution from industrial sources is to know the effluent, stream, groundwater, or sewage standards governing the particular situation.

Before discharging any fluid waste into a waterway, it is essential to comply with the existing regulatory requirements at the particular location. In general, as far as oil content is concerned, the effluent should be free of visible floating oil (not in excess of 15 mg/L). This level assumes a dilution effect in the receiving stream. For example, if the stream is to be used for municipal water supply, its oil content must be kept below 0.2 mg/L. If, on the other hand, the receiving body of water is being used for recreational, agricultural, or industrial uses, as much as 10 mg/L may be acceptable. By the same token, to ensure that there is no interference with planned use of the stream, effluent limits for color, odor, turbidity, dissolved oxygen, heavy metals (such as lead and mercury), phenols, phosphates, suspended solids, and so on must be met.

It is impossible to cover the myriad of regulations here. Instead, this chapter covers general practices aimed at reducing the generation of lubricant wastes while maximizing their useful life.

I. OVERVIEW OF IN-PLANT HANDLING

The objective of proper in-plant handling may be defined as efficiently utilizing petroleum lubricants to prevent them from prematurely becoming waste products. Further, disposal when these products reach the end of their useful life without allowing the waste to become a detriment to the environment is covered in general terms.

Based on 1996 figures, in the United States alone, 2.5 billion gallons of petroleum lubricating and process oils was produced and used as automotive and industrial products. Approximately 1 billion gallons was consumed in the industrial sector as lubricants, hydraulic fluids, process oils, and metalworking fluids. These fluids require recycling or reclamation for continued beneficial use to conserve natural resources or, when no longer suitable, disposal in an approved manner to prevent environmental damage.

ExxonMobil, as a member of the petroleum industry, is keenly aware of its responsibility to utilize, with regard for conservation and environmental quality, one of nature's primary resources. It realizes that this responsibility does not end with ensuring that its own operations use petroleum resources in the best interest of society. But ExxonMobil also desires to assist customers or end users in realizing the maximum utilization from petroleum products and the proper handling and disposal of them once they have served their primary purpose.

An initial step toward efficient usage and waste oil disposal would be the implementation of a lubrication program such as that offered by ExxonMobil. Part of this program would include consultation with ExxonMobil's Engineering Services to ensure proper product selection, maximum product life in service, minimization of leaks, beneficial machine maintenance, optimum drain intervals, and improved handling and storage practices to prevent contamination and spills. It is better not to generate a waste lubricant in the first place than to be obliged to dispose of it later.

Industrial operations and individual plant limits are so widely different that the minimization of waste oil generation and disposal to control pollution in each case constitute unique problems requiring specific solutions. Nevertheless, experience shows that the following concrete suggestions and overall recommendations concerning the handling of lubricants in plants may be adapted to individual cases.

1. Select lubricants, including hydraulic fluids, gear lubricants, metalworking fluids, coolants, and crankcase oils to obtain long service life.

2. Establish a program for good preventive maintenance to keep equipment in good operating condition *(key element)*.
3. Set up good housekeeping procedures.
4. Where feasible, utilize a plant-wide, multimachine circulation system to replace small, single-machine reservoirs.
5. Provide purification equipment for circulation systems to ensure optimum use of recycling where practical.
6. Identify the nature and sources of waste generation and disposal problems.
7. Attack the problem at the source, not at the plant effluent stage.
8. Know local regulations and regulatory agencies.
9. Keep metalworking fluids, solvents, and lubricant streams separated and distinct from each other to prevent complexity in recycling or reclamation.
10. Maintain separate sewer systems for sanitary waste, process water, and storm drainage.
11. Do not use dilution as the solution to pollution.
12. Concentrate waste before treating it for final disposal.

Implementation of a good conservation program can result in a big return on costs of installing such a program. The potential benefits can be found in three categories:

Economic	Less lubricant to be purchased
	Reduced application costs
	Increased machinery availability (longer drains, fewer failures)
	Increased production capacity
	Reduced maintenance costs
Safety	Reduced potential for injury to personnel
	Reduced potential of fires from poor housekeeping
	Lower insurance costs
Environmental	Avoidance of high remediation costs
	Fewer fines
	Improved public opinion

II. PRODUCT SELECTION

The cycle leading to the eventual consumption or disposal of lubricants begins with the purchase and receipt into a plant of these products. It is at this point that we must examine what can be accomplished to maximize use and prevent a new product from prematurely becoming a waste lubricant and a possible pollutant.

The many factors that enter into the selection of the proper product depend on its use as a lubricant, hydraulic fluid, or metalworking fluid. Speed, load, and temperature must be considered (discussed in Chapter 8) in the selection of gear and bearing lubricants, as well as types of metal and severity of operation in machining, and types of engine, speed, and fuel, in crankcase and cylinder lubrication of stationary diesels, gas engines, and gas turbines. These factors and their effects on product selection are well known and usually recognized. In addition, there are other factors that will minimize the waste oil disposal problems in a plant. These factors, which should be considered in the selection process, are long service life, the ability to help control leakage, the ability to minimize contamination effects, compatibility with other products, the compositional make up of the product, the value as a by-product, and ease of disposal. One general rule that should be a guide in product selection to achieve the optimization of in-plant handling is the use of the smallest number of multipurpose premium products that is practical. Product consolidation is a key element in ExxonMobil's lubrication program.

A. Long Service Life

One important characteristic of a lubricant that will do much to minimize the waste oil disposal problem is long service life. It is apparent that the longer the interval between drains in any machine using a closed circulation system for lubrication, the lower the amount of waste oil generated and fewer the shutdowns required for oil changes. Long service life is the result of many factors, including machine and operating conditions, maintenance practices, proper selection of lubricant, and product quality. Product characteristics inherent in the base oil or additive package that enhance long life are as follows:

Chemical stability, to ensure minimum buildup of oxidation products
Resistance to changes in viscosity
Control of acidity and deposit formation
Additive sufficiency, to ensure the needed film strength
Detergency, to prevent deposits and keep system components clean
Viscosity–temperature characteristics
Resistance to depletion of the foregoing additive qualities

All these items lead to selection of the highest quality product to achieve the longest service life. In addition, premium product selection is usually the most economical, for although the initial material price is higher, the cost in man-hours of application, maintenance, and disposal is sharply decreased, and the overall cost of lubrication is lower. For example, synthetic lubricants (Chapter 5) cost more to purchase but can achieve substantially longer service life, resulting in fewer oil drains, increased production capacity of equipment, and less waste disposal. High performance synthetic lubricants are in many cases the most desirable and economical solution for conservation.

B. Compatibility with Other Products

Where the possibility exists of one product mixing with another during use or application, such as the hydraulic fluid in a machine tool with metalworking fluid, the products used for both fluids should be selected so that cross-contamination will minimize the negative effects on performance characteristics. For example, in the instance mentioned, if a chemical coolant type of metalworking fluid can be used as the hydraulic fluid as well, changeout or disposal may not be required because of admixture. Other instances will become apparent in a survey of the types of lubricant used in a plant and the products available from lubricant suppliers. Even at the disposal stage, compatibility of waste oil will permit consolidation without forming complex mixtures, with attendant difficulties of disposal.

C. Value as By-Product

Since eventually most lubricants become waste oils and require disposal, consideration should be given at the selection stage to the value of the waste oil as a by-product. Many lubricants, after satisfying their primary function, can be used for less demanding service such as a fuel or feedstock material for reclamation or re-refining. The ability of a lubricant to serve again as a by-product after initial use should be considered in the selection of a new product, from the points of view of both conservation of natural resources and waste disposal. It must also be recognized that the cost of disposal of a select few lubricants may be more than the original purchase price of those lubricants. Knowing this up front allows for the control of product use and planning for eventual disposal costs.

E. Ease of Disposal

The last factor to be considered in lubricant selection is the ease of disposal of the substance when it has become a waste product. This factor is related to the product type: for example, oil-in-water and water-in-oil emulsions differ in the ease with which the emulsions can be broken. Also, a few select products (e.g., certain metalworking fluids) have been developed that may be disposed of in municipal sewage systems assuming no undesirable contamination. Always check with local authorities and the supplier first. The environmentally aware (EA) lubricants discussed in Chapter 6 are another example of products that generally present fewer problems as pollutants. In general, the value of a by-product increases with the ease of disposal.

Following the selection of products to fulfill the lubricating and other plant requirements, the next area in which premature generation of waste oils and subsequent disposal or pollution control problems can be prevented is the handling, storing, and dispensing of lubricants and associated petroleum products after receipt from the supplier. This subject was covered in Chapter 18.

III. IN-SERVICE HANDLING

Once a lubricant has been charged to a system and until it requires drainage and replacement, much can be done to prevent escape of the lubricant from the system and its premature degradation, requiring early draining. The prevention or minimization of leaks, spills, or drips that can complicate the disposal of lubricants will also reduce the potential for accidental pollution of plant effluent. The elimination of unnecessary contamination through proper system operation and preventive maintenance also is treated, along with in-service purification to prolong the lubricant's life.

A. Reuse Versus All-Loss Systems

The many ways of applying lubricants to bearings, gears, machine ways, cutting tools, and so on vary from hand oilers, and grease guns, through bottle and wick oilers and splash and mist feed to the most sophisticated circulation systems and centralized bulk-fed systems (see Chapters 9 and 18). Some of these applications are all-loss systems in which the lubricant is used in a once-through operation. In some cases, the lubricant is consumed in the process of lubrication. In a two-cycle engine, for example the lubricating oil is mixed with the fuel, is carried off on the product or in exhaust gases, or is collected for disposal. An enclosed lubrication system, such as in gear cases, large engines, turbines, paper machine drying systems, or machine tools, continually reuses the same lubricant until its lubricating characteristics change or are degraded. Enclosed lubrication systems are recommended where practical from the standpoints of both conservation of product and minimization of pollution. These systems are far more economical in terms of costs of material, application labor, housekeeping, and disposal. Further, when these systems are coupled with purification systems, they actually furnish better conditioned lubricant throughout the use period to the machines bearings, gears, and other lubricated components.

Enclosed lubrication systems can be more readily adjusted to prevent excessive lubrication and can be selected to prevent contamination of the lubricating oil or cross-contamination of the lubricant with metalworking fluids, process oils, and materials being fabricated.

Oil mist lubrication systems can be an excellent method of lubrication for specific machine applications while also being quite economical in lubricant consumption. Unless carefully controlled, oil mist lubrication systems may cause fogging and condensation on floors and machine exteriors, while excessive oil mist can possibly cause concern with respect to exposure via inhalation. Proper installation, care, and maintenance are needed here, as well as with many hand-controlled oilers, to ensure proper oil feed rates without any excess lubrication.

B. Prevention of Leaks, Spills, and Drips

Any loss of lubricant from a system or machine component that is not called for in the design and for which a collection system is not provided complicates the disposal problem. It has been estimated that leakage from circulating systems, including hydraulic systems, approximates more than 100 million gallons a year, requiring more than 5.5 million man-hours to provide makeup. With proper leakage control, over 70% of the waste of material and manpower requirements could be reduced or eliminated. In addition to the savings in material and labor, the benefits that accrue from proper leakage control include increased production and decreased machine downtime, prevention of product spoilage and cross-contamination of other lubricant or coolant systems, elimination of safety hazards to plant personnel, and prevention of accidental pollution of the plant's storm, process, or sanitary waters. Also, leaks, spills, and drips seriously complicate the disposal problem because of collection difficulties. Product thus disposed of is lower in by-product value; most often it must be incinerated to ensure disposal in a nonpolluting manner. Further, by cross-contamination, it often forms tight emulsions or complex mixtures with other liquids that are not easily separated. The cost impact (oil only) of even small drips can be seen from the data in Table 19.1.

Table 19.1 Losses by Oil Leaks

	Value of loss[a]					
	In 1 day		In 1 month		In 1 year	
Leakage[b]	Gallons	dollars	Gallons	dollars	Gallons	dollars
One drop in 10 s	0.112	0.45	3.37	13	40	162
One drop in 5 s	0.225	0.90	6.75	27	81	324
One drop/s	1.125	4.50	33.75	135	405	1620
Three drops/s	3.750	15.00	112.50	450	1350	5400
Stream breaks into drops	24.00	96.00	720.00	2880	8640	34560

[a] Based on $4/gal.

[b] Drops are approximately 11/64 in. in diameter.

System leakage control on any machine or complex of machines requires attention to two general classes of joints through which fluid may be lost. These are moving joints (dynamic) and static joints. Moving joints include rod or ram packing, seals for valve stems, pump and fluid motor shaft seals, and in some instances, piston seals. Static joints include transmission lines, pipes, tubing, hoses, fittings, couplings, gaskets, and seals and packing for manifolds, flanges, cylinder heads, and equipment ports (hydraulic pump and motor ports).

Little or no leakage will occur past newly installed seals or packing for moving joints if correct materials and procedures have been used. Some of the causes of leakage in moving joints include improper installation (resulting in seal and packing damage), misalignment, and rough or scarred finishes on rods or shafts. However, both internal leakage (past pistons, vanes, valves, etc.) and external leakage (past rod, shaft, valve stem packing, housings, etc.) may be expected to develop in time, even under normal service conditions. Internal leakage may cause problems in machine operation and loss of lubricant through consumption in the machine. External leakage may cause problems with loss and pollution, as noted earlier. Only a properly planned and executed preventive maintenance program will find and correct such leakage.

Leakage of oil from static type joints may be the result of one or more of several causes:

1. Use of unsuitable types of joint or transmission line
2. Lack of care in preparing joint (machining, threading, cutting, etc.)
3. Lack of care in making up joint
4. Faulty installation, such that joints are loosened by excessive vibration or ruptured by excessive strain
5. Severe system characteristics that subject joints and lines to peak surges of pressure due to ''water hammer''
6. Improper torquing of bolts or fasteners
7. Incompatibility of joint materials with lubricant (takes time to show up)

In addition to slow leakage, a combination of these items may contribute to fatigue failure, line breakage, and loss of large quantities of lubricant and hydraulic fluid.

Several general measures to reduce leakage are as follows.

1. Use joint-type seals and packing material for installation and maintenance work that have proved satisfactory in service.
2. Train maintenance personnel in principles of proper installation of joints, seals, and packing, and maintain surveillance to ensure proper execution of accepted procedures.
3. Minimize the number of connections, and make all lines and connections accessible for checking and maintenance.
4. Design installations to avoid excess vibration, mechanical strain, twisting, and bending.
5. Locate and protect tubing, valves, and other components from damage by shop trucks, heavy work-pieces, or materials handling equipment.
6. Protect finely finished surfaces in contact with seals and packing from abrasive and mechanical damage.

One other area that should be given consideration with respect to preventing spills is the dispensing and draining of lubricant from machine reservoirs. The dispensing of new lubricants has already been discussed (Chapter 18), and the drainage of waste oils is treated later in this chapter (Section VII: Waste Collection and Routing).

C. Elimination of Contamination

It is well known that the largest single cause of waste lubricant generation is contamination. It is estimated that about 70% of the oils are rendered unfit for service because of contamination. Sources of contamination cover a wide range of materials from foreign matter and degradation products formed from the lubricant under action of the lubricating process to cross-contamination through admixture with other process fluids. The result of any one of these contaminating actions can be loss of primary function in the lubricant, with consequent need for reclamation or disposal as a waste oil, a downgrading in value as a by-product, or an increase in complexity of disposal procedures. Therefore, in the interest of minimizing waste and maximizing ease of disposal, contamination of any kind should be eliminated or controlled to the highest degree practical.

1. Central Reservoir Maintenance

Reservoirs for circulating lubricating systems, which include hydraulic and metalworking fluid systems, may be one or a combination of three types: integral with machine and located in the base, separate from the machine but individually associated with it, and centrally located reservoirs serving multiple machines: Whether a central system is economically justified depends on many factors, among which are the size of the facility, the number of different grades or brands of fluids in use, and the facility layout. Obviously in any facility with multimachine lubrication requirements, a central system can be better justified if only one, or at most, two multipurpose lubricants are handled.

All reservoir types can become contaminated in similar fashion, and preventive methods apply equally to each. Dirt, water, dust, lint, and other foreign substances contribute to the formation of emulsions, sludges, deposit, and rust when present in a circulating lubrication system. These materials also detract from the performance of the lubricant, accelerate degradation, and increase the potential for wear or loss of performance of the system components. These contaminants should be removed or the bulk lubricant drained and replaced. Some contaminants present in the waste oil can limit its value as a by-product for use as fuel or for reclamation. Table 19.2 lists numerous contaminants some-

Table 19.2 Some Common Circulating Oil System Contaminants

Air	Packing and seal fragments
Assembly lubes	Paint flakes
Cleaning solvents or solutions	Persistent emulsions
Coal dust	Pipe scale
Dirt	Pipe threading compounds
Drawing compounds	Rust particles
Dust	Rust preventives
Gasket sealants and materials	Water
Grease from pump bearings	Way lubricants
Lint from cleaning rags or waste	Wear particles (metal)
Metal chips	Weld spatter
Metalworking fluids	Wrong oil
Oil-absorbent material	

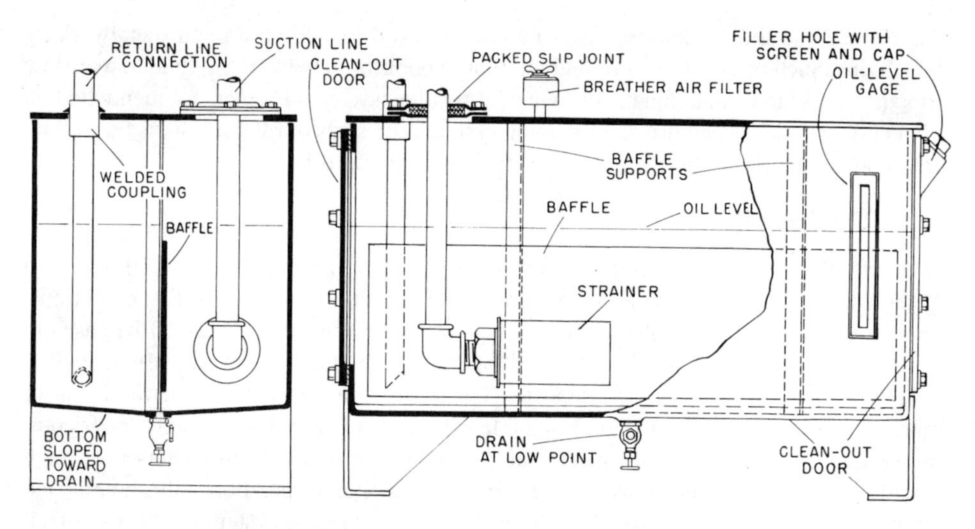

Figure 19.1 Contamination control in central reservoir.

times present in circulating systems. Considering each of these individually may suggest sources of contamination and means to prevent their entry into the system. Contamination prevention of foreign matter involves good system design, good maintenance, and good housekeeping practices. The following general recommendations will help in controlling reservoir contamination by foreign matter (Figure 19.1).

1. The reservoir cover, if of removable type, should fit well and be gasketed and tightly bolted on.
2. Clearance holes in the reservoir cover for suction and drain lines should be sealed, preferably by a compressible gasket and a retainer of the bolted flange type.
3. The oil filler hole should be equipped with a fine mesh screen and dusttight cover.
4. The breather hole should be provided with an air filter and checked regularly.
5. The suction should be equipped with a strainer to prevent the larger particles of dirt and other foreign matter from entering the system and should be inspected regularly.
6. Use of a magnetic pickup in the bottom of the reservoir, a magnetic dipstick, or a magnetic drain plug (not shown in Figure 19.1) will reduce the number of magnetic particles being circulated.

2. Cross-Contamination

Every effort of design and operation must be exerted to prevent contamination of machines that utilize separate systems and products for hydraulic operation, bearing and gear lubrication, and machine tool cooling. For example, hydraulic systems of machine tools are subject to contamination with water-soluble chemical coolants or oil emulsions that can, owing to poor oxidation resistance, chemically active agents, or fatty acids, cause deposits on valves, emulsion formation, and foaming in the hydraulic system. Similarly, in the case of machine tool operation, chips may be allowed to pile up around metalworking fluid drains. Such piles act as dams and may cause pools of metalworking fluid to overflow onto the ways and contaminate the way lubricating system. Further, as mentioned in connection with leaks and drips, lubricating systems and seals of rotating shafts may allow cross-contamination of the metalworking fluid with tramp oil.

3. Proper System Operation

Another type of contamination that will cause a lubricant to become a waste oil is caused by degradation products formed in the lubricant as a result of system operating conditions. For example, excessively high operating temperatures can cause oxidation of the base oil. The oxidation products formed could result in changes in viscosity, an increase in acidity, and the formation of varnish and deposits. The excess temperature may be the result of improper selection of the lubricant, low oil levels, inadequate system capacity, poor cooling, system malfunction, or poor original design. While premium lubricants have high oxidation stability, proper system design, maintenance, and operation will complement this inherent characteristic. Adequate capacity of the lubricant system or the use of oil coolers will help control the bulk temperature during recirculation.

IV. IN-SERVICE PURIFICATION

In circulating systems, the lubricant, hydraulic fluid, or metalworking fluid should be kept as free of contaminants as possible, first, as noted earlier, by preventing entry of contaminants into the system, and second by in-service purification during operation. Purification, depending on the nature and extent of the contaminants, may consist of continuous bypass or full-flow treatment. In many instances, a combination of the two methods is incorporated into the circulating system. In addition, large-capacity critical systems may use portable units for adding oil to the system or for periodic polishing, or independent purification units connected permanently to large reservoirs, or separate batch units to purify or reclaim drained lubricants for reuse.

A. Continuous Bypass Purification

In the continuous bypass system (Figure 19.2), a portion of the oil or coolant delivered by the pump is diverted continuously from the main line for purification. The cleaned oil or coolant is then returned to the system. The remaining unpurified stream is delivered

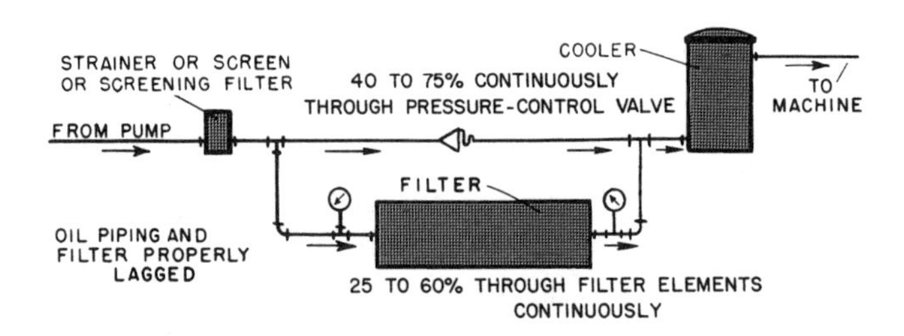

Figure 19.2 Continuous bypass purification.

through the main line to the system. This system of purification is reasonably effective when the rate of contamination is low.

The bypass is limited to applications in which continuous purification of 5–10% of the total pump discharge is sufficient to keep the entire charge in good condition for a satisfactory period.

A throttling orifice to prevent too great a withdrawal of bypass fluid limits flow to the bypass filter. Gradual clogging of the elements reduces the flow of dirty oil through the filter. This condition causes an increase in pressure on a gage on the filter between the orifice and the filter elements. When this gage shows a predetermined increase in pressure, the filter elements should be cleaned or replaced.

B. Continuous Full-Flow Purification

In a continuous full-flow system (Figure 19.3), the entire volume of lubricating oil or coolant is forced through a filter before passing through the cooler to the lubricating system. The oil may pass through the cooler ahead of the filter if the cooled oil is still at a good filtering viscosity.

Clogging of the filter decreases the flow of the fluid through the elements. This condition is indicated when gages on the inlet and outlet of the filter show an increasing drop in pressure across the filter elements. Before this drop becomes excessive, a relief valve in the filter opens and bypasses unpurified fluid around the elements to maintain a full supply of oil or coolant to the system. The filtering elements should be replaced before this has a chance to happen. Filters installed in parallel with the required valving permit the replacement of elements in one filter at a time, without shutdown. This should be done whenever the pressure differential across a filter reaches the value recommended by the filter or machine manufacturer.

C. Continuous Independent Purification

Lubricating oil or coolant is sometimes purified continuously by means of a system entirely independent of the main circulating system. Used oil is drawn from the sump by an independent pump, delivered to a centrifuge (purifier) or fine filter (polishing filter), or both, and after purification, is returned to the system (Figure 19.4).

Although this system would operate normally during system operation, it can be used when the system is shut down, and it can be shut down without system interruption at any time, to allow operators to replace the filter elements or clean the purifier as dictated by the pressure drop across the unit.

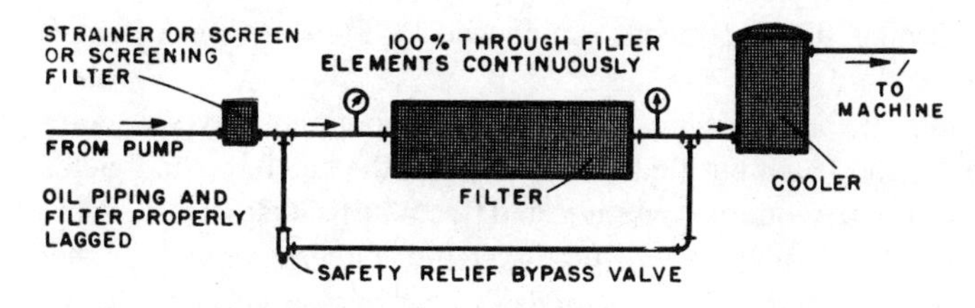

Figure 19.3 Continuous full-flow purification.

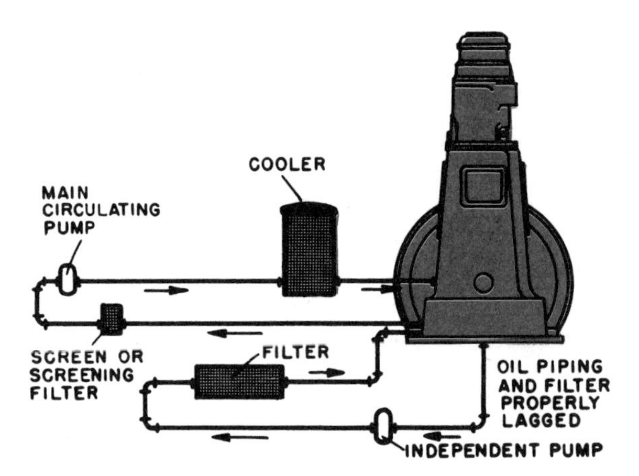

Figure 19.4 Continuous independent purification.

D. Periodic Batch Purification

Periodically, the entire charge of lubricating oil or coolant may be removed from the system for purification. The oil or coolant in this batch is allowed to settle, whereupon it may be reheated and passed through a centrifuge, reclaimer, or other type of purification equipment. Typically, this type of batch purification is used in large-capacity systems such as utility turbine systems, where a single charge of oil can be over 5000 gallons and expected life exceeds 20 years. For example, during a major turbine overhaul or during scheduled downtime, the entire batch of oil may be removed from the system and placed in a holding tank for treatment. This oil is replaced in the system with new oil or oil that has been batch-purified. The oil removed from the system can then be batch-purified on a leisurely basis.

E. Full-Flow and Bypass

In a full-flow and bypass purification system, the dirty fluid from the machine passes through the purification system at a full flow rate, and a portion of the stream delivered to the machine is diverted and returned for further purification. The bypass stream is valved so that when the main stream is shut down, the purification system may continue to operate. Such a system, when utilizing filtration, can provide for gross filtration of the full-flow stream and fine filtration of the bypass stream at a slower flow rate with the combination yielding a cleaner fluid. In addition to fine bypass filtration, the bypass system can be equipped with coalescing filter elements, which can be used to remove small amounts of moisture from the fluid.

A comparison of the four methods (Figure 19.5), full-flow, continuous bypass, combination full-flow and bypass, and batch purification, shows that the batch method yields the cleanest fluid at the least cost but requires downtime and fluctuates widely in efficiency. The combination method, which can function when the system is either operating or shut down and can use more stringent purification methods on the bypass portion without sacrificing operating flow rates, yields the highest overall efficiency.

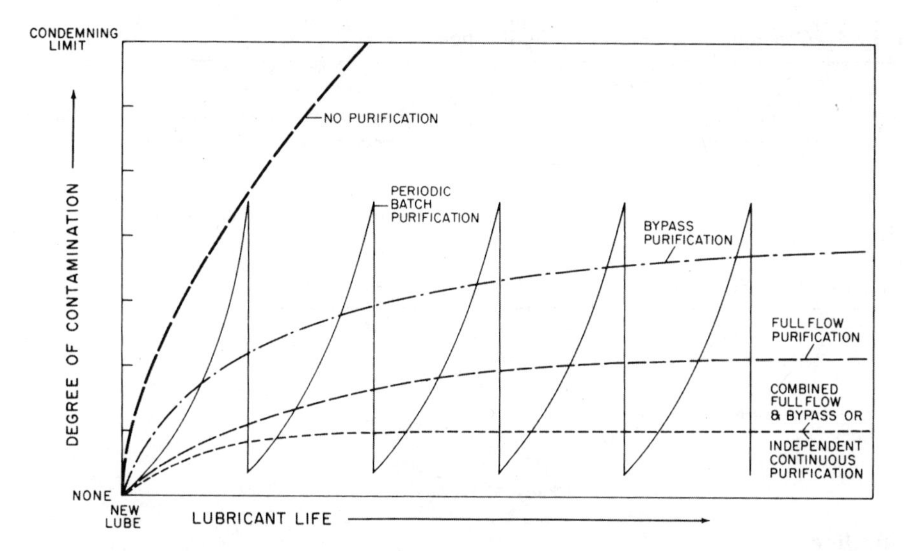

Figure 19.5 Effectiveness of purification methods.

Many industrial oils can be retained in service for long periods before draining by use of in-system or portable purification systems. Other contaminated oils can be reused after independent purification: they can be returned to the system provided they are clean and retain their original performance characteristics, and additives have not been removed by the purification process. If the fluid is not suitable for the original uses, purified lubricants may be used in noncritical gear cases, in all-loss systems such as hand oiling of chains and bar oil for chain saws, or for flushing purposes. Lubricants cross-contaminated with metalworking fluids or vice versa cannot be adequately purified for their original purpose and must, therefore, be drained and used for heating, be re-refined, or be disposed of in another way.

V. PURIFICATION METHODS

The purification methods most commonly used in industrial plants are settling, size filtration, centrifuging, and clay or depth filtration. It is noteworthy that conventional fine filtration will not remove any significant quantities of additives because most of these are soluble in the oil. This applies to the majority of industrial and automotive-type lubricants. In certain instances the use of very fine filtration ($< 3\mu m$) with high (> 100) beta ratios (see Chapter 7, Table 7.3), some additive can be removed from compounded oils such as compressor oil for wet gas conditions or steam cylinder oils. The contaminants removed and the various purification methods are outlined in Table 19.3.

A combination of settling, flotation, and size filtration is often used on the machine or in the circulating system. A central reservoir of proper design will permit settling and may be equipped with size filtration. Clay filtration and centrifuging may be incorporated into the system as separate units. Reclaiming systems have historically been batch operations conducted on the drained lubricant, but recent improvements to reclaiming systems to make them more portable allows them to be rotated from system to system, similar to portable filter units.

Table 19.3 Contaminants Removed by Purification

Materials	Settling	Size filtration	Clay filtration	Centrifuging	Reclaiming	Re-refining
Water	Yes[a]	—[b]	—[b]	Yes	Yes	Yes
Solids	Yes[c]	Yes[c]	Yes[c]	Yes	Yes[c]	Yes
Oxy products	No	No	Yes	No	No	Yes
Fuels, solvents	No	No	No	No	Yes	Yes
Oil additives	No	No	Yes	—[d]	—[d]	Yes

[a] Except water held in tight emulsions.
[b] A little water can be absorbed or held back by the filter material.
[c] Solids will be removed down to the filtration limit (rating).
[d] Some additive may be removed along with the water.

A. Settling

Water and heavy solid contaminants will separate by gravity whenever oil is held in a quiescent state for a suitable period. For separation of solids, water, and oil, a settling system like that shown in Figure 19.6 can be fabricated. Moderate heating (160–180°F) of the oil by a steam coil or electric immersion heater will lower the oil viscosity and thereby accelerate the settling process. Care must be exercised in the selection of heating

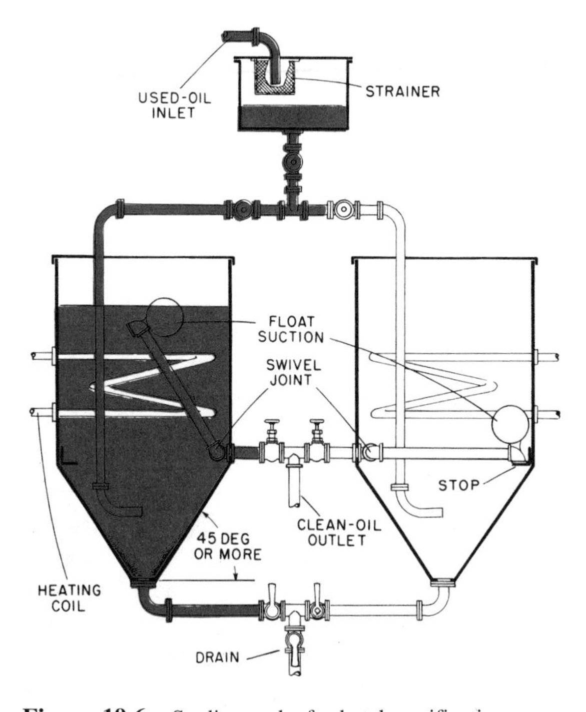

Figure 19.6 Settling tanks for batch purification.

elements to assure that surface temperatures are not high enough to cause actual oxidation or thermal cracking (heating element surface temperatures < 225°F are desirable).

In the dual tanks used in batch purification (Figure 19.6), preliminary straining of gross particulate is achieved in the upper tank, and used oil is introduced into one of the duplicate settling tanks well below any clarified or partially clarified oil present. The used oil is added slowly to prevent agitation of the settled contaminants. The floating suction permits withdrawing clean oil from the surface above a minimum level of water and sludge concentrate in the conical bottom.

B. Filtration

Filtration or straining may be used either as full-flow, continuous bypass, or a combination of these two methods, as discussed earlier. Straining by metal screen or woven wire meshes in cartridges or plates removes only the largest particles, is low in initial cost, and allows maximum flow rates almost independent of fluid viscosity. Conventional filtration will not remove water- or oil-soluble contaminants but, as discussed, coalescing-type filters can be used for the removal of small amounts of water. Any filtration system that removes water can also remove some of the polar rust inhibitors that attach themselves to the water. Filters of the surface, edge, and depth type allow a variation in the size of particle removal, with the surface and depth types suitable for the finer particles. These systems can be high in both initial and operating costs and are adaptable to single- or multimachine systems. When the system is properly designed, little or no downtime is required for servicing. See also the discussion of centrifugal filters (separators) in Section V.F.

C. Size Filtration

Cloth, stainless steel (reusable filters), metal-edge, paper-edge, or surface-type strainers, or depth-type filters can be used to remove particles from the oil by size.

Wire cloth strainers are used to remove only the larger particles—50 μm and larger. Metal-edge-type filters (Figure 19.7) clean by forcing the oil to pass through a series of edge openings formed by a stack of metal wheels or between the turns of metal ribbons. The majority of these units will filter out particles of 90 μm; some will filter particles as small as 40 μm. Paper-edge filters use elements of impregnated cellulose disks or ribbons. Impregnated cellulose disks will filter particles from 0.5 to 10 μm; ribbon elements will filter particles down to approximately 40 μm. Depth-type filters (Figure 19.8) use fuller's earth, felt waste, or rolled cellulose fibers in replaceable cartridges and can filter out particles from 1 to 75 μm in size. Surface-type filters use replaceable resin-treated paper, closely intertwined fiberglass strands (or other synthetic fibers), or woven cloth. Paper elements are available to filter out particles that are smaller than 1 μm; the cloth removes particles no smaller than about 40 μm.

1. Depth-Type Filters

Probably the most versatile of the industrial-type filters are depth-type filters, which can be equipped with cartridges that use a variety of filter media. Commercially available cartridges include the following.

1. An adsorbent-type filter containing fuller's earth in a woven cloth container. Other materials include activated alumina and clay. These materials are recommended for removal of soluble contaminants such as acids, asphaltenes, gums, resins, colloidal particles, and fine solids. An adsorbent filter will also remove polar-type additives but is not

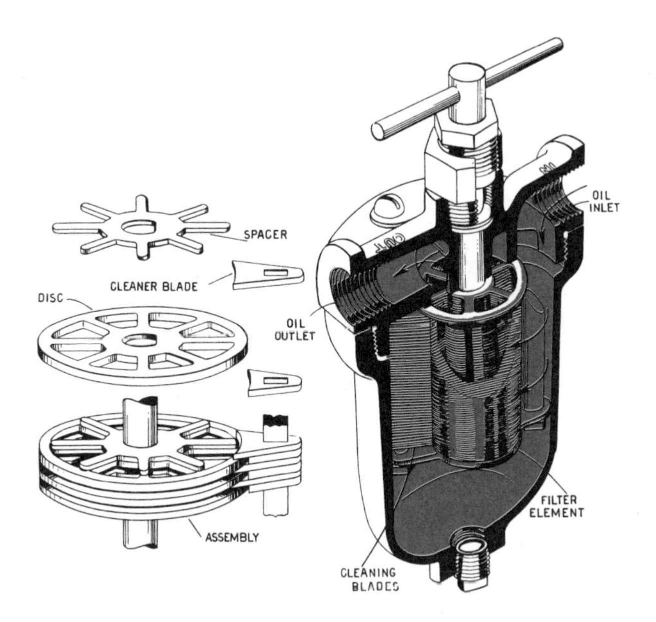

Figure 19.7 Edge-type oil filter.

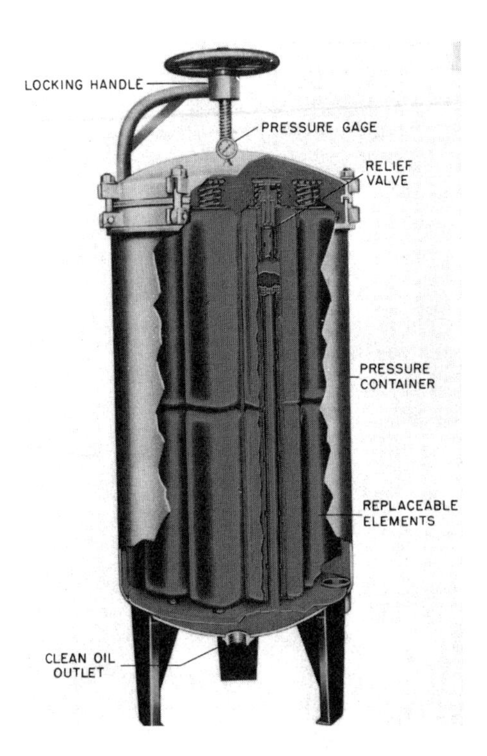

Figure 19.8 Typical depth-type filter.

recommended for general industrial lubricants unless the additive is replaced following purification or the oil is used for purposes not requiring that the additives be removed. Although fuller's earth will remove small amounts of water [≈1 quart (0.95 liter) of water per element], it is not recommended if gross water contamination is prevalent. Typical applications for this type of filter medium are quench oils, transformer oils, and vacuum pump oils. It is also recommended for filtering certain synthetic fluids such as phosphate esters and silicones.

2. A cellulose-type filter element, usually a combination of cotton waste and wood fibers (redwood, excelsior, etc.), is usually recommended for removal of large amounts of gross solids; it will not remove additives or water. The cellulose type retains its resiliency in the presence of water and water emulsions and usually permits higher flow rates. It is recommended for filtering engine oils, hydraulic fluids, turbine oils, fuel oils, and lubricating oils.

3. Resin-impregnated, accordion-pleated paper filter cartridges are recommended for medium contaminant loads, high flow rates, and low pressure drops such as are required by full-flow oil filtration. The cartridges available for particle removal in are sized from <1 μm to >20 μm. These are recommended for most industrial and automotive oil applications.

2. Clay Filtration

Certain clays, such as fuller's earth, will adsorb oil oxidation products, and the depth of clay will screen out fine particles. The fuller's earth may be in cartridge-type (Figure 19.9) units, or it may be mixed with the oil and filtered through size filtration units. The major disadvantage of clay filtration is that certain oil additives may be removed with the contaminants.

D. Multipurpose Purifiers

Filtration, settling, and free water removal are often combined in a single unit, as shown in Figure 19.10. These units have capacities for treating from 50 to over 2600 gal/h (189–9840 L/h). The oil conditioner is a three-compartment unit that provides dry clean oil by one of the following means:

1. Removing free water and coarse solids via horizontal wire screen plates in the precipitation compartment (Figure 19.10A)
2. Removing suspended solids via vertical, cloth-covered, leaf-type filter elements in the gravity filtration compartment (Figure 19.10B)
3. Polishing the oil and stripping free water from moisture vapor (only good for small amounts of moisture) via cellulose filter cartridges in the storage and polishing compartment (Figure 19.10C)

The leaf-type filter elements, which may be removed individually without shutting the system down, have selectivity down to particle sizes below 1 μm. They will not, however, remove rust or oxidation inhibitors except small amounts of polar compounds that may be attached to the moisture that is removed.

E. Centrifugation

Waste lubricating oils can also be purified by centrifugation, a process for separating materials of different densities, whether two liquids, a liquid and a solid, or two liquids

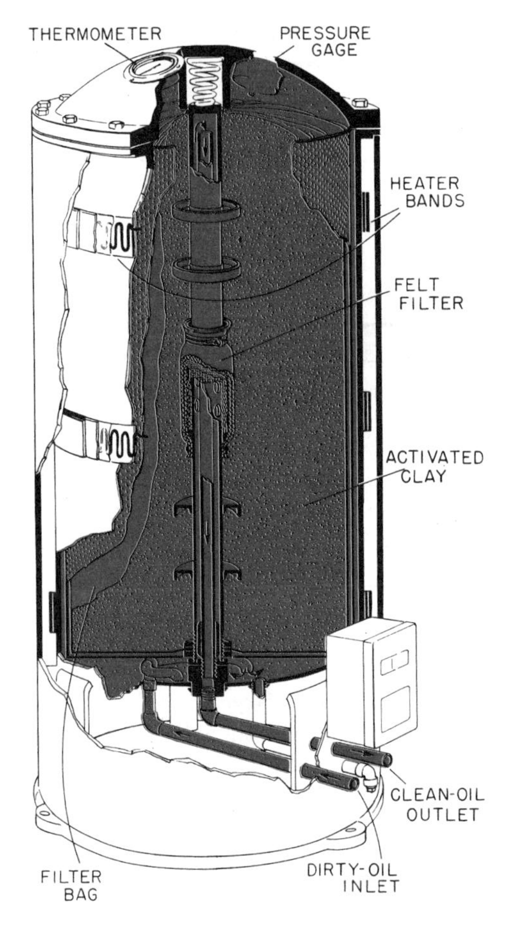

Figure 19.9 Activated-clay-type filter.

and a solid. If the liquids are mutually soluble, the centrifugation process will not be able to separate these materials. A centrifugal force is produced by any moving mass that is compelled to depart from the rectilinear path it tends to follow; the force is exerted in the direction away from the center of the curvature path. The centrifugal force is directly proportional to the mass or specific gravity of a material, and the higher specific gravity, the further the material will travel from the center of rotation. The centrifuge is a machine designed to subject material held in it or being passed through it to relatively high centrifugal force (500–2000 g), with means of collecting the separated materials. A batch centrifuge holds a given quantity of material and must be stopped periodically to discharge the solids and clarified liquid, and be recharged. Continuous centrifuges accept a steady stream of material, apply a centrifugal force, and continuously discharge the separated components. A centrifugal separator or purifier handles a continuous stream of mixed, nonsoluble liquids and effects their separation. A basket centrifuge is designed to hold a mass of material, usually a mixed solid and liquid and, by subjecting the mass to centrifugal force, separates the liquid, which passes through the basket walls, leaving the solid behind.

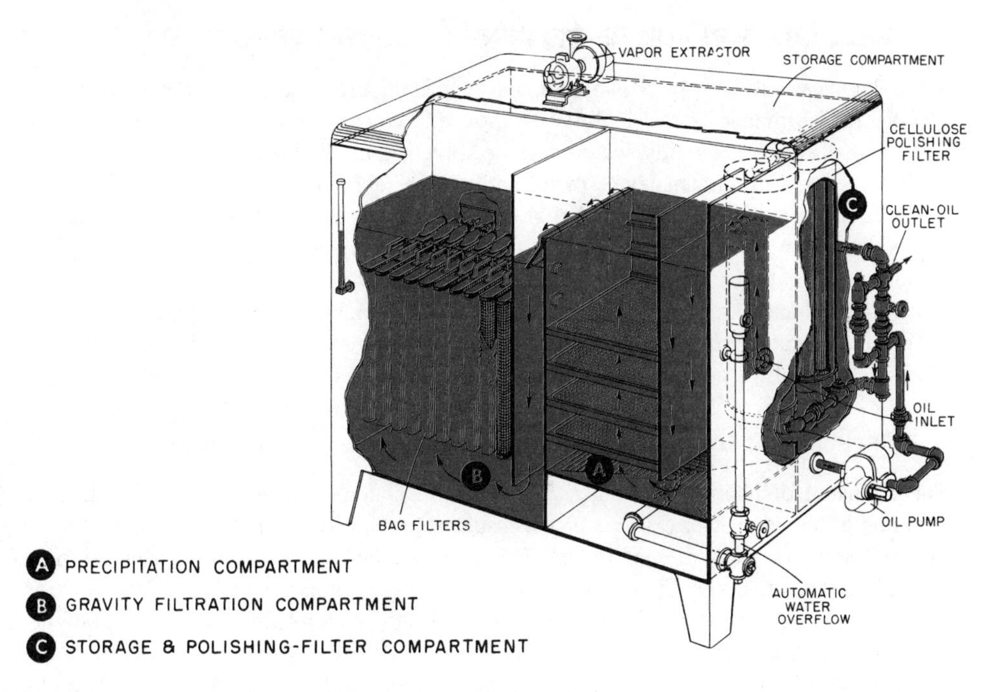

Figure 19.10 Combined filtration and settling purifier.

Centrifugation by disk and tubular centrifuges, operating at high forces, are excellent for removal (at high rates) of the relatively small particles and separation of fluids of close but differing densities. The initial investment and operating costs are relatively high, and fairly large floor space is required for the units with large-volume capacity. Smaller volume capacity units are also available that can be rotated from machine to machine in industrial applications, similar to portable filtering units.

F. Centrifugal Oil Filters (Separators)

A new breed of sidestream purification entered system the market several years ago for application to stationary and mobile internal combustion engines. The centrifugal filters have now expanded their use into industrial applications. Generally small and suited for systems with oil capacities up to about 100 gallons, these units can develop centrifugal forces of about 2000 *g*. With these forces, they have the capability of removing particles down to about 0.5 μm. They are small units, which handle about 2 gal/min in a sidestream and are driven by oil pressure and flow that spins their rotors up to about 6000 rpm. Since they are designed to discharge clean oil directly to the crankcase or reservoir, a simple valve can shut them down for cleaning without shutting down the entire system. Clean-out generally involves the removal of the sludges built up on the walls of the unit. Centrifugal oil filters can remove free water and solids along with minor amounts of degradation products, but they do not remove additive materials other than the minor amounts that have reacted with the contaminants that are removed.

VI. RECLAMATION AND RE-REFINING OF LUBRICATING OILS

The differences between the reclamation and re-refining processes are found in the amount of degradation materials and additives removed. Typically, reclamation units have the capability of removing solids, water, volatile solvents, and fuel dilution products but will not remove any significant levels of oil additives or degradation materials. If the oil to be reclaimed is not severely degraded or contaminated with high flash point materials, it can be reused in its original application or downgraded to a less severe application. Re-refining, on the other hand, strips all the additives and degradation materials from the lubricating oil and essentially produces a base stock, which must be readditized to obtain the desired characteristics. On-site re-refining is not recommended for individual industrial plants because of the complexity of reconstituting additives and economic considerations.

A. Reclamation Units (Oil Conditioners)

The inclusion of heating elements and a vacuum chamber in addition to settling and filtration as shown earlier (Figure 19.10) results in a reclamation unit (sometimes call an oil conditioner or vacuum dehydrator). This unit may be used to remove water and light ends usually due to contamination from solvents and other volatile materials. Reclamation may be either batch or a continuous operation. Units are available in various sizes, ranging from those that are portable, capable of handling from as little as 6 gal/h (22.7 L/h), to large stationary units with capacities as high as 3000 gal/h (11,355 L/h). The portable units can be rotated from one machine to another, similar to the filter carts discussed in Chapter 7, Section IX.B.

In a reclamation unit, the dirty oil is generally prefiltered to remove the bulk of large solids and then passed through heaters to attain the desired temperature. When the correct temperature (around 180°F) is reached, the oil is routed through filters and into a vacuum chamber, where much of the water and volatile materials (solvents, fuels, etc.) can be removed. The high temperature lowers the viscosity, which aids in separation of the water and volatile materials in the vacuum chamber and helps increase the final filterability after leaving the vacuum chamber. The filtering units are generally cascading units starting with large particle size removal capability (40 μm) and going down to 5 μm or less in the final stage. This cascading filtration improves efficiency of filtration and reduces replacement costs. The clean oil is then passed through initial heating section to reclaim the heat for heating of the dirty oil entering the unit, and then the cooled oil is transferred to the operating unit or a holding tank. Reclamation units do not remove additives other than the polar materials that become attached to the water molecules.

B. Re-Refining Units

Re-refining units are similar to reclamation units except that they remove all (or most) of the additives and are more efficient at removing moisture and solvent-type materials. In addition to the prefiltering, heating capability, vacuum dehydration, and degasification, the re-refining unit may contain activated clay, fuller's earth, or activated alumina for removal of additive and degradation materials. In some designs, the clay is mixed with the oil before being passed through the heating elements. The high temperature lowers the oil viscosity, making it easier to filter, and increases the activity of the clay materials. Under controlled vacuum and high temperature, all moisture and the light ends are removed. Since this process removes additive materials, the end product is essentially a

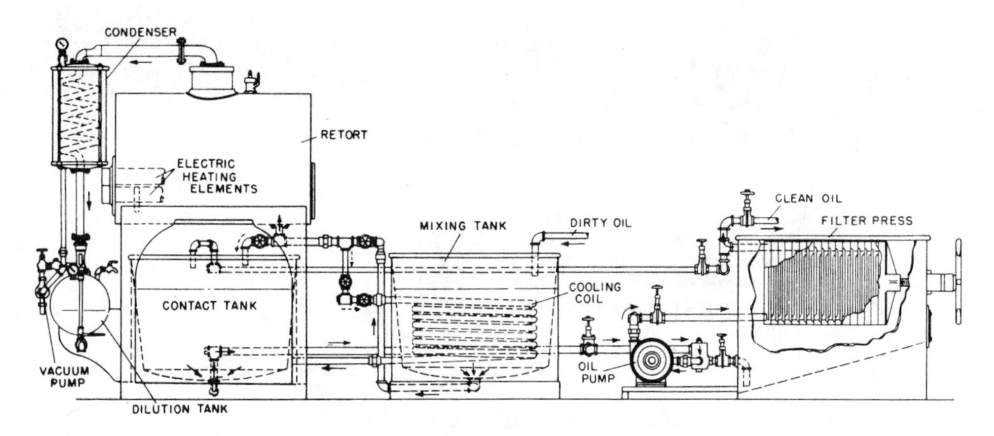

Figure 19.11 Small continuous re-refiner.

base stock and must be reconstituted with additives to provide the desired performance characteristics in the finished lubricant.

A similar process on a continuous basis is conducted in the small oil re-refiner shown in Figure 19.11. Dirty oil is pumped through a heat exchanger and preheater that raises the oil to the proper filtering temperature. The oil passes through a filter bed composed of a filter pad, a bed of activated clay, filter paper, and canvas. The filtered oil passes to a vaporizer, where the temperature is increased. A vacuum is maintained in the vaporizer that aids in the evaporation of water and the distillation of the more volatile portions, if any. The purified oil is cooled in the heat exchanger by the incoming dirty oil. The resulting clarified oil must either be reconstituted with a new additive package or used for less demanding applications.

The unit shown in Figure 19.11 can be equipped with cellulose disks in place of the activated materials for handling additive-based oils, phosphate ester based synthetics containing additives, or other fluids adversely affected by activated clay. The cellulose material will not remove the additive package. At this point, the re-refiner becomes a reclamation unit.

VII. WASTE COLLECTION AND ROUTING

We have discussed the selection, in-service handling, and purification of lubricants for industrial and commercial applications to help assure long life, minimize contamination, and provide appropriate purification in process. Eventually, all lubricants become waste oil and must be drained from the system, collected into waste storage tanks, and prepared for one of several methods of final disposal. All this must be accomplished with two general considerations in mind:

1. Handling the waste oil so that it will have the highest value as a by-product and the optimum stability for disposal
2. Preventing the waste from polluting the process, storm, and sewer systems in the plant and the groundwater and surface waters outside the plant

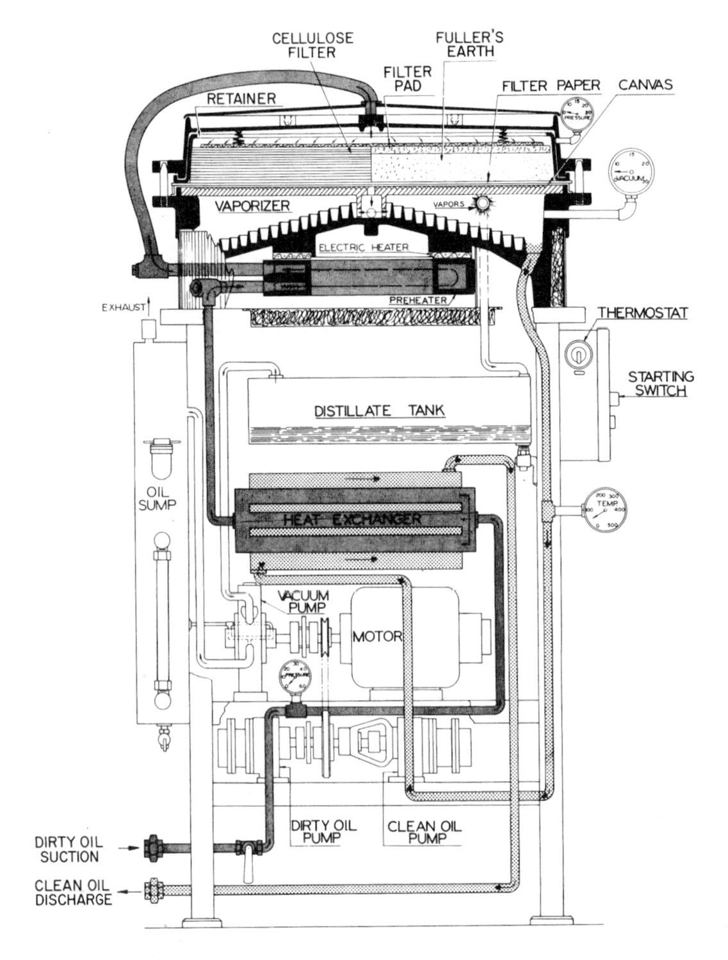

Figure 19.12 Multi-machine collection system.

All this can be accomplished by proper collection and routing systems. A multimachine collection system of intermediate size is shown in Figure 19.12. Alternatively, the system may be simply a combination of sump tank cleaner and centrifuge for renewing and purifying used oil containing sludge, chips, and other solid contaminants, similar to a portable reclamation unit. Most units of these types have their own pumping and discharge system, although some units work on vacuum to pull the oil into the unit and, by means of reversing valves, use air pressure to discharge the oil back to the sump. Waste oil is carted back to be put in holding tanks or drums for final disposal.

Whatever the size of the systems for collection and routing of the waste oil, certain principles should be followed.

1. Separate collection systems should be used for each type of lubricant and separate storage provided until final disposal. Neat oil and lubricants should be kept separate from emulsions, chemical coolants, and so on.

2. Lubricating oils should be kept separate on the basis of degree of contamination even if the type of oil is the same. A relatively clean hydraulic fluid should not be mixed

with waste metalworking fluids or dirty lubricants. Admixture with metalworking fluids or chemical coolants would lower the value of the waste hydraulic fluid as a by-product.

3. Reservoirs should be drained immediately after shutdown while the oil is still at operating temperature and solids are still in suspension. Proper safety precautions must be exercised if the oil is excessively hot. Certain oxidation products also become insoluble in cold oil, which means that if not prevented from doing so, they will precipitate with the solids, remain in the reservoir, and contaminate a new charge.

4. Prevent contamination of process water, storm drains, and sewer systems by waste oil streams. This is a particularly important consideration where floor drains exist and spills of lubricating oil occur. Such contamination is also possible in oil coolers or heaters where leaks in the exchangers can cause contamination of the process stream or cooling water being treated. It is common practice to maintain oil pressures through exchangers at a higher level than water pressure, to prevent the lubricant from contamination with water in the event of leaks.

5. Open troughs or floor gratings and drains to catch leaks, spills, and drips or to route waste oil to holding tanks are not desirable because of the high potential for further contamination and admixture with noncompatible products, making reclamation almost impossible.

Whether individual machine collection systems or plant-wide systems are used will be decided on the basis of economics and feasibility. However, the plant-wide system is desirable. Individual machine systems can be serviced with commercially available oil carts such as shown earlier (Figure 18–32). The unit shown in Figure 18.32 has a capacity of 80 gallons (246 liters) and is equipped with either single-flow or reverse suction rotary hand pumps.

If it is not feasible to connect the individual machines to a waste oil collection system, consideration should be given to a central piping system to carry waste oil to storage, with conveniently located floor drains into which the oil carts could be drained. If such a system is used, it must be designed so that waste oil can be isolated from ordinary floor drainage of water or washing solution. If it is impossible to have a separate oil drainage system, the floor drains in appropriate areas should be equipped with oil traps.

VIII. FINAL DISPOSAL

Lubricants, including metalworking fluids, hydraulic fluids, and gear oils, eventually must be disposed of as waste oil. The method of final disposal will depend on the type of oil, contaminants, and condition of the waste oil. Several generally accepted methods permit disposal with adequate safeguards to prevent water and air pollution:

1. Reclaiming heat value by using as a fuel supplement
2. Re-refining and reconstitution of additives for reuse
3. Incineration
4. Coal or petroleum coke spray, or density control in specific processes where appropriate

One of the presently most feasible and generally acceptable methods of disposing of used lubricating oils is burning as fuel. Federal, state, local, or county regulations should be reviewed before pursuing this option. Testing of the candidate waste oil is recommended to assure that certain hazardous materials do not exist in quantities sufficient to violate clean air standards. Burning waste oils as fuel is highly recommended as an approved method of final disposal for the following reasons.

1. It is the most widely applicable method from the standpoint of type of waste oil and location of generating source.
2. If the burning and heat recovery can be done at the site of the waste oil generation, the process is considered to be recycling.
3. There are practically no volume limitations on the amount of waste oil that may be handled by this method. It is feasible for large, medium, and smaller users of petroleum lubricants.
4. It turns what could be a liability (waste oil ready for disposal) into an asset. Since some contract haulers are charging to remove waste oils, and with high fuel costs, a substantial net value can be realized by using waste oil as fuel.

Further, there is an ever-increasing demand for petroleum products (gaseous and liquid) as an energy source. Therefore, any secondary use of waste oil, subsequent to maximizing its use in its primary function, will recover additional value as a petroleum product and conserve natural resources. Burning of used lubricants as fuel is the simplest of the disposal methods.

While practically all liquid petroleum products may be oxidized into carbon dioxide and water by burning, not all are suitable for use as fuel or for mixing with conventional fuel. Contamination with large amounts of water, low flash petroleum products, excessive sediment, or hazardous materials causes some products to lose their value as fuel and necessitates the use of supplemental fuels to incinerate the waste. This is the reason for proper in-plant handling to ensure a reasonable waste oil by-product, whether the final disposal is by burning as fuel, re-refining, or contract hauling to approved waste collection sites.

Index